SD
409
F63
2005

LIBRARY
NSCC, STRAIT AREA CAMPUS
226 REEVES ST.
PORT HAWKESBURY, NS B9A 2A2 CANADA

Forest Restoration in Landscapes

WWF's Forests for Life Programme

WWF's vision for the forests of the world, shared with its long-standing partner, the World Conservation Union (IUCN), is that "the world will have more extensive, more diverse and higher-quality forest landscapes which will meet human needs and aspirations fairly, while conserving biological diversity and fulfilling the ecosystem functions necessary for all life on Earth."

WWF's approach to forest conservation has evolved over time into a global programme of integrated field and policy activities aimed at the protection, responsible management, and restoration of forests, whilst at the same time working to address the key threats which could potentially undermine these efforts. Those of particular concern to WWF are illegal logging and forest crime, conversion of forests to plantation crops of palm oil and soy, forest fires, and climate change.

The Forests for Life Programme consists of a global network of more than 250 staff working on over 300 projects in nearly 90 countries. Regional forest officers coordinate efforts in each of the five regions, supported by a core team based at WWF International in Switzerland. The programme also draws on the complementary skills and support of partners to help achieve its goals.

WWF and Restoring Forests and Their Functions in Landscapes

WWF has adopted a target for forest restoration: "By 2020, restore forest goods, services, and processes in 20 landscapes of outstanding importance within priority ecoregions to regain ecological integrity and enhance human well-being," which is issued as a challenge to the world.

As its contribution toward the target, WWF is actively developing a portfolio of forest landscape restoration programmes, and also working with governments, international organisations, indigenous peoples, and other communities to pursue its work on forest restoration within a landscape context, by doing the following:

- Initiating and facilitating projects/programmes within landscapes of high restoration priority in WWF Global 200 Ecoregions
- Assisting others, and building local capacity to plan and implement forest restoration interventions
- Developing suitable monitoring tools and techniques to measure progress
- Documenting, exchanging and disseminating lessons learnt and experiences

For more information please see the Web site: http://www.panda.org/forests/restoration/.

Stephanie Mansourian
Daniel Vallauri
Nigel Dudley

Forest Restoration in Landscapes

Beyond Planting Trees

With 28 Illustrations

Stephanie Mansourian
Consultant
WWF International
Avenue Mont Blanc
Gland 1196
Switzerland

Daniel Vallauri
WWF France
6 Rue des Fabres
13001 Marseille
France

Nigel Dudley
Consultant
Equilibrium
47 The Quays
Cumberland Road
BS1 6UQ
United Kingdom

© 1986 Panda symbol WWF

® WWF is a WWF Registered Trademark

Cover Illustrations: Photo Cedar of Lebanon (*Cedrus libani*), tree seedling. Photo Credit: © WWF-Canon/Michael Gunther. Background photo: Mt. Rinjani, Lombok, Indonesia, © Agri Klintuni Boedhihartono.

Library of Congress Control Number: 2005927862

ISBN 0-387-25525-7   Printed on acid-free paper.
ISBN-13: 978-0387-255855

© 2005 Springer Science+Business Media, Inc.
Cite as Mansourian, S., Vallauri, D., Dudley, N., eds. (in cooperation with WWF International) 2005. Forest Restoration in Landscapes: Beyond Planting Trees, Springer, New York.
All rights reserved. This work may not be translated or copied in whole or in part without the written permission of the publisher (Springer Science+Business Media, Inc., 233 Spring Street, New York, NY 10013, USA), except for brief excerpts in connection with reviews or scholarly analysis. Use in connection with any form of information storage and retrieval, electronic adaptation, computer software, or by similar or dissimilar methodology now known or hereafter developed is forbidden.
The use in this publication of trade names, trademarks, service marks, and similar terms, even if they are not identified as such, is not to be taken as an expression of opinion as to whether or not they are subject to proprietary rights.

Printed in the United States of America.   (BS/SBA)

9 8 7 6 5 4 3 2 1

springer.com

Foreword

Is it a sign of the times that last year the Nobel committee chose to award the Nobel Peace prize to Wangari Maathai for having planted 30 million trees? We believe so. We think that while in the 20th century conservation made significant progress on setting up a global protected area network, the 21st century will be a time of forest restoration. The fact that Wangari Maathai is the first African woman to receive such an honourable distinction is in itself a major accomplishment. What is even more remarkable is that, for the first time, this highly esteemed prize, which has long been associated with political feats, was given for an environmental achievement. And not just any environmental achievement, but forest restoration. It is a comfort to see that it is not just us at WWF, the global conservation organisation, who believe forest restoration to be of global significance, but that the Nobel committee is in agreement. The committee members are not the only ones, I should add. In 2003 WWF, IUCN, (the World Conservation Union), and the United Kingdom Forestry Commission launched a global partnership on forest landscape restoration to raise awareness about the importance of the restoration of forests and to invite all decision makers and influential organisations to join in a movement to restore forests. Today this partnership includes governments as diverse as Switzerland, Finland, El Salvador, and Italy, and international organisations such as the United Nations Food and Agriculture Organisation (FAO), the Centre for International Forestry Research (CIFOR), the International Tropical Timber Organisation (ITTO), and it continues to grow.

Too much damage has already been done for us to afford to ignore our dwindling forest resources. If we wait until tomorrow to restore forests, it will be too late. If too little is left, it will take longer, will be more difficult, and will cost much more to begin restoring a healthy forest—and it may also be too late.

At WWF we are aware of this urgency, and with this book we invite practitioners, researchers, and decision makers to join us in doing something practical about our forests. As the Nobel committee has noted, too many wars are fought over dwindling resources. If we do not do something about it, this may well be the new security scourge jeopardising our future and that of our children.

Chief Emeka Anyaoku
President, WWF International

Preface

For WWF, the global conservation organisation, achieving lasting forest conservation requires working on a large scale and integrating global strategies and policies to protect, manage, and restore forests.

In an ideal world, restoration would not be necessary; however, today many forest habitats are already so damaged that their long-term survival, and the ecological services they provide, are in doubt and we urgently need to consider restoration if we are to achieve conservation and sustain the livelihoods of people dependent on nature.

Forest conservation strategies that rely solely on protected areas and sustainable management have proved insufficient either to secure biodiversity or to stabilise the environment. The United Nations Environment Programme now classifies a large proportion of the world's land surface as "degraded," and this degradation is creating a wide range of ecological, social, and economic problems. Forest loss and degradation is a particularly important element in this worldwide problem with annual global estimates of forest loss being as high as 16 million hectares, and those for degradation even higher. Reversing this damage is one of the largest and most complex challenges of the 21st century.

An analysis of the WWF Global 200 ecoregions—those areas of greatest importance for biodiversity on a global scale—demonstrates the problems. For example, over 20 percent of forest ecoregions have already lost at least 85 percent of their forests: sometimes only 1 to 2 percent remains. Deforestation is a key threat to water quality in 59 percent of freshwater ecoregions. Many of the charismatic species that are flagships for conservation (African elephant, Asian elephant, great apes, rhinoceros, giant panda, and tiger) are threatened by forest loss, fragmentation, and degradation.

Forest loss is not only of concern to conservationists. According to the World Bank about 1 billion people in the developing world depend either directly or indirectly on goods and services from the forests, and these provide an essential safety net to many of the world's poorest people.

WWF's mission is to stop degradation on our planet and to achieve a world where humans and nature live in harmony together. Decades of overexploitation have brought us to a world characterised by imbalance: imbalance between rich and poor, imbalance between supply of natural resources and demand for natural resources, imbalance between biodiversity needs and human needs. WWF's approach to forest restoration, in the context of ecoregion conservation, seeks to redress these imbalances in order to restore healthy landscapes that are able to benefit both biodiversity and people.

This book harnesses the expertise of over 70 authors drawing on a wealth of practical experience and a wide range of expertise. It is practical, hands on, and illustrated with numerous examples from across the world. The aim is to synthesise in an easily accessible format the knowledge and expertise that exists and also to highlight areas that need further work. We are hoping to encourage field staff—ours and those of other organisations interested in conservation and development—who are out there dealing with the impacts of forest loss and degradation, to apply landscape-scale forest restoration as an approach to help them meet their conservation goals and our conservation goals.

Dr. Chris Hails
Programme Director, WWF International

Note from the Editors

This book has been designed to help readers understand how forest restoration can be integrated with other aspects of conservation and development in landscapes. Parts A, B, and C introduce the elements for planning and implementing restoration on a broad scale, including a range of social, political, and economic considerations that will influence and that will be influenced by any large-scale restoration effort. Part D focusses on more specific issues, including restoration in different forest habitats and for different reasons.

While we believe that successful restoration generally needs to be planned on a large scale, it will probably be implemented in one or more sites within a landscape, and the book similarly starts with very broad-scale considerations and then focusses increasingly on actions that can be taken at the site. Parts A, B, and C thus provide what could be seen as the foundations, and part D provides some much more specific tools and considerations that are applicable in different situations. We recommend that you read the relevant chapters in part D once you have read all of parts A, B, and C.

The final part (part E) discusses some of the lessons learned to date from practical experiences and recommendations for future work related to forest restoration on a large scale.

Each chapter starts with an introduction to the issue, illustrating it with a series of brief thumbnail examples, showing, where appropriate, both good and bad practice. Some useful tools are then listed followed by a brief description of future work required and finally and importantly a set of references. We cover a vast subject here and each chapter is as a result kept deliberately short, we can only introduce many of the techniques described but have provided detailed sources for those who wish to follow up specific issues in greater detail.

The book includes contributions from a large number of authors. Although we have all been writing within the framework of forest landscape restoration, there are inevitably different nuances in how this should be interpreted and applied. What follows is a set of experienced opinions rather than a rigid blueprint. We will in turn very much appreciate hearing feedback, criticism, and experience from users.

Acknowledgements

The editors would like to thank Mark Aldrich, James Aronson, Chris Elliott, Chris Hails, and Pedro Regato for their emphatic and very welcome support throughout the conception and production of this book. On behalf of WWF International we would also like to thank the 70 authors who donated their expertise, for no payment and under what must have often been a frustratingly tight timetable, to help produce such a comprehensive review of this rapidly emerging field.

The following people have kindly reviewed different sections and chapters and provided us with valuable feedback: Chris Elliott (WWF International), Louise Holloway, Jack Hurd (the Nature Conservancy), Val Kapos (U.N. Environment Programme–World Conservation Monitoring Centre), John Parrotta (U.S. Forest Service), Duncan Pollard (WWF International), Fulai Sheng (Conservation International), P.J. Stephenson (WWF International), and Colin Tingle (NR Group).

A special thank you is due to Tom McShane for taking the time to read and comment on the whole manuscript.

Nelda Geninazzi played an essential role in helping to organise the various editorial meetings, and Katrin Schikorr deserves special mention for helping the editors with references.

The authors would like to specifically thank the following people for contributing in some form or another to their respective chapters: José María Rey Benayas, André Rocha Ferretti, Karen Holl, Ramdan Lahouati, N. Lassettre, Stewart Maginnis, Hal Mooney, Guy Preston, Mohamed Raggabi, Peter Schei, and Kristin Svavarsdottir.

The authors would also like to thank the following agencies and/or institutions for support in projects that have made it possible for them to write their respective chapters: European Life Environment programme "Water and Forest," French Research Ministry, French National Forest Office (ONF) and Water Agency (Agence RMC), the European Commission (EC) (for the project Biodiversity Conservation, Restoration and Sustainable Use in Fragmented Forest Landscapes (BIOCORES), the Long-Term

Ecological Research programme in Puerto Rico, funded by the National Science Foundation, the government of Japan (for the CIFOR/ Japan Research project on lessons from past rehabilitation experiences), and the Generalitat Valenciana and Fundación Bancaja. The authors from CIFOR would like specifically to thank the various research and support staff, as well as workshop and case study participants from the different countries for their invaluable contributions to the project "Review of Forest Rehabilitation Initiatives: Lessons from the Past", which formed the basis for their chapter in this book.

Finally, WWF would like to thank Lafarge for supporting the development of its forest landscape restoration programme.

The book represents a collection of individual essays and are the opinions of the authors and should not be seen as representing opinions from their respective employers or organisations. Needless to say, despite the enormous help we have received in putting this book together, any remaining errors of fact or opinion remain the responsibility of the editors.

Table of Contents

Foreword by Chief Emeka Anyaoku, President, WWF
International v

Preface by Chris Hails, Programme Director, WWF
International vii

Note from the Editors ix

Acknowledgements xi

Acronyms xxi

Contributors' List xxiii

Part A Toward a Wider Perspective in Forest Restoration

Section I Introducing Forest Landscape Restoration

Chapter 1
Forest Landscape Restoration in Context
*Nigel Dudley, Stephanie Mansourian, and
Daniel Vallauri* 3

Chapter 2
Overview of Forest Restoration Strategies and Terms
Stephanie Mansourian 8

Section II The Challenging Context Of Forest Restoration Today

Chapter 3
Impact of Forest Loss and Degradation on Biodiversity
Nigel Dudley 17

Chapter 4
The Impacts of Degradation and Forest Loss on Human
Well-Being and Its Social and Political Relevance for
Restoration
Mary Hobley 22

Chapter 5
Restoring Forest Landscapes in the Face of Climate
Change
    Jennifer Biringer and Lara J. Hansen 31

Section III   Forest Restoration in Modern Broad-Scale Conservation

Chapter 6
Restoration as a Strategy to Contribute to
Ecoregion Visions
    John Morrison, Jeffrey Sayer, and Colby Loucks 41

Chapter 7
Why Do We Need to Consider Restoration in a
Landscape Context?
    Nigel Dudley, John Morrison, James Aronson, and
    Stephanie Mansourian 51

Chapter 8
Addressing Trade-Offs in Forest Landscape Restoration
    Katrina Brown 59

Part B   Key Preparatory Steps Toward Restoring Forests Within a Landscape Context

Section IV   Overview of the Planning Process

Chapter 9
An Attempt to Develop a Framework for
Restoration Planning
    Daniel Vallauri, James Aronson, and Nigel Dudley 65

Section V   Identifying and Addressing Challenges/ Constraints

Chapter 10
Assessing and Addressing Threats in
Restoration Programmes
    Doreen Robinson 73

Chapter 11
Perverse Policy Incentives
    Kirsten Schuyt 78

Chapter 12
Land Ownership and Forest Restoration
    Gonzalo Oviedo 84

Chapter 13
Challenges for Forest Landscape Restoration
Based on WWF's Experience to Date
  Stephanie Mansourian and Nigel Dudley 94

Section VI   A Suite of Planning Tools

Chapter 14
Goals and Targets of Forest Landscape Restoration
  Jeffrey Sayer 101

Chapter 15
Identifying and Using Reference Landscapes for
Restoration
  Nigel Dudley 109

Chapter 16
Mapping and Modelling as Tools to Set Targets, Identify
Opportunities, and Measure Progress
  Thomas F. Allnutt 115

Chapter 17
Policy Interventions for Forest Landscape Restoration
  Nigel Dudley 121

Chapter 18
Negotiations and Conflict Management
  Scott Jones and Nigel Dudley 126

Chapter 19
Practical Interventions that Will Support Restoration in
Broad-Scale Conservation Based on WWF
Experiences
  Stephanie Mansourian 136

Section VII   Monitoring and Evaluation

Chapter 20
Monitoring Forest Restoration Projects in the Context of
an Adaptive Management Cycle
  Sheila O'Connor, Nick Salafsky, and Dan Salzer 145

Chapter 21
Monitoring and Evaluating Forest Restoration Success
  *Daniel Vallauri, James Aronson, Nigel Dudley, and
  Ramon Vallejo* 150

Section VIII Financing and Promoting Forest Landscape Restoration

Chapter 22
Opportunities for Long-Term Financing of Forest Restoration in Landscapes
Kirsten Schuyt 161

Chapter 23
Payment for Environmental Services and Restoration
Kirsten Schuyt 166

Chapter 24
Carbon Knowledge Projects and Forest Landscape Restoration
Jessica Orrego 171

Chapter 25
Marketing and Communications Opportunities: How to Promote and Market Forest Landscape Restoration
Soh Koon Chng 176

Part C Implementing Forest Restoration

Section IX Restoring Ecological Functions

Chapter 26
Restoring Quality in Existing Native Forest Landscapes
Nigel Dudley 185

Chapter 27
Restoring Soil and Ecosystem Processes
Lawrence R. Walker 192

Chapter 28
Active Restoration of Boreal Forest Habitats for Target Species
Harri Karjalainen 197

Chapter 29
Restoration of Deadwood as a Critical Microhabitat in Forest Landscapes
Nigel Dudley and Daniel Vallauri 203

Chapter 30
Restoration of Protected Area Values
Nigel Dudley 208

Section X Restoring Socioeconomic Values

Chapter 31
Using Nontimber Forest Products for Restoring
Environmental, Social, and Economic Functions
Pedro Regato and Nora Berrahmouni 215

Chapter 32
An Historical Account of Fuelwood Restoration Efforts
Don Gilmour 223

Chapter 33
Restoring Water Quality and Quantity
Nigel Dudley and Sue Stolton 228

Chapter 34
Restoring Landscapes for Traditional Cultural Values
Gladwin Joseph and Stephanie Mansourian 233

Section XI A Selection of Tools that Return Trees to the Landscape

Chapter 35
Overview of Technical Approaches to Restoring Tree
Cover at the Site Level
*Stephanie Mansourian, David Lamb, and
Don Gilmour* 241

Chapter 36
Stimulating Natural Regeneration
Silvia Holz and Guillermo Placci 250

Chapter 37
Managing and Directing Natural Succession
Steve Whisenant 257

Chapter 38
Selecting Tree Species for Plantation
Florencia Montagnini 262

Chapter 39
Developing Firebreaks
Eduard Plana, Rufí Cerdan, and Marc Castellnou 269

Chapter 40
Agroforestry as a Tool for Forest Landscape Restoration
Thomas K. Erdmann 274

Part D Addressing Specific Aspects of Forest Restoration

Section XII Restoration of Different Forest Types

Chapter 41
Restoring Dry Tropical Forests
James Aronson, Daniel Vallauri, Tanguy Jaffré, and Porter P. Lowry II 285

Chapter 42
Restoring Tropical Moist Broad-Leaf Forests
David Lamb 291

Chapter 43
Restoring Tropical Montane Forests
Manuel R. Guariguata 298

Chapter 44
Restoring Floodplain Forests
Simon Dufour and Hervé Piégay 306

Chapter 45
Restoring Mediterranean Forests
Ramon Vallejo 313

Chapter 46
Restoring Temperate Forests
Adrian Newton and Alan Watson Featherstone 320

Section XIII Restoring After Disturbances

Chapter 47
Forest Landscape Restoration After Fires
Peter Moore 331

Chapter 48
Restoring Forests After Violent Storms
Daniel Vallauri 339

Chapter 49
Managing the Risk of Invasive Alien Species in Restoration
Jeffrey A. McNeely 345

Chapter 50
First Steps in Erosion Control
Steve Whisenant 350

Chapter 51
Restoring Forests After Land Abandonment
  José M. Rey Benayas 356

Chapter 52
Restoring Overlogged Tropical Forests
  Cesar Sabogal and Robert Nasi 361

Chapter 53
Opencast Mining Reclamation
  José Manuel Nicolau Ibarra and
  Mariano Moreno de las Heras 370

Section XIV Plantations in the Landscape

Chapter 54
The Role of Commercial Plantations in Forest Landscape Restoration
  Jeffrey Sayer and Chris Elliot 379

Chapter 55
Attempting to Restore Biodiversity in Even-Aged Plantations
  Florencia Montagnini 384

Chapter 56
Best Practices for Industrial Plantations
  Nigel Dudley 392

Part E Lessons Learned and the Way Forward

Chapter 57
What Has WWF Learned About Restoration at an Ecoregional Scale?
  Nigel Dudley 401

Chapter 58
Local Participation, Livelihood Needs, and Institutional Arrangements: Three Keys to Sustainable Rehabilitation of Degraded Tropical Forest Lands
  Unna Chokkalingam, Cesar Sabogal, Everaldo Almeida,
  Antonio P. Carandang, Tini Gumartini, Wil de Jong,
  Silvio Brienza, Jr., Abel Meza Lopez, Murniati,
  Ani Adiwinata Nawir, Lukas Rumboko, Takeshi Toma,
  Eva Wollenberg, and Zhou Zaizhi 405

Chapter 59
A Way Forward: Working Together Toward a Vision for
Restored Forest Landscapes
  *Stephanie Mansourian, Mark Aldrich, and
  Nigel Dudley* 415

Appendix
A Selection of Identified Ecological Research Needs
Relating to Forest Restoration 424

Index 427

Acronyms

ACG—Area Conservación Guanacaste
CAP—common agriculture policy
CATIE—Centro Agronómico Tropical de Investigación y Enseñanza
CBD—Convention on Biological Diversity
CBFM—community-based forest management
CDM—clean development mechanism
CEAM—Centro de Estudios Ambientales Mediterráneos (Mediterranean Centre for Environmental Studies)
CIFOR—Centre for International Forestry Research
DFID—U.K. Department for International Development
DG—Directorate General
EC—European Commission
ECCM—Edinburgh Centre for Carbon Management
ERC—ecoregion conservation
EU—European Union
FAO—United Nations Food and Agriculture Organisation
FLO—Fair-Trade Labelling Organisation
FLR—forest landscape restoration
FSC—Forest Stewardship Council
FONAFIFO—Fondo Nacional de Financiamiento Forestal (National Fund for Financing Forestry)
GEF—global environment facility
GIS—geographical information system
GTZ—Deutsche Gesellschaft für Technische Zusammenarbeit (German Company for International Technical Cooperation)
HCVF—high conservation value forest
IAS—invasive alien species
ICDP—Integrated Conservation and Development Programme
IFOAM—The International Federation of Organic Agriculture Movements
IMF—International Monetary Fund
IPF—Intergovernmental Panel on Forests
ITTO—International Tropical Timber Organisation
IUCN—The World Conservation Union
IISD—International Institute for Sustainable Development
IIED—International Institute for Environment and Development
LULUCF—Land Use, Land-Use Change, and Forestry
MOSAIC—Management of Strategic Areas for Integrated Conservation
NTFP—nontimber forest products
NGO—Nongovernmental organisation
ODA—Overseas Development Assistance
PES—payment for environmental services
PRA—participatory rural appraisal
PVA—population viability analysis
RIL—reduced-impact logging
RRA—rapid rural appraisal
REACTION—Restoration Actions to Combat Desertification in the Northern Mediterranean
SAPARD—Special Action for Pre-Accession Measures for Agriculture and Rural Development
SERI—Society for Ecological Restoration International
SDC—Swiss Agency for Development and Cooperation
SEI—Stockholm Environment Institute
SLU—Swedish University of Agricultural Sciences
TDF—tropical dry forests
TNC—The Nature Conservancy
UNCCD—United Nations Convention to Combat Desertification
UNFCCC—United Nations Framework Convention on Climate Change
USAID—U.S. Agency for International Development
WWF—Worldwide Fund for Nature (also known as World Wildlife Fund in North America)

Contributors' List

Mark Aldrich
Manager, Forest Landscape Restoration
Forests for Life Programme
WWF International
Av. Mont Blanc
1196 Gland, Switzerland
E-mail: maldrich@wwfint.org

Thomas F. Alnutt
Senior Conservation Specialist
Conservation Science Programme
World Wildlife Fund–US, Suite 200
Washington, DC 20037
E-mail: tom.allnutt@gmail.com

Everaldo Almeida
CIFOR Regional Office for Latin America
c/o EMBRAPA Amazônia Oriental
Trav. Dr. Enéas Pinheiro s/n
CEP 66.010-080 Belém—Pará, Brazil
E-mail: e.almeida@cgiar.org

James Aronson
Restoration Ecology Group
CEFE (CNRS-U.M.R. 5175)
1919, Route de Mende
F-34293 Montpellier, France
E-mail: james.aronson@cefe.cnrs.fr

Martin Ashby
Sion Chapel
Llanwrin
Powys
SY20 8QH
Wales, United Kingdom
E-mail: martin.ashby@Martin-Ashby.
  demon.co.uk

Eduard Plana
Head of Forest Fire Working Group
Forest Technology Centre of Catalonia
Area de Política Forestal
Pujada del Seminari s/n
Solsona 25280, Spain
E-mail: eduard.plana@ctfc.es

Nora Berrahmouni Corkland Programme
  Coordinator
WWF Mediterranean Programme Office
Via Po 25/C
00198 Rome, Italy
E-mail: nberrahmouni@wwmedpo.org

Jennifer Biringer
World Wildlife Fund–US, Suite 200
Washington, DC 20037
E-mail: jennifer.biringer@wwfus.org

Silvio Brienza, Jr.
Embrapa Amazônia Oriental,
Trav. Dr. Enéas Pinheiro s/n 66095-100
Belém—Pará, Brazil
E-mail: brienza@cpatu.embrapa.br

Katrina Brown
Professor of Development Studies
School of Development Studies
University of East Anglia
Norwich
NR4 7TJ, United Kingdom
E-mail: k.brown@uea.ac.uk

Antonio P. Carandang
Forestry Consultant
Main Street
Marymount Village, Anos
Los Banos,
Laguna 4030, Philippines
E-mail: apc@laguna.net

Marc Castellnou
Forestry Engineer
GRAF-Fire Service
Government of Catalonia
Ctra. Universitat Autònoma, s/n
08290 Cerdanyola del Vallès (Vallès
  Occidental)
Spain
E-mail: incendis@yahoo.com

Rufi Cerdan
Dr. in Geography
Autonomous University of Barcelona
Campus de la UAB, Edifici B
08193 Bellaterra
Spain
E-mail: rufi.cerdan@uab.es

Soh Koon Chng
Communications Manager
WWF International
Av. Mont Blanc
1196 Gland, Switzerland
E-mail: skchng@wwfint.org

Unna Chokkalingam
Scientist
Environmental Services and Sustainable Use
  of Forests Programme
Centre for International Forestry Research
  (CIFOR)
P.O. Box 6596 JKPWB
Jakarta 10065, Indonesia
E-mail: u.chokkalingam@cgiar.org

Nigel Dudley
Consultant
Equilibrium
47 The Quays
Cumberland Road
Bristol
BS1 6UQ, United Kingdom
E-mail: equilibrium@compuserve.com

Simon Dufour
PhD student
CNRS UMR 5600
18 rue Chevreul
69362 Lyon Cedex 07, France
E-mail: sim_dufour@yahoo.fr

Chris Elliott
Director, Forest Programme
WWF International
Av. Mont Blanc
1196 Gland, Switzerland
E-mail: celliott@wwfint.org

Thomas K. Erdmann
Regional Coordinator
ERI Madagascar Project
c/o Development Alternatives, Inc.
7250 Woodmont Ave.
Suite 200
Bethesda, MD 20814
E-mail: tom_erdmann@dai.com

Alan Watson Featherstone
Trees for Life
The Park
Findhorn Bay
Forres IV36 3TZ
Scotland, United Kingdom
E-mail: trees@findhorn.org

Don Gilmour
Environmental Consultant
42 Mindarie Cres Wellington Point
4160 Queensland, Australia
E-mail: gilmour@itxpress.com.au

Manuel Guariguata
Environmental Affairs Officer
United Nations Environment Programme
Secretariat of the Convention on Biological
  Diversity
413 St. Jacques, Suite 800
Montreal, Quebec, Canada
E-mail: manuel.guariguata@biodiv.org

Tini Gumartini
Environmental Services and Sustainable Use
  of Forests Programme
Centre for International Forestry Research
  (CIFOR)
P.O. Box 6596 JKPWB
Jakarta 10065, Indonesia
E-mail: gumartini@cgiar.org

Lara J. Hansen
Chief Scientist for Climate Change
WWF
1250 24th Street NW
Washington, DC 20016
E-mail: lara.hansen@wwfus.org

Mary Hobley
Consultant
Glebe House
Chard Street
Thorncombe
Chard TA20 4NE, United Kingdom
E-mail: mary@maryhobley.co.uk

Marja Hokkanen
Metsähallitus
Natural Heritage Services
P.O Box 94
01301 Vantaa, Finland
E-mail: marja.hokkanen@metsa.fi

Silvia Holz
Ph.D. candidate,
National University of Buenos Aires
Departamento de Ecología, Genética y
    Evolución
(4to Piso, Pabellón II).
Facultad de Ciencias Exactas y Naturales
Güiraldes 2620. Ciudad Universitaria
CP: 1428, Ciudad de Buenos Aires, Argentina
E-mail: silviaholz@yahoo.com.ar

José Manuel Nicolau Ibarra
Profesor Titular de Ecología
Departamento de Ecología
Universidad de Alcalá
28871 Alcalá de Henares
Madrid, Spain
E-mail: josem.nicolau@uah.es

Tanguy Jaffré
Directeur de Recherche de l'IRD
Laboratoire de Botanique et d'Ecologie
    Végétale
Centre IRD
BP A5
F-98848 Noumea Cedex
Nouvelle-Calédonie, France
E-mail: jaffre@noumea.ird.nc

Wil de Jong
Professor
Japan Centre for Area Studies, National
    Museum of Ethnology
10-1 Senri Expo Park, Suita
Osaka 565-8511, Japan
E-mail: wdejong@idc.minpaku.ac.jp

Scott Jones
Forests for People Group
Centre for International Development and
    Training
University of Wolverhampton
Telford Campus
TF2 9NT, United Kingdom
E-mail: tiger.moth@ntlworld.com

Gladwin Joseph
Director and Fellow
Ashoka Trust for Research in Ecology and the
    Environment
Hebbal
Bangalore 560024, India
E-mail: gladwin@atree.org

Harri Karjalainen
Head, Forest Programme
WWF Finland
Lintulahdenkatu 10
00500 Helsinki, Finland
E-mail: harri.karjalainen@wwf.fi

David Lamb
School of Integrated Biology
University of Queensland
Brisbane, 4072, Australia
E-mail: d.lamb@botany.uq.edu.au

Abel Meza Lopez
Centre for International Forestry Research
    (CIFOR)
Apdo. 558, Carretera Fdco. Basadre km 4,200
Pucallpa, Peru
E-mail: cifor-peru@cgiar.org;
    a.meza@cgiar.org

Colby Loucks
Conservation Science Programme
World Wildlife Fund–U.S., Suite 200
Washington, DC 20037
E-mail: colby.loucks@wwfus.org

Porter P. Lowry II
Curator and Head, Africa and Madagascar
  Department
Missouri Botanical Garden
P.O. Box 299
St. Louis, Missouri 63166-0299,
and
Département Systématique et Evolution,
  Muséum National d'Histoire Naturelle
C.P. 39
57 rue Cuvier
75231 Paris CEDEX 05, France
E-mail: Pete.Lowry@mobot.org

Stephanie Mansourian
Consultant—WWF International
10 rte de Burtigny
1268 Begnins, Switzerland
E-mail: stephanie.mansourian@worldcom.ch

Jeffrey A. McNeely
Chief Scientist
IUCN—The World Conservation Union
Rue Mauverney, 28
1196 Gland, Switzerland
E-mail: jam@iucn.org

Florencia Montagnini
Professor in the Practice of Tropical Forestry
Yale University
School of Forestry and Environmental Studies
370 Prospect St.
New Haven, CT 06511
E-mail: florencia.montagnini@yale.edu

Peter Moore
Fire Management and Policy Specialist
Metis Associates—Strategic Analysts
P.O. Box 1772
Bowral NSW 2576, Australia
E-mail: metis@metis-associates.com

Mariano Moreno de las Heras
PhD Student
Departamento de Ecología
Edificio de Ciencias
Universidad de Alcalá
28871 Alcalá de Henares
Madrid, Spain
E-mail: mariano.moreno@uah.es

John Morrison
Deputy Director
Conservation Science Programme
World Wildlife Fund–U.S., Suite 200
Washington, DC 20037
E-mail: john.morrison@wwfus.org

Murniati
Forestry Research and Development Agency
  (FORDA)
Jalan Gunung Batu No. 5
Bogor, Indonesia
E-mail: murniati@forda.org

Robert Nasi
Principal Scientist
Programme on Environmental Services and
  Sustainable Use of Forests
Centre for International Forestry Research—
  CIRAD
Campus International de Baillarguet TA 10/D
34398 Montpellier Cedex 5, France
E-mail: r.nasi@cgiar.org

Ani Adiwinata Nawir
Scientist
Forests and Livelihoods Programme
Centre for International Forestry Research
  (CIFOR)
P.O. Box 6596 JKPWB
Jakarta 10065, Indonesia
E-mail: a.nawir@cgiar.org

Adrian Newton
Senior Lecturer
School of Conservation Sciences
Bournemouth University
Talbot Campus
Fern Barrow
Poole
Dorset BH12 5BB, United Kingdom
E-mail: anewton@bournemouth.ac.uk

Nguyen Thi Dao
Annamites Ecoregion Conservation Manager,
  Vietnam
WWF Indochina Programme
40 Cat Linh
Ba Dinh District
Hanoi, Vietnam
E-mail: dao@wwfvn.org.vn

Sheila O'Connor
Director, Conservation Measures and Audits
WWF International
39 Stoke Gabriel Rd.
Galmpton nr Brixham
Devon TQ5 0NQ, United Kingdom
E-mail: soconnor@wwfint.org

Jessica Orrego
Forestry Project Manager
Edinburgh Centre for Carbon Management
Tower Mains Studios
18F Liberton Brae
Edinburgh, EH16 6AE, United Kingdom
E-mail: jessica.orrego@eccm.uk.com

Gonzalo Oviedo
Senior Advisor, Social Policy
IUCN–The World Conservation Union
28 Rue Mauverney
1196 Gland, Switzerland
E-mail: gonzalo.oviedo@incn.org

Jussi Päivinen
Metsähallitus, Natural Heritage Services
P.O. Box 36
40101 Jyväskylä, Finland
E-mail: jussi.paivinen@metsa.fi

Hervé Piégay
Researcher
CNRS UMR 5600
18 rue Chevreul
69362 Lyon Cedex 07, France
E-mail: piegay@univ-lyon3.fr

Guillermo Placci
Consultant
Constitución 237
5800—Río Cuarto, Cba, Argentina
E-mail: guillermoplacci@ciudad.com.ar

Gérard Rambeloarisoa
Forest Programme Officer
WWF Madagascar
WWF Madagascar and West Indian Ocean
  Programme Office
B.P. 738
Antananarivo 101, Madagascar
E-mail: grambeloarisoa@wwf.mg

Pedro Regato
Head, Forest Programme
WWF Mediterranean Programme Office
Via Po 25/C
00198 Rome, Italy
E-mail: pregato@wwfmedpo.org

José M. Rey Benayas
Dpto. de Ecología
Edificio de Ciencias
Universidad de Alcalá
28871 Alcalá de Henares, Spain
E-mail: josem.rey@uah.es

Doreen Robinson
Biodiversity and Natural Resources Specialist
USAID
1300 Pennsylvania Ave. NW
Ronald Reagan Building 3.08
Washington, DC 20523-3800
E-mail: drobinson@usaid.gov

Triagung Rooswiadji
Programme Manager
WWF Indonesia's Nusa Tenggara Programme
Jl. DODIKLAT No. 2
Kelurahan Oebubu
Kupang, NTT 8500, Indonesia
E-mail: triagung@kupang.wasantara.net.id

Lukas Rumboko Wibowo
Forestry Research and Development Agency
  (FORDA)
Jalan Gunung Batu No. 5
Bogor, Indonesia
E-mail: lukas_19672000@yahoo.com

Cesar Sabogal
Senior Scientist, Tropical Silviculture and
  Forest Management
CIFOR Regional Office for Latin America
c/o EMBRAPA Amazônia Oriental
Trav. Dr. Enéas Pinheiro s/n,
CEP 66.010-080 Belém
Pará, Brazil
E-mail: c.sabogal@cgiar.org

Nick Salafsky
Co-Director
Foundations of Success
4109 Maryland Ave.
Bethesda, MD 20816
E-mail: Nick@FOSonline.org

Daniel W. Salzer
Conservation Measures Manager
The Nature Conservancy
Conservation Measures Group
821 SE 14th Ave.
Portland, OR 97214
E-mail: dsalzer@tnc.org

Jeffrey Sayer
Senior Advisor
WWF International
Av. Mont Blanc
1196 Gland, Switzerland
E-mail: jsayer@wwfint.org

Kirsten Schuyt
Resource Economist
WWF International
Av. Mont Blanc
1196 Gland, Switzerland
E-mail: Kschuyt@wwfint.org

Sue Stolton
Consultant
Equilibrium
47 The Quays, Cumberland Road
Bristol
BS1 6UQ, United Kingdom
E-mail: equilibrium@compuserve.com

Takeshi Toma
Senior Scientist
Environmental Services and Sustainable Use of Forests Programme
Centre for International Forestry Research (CIFOR)
P.O. Box 6596 JKPWB,
Jakarta 10065, Indonesia
E-mail: t.toma@cgiar.org
Present address:
Associate Research Coordinator
Research Planning and Coordination Division, Forestry and Forest Products Research Institute (FFPRI)
1 Matsunosato, Tsukuba, Ibaraki 305-8687, Japan
E-mail: toma@affrc.go.jp

Daniel Vallauri
WWF France
6 Rue des Fabres
13001 Marseille, France
E-mail: dvallauri@wwf.fr

Ramon Vallejo
CEAM,
Parque Tecnológico,
Ch. Darwin 14
E-46980 Paterna, Spain
E-mail: vvallejo@ub.edu

Lawrence R. Walker
Professor of Biology
Department of Biological Sciences
University of Nevada, Las Vegas
Box 454004
4505 Maryland Parkway
Las Vegas, NV 89154-4004
E-mail: walker@unlv.nevada.edu

Steve Whisenant
Professor and Department Head
Department of Rangeland Ecology and Management
2126-TAMU
Texas A&M University
College Station, TX 77843-2126
E-mail: rangerider@mac.com

Eva Wollenberg
Senior Scientist
Forests and Governance Programme
Centre for International Forestry Research (CIFOR)
P.O. Box 6596 JKPWB
Jakarta 10065, Indonesia
E-mail: L.wollenberg@cgiar.org

Zhou Zaizhi
Research Institute of Tropical Forestry
Chinese Academy of Forestry
Longdong, Guangzhou 510520, China
E-mail: zzzhoucn@21cn.com

Part A
Toward a Wider Perspective in Forest Restoration

Section I
Introducing Forest Landscape Restoration

1
Forest Landscape Restoration in Context

Nigel Dudley, Stephanie Mansourian, and Daniel Vallauri

Key Points to Retain

Forest landscape restoration is grounded in ecoregion conservation and is defined as a planned process that aims to regain ecological integrity and enhance human well-being in deforested or degraded landscapes.

Such an approach helps achieve a balance between human needs and those of biodiversity by restoring a range of forest functions within a landscape and accepting the trade-offs that result.

1. Background and Explanation of the Issue

People have been actively using forests since long before the beginning of history. The oldest known written story, the *Epic of Gilgamesh* recorded on 12 cuneiform tablets in Assyria in the seventh century B.C., includes reference to the problems of forest loss. The need for good tree husbandry was stressed in Virgil's pastoral poem *The Georgics* in 30 B.C., written to promote rural values within the Roman Empire. The oldest records of forest management in the world have been kept without a break for 2000 years in Japan, relating to forests managed to produce timber for Shinto temples. The need for large-scale restoration has also been recognised for centuries; for example, the English pamphleteer John Evelyn wrote a tract calling for major tree planting during the time of Queen Elizabeth I in the 1600s. In more recent times, forest departments around the world have developed major efforts at reforestation in Europe, eastern North America, Australia, New Zealand, and increasingly in parts of the tropics.[1] In the last 20 years, hundreds of aid and conservation projects have promoted and carried out tree planting schemes and the development of tree nurseries, aimed at both supplying goods such as fuelwood and at restoring ecological functions and protecting biodiversity. Following the Society for Ecological Restoration International (SERI) and its chapters around the world, the scientific knowledge on ecological restoration has been conceptualised and applied to many different types of ecosystem, including forest landscapes. Good books have already been published.[2] Why then do we need another book about restoration?

The arguments for forest restoration are becoming more compelling. Forest loss and degradation is a worldwide problem, with net annual estimates of forest loss being 9.4 million hectares throughout the 1990s[3] and those for degradation uncalculated but universally agreed to be even higher. The most severe losses are currently concentrated mainly, although not exclusively, in the tropics, with

---

[1] For an overview see Perlin, 1991.
[2] Perrow and Davy; 2002, SERI, 2002; Whisenant, 1999.
[3] FAO, 2001.

the temperate countries gradually recovering forest area if not necessarily quality after severe deforestation in the past. As well as creating acute threats to forest dependent biodiversity, the decline in global forests also has a series of direct social and economic costs because of the role of forests in supplying timber and many important nontimber forest products along with a wide range of environmental service such as the stabilisation of soils and climate. Forest loss and degradation has already led to the extinction of species, has altered hydrological regimes and damaged the livelihoods of millions of people—mainly amongst the poorest on the earth—who rely on forests for subsistence. In many areas, protecting and managing the remaining forests are no longer sufficient steps in themselves to ensure that forest functions are maintained, and restoration is already an essential third component of any management strategy.

Unfortunately, many existing restoration projects have partially or completely failed, often because the trees that they sought to establish have not survived or have been rapidly destroyed by the same pressures that have caused forest loss in the first place. Anyone working regularly in the tropics becomes accustomed to finding abandoned tree nurseries, often with their donor organisations' signboards still in place, the paint gradually peeling away. Even when crops of trees have survived to maturity, they have not necessarily been welcomed, as evidenced by the widespread controversy over afforestation with exotic monocultures of conifers in much of western Europe[4] and the increasingly bitter debates about tree plantations in the tropics.[5]

There has also often been a mismatch between social and ecological goals of conservation; either restoration has aimed to fulfil social or economic needs without reference to its wider ecological impacts, or it has had a narrow conservation aim without taking into account people's needs.

A number of consequent problems can be identified. Most restoration to date has been site-based, aiming to produce one or at most a limited number of goods and services. Projects have often sought to encourage and sometimes impose tree planting without understanding why trees disappeared in the first place and without attempting to address the immediate or underlying causes of forest loss.[6] Projects have also relied heavily on tree planting, which is often the most expensive way of reestablishing tree cover over a large area, frightening off governments, donors, and nongovernmental organisations. Because restoration takes time, it is essential to think and plan long term. Unfortunately, short-term political interests often supersede longer term priorities, creating simplistic approaches.

The above reservations are not to underestimate the major steps that have been made in understanding the ecological and social aspects of restoration, many of which are summarised in this book. Criticising after the event is always easy, and we also recognise the very real benefits that have accrued from successful restoration projects. Nonetheless, we are far from alone in believing that some new perspectives are needed in addressing the current restoration challenge. Perhaps the most important of these relates to working on a broader scale, along with all the implications that this has.

1.1. Taking a Broader Approach

An increasing number of governmental and nongovernmental conservation institutions have recognised that in order to achieve lasting conservation impacts it is necessary to work on a larger scale than has been the case in the past. Although there are a number of ways of defining useful ecological units for planning conservation, the concept of the *ecoregion* is increasingly being adopted, including by WWF, the global conservation organisation. An ecoregion is defined as a large area of land or water that contains a geographically distinct assemblage of natural communities that share a large majority of their species and ecological dynamics, share similar environmental conditions, and

---

[4] Tompkins, 1989.
[5] Carrere and Lohmann, 1996.
[6] Eckholm, 1979.

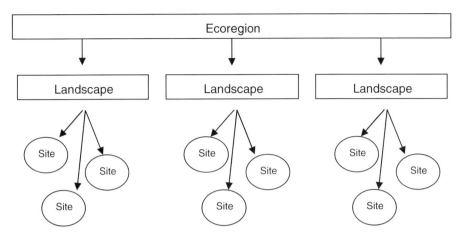

FIGURE 1.1. At the ecoregional scale, ecoregion visioning can help to identify a series of priority landscapes. At the landscape level, assessment and negotiation can help to identify agreed forest functions to be restored, leading to a number of actions at individual sites within the landscape. All these fit within the landscape goals for restoration, which themselves contribute to the ecoregion vision.

interact ecologically in ways that are critical for their long-term persistence. Ecoregions are suitable for broad-scale planning, which usually includes the identification of a few smaller priority landscapes that are particularly important from a conservation perspective, themselves composed of numerous sites with different management regimes or habitats (see chapter "Restoration as a Strategy to Contribute to Ecoregion Visions").

As used here (Fig. 1.1), landscapes are generally smaller than ecoregions, and typically a number of important "conservation landscapes" have been identified within ecoregions during planning processes. But the key point here is that landscapes are bigger than single sites and therefore almost always encompass a range of different management approaches.

Coming from a conservation organisation, this book is biased toward ecological and biodiversity issues. However, forests have social and economic functions as well, and restoration efforts often need to address many needs at once. This may not be possible within a single site; it is, for example, difficult to create a large harvest of industrial timber or firewood in an environment that is also suitable for specialised or sensitive wildlife species. One important reason for shifting the focus to a landscape scale is that it is hoped this can provide a broad enough area to plan a suite of restoration activities that could meet multiple needs and to negotiate the compromises and trade-offs that such a mosaic entails. The aims of forest landscape restoration have therefore always transcended conservation to embrace development as well, and we have invited a number of experts to provide a parallel set of social tools and approaches within the current volume. We believe that successful restoration on a broad scale relies on getting the right mix between social and environmental needs; this is a fundamental part of the process and not an optional extra.

Accordingly, in 2000, WWF and IUCN, the World Conservation Union, brought together a range of experts from different organisations, different regions, and different disciplines to agree on a definition for forest landscape restoration[7]: "A planned process that aims to regain ecological integrity and enhance human well-being in deforested or degraded landscapes." This definition and approach lies at the heart of the current book. "Ecological integrity" is described by Parks Canada as a state of ecosystem development that is characteristic of its geographic location, containing a full range of native species and supportive processes that are present in viable numbers. "Well-being" embraces the factors that make

[7] WWF and IUCN, 2000.

human life comfortable, such as money, peace, health, stability, and equable governance.

1.2. What Is Special About Forest Restoration in a Landscape?

Restoring the complexity of a small patch of forests is in itself an achievement. However, a greater challenge lies in restoring a matrix of forests within larger areas—landscapes—to meet different needs. At this greater spatial scale, different influences, pressures, stakeholders, and habitats coexist, which in some ways increases the challenges of restoration. However, the landscape scale also provides enough space to plan and implement restoration to meet multiple needs.

Conservation priorities therefore must be balanced with other aspects of sustainable development. Specific uses and priorities may have to be focussed on part of the forest landscape, and the resulting trade-offs negotiated and agreed to by a wide range of stakeholders. The resulting task is generally too complex to be solved solely by site-based approaches focussing on a narrow range of benefits from individual forests. Achieving a balance between the various goods and services required from restored forest ecosystems requires conceptualisation, planning, and implementation on a broader scale.

It also requires deciding where forest is and is not needed. Aiming at restoring forest functions does not necessarily mean restoring forest across the whole landscape; this is often impossible in a crowded world with many competing claims on land. Rather, it entails identifying those areas where forests are most useful, from a variety of social and ecological perspectives, and further identifying what type of forest is likely to be most useful in a particular location. Whilst from a conservation perspective a high degree of naturalness is often important, this may not be the case for social or economic uses. Even in the parts of the landscape that are "specialised" in conservation, sometimes cultural landscapes are desired either because they have been in place for so long that remaining biodiversity has adapted to these conditions or because there is not sufficient space for a fully functioning natural system (for instance, with respect to the way that the forest changes and regenerates over time).

Forests managed for social needs may have different priorities. Sometimes these overlap with conservation requirements—for instance some forests managed for nontimber forest products can be extremely rich in biodiversity—in other cases they do not. Seeking a balance at a landscape scale is more important than trying to make sure that every scrap of forest fulfils every possible role. Broad-scale restoration in most cases, therefore, has to address multiple, sometimes competing, needs that will themselves entail different types of forests (perhaps ranging from natural forests to plantations) and sometimes also including quite specific requirements such as particular nontimber forest products required by local communities or maintenance of water quality in a certain watershed. Such *multifunctional landscapes* by their nature need to be planned and implemented on a far broader scale than an individual forest patch.

2. Conclusion

For foresters, restoration traditionally meant establishing trees for a number of functions (wood or pulp production, soil protection). For many conservationists restoration is either about restoring original forest cover in degraded areas or about planting corridors of forest to link protected areas. For many interested in social development, the emphasis will instead be on establishing trees that are useful for fuelwood, or fruits, or as windbreaks and livestock enclosures. The sad fact is that all too many restoration projects do not bother to find out what local people really want at all; if they do, then a collection of different and often opposing or mutually exclusive wants and desires emerge. There is still a lot to be learned and disseminated about reconciling nature and human needs, and about planning restoration areas within larger scales in order to return as wide a range of forest functions as possible. This requires the ability to work across disciplines, including agriculture, forest-compatible income-generation activities, forestry, and addressing water issues as well as specific social issues. It also, perhaps even more importantly, requires finding out how to bring the people most affected into the debate, not as a matter of duty or because funding agencies expect it but because this is vital and necessary for both nature and human well-being.

Through ecoregion conservation, WWF has learned that working on a large scale is complex, costly, and time-intensive; however, it is also a more sustainable way of addressing conservation than through small, often unrelated projects. This approach is also a challenge for restoration.

References

Carrere, R., and Lohmann, L. 1996. Pulping the South: Industrial Plantations and the World Paper Economy. Zed Books and the World Rainforest Movement, London and Montevideo.

Eckholm, E. 1979. Planting for the Future: Forestry for Human Needs. Worldwatch Paper number 26. Worldwatch Institute, Washington, DC.

FAO. 2001. Global Forest Resource Assessment 2000: Main Report. FAO Forestry Paper 140. Food and Agriculture Organisation of the United Nations, Rome.

Perlin, J. 1991. A Forest Journey: The Role of Wood in the Development of Civilisation. Harvard University Press, Cambridge, MA, and London.

Perrow, M.R., and Davy, A.J. 2002. Handbook or Ecological Restoration, vol. 1 and 2. Cambridge University Press, Cambridge, UK.

Society for Ecological Restoration International. Science and Policy Working Group. 2002. The SER Primer on Ecological Restoration, www.ser.org.

Tompkins, S. 1989. Forestry in Crisis: The Battle for the Hills. Christopher Helm, London.

Whisenant, S.G. 1999. Repairing Damaged Wildlands—a Process-Oriented, Landscape-Scale Approach. Cambridge University Press.

WWF and IUCN. 2000. Minutes, Restoration workshop, Segovia, Spain (unpublished).

2
Overview of Forest Restoration Strategies and Terms

Stephanie Mansourian

Confusion reigns as the term restoration is used indiscriminately, with no consensus even among practitioners in its meaning.

Stanturf and Madsen, 2002

Key Points to Retain

There are numerous terms promoting different strategies when dealing with forest restoration, which could be a source of confusion.

WWF is implementing forest landscape restoration (FLR) as an integral component of the conservation of large, biologically important areas such as ecoregions, along with protection and good management.

Forest landscape restoration is an approach to forest restoration that seeks to balance human needs with those of biodiversity, thus aiming to restore a range of forest functions and accepting and negotiating the trade-offs between them.

While the challenge of restoration on a large scale is greater than at individual sites, it is accepted nowadays that the effectiveness of forest restoration and its chances of sustainability are both much greater on a large scale.

Forest landscape restoration aims to achieve a landscape containing valuable forests, rather than returning forest cover across an entire landscape.

1. Background and Explanation of the Issue

When forests are lost or degraded, we lose far more than just the trees that they contain. Forests provide a large number of goods and services, including habitat for species, homeland for indigenous peoples, recreational areas, food, medicines, and environmental services such as soil stabilisation. And as forest areas are reduced, pressure on remaining forests increases.

Efforts at reversing this trend have had only limited success. For many, restoration signifies large-scale afforestation or reforestation (mainly using fast growing exotic species), which have only limited conservation benefits. This has been the approach taken by many governments that are seeking to support a timber industry or create jobs or, equally, those who have taken a simplistic approach to flood or other disaster mitigation. On the other hand, some have sought to re-create original forests, a near-impossible feat in areas where millennia of human intervention have modified the landscape and local conditions.

Many different terms are used to describe these different approaches and can result in some confusion or misconceptions.[8] We attempt here to cover most of the terminology used in English taken from the Society for Ecological Restoration International (SERI), which has

[8] Ormerod, 2003.

made the best attempt at cataloguing and defining these different terminologies and concepts. It must be noted that this complexity is also apparent and sometimes exacerbated when translating these terms into other languages.

2. Examples

We present below a number of terms that have been defined recently by SERI in its "The SER Primer on Ecological Restoration."[9]

2.1. Ecological Restoration

Ecological restoration is defined as the process of assisting the recovery of an ecosystem that has been degraded, damaged, or destroyed. It is an intentional activity that initiates or accelerates the recovery of an ecosystem with respect to its health, integrity, and sustainability.

Example 1: In 2000, in an attempt to re-create a native wild wood, the Scottish nongovernmental organisation (NGO), Borders Forest Trust, together with many partners, bought a 600-hectare plot of land, Carrifran, in the Southern Uplands of Scotland in order to restore its original forest. Thanks to fossil pollen buried deep in peat, it was possible to identify the nature of the variety of species previously found on this now near-denuded site and therefore to develop a restoration plan that aimed to re-create the species' mix that had occurred in the past. Thousands of native tree seeds from surviving woodland remnants in the vicinity were collected. A total of 103.13 hectares (165,008 trees) have been planted at Carrifran since the start of the project. The upper part of the site is being allowed to regenerate naturally.[10]

2.2. Rehabilitation

Rehabilitation emphasises the reparation of ecosystem processes, productivity, and services, whereas the goals of restoration also include the reestablishment of the preexisting biotic integrity in terms of species' composition and community structure.

Example 2: Bamburi Cement's quarries in Mombasa (Kenya) were once woodland expanses covering 1,200 hectares.[11] Starting in 1971, experiments began with the rehabilitation of the disused quarries. In the face of badly damaged soils, three tree species proved capable of withstanding the difficult growing conditions: *Casuarina equisetifolia, Conocarpus lancifolius*, and the coconut palm. The Casuarina is nitrogen fixing and is drought and salt tolerant, enabling it to colonise areas left virtually without soil. The Conocarpus is also a drought-, flood-, and salt-tolerant swamp tree. The decomposition of the Casuarina leaf litter was initially very slow due to a high protein content, thus impeding the nutrient cycling process, although this problem was overcome by introducing a local red-legged millipede that feeds on the dry leaves and starts the decomposition process. Today this area contains more than 200 coastal forest species and a famous nature trail, attracting 100,000 visitors a year since opening in 1984.

2.3. Reclamation

Reclamation is a term commonly used in the context of mined lands in North America and the United Kingdom. It has as its main objectives the stabilisation of the terrain, assurance of public safety, aesthetic improvement, and usually a return of the land to what, within the regional context, is considered to be a useful purpose.

Example 3: A large open-cut bauxite mine at Trombetas in Pará state in central Amazonia is located in an area of relatively undisturbed evergreen equatorial moist forest. A reclamation programme has been developed to restore the original forest cover as far as possible. The project has treated about 100 hectares of mined land per year for the last 15 years. First, the mined site was levelled and topsoil replaced to a depth of about 15cm using topsoil from the site that was removed and stockpiled (for less

---

[9] SERI, 2002.
[10] www.carrifran.com.
[11] Baer, 1996.

than 6 months) prior to mining. Next, the site was deep-ripped to a depth of 90 cm (1-m spacing between rows). Trees were planted along alternate rip lines at 2-m spacings (2500 trees per hectare) using direct seeding, stumped saplings, or potted seedlings. Some 160 local tree species were tested for their suitability in the programme, and more than 70 species from the local natural forests are now routinely used. After 13 years most sites have many more tree and shrub species than those initially planted because of seeds stored in the topsoil or colonisation from the surrounding forest. Not surprisingly, the density of these new colonists is greater at sites near intact forest, but dispersal was evident up to 640 m away from old-growth forest. The new species, most of which have small seed, have been brought to the site by birds, bats, or terrestrial mammals.[12]

2.4. Afforestation/Reforestation

Afforestation and reforestation refer to the artificial establishment of trees, in the former case where no trees existed before. In addition, in the context of the U.N.'s Framework Convention on Climate Change (UNFCCC) and the Kyoto protocol, specific definitions have been agreed on reforestation and afforestation.[13] Afforestation is defined by the UNFCCC as "the direct human-induced conversion of land that has not been forested for a period of at least 50 years to forested land through planting, seeding, and/or human induced promotion of natural seed sources."

Example 4: During the middle years of the 20th century, very large areas of long-deforested land were planted in Scotland by the state forestry body, initially as a strategic resource. In contrast to the Borders Forest Trust project described above, these efforts made no attempt to re-create the original forest, instead using exotic monocultures, mainly of Sitka spruce from Alaska (*Picea sitchensis*) or Norway spruce (*Picea abies*) from mainland Europe. Planting was generally so dense that virtually no understorey plant species developed.

Reforestation is defined by the UNFCCC as "the direct human-induced conversion of non-forested land to forested land through planting, seeding, and/or the human-induced promotion of natural seed sources, on land that was forested but that has been converted to non-forested land."

Example 5: In Madagascar, large plantation projects were planned in the early 1970s to supply a paper mill on the "Haut Mangoro." By 1990 about 80,000 hectares had been planted, 97 percent of which was *Pinus spp*. This project created significant social and political tensions, as the local population systematically opposed a project that it felt was not providing much benefit.[14]

2.5. What Is WWF's Definition?

In 2000 WWF and IUCN, the World Conservation Union, were asking the questions: What is meant by forest restoration? How can we achieve lasting and successful forest restoration in our ecoregional programmes? The two organisations felt that a suitable definition and typology of restoration were needed. In particular, given the large-scale conservation work that the organisations were engaging in, it was felt that there was still a gap in knowledge and in approaches to forest restoration. Notably, how does forest restoration relate to plantations, agroforestry, secondary forests, biological corridors, and single trees in the landscape?

In July 2000 WWF and IUCN brought together a number of regional conservation staff, foresters, economists, and other professionals to help them take restoration forward. They defined the term *forest landscape restoration* as "a planned process that aims to regain ecological integrity[15] and enhance human well-being[16] in deforested or degraded landscapes."

[12] Lamb and Gilmour, 2003.
[13] UNFCC, 2003.

[14] Faralala, 2003.
[15] Ecological Integrity, for WWF and IUCN, is "maintaining the diversity and quality of ecosystems, and enhancing their capacity to adapt to change and provide for the needs of future generations."
[16] Human well-being, for WWF and IUCN, is "ensuring that all people have a role in shaping decisions that affect their ability to meet their needs, safeguard their livelihoods, and realise their full potential."

The key elements of FLR are as follows:

- It is implemented at a landscape scale rather than a single site—that is to say, planning for forest restoration is done in the context of other elements: social, economic, and biological, in the landscape. This does not necessarily imply planting trees across an entire landscape but rather strategically locating forests and woodlands in areas that are necessary to achieve an agreed set of functions (e.g., habitat for a specific species, soil stabilisation, provision of building materials for local communities).
- It has both a socioeconomic and an ecological dimension. People who have a stake in the state of the landscape are more likely to engage positively in its restoration.
- It implies addressing the root causes of forest loss and degradation. Restoration can sometimes be achieved simply by removing whatever caused the loss of forest, (such as perverse incentives and grazing animals). This also means that without removing the cause of forest loss and degradation, any restoration effort is likely to be in vain.
- It opts for a package of solutions. There is no single restoration technique that can be applied to all situations. In each case a number of elements need to be covered, but how to do that depends on the local conditions. The package may include practical techniques, such as agro-forestry, enrichment planting, and natural regeneration at a landscape scale, but also embraces policy analysis, training, and research.
- It involves a range of stakeholders in planning and decision making to achieve a solution that is acceptable and therefore sustainable. The decision of what to aim for in the long term when restoring a landscape should ideally be made through a process that includes representatives of different interest groups in the landscape in order to reach, if not a consensus, at least a compromise that is acceptable to all.
- It involves identifying and negotiating trade-offs. In relation to the above point, when a consensus cannot be reached, different interest groups need to negotiate and agree on what may seem like a less than optimal solution if taken from one perspective, but a solution that when taken from the whole group's perspective can be acceptable to all.
- It places the emphasis not only on forest quantity but also on forest quality. Decision makers often think predominantly about the area of trees to be planted when considering restoration, yet often improving the quality of existing forests can yield bigger benefits for a lower cost.
- It aims to restore a range of forest goods, services, and processes, rather than forest cover per se. It is not just the trees themselves that are important, but often all of the accompanying elements that go with healthy forests, such as nutrient cycling, soil stabilisation, medicinal and food plants, forest-dwelling animal species, etc. Including the full range of potential benefits in the planning process makes the choice of restoration technique, locations, and tree species much more focussed. It also allows more flexibility for discussions on trade-offs with different stakeholders, by providing a diversity of values rather than just one or two.

Forest landscape restoration goes beyond establishing forest cover per se. Its aim is to achieve a landscape containing valuable forests, for instance partly to provide timber, partly mixed with subsistence crops to raise yields and protect the soils, as well as partly improving biodiversity habitat and increasing the availability of subsistence goods. By balancing these within a landscape, WWF believes that it is possible to enhance the overall benefits to people and biodiversity at that scale.

3. Outline of Tools

Broad definitions and explanations of what restoration entails can be found in most conservation and forestry institutions. Nonetheless, little of this has reached the field. Because of its complexity, large-scale restoration requires a mixture of responses from practical to political and many practitioners are at a loss as to where to begin.

Some practical guidance is available:

- The Society for Ecological Restoration (SERI) have developed guidelines for restoration (see *Guidelines for Developing and Managing Ecological Restoration Projects*, 2000, at www.ser.org).
- The International Tropical Timber Organisation (ITTO) developed some guidelines[17] on the restoration, management, and rehabilitation of degraded and secondary tropical forests.
- The International Union of Forest Research Organisations (IUFRO) runs a special programme on correct usage of technical terms in forestry called SilvaVoc, available on its Web site: www.iufro.org/science/special/silvavoc/.
- The Nature Conservancy (TNC)[18] has identified some guidance on when and where to restore (see *Geography of Hope Update, When and Where to Consider Restoration in Ecoregional Planning* at www.conserveonline.org).
- In 2003, IUCN and WWF published a book, by David Lamb and Don Gilmour,[19] *Rehabilitation and Restoration of Degraded Forests*, which covers site-based techniques to restoration (summarised in a paper in this manual) but also highlights some of the gaps.
- Cambridge Press has produced a *Handbook of Ecological Restoration*,[20] which is a two-volume handbook containing a large amount of material on the diverse aspects of restoration.

It should also be noted that a number of state forest services and the U.S. Department of Agriculture have produced guidelines for planting trees. However, while these guidelines may have some applicability for very specific cases (issues dealing with one or another specific species), they are of limited value for restoration within ecoregions or large and biologically and structurally complex areas.

Tools available to address specific elements of restoration are summarised in other chapters of this manual.

[17] ITTO, 2002.
[18] TNC, 2002.
[19] Lamb and Gilmour, 2003.
[20] Perrow and Davy, 2002.

4. Future Needs

In the context of terminology related to restoration, given the flurry of interest, concepts, and definitions being touted, there is a need for

- a set of widely accepted definitions (such as those of SERI) to be used more systematically and rigorously;
- efforts and resources to be more focussed on the "doing" than on the "defining";
- greater exchanges, debates, and sharing of experiences in order to disseminate the accepted concepts and the positive experiences; and
- the accepted definitions in the restoration field to be shared with other relevant expert groups, such as development workers, foresters, extension officers, etc.

References

Baer, S. 1996. Rehabilitation of Disused Limestone Quarries Through Reafforestation (Baobab Farm, Mombasa, Kenya). World Bank/Unep Africa Forestry Policy Forum, Nairobi, August 29–30, 1996.

Faralala. 2003. Rapport de Reconnaissance dans Cinq Paysages Forestiers. WWF, Madagascar.

ITTO Policy Series No. 13. 2002. Guidelines on the Restoration, Management and Rehabilitation of Degraded and Secondary Tropical Forest. Yokohama, Japan.

Lamb, D., and Gilmour, D. 2003. Rehabilitation and Restoration of Degraded Forests. IUCN, Gland, Switzerland and Cambridge, UK, and WWF, Gland, Switzerland.

Ormerod, S.J. 2003. Restoration in applied ecology: editor's introduction. Journal of Applied Ecology 40:44–50.

Perrow, M., and Davy A., eds. 2002. Handbook of Ecological Restoration. Cambridge University Press, Cambridge, England.

Society for Ecological Restoration International. Science and Policy Working Group. 2002. The SER Primer on Ecological Restoration, www.ser.org/.

Stanturf, J.A., and Madsen, P. 2002. Restoration concepts for temperate and boreal forests of North America and Western Europe. Plant Biosystems 136(2):143–158.

The Nature Conservancy (TNC). 2002. Geography of Hope Update: When and Where to Consider Restoration in Ecoregional Planning. www.conserveonline.org.

United Nations Framework Conference on Climate Change (UNFCCC) Subsidiary Body for Scientific and Technological Advice. 2003. Land Use, Land-Use Change and Forestry: Definitions and Modalities for Including Afforestation and Reforestation Activities Under Article 12 of the Kyoto Protocol. Eighteenth session, Bonn, June 4–13, 2003.

Section II
The Challenging Context of Forest Restoration Today

3
Impact of Forest Loss and Degradation on Biodiversity

Nigel Dudley

> ## Key Points to Retain
>
> Assessment of current forest condition is a necessary precursor to restoration.
>
> Ecological assessments should consider issues related to biodiversity, level of naturalness, and more generally ecological integrity.
>
> A number of assessment tools exist, for national, landscape, and site-level assessments. They include: at national scale, frontier forests; at landscape scale, forest quality assessment; and a number of site-level tools including High Conservation Value Forest assessments.

1. Background and Explanation of the Issue

1.1. The Need for Assessment and Likely Impacts of Forest Loss

Assessment of forest condition is an important precursor to the planning and implementation of restoration programmes. Restoration is a process that in the case of forests generally aims at rebuilding the ecosystem to some earlier or more desirable stage. There is widespread recognition of the need for restoration; for example, in its Programme of Work on Protected Areas the Convention on Biological Diversity advises governments to "rehabilitate and restore habitats and degraded ecosystems, as appropriate, as a contribution to building ecological networks, ecological corridors and/or buffer zones." Given limited time and resources, restoration must be strategic, focussing on forests that have the highest importance to biodiversity or to society, and considering the four goals of conservation biology: representation, maintenance of evolutionary/ecological processes, maintenance of species, and conservation of large habitat blocks. Reasonably fine-scale analyses are needed to choose specific sites where restoration might bring the highest benefits. From a conservation perspective, this means evaluating the impacts of forest loss, including analysis of biodiversity, authenticity, and ecological integrity.

Impacts on biodiversity: Complete forest loss has the clearest impact on biodiversity, with most forest-dwelling species unable to live in habitats that replace forests. However, it is harder to measure the impacts of changes such as fragmentation and loss of microhabitats. Management often simplifies forests, reducing biodiversity and age range; as older and dead trees disappear, so do many associated species. Conversely, pioneer or weed species may increase. Biodiversity monitoring is costly, and our knowledge of many forest ecosystems is still incomplete. One concept that has gained increasing recognition in the last few years is that of *critical thresholds* for particular species, that is, the population level below which further decline and eventual extirpation or extinction is likely, and where these thresholds are known they can play a key role in monitoring impacts and planning restoration strategies.

Impacts on authenticity or naturalness: On an ecosystem scale, measuring impacts on overall naturalness of forests is easier than surveying biodiversity and acts as a partial surrogate; generally the greater the naturalness of a forest, the more of its original constituent species are likely to survive. Worldwide forest authenticity is declining fast. In most West European countries less than 1 percent of forests are classified by the United Nations as "undisturbed."[21] A growing proportion of forests in Africa, the Pacific, and the Amazon have been logged at least once.

Ecological integrity: This concept covers many of the above issues. It is defined by Parks Canada as "a condition that is determined to be characteristic of its natural region and likely to persist, including abiotic components and the composition and abundance of native species and biological communities, rates of change, and supporting processes."[22]

Evaluation of options for restoration should also consider the reasons why forest loss or degradation have occurred. Many restoration programmes fail because the pressures that caused deforestation are not addressed, and restored forests suffer the same fate as the original forests. If population or economic pressures mean that there is insufficient fuelwood, then planted trees will be burned long before they have a chance to mature and reach a useful size. On the other hand, understanding the nature of the pressures and working with local communities to plan restoration in ways that are mutually beneficial increases the chances of restoration succeeding. Assessment needs to address several different aspects:

- Impacts of forest loss and degradation on biodiversity, naturalness, and ecological integrity;
- Some of the key factors causing change;
- Changes in biodiversity, naturalness, and ecological integrity following restoration interventions.

Whilst the first two can be assessed through single surveys, assessment of trends implies the need for a monitoring system.

2. Examples

2.1. New Caledonia

In New Caledonia the overall loss of forests creates a critical threat to biodiversity and ecological integrity. Today only 2 percent of the dry forest remains in the island, in scattered fragments of 300 hectares or less, leading to extreme threats to the remaining biodiversity. Over half of the 117 dry forest plant species assessed by the IUCN Species Survival Commission are threatened, and it is likely that several have already gone extinct. For example, the tree *Pittosporum tanianum* was discovered in 1988 on Leprédour Island in an area that has been devastated by introduced rabbits and deer, declared extinct in 1994, and rediscovered in 2002. This level of damage suggests an urgent need for both restoration of forest cover and a carefully designed series of interventions to protect and allow the spread of species that may already be at critically low levels.[23]

2.2. Western Europe

Changes in management and human disturbance have reduced near-natural forests to less than 1 percent of their original area in most western European countries, despite an expanding forest estate. In Europe as a whole, almost nine million hectares are defined as "undisturbed by man," but most of this exists in the Russian Federation and Scandinavia; Sweden records 16 percent of its forest as natural, Finland 5 percent, and Norway 2 percent. In most of Europe the proportion is usually from zero to less than 1 percent; for instance, Switzerland records 0.6 percent.[24] Even in forest-rich countries like Finland and Sweden, many forestd-welling species are threatened because the forests contain only a proportion of the

---

[21] UNECE and FAO, 2000.
[22] Parks Canada, undated.
[23] Vallauri and Géraux, 2004.
[24] UNECE and FAO, 2000.

expected habitats and ecosystem functions. Here the challenge is less to recover forest area (although this may sometimes be important) than to restore natural ecosystem processes and microhabitats. Specific monitoring criteria are needed and these have started to be developed, for instance by the Ministerial Conference on the Protection of Forests in Europe.[25]

2.3. Brazilian Atlantic Forests

In the Atlantic forest of Brazil, forest loss and fragmentation are combining to threaten endemic species. Although international attention tends to focus on threats to the Amazon, the Atlantic forests of Brazil have undergone far more dramatic losses. The forests have already been reduced to just 7 percent of their original size, and the associated threats to biodiversity are increased because the remaining areas are fragmented and the populations are genetically isolated. The area is home to many endemic species, including some of the 19 resident primates and 92 percent of amphibian species found there. Attention has focussed particularly on the golden lion tamarins (*Leontopithecus rosalia*), which now inhabit less than 2 percent of their original range. Their population is currently around 1000, up from little more than 200 twenty years ago following a major conservation effort. However, population is still believed to be below long-term viability, and subpopulations are isolated in remaining forest fragments. Restoration efforts, therefore, focus particularly in reconnecting the remaining forest fragments of high biological importance.

2.4. Uganda

In Uganda loss of connectivity is separating populations of mountain gorillas even in areas with relatively high forest cover. The world's remaining mountain gorillas (*Gorilla beringei beringei*) live in isolated rain forests in the mountains on the borders of Uganda, Rwanda, and the Democratic Republic of Congo, with half of the world's known population, 350 individuals, in Bwindi Impenetrable Forest Reserve in Uganda. Another major population is in the Virunga volcanoes area, some of which is in Mgahinga National Park. Neither of these populations is considered large enough to be genetically secure over time, but both reserves are also thought to be reaching their natural carrying capacity. Linking the two populations is important for their long-term survival, but the intervening land has all been converted to agriculture, and any restoration efforts will need a long period of planning and negotiation (information from park staff in Bwindi).

Understanding of what has been lost, and what is at risk of being lost, should be the basis for any forest restoration that has biodiversity conservation amongst its aims. This needs to be augmented with an understanding of what type or quality of forest is needed to maintain biodiversity. If the key issue is connectivity for large mammals and birds, for example, managed secondary forests or even plantations or shade-grown coffee may be suitable. If the threats are more generally to forest biodiversity, restoration efforts should probably be aimed at creating a forest as near to natural as possible.

3. Outline of Tools

Detailed biodiversity surveys are expensive and rely on a high level of expertise. Methodologies for achieving these have become increasingly sophisticated, and a number of short cuts have been developed where time and money are limited.

3.1. National Level Surveys

National level surveys can help identify the scale of the problems and the locations of valuable remaining forest habitat, which should usually serve as the starting point for restoration efforts. The U.N. Economic Commission for Europe and the Food and Agriculture Organisation asked countries to report on the proportion of their forest that was "undisturbed by man," taken here to mean left without management interventions for at least 200 years. This has created a fairly crude but effective

---

[25] Ministerial Conference on the Protection of Forests in Europe, 2002.

international database for many of the temperate countries, but as yet no similar exercise has been attempted in the tropics. It also does not create a very useful way of measuring progress in restoration. Some individual countries (e.g., Austria, France, and the U.K.) have also carried out detailed surveys of ancient forest.

3.2. High Conservation Value Forests (HCVF)

This is a WWF/ProForest methodology for identifying the forests of the highest conservation and social value in a landscape, drawing on six different types of HCVF: (1) forest areas containing globally, regionally, or nationally significant concentrations of biodiversity values (e.g., endemism, endangered species, refugia); (2) forest areas containing globally, regionally, or nationally significant large landscape level forests, where viable populations of most if not all naturally occurring species exist in natural patterns of distribution and abundance; (3) forest areas that are in or contain rare, threatened, or endangered ecosystems; (4) forest areas that provide basic services of nature in critical situations; (5) forest areas fundamental to meeting basic needs of local communities; and (6) forest areas critical to local communities' traditional cultural identity.[26] Although designed initially for site-level assessments, a landscape-scale methodology is being developed.

3.3. Forest Quality Assessment

WWF and IUCN have developed an approach to landscape assessment of forest quality using indicators to map social and ecological values, including identifying different elements of naturalness or authenticity, drawing on the following: composition, pattern, ecological functioning, process, resilience, and area (also see "Restoring Quality in Existing Native Forest Landscapes"). Assessment is based on a seven-stage process: identification of aims, selection of the landscape, selection of a toolkit (relevant indicators), collection of information about each indicator, assessment, presentation of results, and incorporation into management. Information is collected through primary research, literature review, and interviews. The extent to which assessment is a participatory process can change depending on the situation and aims.[27]

3.4. Frontier Forest Analysis

Frontier forest analysis is a World Resources Institute/Global Forest Watch approach[28] that defines frontier forests as free from substantial anthropogenic fragmentation (settlements, roads, clearcuts, pipelines, power lines, mines, etc.); free from detectable human influence for periods that are long enough to ensure that it is formed by naturally occurring ecological processes (including fires, wind, and pest species); large enough to be resilient to edge effects and to survive most natural disturbance events; containing only naturally seeded indigenous plant species; and supporting viable populations of most native species associated with the ecosystem.[29] It is mainly used at a national scale.

3.5. Site-Scale Survey Methods

A wide range of survey methods exist including some that have specifically been developed to facilitate rapid surveys for conservation practitioners, amongst these are the Rapid Ecological Assessment methodology developed by The Nature Conservancy.[30] Increasingly surveys by outside experts are being augmented by interviews and collaboration with local communities, which often have great understanding of population levels of key plants and animals; these sources are usually referred to as traditional ecological knowledge.

4. Future Needs

Despite expertise in survey methods, there is still much to be learned about accurate ways

---

[26] Jennings et al, 2003.
[27] Dudley et al, in press.
[28] Bryant et al, 1997.
[29] Smith et al, 2000.
[30] Sayre et al, 2002.

of monitoring of both biodiversity and, more critically, ecological integrity that would allow proper assessment of restoration outcomes over time and thus help set realistic goals for restoration. In general, quick and cost-effective methods of monitoring the impacts of restoration on biodiversity and ecology are still required in many ecosystems.

References

Bryant, D., Nielsen, D., and Tangley, L. 1997. The Last Frontier Forests: Ecosystems and Economies on the Edge. World Resources Institute, Washington, DC.

Dudley, N., Schlaepfer, R., Jackson, W., and Jeanrenaud, J. P. In press. A Manual on Forest Quality.

ECE and FAO. 2000. Forest Resources of Europe, CIS, North America, Australia, Japan and New Zealand. U.N. Regional Economic Commissions for Europe and the Food and Agriculture Organisation, Geneva and Rome.

Jennings, S., Nussbaum, R., Judd, N., et al. 2003. The High Conservation Value Toolkit. Proforest, Oxford (three-part document).

Ministerial Conference on the Protection of Forests in Europe. 2002. Improved Pan-European Indicators for Sustainable Forest Management: as adopted by the MCPFE expert level meeting, October 7–8, 2002, Vienna, Austria.

Parks Canada. Undated. http://www.pc.gc.ca/progs/np-pn/eco_integ/index_e.asp.

Sayre, R., et al. 2002. Nature in Focus: Rapid Ecological Assessment. The Nature Conservancy and the Island Press, Covelo and Washington, DC.

Smith, W., et al. 2000. Canada's Forests at a Crossroads: An Assessment in the Year 2000. Global Forest Watch, World Resources Institute, Washington, DC. See also the Global Forest Watch Web site: http://www.globalforestwatch.org.

Vallauri, D., and Géraux, H. 2004. Recréer des forêts tropicales sèches en Nouvelle Calédonie. WWF France, Paris.

4
The Impacts of Degradation and Forest Loss on Human Well-Being and Its Social and Political Relevance for Restoration

Mary Hobley

Forests: "the poor man's overcoat" (Westoby, 1989).

Forests have an important role to play in alleviating poverty worldwide in two senses. First, they serve a vital safety net function, helping rural people avoid poverty, or helping those who are poor to mitigate their plight. Second, forests have untapped potential to actually lift some rural people out of poverty (Sunderlin et al, 2004).

Key Points to Retain

Poor people rely on forests as a safety net to avoid or mitigate poverty and sometimes as a way to lift themselves out of poverty.

It is important to recognise different levels of poverty and different types of dependence on forests when trying to understand the likely social implications of forest restoration.

A series of tools and questions exist that can help to identify potential benefits from restoration, although these need to be used with care to avoid overlooking some of the poorest members of society.

1. Background and Explanation of the Issue

For many millions of people forests and forest products and services supply both direct and indirect sources of livelihood, providing a major part of their physical, material, economic, and spiritual lives[31]). The World Bank has estimated that 90 percent of the world's 1.2 billion poorest people depend on forests in some way or another. Forest areas often coincide with areas of high poverty incidence and livelihood dependence on forests. They often occur in remote rural areas with poor infrastructure and limited access to markets and other basic services; the livelihood options in such areas are highly circumscribed. The challenge facing many communities is not just the restoration of trees in their landscape but the growth of a political and social landscape that facilitates their ability to make choices to secure their livelihoods.

In this section we consider the impacts of forest loss and degradation on human well-being. At the most simple level the first question must be: impact on whom? This is an important point because degradation and loss of resources affects people in different ways. To explore this question we need to unpick the concept of well-being and then look at the ways in which forests and people are intertwined. The major focus of this section, however, is on those who are most adversely affected by changes in forest cover and quality—the poor, and in particular those living in forest areas. The second question to ask is why deforestation and degradation happen, since understanding the

[31] Byron and Arnold, 1997.

4. The Impacts of Degradation and Forest Loss on Human Well-Being

answers to this question provides answers to whom it impacts on. As part of this process we need to set out the major concepts and terms that support this understanding. These are *deforestation* and *degradation, well-being, livelihoods, people*, and *impact*.

The drivers of forest loss and degradation are complex and variable, moving from the extreme of deforestation for other land uses to more subtle forms of degradation through multiple overuse, either happening slowly or more rapidly depending on the pressures driving change. Who drives the changes in the forests and who benefits from them also helps to determine the impacts. These are not simple events and do not have simple causal consequences. For example, one person's loss as a result of forest degradation may be another person's gain if for instance opportunities to farm land are opened up. Timber companies benefit from timber extraction but generally the capture of benefits at the local level is very weak and the local social and environmental costs of logging are high.

Following Wunder[32] and the U.N. Food and Agricultural Organisation, *deforestation* (or forest loss) is defined as a radical removal of vegetation to less than 10 percent crown cover. For local people deforestation can be catastrophic, as in the case of large-scale clearfelling by an outside agency that destroys resources without offering any alternatives, or in other cases it can be the planned precursor to an alternative land use system such as farming, which in terms of livelihood outcomes may provide more secure alternatives than that offered by the forest.

Degradation is taken to mean a loss of forest structure, productivity, and native species' diversity. A degraded site may still contain trees or forest but it will have lost its former ecological integrity.[33] Degradation is a process of loss of forest quality that is in practice often part of the chain of events that eventually leads to deforestation.

Impact: "Impact concerns the long-term and sustainable changes introduced by a given intervention in the lives of beneficiaries. Impact can be related either to the specific objectives of an intervention or to unanticipated changes caused by an intervention; such unanticipated changes may also occur in the lives of people not belonging to the beneficiary group. Impact can be either positive or negative, the latter being equally important to be aware of."[34]

Well-being is a concept used to describe all elements of how individuals experience the world and their capacities to interact, and includes the degree of access to material income or consumption, levels of education and health, vulnerability and exposure to risk, opportunity to be heard, and ability to exercise power, particularly over decisions relating to securing livelihoods.[35] When used in connection with livelihoods it becomes a powerful concept for considering the effects of change on all aspects of the lived experience of an individual.

A useful definition of *livelihoods* is as follows: "People's capacity to generate and maintain their means of living, enhance their well-being and that of future generations. These capacities are contingent upon the availability and accessibility of options which are ecological, economic, and political and which are predicated on equity, ownership of resources, and participatory decision making."[36]

The individual experience of well-being varies along a continuum, with ill-being at one end and well-being at the other, and is not static; it can vary during an individual's life cycle. Those classified as extreme poor often suffer ill-being, particularly expressed through high degrees of exposure to vulnerability and risk, whereas those who can be classified as improving poor generally experience higher levels of well-being. It is important to be able to differentiate among people's vulnerabilities in order to understand the differential effects that forest loss and degradation may have.

One of the most important issues to consider when looking at the effects of a change in access to or availability of forest products and services is a household's exposure to vulnera-

---

[32] Wunder, 2001.
[33] Lamb and Gilmour, 2003: 4.
[34] Blankenberg, 1995.
[35] World Bank, 2001:15.
[36] de Satgé, 2002:4.

bility and risk. It is clear that households and individuals within households experience different levels of vulnerability and exposure to risk. This is particularly important in the assessment of the effects of forest quality change, as it has differential impacts within and between households.

There are two main ways in which forests impact on livelihoods and reduce vulnerability:

- as a *safety net* helping rural people avoid poverty and helping those who are poor to mitigate their poverty;
- through their potential to lift some people out of poverty.

For the sake of understanding the likely impacts of forest loss or restoration, it is useful to define *people* in terms of their vulnerability and their relationships with forests and forest products (see Table 4.1 for examples of impacts of degradation and deforestation on these different groups):

- Extreme poor with very little or no capability for social mobilisation
- Coping poor with little capability for social mobilisation
- Improving poor with some capability for social mobilisation

This typology helps to underline the importance of understanding the social situation of households and individuals. Attempts to address restoration in a social context, without recognising the differences that degrees of poverty have on people's relative vulnerability and opportunities, most often at best ignore those in extreme poverty and at worst exacerbate their condition.

Also important in this context are the different relationships that people have with forests which can usefully be categorised as[37]:

- hunters and gatherers,
- shifting cultivators,
- farming communities with inputs from the forest, and
- livelihoods based on commercial forest product activities.

Poverty is not a uniform experience for these four types of forest-related people, and neither is it possible to say, for example, that all shifting cultivators are extremely poor or that all farming communities are "improving poor." This makes it even more difficult to generalise about the impacts that forest change will have on individual livelihoods. Within the same community, dependence on forests and wildlands will vary, although generally the extremely poor will be the most dependent on the resources from natural habitats and the improving poor will be less dependent. However, those whose livelihoods are most interlinked with the forest resource, such as hunter-gatherer groups and shifting cultivators, are those who are the most vulnerable to any changes in that resource and are also the least able to move into other livelihood options.

It should be noted that these are by no means static categories; they change as the local and national environment changes. For example, increasing market penetration has profound effects on the choices or enforced changes that people have to make in their livelihood base. The key point to recognise here is the diversity of the types of relationships that people have with forests and therefore the diversity of impacts that changes in forests and associated landscapes might have on the livelihoods of those living in and around them.

1.1. Relationships to the Forest

It is also important to move away from a broad-brush consideration of communities to recognition of differences between individual households and categories of well-being.[38] Many people assume that communities have common interests or, where they are conflicting, that disagreements could be resolved by working with the different interest groups, but this is not always the case. This becomes particularly important when considering the impacts of changes in forest cover and quality and how this is experienced by different households. For some of the most dependent people,

---

[37] Byron and Arnold, 1997.

[38] de Satgé, 2002.

TABLE 4.1. Examples of impacts of deforestation and degradation.

Process	Product	Extreme poor	Impacts on people — Coping poor	Improving poor
Deforestation	Conversion of forests to agriculture	Lose access to forest resources. Will not obtain land for agriculture as generally do not have the power to acquire the land. May be labourers for others but generally too marginalised	Lose access to safety net functions of forest resources. May become labourers for others on converted forest land	Lose access to safety net functions of forest resources; may acquire land under clearance as have better access to influence local decision making
Degradation	Foods: variety to diets, palatability, meet seasonal dietary shortfalls, snack food, emergency foods during flood, famine, war, etc. Fuels: firewood, charcoal growing importance for urban as well as rural energy needs Medicines: range of traditional plant medicines essential to those in remote rural areas distant from other medical services	Diminishing access to foods, fuels, and medicines make their livelihoods even more insecure and more vulnerable to hazards; in areas of high forest cover this group in particular is highly forest resource dependent and most particularly affected by changes in access or reduction in quality of forest; this range of products needs little or no capital investment and is therefore more readily accessible to the extreme poor	The importance of this range of products to the coping poor is two fold: (1) as a safety net, and (2) as an income earner to contributing household economies; for women, these are often the only source of income that they are allowed to access and so although a small proportion of overall household income, they are of high gender significance	With a more diverse livelihood portfolio with more assets and opportunities for diversifying, this group is not so vulnerable to changes in forest condition; it is more able to access alternatives to the forest products; nonetheless, its need for the safety net functions of the forest remains, and without it these households could become more vulnerable and less resilient to shocks
	Timber	Reduced access to timber usually has little impact on this group because they have little power to control access to high value resources; benefits of timber are mostly captured by the elites often in urban centres	This group, as for the extreme poor, is unlikely to benefit in any direct way from the economic benefits of timber harvesting; although because of their better social networks and levels of well-being they may have more opportunity to be labourers for timber contractors	With greater ability to take risk and invest in some relatively low-cost technology such as chain saws, this group can access some limited benefits from timber harvesting; being better socially networked, this group is more likely to be engaged as timber harvesters
	Environmental services	Across all groups the environmental functions of forests are important for maintaining water supplies, inputs to agricultural productivity through improving soil fertility, and providing the range of biodiversity necessary to maintain a robust local ecosystem. Degradation of environmental services is again most acutely felt by those who have no other options		For this group their more diverse portfolio and higher levels of risk-taking capacity means that they are more resilient to minor changes in environmental services.

Adapted from work by Brocklesby (2004) and Hobley (2004) differentiating between forms of poverty dependent on vulnerability and capability to have a voice.

forest change can be devastating, whereas for others with a broader livelihood portfolio that includes only limited dependence on the forests, changes in forest quality and extent may only have relatively minor effects. In such cases, responses to forest restoration will also be different between individual households in a community. The importance of a broad-based and carefully structured participatory process, linked to social mobilisation and including attempts to build the capacity of different social groups to have a voice, cannot be underestimated.

For some of the poorest rural peoples there is extreme forest dependence, but for others who are not so poor (the "coping" poor), the use of forests is indirect and more often is a means of poverty prevention, providing important seasonal safety nets. This latter role is often transitory as poor people build other assets to move out of poverty. It is rarely the case that forests themselves are the means to poverty reduction. However, what happens to the forests, their products and services, does have a profound impact on people's livelihoods, particularly when this is linked with the effects on other land uses such as grazing and agriculture.

Risk and uncertainty are universal characteristics of life in rural areas. Sources of risk include natural hazards like drought and flood, commodity price fluctuations, illness and death, changing social relationships, unstable governments, and armed conflicts. Some risky events like drought or flood simultaneously affect many households in a community or region. Other risky events, like illnesses, are household-specific and again have differential effects depending on the overall robustness of a particular household and its livelihood strategies. Catastrophic forest loss, for example through fire or clear-felling, thus affects whole communities, but the intensity of the effects are not necessarily uniform.

It is not only total forest loss that leads to negative impacts on well-being. For example, loss of particular nontimber forest products (NTFPs) from a surviving forest can be equally catastrophic to those households who have based their livelihoods around the use and sale of these products. Changes in market conditions, including in particular the recognition of the value of an NTFP on national and international markets, can disadvantage the very poor as the elites seize control of valuable natural resources and dominate market access.

1.2. Implications of Differential Social Impacts for Forest Restoration

1.2.1. Guiding Questions for Restoration

Forests can affect livelihoods in two principal ways that must be considered when any landscape restoration is under consideration[39]:

- Poverty avoidance or mitigation, that is, where forest resources serve a safety net function, or as a gap filler, including as a source of petty cash
- Poverty elimination, that is, where forest resources help lift a household out of poverty by functioning as a source of savings, investment, accumulation, asset building, and permanent increases in wealth and income

When restoration is planned to ameliorate the impacts of forest changes on the well-being of target groups a set of questions can help to guide responses as to the nature and extent of restoration required.[40] The usefulness of such questions depends to a large extent on the way in which they are asked. It is important to use participatory processes that lead to people being able to influence decisions about land use and control the outcomes of these decisions, but processes must also allow space for the voices of the extreme poor to be heard as well as those of the more articulate and much less vulnerable poor and wealthier groups:

What is the frequency or timing of use of forest products and the extent to which a household's labour is allocated to these activities?
What is the role of forest products in household livelihood systems? What is their importance as a share of household inputs, and in

---

[39] Sunderlin et al, 2004:1.
[40] Byron and Arnold, 1997.

meeting household livelihood strategy objectives?

What is the impact of reduced access to forests? Does the forest serve as a (critical) economic and ecological buffer for its users, or are there alternatives, such as trees outside forests or non–forest/tree sources of needed inputs and income?

What is the likely future importance of forest products? Do users face a growing or declining demand for forest products, or the potential for expanded or decreased involvement in production and trade in forest products?

2. Examples

Undoubtedly forest degradation and loss has major livelihood and well-being impacts for many people, from those with secure livelihoods to the extreme poor. It is therefore particularly important to understand the differential effects of forest change and the implications for livelihoods and livelihood options.

Byron and Arnold[41] provide a useful categorisation that aids this understanding and directs practical intervention. Clearly there is no general solution that can be applied across all situations. Any support to forest landscape restoration must be based on a careful assessment that "covers the range of the relationships between the people and the forests which they use and/or manage, the current limitations to their livelihoods, and the potentials and desires for change." They outline five generalised (and potentially overlapping) situations:

1. *Forests continue to be central to livelihood systems.* Local people are or should be the principal stakeholders in these forest areas. Meeting their needs is likely to be the principal objective of forest management and restoration, and this should be reflected in control and tenure arrangements (also see "Land Ownership and Forest Restoration").

2. *Forest products play an important supplementary and safety net role.* Users need security of access to the resources from which they source these products, but are often not the only users in that forest area. Forest management and control is likely to be best based on resource-sharing arrangements among several stakeholder groups. Successful restoration activities need to recognise and be planned with respect to these roles. Examples across the world include joint forest management in India and collaborative management in Ghana, where the state and local forest users share both in management decisions and in the benefits of forest products, which provide incentives to both partners to manage the forests for a range of benefits. However, in many cases the state is still reluctant to allow these agreements to cover high value forests, retaining control and access to the benefits and restricting local access to the forests and its products.[42] Community forestry in the hills of Nepal is widely cited as a successful example of transfer of control of management and benefits to local communities; again, however, the government has demonstrated its reluctance to extend management authority to the high value forests of the lowlands.

3. *Forest products play an important role but are more effectively supplied from nonforest sources.* Management of a proportion of the forests needs to be geared towards agro-forest structures, and control and tenure need to be consistent with the individual rather than the collective forms of governance that this shift is likely to require. Examples of these situations abound: PASOLAC (Programa para la Agricultura Sostenible en las Laderas de América Central) in Central America has been working with communities living in areas of high environmental degradation and insecurity to reduce their vulnerability to extreme natural events. This programme supports farmers to identify their own training requirements, provides financial and in-kind compensation for the management and maintenance of natural resources and their services and works to develop the integration of farmers and forest products into local markets. This integrated

---

[41] Byron and Arnold, 1997.

[42] Arnold, 2001; Molnar et al, 2004.

approach "combining improvements in human and social capital with advances in locally adapted resource management techniques and the creation of financial instruments"[43] is an important combination and an interesting progression away from approaches that have generally limited their support to more technically based interventions.

4. *Participants need help in exploiting opportunities to increase the benefits they obtain from forest product activities.* Constraints in the way of smallholders' access to markets need to be removed. Improved access to credit, skills, marketing services etc., may be required. A good example of the increasing experience with this type of support is provided by the PROCYMAF project (Proyecto de Conservación y Manejo Sostenible de Recursos Forestales) in Mexico. It has focussed on strengthening producer organisations and overcoming value chain "gaps."[44] This support is packaged with the supply of business services, which develop the skills of producer organisation leaders and members. A range of other programmes across the world are focussing on the better harvesting and marketing of a wide variety of NTFPs through understanding value chains and developing producer skills at entering markets in a more informed and secure environment.

5. *Participants need help in moving out of dead-end forest product activities.* An important example of this is firewood collection for sale in the market, often conducted by women who say they would rather be employed in other easier activities that are not so physically burdensome and poorly paid. It is often an activity of last resort and does not lead to opportunity to move out of these poverty confining conditions.

3. Outline of Tools

Baseline assessment: To build understanding of people's livelihoods and well-being, exposure to risk, and vulnerability, there are a range of tools that have been gathered under the umbrella of livelihoods analysis. These include survey methodologies and participatory appraisal approaches and are discussed in other chapters in this book. A useful guide to the range of tools and their applications can be found on Web sites including www.livelihoods.org. With this baseline assessment, it is then possible to begin to work with local people to identify different approaches to support their relationships with forests and forest products. It can be used as the basis for implementation and for later evaluation to assess the degree of change in exposure to risk and reduction in vulnerability as a result of livelihood interventions.

Tools for engagement: Voice, as has already been discussed, is an essential element of changing relationships and shifting power. Building poor people's capabilities to be able to influence decisions and policy is a key part of any restoration effort. Participatory tools and social mobilisation approaches are all used to build people's capabilities, but often voice is most strongly developed as poor people's livelihoods become more secure.

Community-based cost-benefit analysis: For communities, changing their use of forests and forest lands depends very much on individual and collective cost-benefit analyses. Communities are likely to be prepared to manage forests only if they offer greater benefit than under other uses of the land on which the forests grow. Such analyses are an essential part of any landscape restoration initiative because unless these costs and benefits are understood and factored into the process, initiatives will fail where perceived costs of maintaining the forest outweigh the tentative benefits. This is where ecosystem service payment schemes become an important part of the analysis and where it will be important to change local incentives and attitudes toward forests.[45] Additionally, focus on market access is critical where poor access and low values for forest products act as major barriers and disincentives.

---

[43] IISD et al, 2003.
[44] Scherr et al, 2003.
[45] Arnold, 2001.

Facilitating access to green markets: Providing mechanisms and funds that allow local people to access markets for ecosystem services such as watershed protection, biodiversity protection, etc., is another important element of changing the relationship between people's livelihoods and the forest resource. Forest certification can also be used to help forest managers to access higher value markets. There are some successful experiences with community-based certification in Latin America,[46] although the certification costs are often very high for small community groups and much more still needs to be done to provide standards that facilitate access of community managed natural timber into the green markets.

Securing tenure and management rights: Clearly tenure or at least long-term management rights are important elements in any forest restoration effort. There are now many models of communities that own forests with evidence of the incentives this creates for wise management. Tenure is often highly contested and requires careful work with governments to build an environment in which it is possible to shift tenure patterns. Often this requires significant evidence that changing tenure arrangements does lead to fundamental environmental and social benefits.

4. Future Needs

In any process of restoration, and perhaps particularly restoration projects driven by conservation concerns, some key messages need to be incorporated into the planning and implementation of any programme:

- Recognition of the differential importance of forests, products, and services on different people and therefore the differential impacts of changes in forest quality and extent;
- Recognition of the role of forests in poverty prevention as well as poverty reduction;
- The need to involve people in the decision-making process to build voice and capacity to articulate voice in an institutional and political environment that is able to respond to these voices;
- Recognition of the need to support the building of livelihoods that reduce people's exposure to risk and remove vulnerabilities;
- Recognition that forests alone do not necessarily move people out of poverty but actually can secure them in poverty;
- Support to decentralised service provision that can be socially responsive and tailored to particular ecological and economic conditions[47];
- Impacts of restoration also need to be carefully considered. Just as the impacts of degradation are not equally felt across livelihood groups, it is the case with restoration. Restoration of forest cover for some may have negative livelihood implications. Often the beneficiaries of restoration are not those living locally to the forest but are downstream users of services, therefore, the distribution of costs and benefits of restoration need to be carefully considered.

References

Arnold, M. 2001. 25 Years of Community Forestry. FAO, Rome.

Blankenberg, F. 1995. Methods of Impact Assessment Research Programme, Resource Pack and Discussion. Oxfam UK/I and Novib, the Hague.

Brocklesby, M.A. 2004. Planning against risk: tools for analysing vulnerability in remote rural areas. Chars Organisational Learning Paper 2, DFID, London, www.livelihoods.org.

Byron, N., and Arnold, M. 1997. What futures for the people of the tropical forests? CIFOR working paper No 19. CIFOR, Bogor, www.cifor.cgiar.org.

de Satgé, R. 2002. Learning about livelihoods: insights from Southern Africa. Periperi Publications, South Africa and Oxfam Publishing, Oxford.

Hobley, M. 2004. The Voice-responsiveness framework: creating political space for the extreme poor. Chars Organisational Learning Paper 3, DFID, London, www.livelihoods.org.

IISD, SEI, IUCN, and Intercooperation. 2003. Livelihoods and climate change: increasing the

---

[46] Molnar et al, 2004.

[47] Ribot, 2002.

resilience of tropical hillside communities through forest landscape restoration. Information Paper 2 IUCN and SDC, www.iucn.org/themes/ceesp/index.html.

Lamb, D., and Gilmour, D. 2003. Rehabilitation and Restoration of Degraded Forests. IUCN and WWF, Gland Switzerland and Cambridge, UK.

Molnar, A., Scherr, S.J., and Khare, A. 2004. Who conserves the world's forests? Community-driven strategies to protect forests and respect rights. Forest Trends, and Ecoagriculture Partners, Washington, DC, www.forest-trends.org.

Ribot, J.C. 2002. Democratic Decentralisation of Natural Resources: Institutionalising Popular Participation. World Resources Institute, Washington, DC.

Scherr, S.J., White, A., and Kaimowitz, D. 2003. Making markets work for forest communities. International Forestry Review 5(1):67–73.

Sunderlin, W.D., Angelsen, A., and Wunder, S. 2004. Forests and poverty alleviation. CIFOR, Bogor, www.cifor.cgiar.org.

Westoby, J. 1989. Introduction to World Forestry. Basil Blackwell, Oxford.

World Bank. 2001. World Development Report 2000–2001. World Bank, Washington.

Wunder, S. 2001. Poverty alleviation and tropical forests—what scope for synergies? World Development 29(11):1817–1833.

Additional Reading

Forestry Research Programme (FRP). 2004. Community forestry gets the credit. Forestry Research Programme Research Summary 006, FRP, Kent.

5
Restoring Forest Landscapes in the Face of Climate Change

Jennifer Biringer and Lara J. Hansen

Key Points to Retain

Climate change increases the need for restoration, both to help forest systems to manage existing changes and to buffer them against likely changes in the future by increasing areas of natural, healthy forest systems.

Care needs to be taken to avoid oversimplistic reliance on forests for carbon sequestration, and attempts at restoration to increase carbon storage must be assessed carefully to judge their true worth.

Tools such as vulnerability analyses can help to design effective restoration strategies, which are likely to include reduction of fragmentation, increasing connectivity, development of effective buffer zones, and maintenance of genetic diversity.

1. Background and Explanation of the Issue

Climate change is arguably the greatest contemporary threat to biodiversity. It is already affecting ecosystems of all kinds and these impacts are expected to become more dramatic as the climate continues to change due to anthropogenic greenhouse gas emissions into the atmosphere, mostly from fossil fuel combustion. While restoration is made more difficult by climate change, it can conversely be seen as a possible adaptive management approach for enhancing the resilience of ecosystems to these changes.

Climate change will result in added physical and biological stresses to forest ecosystems, including drought, heat, increased evapotranspiration, altered seasonality of hydrology, pests, disease, and competition; the strength and type of effect will depend on the location. Such stresses will compound existing nonclimatic threats to forest biodiversity, including overharvesting, invasive species, pollution, and land conversion. This will result in forest ecosystems changing in composition and location. Therefore, in order to increase the potential for success, it will be necessary to consider these changes when designing restoration projects.

On the other hand, restoration projects can also be viewed as a key aspect of enhancing ecosystem resilience to climate change. Human development has resulted in habitat loss, fragmentation, and degradation. A first step in increasing resilience to the effects of climate change is enhancing or protecting the ecosystem's natural ability to respond to stress and change. Research suggests that this is best achieved with "healthy" and intact systems as a starting point, which can draw on their own internal diversity to have natural adaptation or acclimation potential,[48] and therefore greater resilience. Any restoration activities that enhance the ecological health of a system can

---
[48] Kumaraguru and Beamish, 1981; McLusky et al, 1986.

thus be seen as creating or increasing the potential buffering capacity against negative impacts of climate change. It should be mentioned that there are obvious limits to the rate and extent of change that even a robust system can tolerate. As a result it is only prudent to conduct restoration for enhancing resilience in tandem with efforts to reduce greenhouse gas emissions, the root cause of climate change.

For many with a forestry background, carbon dioxide sequestration might seem a concomitant advantage to restoration projects, which can aid in reduction of atmospheric concentrations of greenhouse gases. While forests do hold carbon, and their loss does release carbon, their long-term capacity to act as a reliable sink in the face of climate change, especially for effective mitigation, is not a foolproof strategy (for more on carbon sequestration projects, see "Carbon Knowledge Projects and Forest Landscape Restoration"). Where restoration is promoted with a focus on capturing carbon, an analysis of climate change impacts should be integrated into project planning to determine whether there really are net sequestration benefits. Increased incidence of forest fires as a result of warming and drying trends, for example, could outweigh any efforts to reduce carbon emissions. Case studies of successful resilience-building efforts are not yet plentiful, due to relatively recent revelations about the scale and impact that climate change will have on ecosystems. However, the global temperature has risen 0.7°C as atmospheric concentrations have risen[49] and extinctions and large-scale ecosystem changes are expected. A number of forest types are already being negatively impacted, including tropical montane cloud forests, dry forests, and forests in the boreal zone, and climate-related extinctions are already thought to have occurred, for example amongst amphibians. Along the coasts, the rising sea level is increasing the vulnerability of mangroves. Restoration as a means to ensure healthy ecosystem structure and function will have a large part to play in adapting ecosystems to these broad-scale changes. See Box 5.1 for more in-depth exploration of these topics.

2. Example: Mangrove Restoration as an Adaptive Management Strategy

Mangroves provide a concrete example of how restoration can be used as a tool to help enhance resistance and resilience to climate change. Mangroves are clearly vulnerable to rising sea levels, which will change sediment dynamics, cause erosion, and change salinity levels. The rate of sediment buildup, which is the backbone of mangrove survival, is expected to take place at only half the pace of sea-level rise in many places, and mangrove survival will therefore require active restoration. Another aspect of mangroves that makes them an ideal testing ground for restoration is their relative ecological simplicity. Furthermore, the relationship between human and ecological vulnerability to climate change is relatively clear. Low-lying coastal areas, particularly those in tropical Africa, South Asia, and the South Pacific, are predicted to experience among the most severe consequences of global climate change.[50] As these are among the most populous areas across the globe, the livelihoods of many coastal communities that depend on mangrove resources for wood and shrimp farming, will be increasingly tied to their vulnerability to climate change.

Mangrove restoration can do much to limit or delay the negative effects of climate change on associated human and natural communities. Mangroves play an integral role in coastal ecosystems as the interface among terrestrial, freshwater, and marine systems. They are extensively developed on sedimentary shorelines such as deltas, where sediment supply determines their ability to keep up with sea-level rise. They afford protection from dynamic marine processes to both terrestrial and estuarine systems, preventing erosion and chaotic mixing. They also act locally to filter water. Mangrove forests protect sea grass beds and coral reefs from deposition of suspended matter that is transported seaward by rivers and

---

[49] Hansen et al, 2003.

[50] IPCC, 2001.

Box 5.1. Framework for Understanding Intersection of Resilience-Building and Forest Restoration and Protection

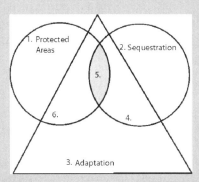

1. **Protection**: For some forests protection alone will not increase resilience to climate change. Many tropical montane cloud forests provide a case in point. Australia's Wet Tropics World Heritage Area is expected to experience a 50% reduction in habitat with warming of 1 degree Celsius, which will leave amphibians and other cool-adapted species no upland migration options as conditions become warmer and drier.

2. **Sequestration via restoration**: Many examples exist where the planting of trees stores carbon but is not coordinated with conservation or resilience-raising advantages. Nonnative trees, such as Eucalyptus, are often planted solely for the carbon benefit, though the planting may cause degradation of the landscape, and thus not provide a buffer against climate change.

3. **Resilience/adaptation**: Restoration is but one of the many types of management options that increase resilience. For example, actions that respond to changing dynamics such as insect infestations and changing fire patterns are aspects of good forestry that will receive special attention with the advent of climate change. Activities that increase the efficiency of resource use will also increase resilience. In Cameroon, mangroves are being aided by increasing the efficiency of wood-burning stoves so that 75 percent less mangrove wood is needed for cooking, thereby increasing the resilience of the system by reducing harvest levels. Such actions decrease degradation of the mangrove and raise the probability that it will be equipped to respond to the effects of climate change.

4. **Sequestration and resilience/adaptation**: Restoration and resilience go hand in hand when the impacts of climate change are taken into account in project planning. Whether passive or active restoration, activities target those areas that will be more suitable to climate change, and encourage use of species that will be hardier under new climatic conditions (successful seed dispersers, for example).

5. **Intersection of protection, sequestration, and resilience/adaptation**: Creating buffer zones through restoration can increase the resilience of protected areas to the impacts of climate change while at the same time sequestering carbon. This scenario is similar to the one above, except that restoration is focussed on increasing the resilience of protected areas by expanding boundaries to increase suitable habitat under changing climatic conditions.

6. **Protection and adaptation**: Protection can lead to increased resilience to the impacts of climate change, where suitable habitat is intact, and the expansion of boundaries is possible to accommodate species' needs with a changing climate. A successful protected area system includes identification and conservation of mature forest stands, functional groups and keystone species, and climate refugia.

provide nursery habitat for many fish species. Deteriorating water quality and coastal degradation are anticipated to be magnified by climate change. Globally, however, many mangrove systems have already been degraded and destroyed. Loss of these buffering systems precludes any protection they might afford. This has been recognised for some time, and many individual projects have attempted to rebuild mangrove systems. However, in the past, the emphasis of mangrove restoration projects has been on planting trees, and this has led to poor survival rates, such as in West Bengal, India, where survival rates in some projects were reported as low as less than 2 percent.[51]

New approaches are therefore required. In addition, simply restoring a mangrove where it has been degraded will not necessarily be enough in the face of climate change. Restoration in an environment where the climate is rapidly changing will require taking into account a few additional elements as opposed to restoration in a stable context. Before starting a restoration programme, two additional steps are required: (1) assess the cause of mangrove loss and evaluate how to remove those causes if possible; and (2) take into account the added complexity relating to how climate change will affect the system: in this case primarily through sea-level rise.

A large-scale mangrove restoration effort in Vietnam has demonstrated that this approach to mangrove management can benefit local resource users and enhance protection from storm surge and sea-level rise.[52] The restoration project in this region has planted more than 18,000 hectares of mangrove along 100 kilometres of coastline. In addition to creating a more stable coastline capable of surviving changing marine conditions, harvestable marine resources are also increasing in number.

Understanding the hydrology (both frequency and duration of tidal flooding) is the single most important factor in designing successful mangrove restoration projects.[53] Incorporating projections of sea-level rise into project design will be necessary so that mangroves are planted or are allowed to colonise naturally or regenerate (this takes 15 to 30 years where stresses leading to degradation are no longer present) in areas that will be more hospitable in the future. If the shoreline is moving, for instance, mangroves may need to be restored some distance from their original location.

3. Outline of Tools

This section offers a framework for integrating knowledge about climate change to forest managers who are considering restoration. It is based on an understanding of how adaptation (in this case to climate change) needs to be integrated with both restoration and protection, as outlined in Box 5.1 above.

3.1 Vulnerability Analysis

To understand how climate change will affect an existing forest system, an analysis of the vulnerability of the defined area can be undertaken. As a first stop, climate change impacts on the major forest types are presented in WWF's *Buying Time: A User's Manual for Building Resistance and Resilience to Climate Change in Natural Systems*,[54] with examples from many different regions collected from the literature. For more specific information on a particular site, a literature search may identify whether a vulnerability analysis has been made of the project area in question.

If limited information on climate change impacts exists for the selected site, a vulnerability analysis can be commissioned to feed into project design activities. An expert conversant in climate change science as well as biological science for the region can piece together a picture of regional vulnerability that will help to guide project activities so that they can take account of likely alterations in environmental conditions as the climate changes. At a large

---

[51] Sanyal, 1998.
[52] Tri et al, 1998.
[53] Lewis and Streever, 2000.

[54] Hansen et al, 2003 (available on www.panda.org).

scale, major shifts in biome types can be projected by combining biogeography models such as the Holdridge Life Zone Classification Model with general circulation models (GCMs) that project changes under a doubled CO_2 scenario. Biogeochemistry models simulate the gain, loss, and internal cycling of carbon, nutrients, and water-impact of changes in temperature, precipitation, soil moisture, and other climatic factors that give clues to ecosystem productivity. Dynamic global vegetation models integrate biogeochemical processes with dynamic changes in vegetation composition and distribution. Studies on particular species comparing present trends with paleo-ecological data also provide indications for how species will adapt to climate change.[55]

A vulnerability analysis can help to assess what systems or aspects of the systems have greater resilience and resistance to climate change impacts. This type of information can help to identify sites that have greater long-term potential as ecosystem "refugia" from climate change impacts. Some refugia exist due to their unique situational characteristics, but their resilience could be enhanced by management and restoration.

3.2 Restoration as a Resilience/Adaptation Strategy

After completing a vulnerability analysis to determine how a forest system may be impacted by changing climatic conditions, the next step is to look at the range of adaptation options available in order to promote resilience. An effective vulnerability analysis will determine which components of the system—species or functions, for example—will be most vulnerable to change, together with consideration of which parts of the system are crucial for ecosystem health. An array of options pertinent to adapting forests to climate change are available, both to apply to forest communities at high risk from climate change impacts as well as for those whose protection should be prioritised given existing resilience. Long-term resilience of species will be enabled where natural adaptation processes such as migration, selection, and change in structure are allowed to take place due to sufficient connectivity and habitat size within the landscape.

Restoration can provide a series of critical interventions to reduce climate change impacts.[56] Basic tenets of restoration for adaptation include working on a larger scale to increase the amount of available options for ecosystems, inclusion of corridors for connectivity between sites, inclusion of buffers, and provision of heterogeneity within the restoration approach. Key approaches are as follows:

Reduce fragmentation and provide connectivity: Noss[57] provides an overview of the negative effects of ecosystem fragmentation, which are abundantly documented worldwide. "Edge effects" threaten the microclimate and stability of a forest as the ratio of edge to interior habitat increases. Eventually, the ability of a forest to withstand debilitating impacts is broken. Fragmentation of forest ecosystems also contributes to a loss of biodiversity as exotic, weedy species with high dispersal capacities are favoured and many native species are inhibited by isolation. Restoration strategies should therefore often focus first on those areas where intervention can connect existing forest fragments into a more coherent whole.

Provide buffer zones and flexibility of land uses: The fixed boundaries of protected areas are not well suited to a dynamic environment unless individual areas are extremely large. With changing climate, buffer zones might provide suitable conditions for species if conditions inside reserves become unsuitable.[58] Buffer zones increase the patch size of the interior of the protected area and overlapping buffers provide migratory possibilities for some species.[59] Buffer zones should ideally be large, and managers of protected areas and surrounding lands must demonstrate considerable flexibility by adjusting

---

[55] Hansen et al, 2001.

[56] Biringer, 2003; Noss, 2001.
[57] Noss, 2000.
[58] Noss, 2000.
[59] Sekula, 2000.

land management activities across the landscape in response to changing habitat suitability. A specific case for a buffer zone surrounding tropical montane cloud forests can be made based on research that shows that the upwind effects to deforestation of lowland forests causes the cloud base to rise.[60] Restoring forest around protected areas, for example to supply timber through continuous cover forestry, or for nontimber forest products, watershed protection, or as recreational areas, could help maintain the quality of the protected area in the face of climate change.

Maintain genetic diversity and promote ecosystem health via restoration: Adaptation to climate change via selection of resilient species depends on genetic variation. Efforts to maintain genetic diversity should be applied, particularly in degraded landscapes or within populations of commercially important trees (where genetic diversity is often low due to selective harvesting). In such places where genetic diversity has been reduced, restoration, especially using seed sources from lower elevations or latitudes, can play a vital role in maintaining ecosystem resilience.[61] Hogg and Schwarz[62] suggest that assisted regeneration could be used in southern boreal forests in Canada where drier conditions may decrease natural regeneration of conifer species. Similarly, genotypes of beach pine forests in British Columbia may need assistance in redistributing across the landscape in order to maintain long-term productivity.[63] In addition, species that are known to be more resilient to impacts in a given landscape can be specifically selected for replanting. For example, trees with thick bark can be planted in areas prone to fire to increase tree survival during increased frequency and severity of fires.[64]

---

[60] Lawton et al, 2001.
[61] Noss, 2000.
[62] Hogg and Schwarz, 1997.
[63] Rehfeldt et al, 1999.
[64] Dale et al, 2001.

4. Future Needs

Documentation of the role restoration plays in building resilience to climate change is in its infancy. Although field projects are beginning to test restoration as a resilience-building tool, we are far from definitive guidance. Unfortunately, this is the nature of the practice of conservation; decisions based on best knowledge need to be made now while we continue to gather more information. Otherwise, opportunities will be lost.

To meet these needs we propose additional field projects to test, confirm, and develop restoration's role in building resilience to climate change. This needs to be conducted across different forest types with as much replication as possible. A strong monitoring component is necessary for any such project, especially given the complex relationships between species' structure, composition, and functioning on which climate change is unfolding. The results of monitoring will also enable lessons to be drawn from resilience-building efforts, and to compare these with similar "control" landscapes or other resilience-building projects in different regions with similar habitat type.

Ideally, resilience-building management strategies will serve as another layer in a comprehensive forest management plan that has as its objective the overall health of the forest ecosystem. For example, many WWF ecoregional visions are adding vulnerability to climate change as another component that will drive conservation decisions. Such anticipatory resilience-building plans take climate change into account during the planning process, and will better ensure synergies with other management priorities. A number of scientific, governmental institutions and non-governmental organisations (NGO) are acquiring expertise in the area of climate change impacts and adaptation/resilience. It will be fruitful to seek partnerships with these institutions at the beginning of any restoration project to analyse climate impacts and proposed restoration activities.

References

Biringer, J. 2003. Forest ecosystems threatened by climate change: promoting long-term forest resilience. In: Hansen, L.J., Biringer, J.L., and Hoffman, J.R. eds. Buying Time: A User's Manual for Building Resistance and Resilience to Climate Change in Natural Systems. WWF, Washington, pp. 41–69. (Also online at www.panda.org/climate/pa_manual)

Dale, V., Joynce, L., McNurlty, S., et al. 2001. Climate change and forest disturbances. Bioscience 51(9): 723–734.

Hansen, A., Neilson, R., Dale, V., et al. 2001. Global change in forests: responses of species, communities, and Biomes. Bioscience 51(9):765–779.

Hansen, L.J., Biringer, J.L., and Hoffman, J.R. eds. 2003. Buying Time: A User's Manual for Building Resistance and Resilience to Climate Change in Natural Systems. WWF, Washington, 242 pages. (Also online at www.panda.org/climate/pa_manual.)

Hogg, E., and Schwarz, A. 1997. Regeneration of planted conifers across climatic moisture gradients on the Canadian prairies: implications for distribution and climate change. Journal of Biogeography 24:527–534.

Intergovernmental Panel on Climate Change (IPCC). 2001. Impacts, Adaptations and Vulnerability. Working Group II, Third Assessment Report. Cambridge University Press, Cambridge, UK, 1032 pages.

Kumaraguru A.K., and Beamish, F.W.H. 1981. Lethal toxicity of permethrin (NRDC 143) to rainbow trout, *Salmo gairdneri*, in relation to body weight and water temperature. Water Research 15:503–505.

Lawton, R., Nair, U., Pielke, R., and Welch, R. 2001. Climate impact of tropical lowland deforestation on nearby montane cloud forests. Science 294 (5542):584–587.

Lewis, R., and Streever, B. 2000. Restoration of mangrove habitat. WRP Technical Notes Collection (ERDC TN-WRP-VN-RS-3.2), U.S. Army Engineer Research and Development Center, Vicksburg, MS. www.wes.army.mil/el/wrp.

McLusky, D.S., Bryant, V., and Campbell, R. 1986. The effects of temperature and salinity on the toxicity of heavy metals to the marine and estuarine invertebrates. Oceanography and Marine Biology Annual Review 24:481–520.

Noss, 2000. Managing forests for resistance and resilience to climate change: a report to World Wildlife Fund U.S., 53 pages.

Noss, R. 2001. Beyond Kyoto: forest management in a time of rapid climate change. Conservation Biology 15(3):578–590.

Rehfeldt G., Ying, C., Spittlehouse D., and Hamilton, D., Jr. 1999. Genetic response to climate in *Pinus contorta*: niche breadth, climate change and reforestation. Ecological Monographs 69(3):375–407.

Sanyal, P. 1998. Rehabilitation of degraded mangrove forests of the Sunderbans of India. Programme of the International Workshop on the Rehabilitation of Degraded Coastal Systems. Phuket Marine Biological Center, Phuket, Thailand, January 19–24, p. 25.

Sekula, J. 2000. Circumpolar boreal forests and climate change: impacts and managerial responses. An unpublished discussion paper prepared jointly by the IUCN Temperate and Boreal Forest Programme and the IUCN Global Initiative on Climate Change.

Tri, N.H., Adger, W.N., and Kelly, P.M. 1998. Natural resource management in mitigating climate impacts: the example of mangrove restoration in Vietnam. Global Environmental Change 8(1): 49–61.

Additional Reading

Krankina, O., Dixon, R., Kirilenko, A., and Kobak, K. 1997. Global climate change adaptation: examples from Russian boreal forests. Climatic Change 36(1–2):197–215.

Section III
Forest Restoration in Modern Broad-Scale Conservation

6
Restoration as a Strategy to Contribute to Ecoregion Visions

John Morrison, Jeff Sayer, and Colby Loucks

Key Points to Retain

Ecoregion conservation is a large-scale, long-term, and flexible concept whose purpose is to meet the four goals of biodiversity conservation: representation, maintenance of evolutionary processes, maintenance of viable populations, and resilience.

In degraded landscapes and ecoregions restoration goals and strategies will be critical to the success of an ecoregion vision.

But as restoration can be energy intensive, its role must be defined in the context of quantifiable goals related to the four larger goals of biodiversity conservation.

1. Background and Explanation of the Issue

Most people are aware of the global reduction in forest cover as a result of ever-increasing human domination of the planet. The impacts are felt on biodiversity and on people as shown in the previous chapters of this book. A natural reaction to this forest loss is to engage in forest restoration activities.

Across the planet, conservationists are working to increase overall forest coverage using a variety of strategies. In some cases this includes attempting to intensify agriculture so that it requires less land, focussing on value over volume in wood products, and concentrating production in (native) plantation forests. Another strategy is to de-intensify agricultural uses and promote a mosaic of natural and anthropogenic elements, allowing native species and communities to fill in around our use of the landscape, and provide necessary ecosystem services to operate more freely.

In any case, the competition for land among a range of interests and stakeholders necessitates that all forest conservation activities, including forest restoration, be strategic and for a specific purpose(s), be it conservation or otherwise. This strategic focus should ideally be identified through a participatory process that leads to a long-term "vision" for the desired future state of the area. Increasing the quality and quantity of forest cover is an important general goal for conservation, both for ecosystem services (watershed protection, climate regulation, etc.) and for the needs of those species that depend on forests. However, due to the intense competition for land between the forces of development and conservation, efficiency in how and where forest restoration occurs is critical. In other words, while increased tree cover will nearly always be beneficial from a conservation perspective, if possible, restoration efforts should be focussed in such a way that multiple conservation and social goals are reached (also see sections "Restoring Ecological Functions" and "Restoring Socioeconomic Values"). Meeting both conservation and social goals simultaneously maximises the chances that the activities will be sustainable and that they will have local support. An example of this integration is provided by the activities in the Upper Paraná Atlantic Forest. Within this ecoregion forest patch connectivity is being improved through the incorporation of native plants that can also be sustainably used by local people (see case study "Finding Economically Sustainable Means of Preserving and Restoring the Atlantic Forest in Argentina").

What are the primary conservation goals that we should be trying to achieve?

1.1. The Four Goals of Biodiversity Conservation and Ecoregion Conservation[65]

The goals of biodiversity conservation and ecoregion conservation are as follows:

1. Representation of all distinct natural communities within conservation landscapes and protected areas' networks
2. Maintenance of ecological and evolutionary processes that create and sustain biodiversity
3. Maintenance of viable populations of species
4. Conservation of blocks of natural habitat large enough to be resilient to large-scale disturbances and long-term changes

Because these conservation goals often operate over large spatial and temporal scales, the design of conservation programmes "requires a perspective that spans nations and centuries."[66] Large-scale conservation initiatives have become standard in a number of conservation organisations over the last decade. This evolution is seen as a reaction to the often disjointed, isolated, and nonstrategic activities that once characterised site-level conservation. While site-level conservation will always be an important and, many would argue, the most important scale of conservation intervention, site-level activities can be planned in the context of larger scale (landscape and ecoregion) visions. The thinking behind using large biogeographic units as the framework in which to achieve conservation goals is that natural communities, species, and even human threats to biodiversity move and operate at large scales, often irrespective of political boundaries. Actions conceived at the same scale as the ecological entities and processes that the actions are trying to protect should be more robust and efficient than uncoordinated efforts at a site scale. At WWF, the global conservation organisation, this evolution has taken the form of Ecoregional Conservation (ERC). Ecoregion conservation is really a philosophy that espouses using large, biogeographically defined units as an arena within which to achieve the four goals of conservation outlined above. The actual process of ecoregion conservation planning has followed a number of paths, generally relying on experts, computer algorithms, or even a mixture of the two to identify conservation priorities.

A range of spatial scales has been addressed to date, under the heading of "ecoregion conservation." A system of ecoregional boundaries of the world has been stitched together by WWF.[67] This system is also used by the Nature Conservancy. Conservation effort is not applied equally across this system. WWF has defined 825 terrestrial ecoregions (Fig. 6.1), of which a large proportion is forest ecoregions of various subtypes (tropical dry, tropical moist, temperate moist, etc.). A further analysis by WWF identified 237 groupings of these terrestrial ecoregions as being of particular importance to conservation and named these the Global 200 Ecoregions—it is usually these Global 200 ecoregions that are the focus of WWF Ecoregion Action Programmes.[68] In the process of analysing ecoregions, "priority areas" or "priority landscapes" are often identified that become the subject of further conservation planning and initiatives. Thus the general hierarchical spatial scale, from largest to smallest, is Global 200 ecoregion, terrestrial ecoregion, and priority landscape—but this is not a steadfast rule,

---

[65] Noss, 1992.
[66] Scott et al, 1999.
[67] Olson et al, 2001.
[68] Olson and Dinerstein, 1998.

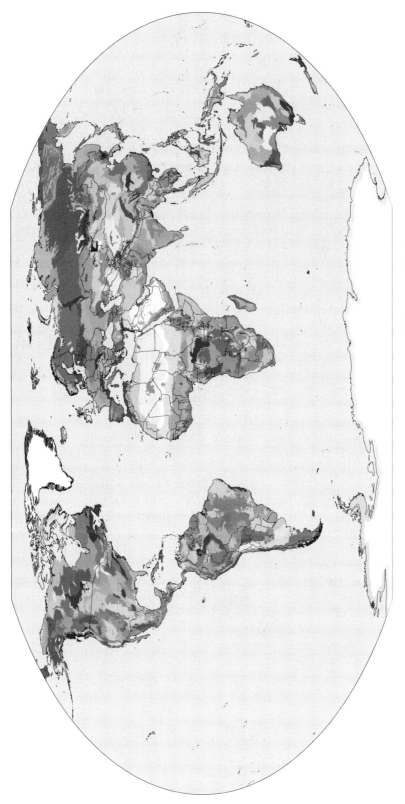

FIGURE 6.1. Terrestrial ecoregions of the world. (*Source:* WWF.)

and there are very small ecoregions (tens of km^2) and very large priority landscapes (thousands of km^2). Most of the principles discussed below hold for a range of scales, from the landscape to the ecoregion.

1.2. Protect, Manage, and Restore

More than likely, any comprehensive conservation strategy in an ecoregion will involve a combination of protection, management, and restoration, plus the abatement/amelioration of threats. The relative proportion of each strategy that is appropriate is a function of both the overall conservation status of the ecoregion, and the location in the ecoregion—and this will change over time. For example, restoration is not necessarily an appropriate strategy in all ecoregions or landscapes. One can imagine that restoration may not currently be the highest priority in those ecoregions that are composed mostly of wilderness or large forest blocks, such as in the Amazon. A primary output of many ecoregional visions is a map of priority areas, where conservation activities are more focussed than in the surrounding matrix of the ecoregion. Yet even in the matrix, some proportion of protection, management, and restoration activities will be appropriate, and in the case of the wilderness ecoregions mentioned above, over the long-term, restoration may rise in priority in those ecoregions as more comprehensive protection and better management are instituted.

From a conservation standpoint, the decisions about how much protection, management, and restoration will be a natural consequence of attempting to achieve the above four conservation goals in a strategic fashion in an ecoregion or a landscape within that ecoregion. Is there enough of a given target habitat present in the ecoregion or landscape to meet representation objectives that we can simply protect a (greater) proportion of it? Or will some areas containing that habitat need active or passive restoration in order to meet the prescribed target for that habitat? Can existing multiuse buffer zones of forest simply be managed in their current state to provide landscape connectivity, or will some areas need to be rehabilitated to restore connectivity?

Forest "restoration" activities range from active planting, to management (e.g., invasive species' removal), to more passive restoration (creating the conditions that will allow natural processes to regenerate high-quality forest). Because active restoration is so resource intensive, it should generally be the last option selected to meet a conservation objective. The key point is that from a conservation perspective restoration activities should not be undertaken for the sake of restoration; rather, the activity should be a strategic response to a specific need identified during the formation of conservation goals. The Forests of the Lower Mekong ecoregion has endeavoured to find the right balance of protection, management, and restoration—all stemming from the conservation goals highlighted during the ecoregional vision process.

2. Examples: Restoration and the Four Conservation Goals

Conceptually, it is a relatively simple matter to decide whether restoration is necessary or not. By selecting conservation targets that are applicable to the aforementioned four goals of conservation, it should quickly become clear whether or not the relevant ecoregion or priority landscape still contains the necessary components to satisfy all four goals. If there are elements missing or the ecoregion/landscape is too fragmented, some restoration is probably necessary. At the basic level of the four conservation goals, the following discussion illustrates how the need for restoration can be identified.

2.1. Representation

Conservationists need to represent all natural communities in some sort of a conservation network, which is generally a mix of different levels of protection. It is important that the mix of natural communities is one that has existed before a major disturbance rather than the existing mix. But all of these original communities may no longer be present in the quantity and quality necessary, and that is where the potential application of restoration comes in. This is especially true during periods of climate change when species will need to move in response to changing conditions.

One of the first steps in any conservation planning initiative is to obtain or develop a map of historic (sometimes called "potential") natural community types across the entire ecoregion/priority landscape. A number of coverages may suffice for this purpose, including historic vegetation maps, potential vegetation maps, or maps of plant communities or ecosystems. In the case where land conversion has made this task impossible, maps of environmental domains, which are unique combinations of substrate (soils or geology), elevation, and climate classifications, may be developed. If these environmental domains are carefully developed, they should represent unique environmental classes that correlate with the species living in them.

It is common practice for a target level of representation to be chosen for each natural community type (or environmental domain). This is not always easy, but endeavouring to determine what these levels should be (preferably on an individual habitat-by-habitat basis rather than a blanket prescription) is one of the highest callings of a conservation biologist. It is altogether appropriate to begin with coarse estimates that can be improved over time. Custom representation targets are preferable to blanket prescriptions. Once an appropriate level of representation of each historic natural community is decided (20 percent, 30 percent, 50 percent, etc.), it may be discovered that less intact habitat of a particular type(s) remains than the target representation amount. This is a sign that some restoration is in order. Madagascar and the dry forests of New Caledonia are prime examples—forest conversion has proceeded so far in these ecoregions that forest restoration is required to meet the most basic habitat representation goals.

It should also be noted that each natural community is itself made up of seral stages, and the appropriate mix of seral stages, or more likely the allowable ranges of seral stages, corresponding to a natural range of variation, must be specified. The ability of a natural community type to support a natural range of seral stages must be protected, or if necessary enhanced, and this may also require some forest restoration activities. An example is the relative lack of primary, or old-growth forest, in many temperate forest ecoregions compared to historic levels. Efforts to increase the proportion of late seral stages are an appropriate application of forest restoration in this case.

Many ecoregional programmes, especially those in developed or densely populated countries, have found that the amount of lowland and riparian communities are in short supply—they have already been converted for human uses. Clearly in such situations, restoration will necessarily be an important component of the overall conservation strategy if representation targets are to be met.

2.2. Viable Populations

The idea behind this goal is that all species should have conserved viable populations, but in practice it is never possible to plan for all species (if for no other reason than that all species are never really identified). During any large-scale conservation initiative, therefore, focal species are selected for special attention. Focal species are chosen because they are "keystone," highly threatened endemics, habitat specialists, or because they are very "area-sensitive" and act as umbrellas for a number of species with smaller area requirements. The number of focal species chosen will vary from ecoregion to ecoregion, and certainly from priority landscape to priority landscape, but is generally a manageable number of five to 20 species from the above categories.

After determining what the list of focal species is, the next step is to determine the number of breeding individuals that represent a viable population, or potentially a viable subpopulation in the case of a priority landscape. This is not a trivial determination, and there is an extensive literature discussing rules of thumb for the number of breeding individuals that constitutes a viable population—with little consensus. In some cases a species-specific and resource-intensive population viability analysis (PVA) will be necessary. If a viable population estimate is difficult to come by or there are severe limits to the number of individuals that are possible, the bottom line is that a target level should be chosen that represents the largest conceivable achievable population level.

For restoration purposes, the specific needs of each focal species must be analysed individually. A number of related metrics, including minimum patch size, connecting patches to enlarge the effective habitat area or feature (breeding, feeding, or nesting areas/cavities), corridor width, specific habitat requirements (plant species), access to water, etc. must be considered. During the course of the analysis to determine the habitat and total area requirements for each species, it should quickly become clear if there is not enough habitat necessary for a viable population of a particular species—and restoration will be necessary. This is frequently the case in those ecoregions that have been highly degraded.

The reconnection of now disjunct habitat patches is a common application of forest restoration activities. This is the focus of the current work in the Terai Arc in the Eastern Himalayas: reconnecting 10 protected areas by encouraging the growth of community-managed forests (Fig. 6.2). Tigers are loath to cross more than $5\,km^2$ of nonhabitat, but the existing protected areas are not large enough to maintain viable populations of tigers. Some mixing of the respective populations is desirable. Therefore, community forests are being encouraged where gaps in forest cover are noted between the existing protected areas. This will allow tigers, greater-one horned rhinoceroses, and Asian elephants to disperse between patches of prime habitat. Restoration is an important activity in other fragmented ecoregions that still contain large carnivores, including for jaguars in South America's Atlantic Forest and for wolves and grizzly bears

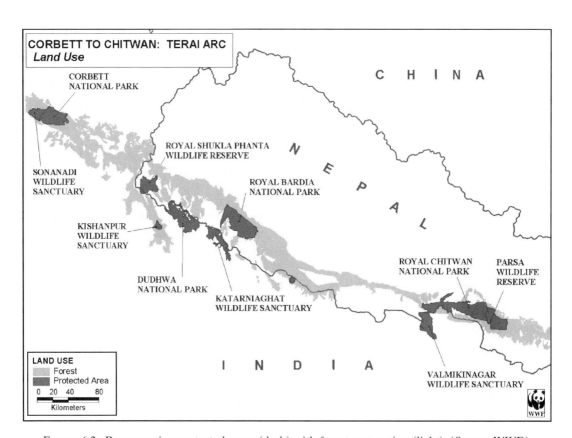

FIGURE 6.2. Reconnecting protected areas (dark) with forest restoration (light). (*Source*: WWF.)

in the ecoregions of the Northern Rockies of North America.

2.3. Ecological and Evolutionary Processes

The many evolutionary and ecological processes that create and sustain biodiversity are complex, and often poorly understood. Gene flow, migration, pollination, seed dispersal, predator–prey dynamics, and nutrient cycling are some of the many that should be considered when a conservation plan is developed. All of these processes can potentially benefit from restoration activities, because many species (and the processes that they are involved in) will respond positively to restored forest quality, but some of them will benefit more obviously than others. Gene flow and migration can directly benefit from restored forest corridors, as in the above examples. Likewise, if key processes such as pollination or seed dispersal are threatened by insufficient forest area to support the species that are performing these functions, restoration activities would be appropriate.

In some regions, reduced forest cover threatens to throw the area into a not-easily-reversible regional climatic shift. Restoration of forest cover (that simultaneously meets finer scale representation targets and is configured to maximise forest block size for area-sensitive species) would be a high priority activity.

The Terai Arc is also a good example for this set of conservation goals. By reconnecting disjunct forest patches and thus tiger subpopulations, the ecological processes of subadult dispersal, gene flow, and restoration of predator–prey dynamics can be restored. Because systems with large predators are often dominated top-down forces (in this case elephants and tigers), the reintroduction of tigers and elephants across the entire landscape will help put a number of natural ecological processes back into a more natural dynamic balance. However, the needs of finer-scale habitat specialists (particularly for breeding or feeding) within the larger area should not be overlooked.

2.4. Environmental Change

Planning for inevitable environmental change (even without the additional spectre of anthropogenic climate change) is a key precept in conservation. Ecological systems are by their very nature dynamic, and it is important to incorporate large habitat areas and sufficient connectivity between habitat areas in order to build resiliency into the protected area network. Increased connectivity is the main option available to conservation planners trying to anticipate the effects of anthropogenic climate change. Species' ranges are already beginning to shift in latitude and altitude; this is true not only for animals but for plant species as well. Again, reconnecting now disjunct habitat patches through restored forest corridors is an appropriate application for forest restoration activities to help migration to keep pace with changing conditions. In addition, managing the landscape in such a way that it provides more flexibility for species and gene flow in times of stress is an important element of restoration.

This connectivity strategy will be important for every ecoregion across the planet to consider. Ecoregions likely to be faced with this threat in the near term are tropical montane ecoregions that contain significant topographic relief. Climatological changes are concentrated in narrow bands, and maintaining altitudinal connectivity will be critical for allowing habitats to shift in response to changing temperature and moisture regimes.

Restoration activities are important for all ecoregions where human activities have fragmented the ecoregion, and this includes most ecoregions. Rising temperatures and changing precipitation patterns will cause natural communities to shift latitudinally and altitudinally. Without restoration to reconnect fragmented habitat patches with corridors, natural communities will have great difficulty shifting across human-dominated landscapes. A more specific example of the need for restoration will be in tropical coastal ecoregions with mangroves. As sea level continues to rise, mangrove belts will tend to shift inland (Fig. 6.3). However, if the landward edge of the mangrove belt has been degraded, which it commonly is, space and

FIGURE 6.3. Mangrove belts along coastal areas are expected to shift inland with rising sea levels. (Photo © John Morrison.)

restoration activities will be necessary to allow the continued persistence of the mangroves, and with them the important ecological (and social) functions they perform.[69]

2.5. Deciding Where to Do Restoration When There Are Choices

In the preceding discussion, the need for restoration fell into two broad categories: increasing the area of a particular forest type for representation or for particular species/processes, and restoring particular landscape features, especially corridors, which allow specific ecological processes to operate. Sometimes there are choices of where restoration is most appropriate. All other things being equal, it is generally easier to restore the less degraded example of a forest type, since less effort or time will be required. All other things are rarely equal, however. How does one decide which semi-irreplaceable example of a forest type to restore if there are several choices? Obviously, many factors must often be weighed.

The first step is to be clear about the end objective(s). For example, is primary forest the only possible objective, or would secondary forest do just as well (or even better) for the focal species being considered? Factors to consider when determining which area to restore are the following:

- The current condition of the forest area in question—how much effort/time is required to restore?
- Proximity to other viable habitats, to allow species to disperse or facilitate later reconnection
- Proximity to the existing or anticipated urban frontier

This last bullet point highlights an entire class of information that can help to assure that restoration activities (and in fact any conservation activities) have the greatest chance of success. The mapping of human population density, distance from access corridors, government capacity, ethnic stability and homogeneity, and similar factors can help a project see where the threats and opportunities lie across the ecoregion or landscape. Additionally, the incorporation of socioeconomic information and consultation will help to assure that restoration activities undertaken for ecological reasons will also benefit local people either through ecological services or even through employment in restoration activities.

3. Outline of Tools

As already noted, ecoregion conservation in the WWF network is more of a philosophy than a particular methodology, and a number of methodologies have been used to achieve the four goals of conservation. This is altogether appropriate, since there is a great variety of

[69] Noss, 2001.

data availability, social structures, infrastructure, and professional capacity in the ecoregions across the planet. There is no tool especially tailored to help set restoration priorities. These priorities should emerge from a generic comprehensive planning process.

A full discussion of the tools available for ecoregional conservation planning is beyond the scope of this paper. Some of the primary tools include:

- WWF's approaches to ecoregion conservation,[70] including specific advice about actions in priority conservation landscapes[71] and case studies[72] and a detailed guide to implementation within ecoregions[73]
- The Nature Conservancy's approach to ecoregion conservation[74]
- Systematic conservation planning approaches as developed in New South Wales, Australia[75]

The use of a geographic information system (GIS) is practically mandatory when considering spatial planning for conservation. The GIS allows spatial maps to display conservation options, and more powerfully, allows the user to combine biological and socioeconomic information to analyse ways of meeting conservation goals at the least socioeconomic "cost." Additional tools that work alongside and with a GIS are decision support software tools, which allow numerous competing variables to be combined. Depending on the particular tool used, a single best conservation configuration may be generated or a range of choices can be portrayed. In some of these tools, once a decision is made regarding a particular portion of the landscape, the entire study area can be recalculated to portray the next best options.

4. Future Needs

Further development is needed for tools to prioritise restoration needs. Current decision support tools are able to identify remaining habitat for inclusion in protected area networks, and these tools can be used to work with maps of previously existing potential vegetation. However, further refinement of these tools and associated techniques to identify areas that could be restored to meet representation goals is needed.

References

Dinerstein, E., Powell, G., Olson, D., et al. 2000. A workbook for conducting biological assessments and developing biodiversity visions for ecoregion-based conservation. World Wildlife Fund, Washington, DC. http://www.worldwildlife.org/science/pubs2.cfml.

Groves, C.R., Valutis, L.L., Vosick, D., et al. 2000. Designing a geography of hope: a practitioner's handbook to ecoregional conservation planning. The Nature Conservancy, Arlington, VA. www.conserveonline.org.

Loucks, C., Springer, J., Palminteri, S., Morrison, J., and Strand, H. 2004. From the Vision to the Ground: A Guide to Implementing Ecoregion Conservation in Priority Areas. World Wildlife Fund, Washington, DC.

Margules, C.R., and Pressey, R.L. 2000. Systematic conservation planning. Nature 405:243–253.

Noss, R.F. 1992. The wildlands project: land conservation strategy. Wild Earth (Special issue) 10–25.

Noss, R.F. 2001. Beyond Kyoto: forest management in a time of rapid climate change. Conservation Biology 15(3):578–590.

Olson, D.M., and Dinerstein, E. 1998. The global 200: a representation approach to conserving the earth's most biological valuable ecoregions. Conservation Biology 12:502–515.

Olson, D.M., Dinerstein, E., Wikramanayake, E.D., et al. 2001. A new map of life on earth. BioScience 15:933–938.

Palminteri, S. 2003. Ecoregion conservation: securing living landscapes through science-based planning and action. A users guide for ecoregion conservation through examples from the field (draft). CD-Rom. World Wildlife Fund US, Washington, DC.

Scott, J.M., Norse, E.A., Arita, H., et al. 1999. The issue of scale in selecting and designing biological reserves. In: Soule, M.E., Terborgh, J. Continental Conservation; Scientific Foundations of Regional Reserve Networks. Island Press, Washington, DC.

WWF. 2003. Ecoregion Action Programmes A Guide for Practitioners. WWF International, Gland, Switzerland.

---

[70] Dinerstein et al, 2000.
[71] Loucks et al, 2004.
[72] Palminteri, 2003.
[73] WWF, 2003.
[74] Groves et al, 2000.
[75] Margules and Pressey, 2000.

Additional Reading

International Tropical Timber Organisation. 2002. ITTO Guidelines for the Restoration, Management, and Rehabilitation of Degraded and Secondary Tropical Forests. ITTO Policy Development Series No. 13, Yokohama, Japan.

Moguel, P., and Toledo, V.M. 1999. Biodiversity conservation in traditional coffee systems of Mexico. Conservation Biology 13:11–21.

Pimentel, D., Stachow, U., Takacs, D.A., et al. 1992. Conserving biological diversity in agricultural forestry systems: most biological diversity exists in human-managed ecosystems. Bioscience 42: 354–362.

Victor, D.G., and Ausubel, J.H. 2000. Restoring the forest: skinhead earth? Foreign Affairs 79(6):127–144.

7
Why Do We Need to Consider Restoration in a Landscape Context?

Nigel Dudley, John Morrison, James Aronson, and Stephanie Mansourian

Key Points to Retain

Restoration is already needed in many important forest ecosystems because loss and degradation have proceeded to a point where the ecosystem is no longer sustainable in the long term.

Approaching restoration on a landscape scale means addressing conservation issues while considering social concerns, at a scale where optimisation and trade-offs are easier to agree on than at the site level.

Most current restoration activities tend too often to focus on one or two benefits and miss the wider picture.

Tools are starting to be developed that help to negotiate realistic mixes of management actions, including a suite of restoration activities, and biodiversity protection, at the full landscape scale.

1. Background and Explanation of the Issue

The landscape is the spatial and ecological scale at which the range of different ecological, social, and economic needs and desires of stakeholders can best be discussed, compared, and integrated.

1.1. Why Restore?

Conservation strategies that rely solely on protected areas and sustainable management have proved insufficient either to secure biodiversity or to stabilise the environment. The United Nations Environment Programme now classifies a large proportion of the world's land surface as "degraded," and reversing this damage is one of the largest and most complex challenges of the 21st century. Habitat loss is already so severe that conservation programmes need to include restoration if they are to deliver long-term success. Analysis of the WWF Global 200 ecoregions—identified as those of the highest conservation importance—demonstrates the problems. Over 80 percent of the G200 forest ecoregions need restoration in at least parts of their area; deforestation is a key threat to water quality in 59 percent of G200 freshwater ecoregions, and three quarters of G200 mangrove ecoregions are under threat.[76] Even where forest is stable or increasing, parallel losses of forest quality create the need for restoration. In Western Europe, for instance, research by the United Nations Economic Commission for Europe found that most countries had less than 1 percent of their forests surviving in an unmanaged state.[77]

Forest loss is not only of concern to conservationists. The United Nations estimates that 60 million people are directly dependent on forest

[76] Dudley and Mansourian, 2000.
[77] Dudley and Stolton, 2004.

resources including many of the poorest people. A far larger number are indirectly dependent, for example, on environmental services from forests such as soil and watershed protection. Forests also provide a wealth of recreational, spiritual, and aesthetic services.

1.2. Why Landscapes?

Many restoration efforts have ended in failure (see "Forest Landscape Restoration in Context"). Some of the reasons for this relate to their limited scope, their lack of engagement with local people and other stakeholders' interests and needs, their short-term nature, and their failure to address underlying causes of forest loss and degradation. In the last decade or so it has become increasingly clear to conservationists that developmental and socio-economic concerns cannot be overlooked if conservation is to be successful. Conservation activities, therefore, inevitably take place alongside other aspects of sustainable development, and a landscape approach can help to embrace both aspects of conservation and development. Because the restoration of forests in landscapes aims to repair and recover forest products and services that are valuable to people, it has a key role to play in development programmes. Balancing competing ecological and social needs is always difficult, but is most likely to succeed if we work on a large enough area to encompass two or more interactive ecosystems, as well as different landscape units with different land uses by local people. This facilitates negotiation and trade-offs among different demands.

Thus, rather than relying on a series of individual projects attempting to restore individual forest values, at the landscape scale it becomes possible to attempt the integration of these projects. Where successful, the net result should be much more than the sum of individual site-based restoration actions. Achieving a balance between the various goods and services required from restored forest ecosystems requires conceptualisation, planning, and implementation on a broader scale. It also assumes some negotiations and trade-offs among the various stakeholders involved to identify those restoration actions that have enough of a groundswell of support to be likely to succeed. A landscape or ecoregion approach also allows forest restoration to be fully integrated with protection and sustainable management of forest.

From the perspectives of biodiversity, long-term viability and ultimately social and economic values, approaches to restoration need to focus on forest functions and ecological processes. A key concern in many restoration projects is increasing the size of core areas of forest habitat. However, where space is limited by competing land uses, many functions of a large forest can be simulated by increasing connectivity between patches of forest by biological corridors and ecological stepping stones (patches of habitat that can provide "way stations" for migrating or mobile species). Increasing the values of existing forests, for example by changing management or decreasing interference, can also play a vital role in restoration. The landscape scale also allows us to consider the links between different habitat types. The interface between habitats may be abrupt (particularly in managed landscapes) or gradual, and they will have a varying ability to allow dispersal and interchange of species (see "Restoring Tropical Montane Forests"). Increasing the permeability of habitat boundaries to genetic interchange may be as important as specific habitat creation such as biological corridors.

1.3. Protect, Manage, Restore in a Landscape

The result of integrating efforts to restore multiple functions at a landscape scale often resembles a mosaic, where protected areas, other protective forests, and various forms of use and management are combined, depending on existing and evolving needs, legislative constraints, and land ownership patterns. Restoration becomes a management option that can be used within any part of the landscape to contribute to the overall long-term aims for the landscape. Agreeing on the mosaic and balancing different social, economic, and environmental needs on a landscape scale requires careful planning and negotiation.

A landscape approach recognises that overall landscape values and services are more important than individual sites, and that in a world of competing interests, conservation aims need to be integrated with those of, for example, poverty alleviation, human health, and other legitimate forms of social and economic development and welfare. Conservation cannot, or should not, take place divorced from issues relating to human well-being, and people working for conservation are usually also concerned about social justice and sustainable development. The appropriate approach, therefore, is to identify where and how these different but overlapping interests can best be integrated into a multifunctional landscape. Such integration will necessarily include negotiation and trade-offs.

1.4. The Process of Restoring Forest Functions in a Landscape

Deciding what forms of restoration to apply requires a suite of different activities, including careful analysis of what is needed, assessment of what is possible, and agreement amongst relevant stakeholders about the aims of restoration and the appropriate actions to undertake. It is axiomatic of forest landscape restoration that in most cases we are not looking at a single project or a single forest use, but rather at a range of different restoration efforts that will, as far as is feasible, be coordinated and complementary. The extent to which this is attainable in practice depends on the willingness of different groups of stakeholders to cooperate, the negotiation skills of those involved, and hard-to-define issues such as ownership patterns and other demands on the landscape. In areas where much of the land is in private ownership, many "common goods" including conservation can only be addressed through voluntary agreements, land purchase, or overarching policy decisions, and all of these options are slow and laborious to achieve in most situations.

2. Examples

Some examples show how different countries or regions have approached issues of restoration and how different priorities have shaped and in some cases distorted options for restoring a balanced forest mosaic.

2.1. Switzerland: Restoration for Environmental Services but with Additional Economic and Biodiversity Values

Following severe erosion and flooding problems in the past resulting from historical deforestation, during the 19th and 20th centuries Switzerland devised a system of continuous cover forestry to protect slopes and provide resources and fuel. The government has one of the few forest policies that explicitly rank social and protective functions above commercial functions. The country has 1,204,047 hectares of forest and woodland, covering 29 percent of the country.[78] Trees within managed forests are generally native and around 60 percent are conifers, with almost half the growing stock being Norway spruce. Although forest management is less intensive than in many European countries on a stand level, it affects virtually the entire forest area, and there are very few old-growth forests. Around 0.5 percent of forests are in natural forest reserves. Landscape-scale planning has played a critical role in identifying where best to restore forests, with an emphasis being placed on avalanche control, stabilisation of slopes, provision of local firewood, and biodiversity conservation.[79]

2.2. Guinea: Traditional Management Including Forest Restoration

Careful research with villages on the forest-savannah interface in Guinea, in West Africa, found that rather than contributing to deforestation as was once thought, local communities were actually planting and tending forest patches. Once villages were abandoned (a periodic response to declining soil fertility so that communities moved every few decades), such forests tended to decline and disappear as a

[78] Holenstein, 1995.
[79] McShane and McShane-Caluzi, 1997.

result of increased grazing pressure from savannah herbivores. New areas were chosen on the basis of past use and where fertility was likely to have recovered, thus focussing on different parts of the landscape at different times to ensure long-term continuity. Villagers established forest patches on the edge of the grassland to provide needed nontimber forest products and protected these from fire and grazing.[80]

2.3. United Kingdom: Plantations Replacing Natural Forests and Dominating the Landscape

Following the First World War, concern about lack of timber led to the establishment of the Forestry Commission, which was provided with considerable funds and political power to undertake compulsory purchase, to establish fast-growing plantations of trees. The emphasis was on conifers, particularly Sitka spruce (*Picea sitchensis*) from Alaska. Many of these plantations were established on upland grazing areas (which were originally forested but had lost their tree cover, in some cases centuries before). Some plantations were also established on the site of native woodland, which was occasionally cleared with herbicides, and in northeast Scotland on moor that had never contained trees. Whilst the planting was successful in creating a strategic reserve, it led to resentment about loss of access, native woodlands, and other natural habitats, and a limited range of forest functions. Dense forest created access problems and the abrupt boundaries between this and other habitat limited usefulness for biodiversity. Planning was usually at site rather than landscape scale. From the 1980s onward, the commission started revising its aims, increasing native planting and playing a more general stewardship role in land management; experiments are also taking place in returning woodland areas to local community control.[81]

2.4. Costa Rica: Shade-Grown Coffee as a Linking Habitat in Fragmented Landscape with a High Population Density

Although Costa Rica still contains large areas of native forests, some forest ecosystems have declined to a fraction of their former size and are no longer ecologically viable, particularly in Talamanca and Guanacaste. In the former area, The Nature Conservancy (TNC) has been working with local communities to link remaining forest fragments to allow access for birds. Because pressure on land was too intense to allow space for native woodland as such, shade grown cacao and coffee production was encouraged and supported, planned at a landscape scale to link remaining forest fragments. While far from a natural woodland, the trees shading coffee provide habitats to allow passage for rare birds, thus allowing them to form viable populations.[82]

The above cases illustrate only a fraction of the possible examples. They show that in most places where restoration is encouraged, its purpose is generally fairly narrow (also see "Goals and Targets of Forest Landscape Restoration"): erosion control, strategic reserves, etc. If other benefits accrue, it has sometimes been fortuitous. One of the key aspects of forest landscape restoration is to reduce the elements of chance and increase the sophistication of restoration planning.

3. Outline of Tools

3.1. Ecoregional Planning Tools

A wide range of possible tools exist to plan regional scale forest cover and management (see also previous chapter). Among the most popular are the following:

- Ecoregional workshops: used to help establish a vision for an ecoregion, prioritise actions and conservation landscapes, and develop strategies

---

[80] Fairhead and Leach, 1996.
[81] Garforth and Dudley, 2003.
[82] Parrish et al, 1999.

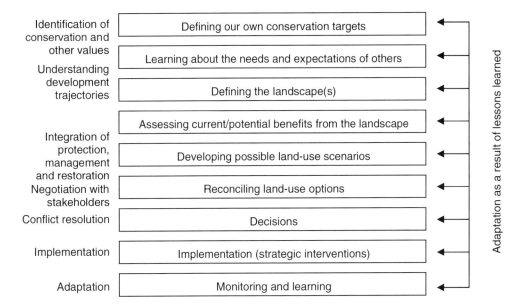

FIGURE 7.1. Protect–manage–restore approach.

- Computer-aided design packages: including those involved in the development of systematic conservation planning
- Conservation by design: developed by TNC, using a five-step process (identifying targets, gathering information, setting goals, assessing viability, assembling portfolios) and the 5-S framework (systems, stresses, sources, strategies, success)

There are many other examples; a selection are available on the Web-based Earth Conservation Toolbox.[83]

3.2. Protect, Manage, Restore

WWF[84] and IUCN have developed a number of landscape approaches to help address this kind of broadscale decision making, and these or similar exercises could provide help in determining where restoration could be used most effectively. An outline of one approach is shown diagrammatically in Figure 7.1 (also see Box 7.1 for the detailed steps):

3.3. Implementing Conservation in Priority Areas

WWF also has a science-based methodology for continuing ecoregion planning inside priority conservation landscapes, containing a set of guidelines to develop and implement a conservation landscape, which could be used to include restoration issues.[85]

3.4. Reference Forests

Restoration for conservation usually involves trying to regain something as similar to a native forest as possible (for more, see "Identifying and Using Reference Landscapes for Restoration").

---

[83] www.earthtoolbox.net.
[84] Aldrich et al, 2004.
[85] Loucks et al, 2004.

Box 7.1. The stages in a protect–manage–restore process

- ✓ **Defining our own conservation targets:** As stakeholders, conservation organisations need to start with some ideas of the landscape mix that they are aiming for, including ideas about geographical areas and ecological processes of primary interest. Reaching these targets will require a mix of protection, management, and restoration.

- ✓ **Learning about the needs and expectations of others:** At an early stage it is important to get an initial idea about the other key stakeholders and their relationships, what they need and want, and what they are planning. While the focus will be on economic or development issues, culture, history, expectations within society, level of development, and spiritual needs are all important.

- ✓ **Defining the landscape(s):** The concept of "landscape" has many different meanings; a conservation programme will usually work within a predetermined "conservation landscape," but it is important to identify any "cultural landscapes" nested within or overlapping the conservation landscape: e.g., a village, land used by nomadic pastoralists, or a timber concession.

- ✓ **Assessing current/potential benefits from the landscape:** The next stage involves assessment to identify lost, current, and potential future values from the landscape. While conservationists tend to focus on biodiversity, assessment also takes full account of social, cultural, and economic values. The extent to which this is a *participatory process* can be decided on a case-by-case basis. Including stakeholders also means that assessment is part of the negotiation process.

- ✓ **Developing land-use scenarios:** Integration of potential conservation and development actions to develop scenarios including a combination of elements such as protected areas; other protected forests (set asides, watershed protection etc); well-managed forests; areas needing restoration; and other compatible and competing land uses. All these factors interact. What mosaic will work best? Are we looking at one "master plan" or a pattern that emerges gradually over time?

- ✓ **Reconciling land use options:** The approach is predicated on the idea that trades-offs among social, economic, and environmental values are often essential and are acceptable if overall values are maintained or enhanced within the landscape.

- ✓ **Decisions:** In some situations government(s), nongovernmental organisations, corporate interests, and communities may agree on a package of actions within one action plan. In many other cases, negotiations are likely to be continuing and sporadic. Here it is unlikely that a single master plan could be agreed; rather, decisions will be over smaller parcels of land within a framework that will continue to evolve.

- ✓ **Implementation (strategic interventions):** Some of the resulting actions will take place at the site level and may involve creating the right conditions for natural regeneration, selective tree planting to reconnect forest fragments, or community initiatives to improve fire management. Other interventions may be necessary at a landscape or even larger scale, e.g., working with governments to realign reforestation programmes.

- ✓ **Monitoring and learning:** Much of what we will be attempting with the landscape approach is quite new, and therefore it is especially important to ensure that progress is monitored effectively and that

> lessons are both used to improve programmes as they develop and are also transmitted around and beyond the immediate conservation programme. At a larger scale, combining monitoring of many individual projects, along with some additional indicators that transcend individual project work, will be needed to measure progress over the whole landscape.

3.5. Gap Analysis

Several methodologies exist for identifying gaps in existing forest systems. For example, a WWF Canada methodology used enduring landform features to identify likely past vegetation,[86] while another developed by the United Nations Environment Programme-World Conservation Monitoring Centre(UNEP-WCMC) used analysis of current forest cover.[87]

4. Future Needs

Although restoration needs are increasingly being addressed within broader-scale conservation, they generally remain less well supported in terms of approaches and methodologies than, for example, planning of protected areas. These needs include the following:

Prioritisation: There is a need for better tools for prioritisation of areas for restoration, for example to balance the importance of connectivity with core areas, identification of microhabitat gaps in current forest cover, calculation of minimum viable areas, etc.

Decision support: Methodologies are needed for balancing social and ecological values, including participatory methods.

Incorporating a range of management schemes into existing decision support tools: Currently, decision support tools consider an area either protected, or not, based on the input of the user. More sophisticated tools are needed that can handle a wider range of "protection" schemes (e.g., sustainably managed forests).

There is also the need for some degree of advocacy and explanation, to encourage those involved in broad-scale planning to consider restoration, particularly in the case of restoring forest quality. Some of these tools are being developed during current forest landscape restoration projects, but it is still too early to judge their success.

References

Aldrich, M., et al. 2004. Integrating Forest Protection, Management and Restoration at a Landscape Scale. WWF, Gland, Switzerland.

Dudley, N., and Mansourian, S. 2000. Forest Landscape Restoration and WWF's Conservation Priorities. WWF International, Gland, Switzerland.

Dudley, N., and Stolton, S. 2004. Biological diversity, tree species composition and environmental protection in regional FRA-2000. Geneva Timber and Forest Discussion Paper 33. United Nations Economic Commission for Europe and Food and Agricultural Organisation of the United Nations, Geneva.

Fairhead, J., and Leach, M. 1996. Misreading the African Landscape: Society and Ecology in a Forest-Savanna Mosaic. Cambridge University Press, Cambridge, UK.

Garforth, M., and Dudley, N. 2003. Forest Renaissance. Published in association with the Forestry Commission and WWF UK, Edinburgh and Godalming.

Holenstein, B. 1995. Forests and Wood in Switzerland. Federal Office of Environment, Forests and Landscape. Swiss Forest Agency, Bern.

Iacobelli, T., Kavanagh, K., and Rowe, S. 1994. A Protected Areas Gap Analysis Methodology: Planning for the Conservation of Biodiversity. World Wildlife Fund Canada, Toronto.

Loucks, C., Springer, J., Palminteri, S., Morrison, J., and Strand, H. 2004. From the Vision to the

---

[86] Iacobelli et al, 1994.
[87] UNEP-WCMC, 2002.

Ground: A Guide to Implementing Ecoregion Conservation in Priority Areas. WWF-US, Washington, DC.

McShane, T.O., and McShane-Caluzi, E. 1997. Swiss forest use and biodiversity conservation. In Freese, C.H., ed. Harvesting Wild Species: Implications for Biodiversity Conservation. John Hopkins University Press, Baltimore and London, pp. 132–166.

Parrish, J.D., Reitsma, R., and Greenberg, R., et al. 1999. Cacao as Crop and Conservation Tool in Latin America: Meeting the Needs of Farmers and Biodiversity. Island Press/America Verde Publications, The Nature Conservancy, Arlington, Virginia.

UNEP-WCMC. 2002. European forests and protected areas gap analysis 2002. http://www.unep-wcmc.org/forest/eu_gap/index.htm.

8
Addressing Trade-Offs in Forest Landscape Restoration

Katrina Brown

Key Points to Retain

In questions of land management and natural resource allocation it will nearly always be impossible to satisfy all stakeholders and there will necessarily be winners and losers.

Applying the concept of multifunctionality can help to allow different forest functions to coexist, meeting a wider range of different stakeholder groups' interests.

Capacity needs to be created among conservationists to engage stakeholders in constructive trade-off discussions and to deal with the outcomes of these.

1. Background and Explanation of the Issue

In most of the places where forest restoration is being considered, from the perspective of either conservation or development, the landscape is already inhabited. Furthermore, the resident or transient populations are unlikely to be a single homogeneous entity. Therefore, forest restoration involves many different stakeholder groups with their own wants and needs.[87a] Agreeing what the restoration priorities should be within a given landscape will consequently necessitate negotiating trade-offs among a range of stakeholders.

1.1. Win–Win Situations

It is often assumed that with enough discussion and compromise, questions of land management and natural resource allocation can be agreed to in ways that satisfy everyone—in this case that a sufficient number and variety of forest functions can be restored in a landscape to satisfy all stakeholder groups: so-called win–win situations. The question of how to attain such win–win situations has been addressed by many integrated conservation and development projects, and the consensus seems to be that in most real-life situations it will be impossible to satisfy everybody and there will necessarily be winners and losers.[88] From our perspective, some people will stand to gain more from the restored functions of a forest, for example with increased availability of fuelwood or salable products, while others will lose for instance, through access or grazing rights. The realistic aim of a negotiated process is to minimise the losses and to ensure that these do not fall disproportionately on those already amongst the poorest or otherwise disadvantaged. Indeed, raising false assumptions that careful planning and participatory processes can deliver win–win results, and an accompanying failure to deal with necessary trade-offs are often major sources of conflict, because people have their expectations raised and then not met.

[87a] Sheng (no year).

[88] McShane and Wells, 2004.

1.2. Identifying Stakeholders

The need for trade-offs arises because different stakeholder groups have different expectations or needs from a landscape. To understand trade-offs when dealing with a restoration programme in a landscape, the first step is to identify all the stakeholders. Often stakeholders are characterised by their degree of influence and importance.[89] The results of such an analysis can be categorised into primary stakeholders, secondary stakeholders, and external stakeholders. Primary stakeholders have little influence on the outcomes but they have the most to lose from management decisions. A primary stakeholder could be a farmer, a fisher, or a forest-dweller. Secondary stakeholders are often managers or decision makers, and they are the ones charged with implementing the decision, although the outcomes do not impact directly on them. External stakeholders are those who can significantly influence the outcome even if they are located far away, typically international nongovernmental organisations (NGOs). Many more complex stakeholder categories have been suggested, but these three capture the main groupings. Depending on the objectives of the trade-off process, stakeholder analysis can be critical in identifying *who* to include and perhaps *how* to engage them.

1.3. Brokering a Satisfactory Outcome

The next requirement in an equitable trade-off process is to allow genuine discussion on trade-offs between different stakeholders. There is usually a need for someone to help facilitate this process, ideally a person without a stake (perhaps a trusted outsider) who can act as an "honest broker."[90] The role of the broker is to encourage an open discussion and to help facilitate a process whereby different stakeholders feel that they are gaining something from the process, even if that may mean also agreeing to some sacrifices. For instance, shifting cultivators may need to modify their approach to farming, but in return they may gain legitimate access to nontimber forest products located in the landscape. Frequently, conservation or development organisations like to consider themselves as "neutral brokers," yet the reality is that they also have a position and an interest. Conservation organisations are stakeholders just like any other, with a particular vision that will sometimes be in competition with other legitimate economic and social "visions," and conservationists are therefore unlikely to get everything that they want.[91] "Valid processes require much more time, patience and sensitivity to local cultures than most outside experts are prepared to allocate. Neutral facilitation and explicit recognition of the trade-offs between the interests of different stakeholders are important ingredients of success."[92]

1.4. The Concept of Multifunctionality

When negotiating trade-offs in attempting to restore forest functions in a landscape, the concept of "multifunctionality" is important. If one stakeholder group, for instance biologists, is the only one deciding on the restoration outcomes of a given landscape, it may be that an ideal landscape for that group is one containing pristine habitat for all identified species in the given area. On the other hand, if the single stakeholder is a plantation company, it may be that its vision for the main function to restore in the landscape is that of productive monoculture plantations bringing in money from pulp and paper. For a poor local family, the main function it may be interested in restoring might be fuelwood. Applying the concept of multifunctionality can help to allow these different functions to coexist, meeting a wider range of different stakeholder groups' interests.

1.5. Types of Trade-Offs[93]

Restoring a landscape intentionally to meet a range of functions requires negotiating trade-offs. There are different types of trade-offs:

---

[89] Brown, 2004.
[90] Franks, 2004.
[91] Aldrich et al, 2003.
[92] Sayer et al, 2003.
[93] Brown, 2004.

- Trade-offs between different interest priorities, as per the example above
- Trade-offs between short and long-term horizons
- Trade-offs between different spatial scales, notably sites and landscapes
- Trade-offs between different sections of society and biodiversity conservation, typically farmers or plantation owners and conservation NGOs
- Trade-offs between different aspects of biodiversity, as it may not always be possible to restore a landscape to secure all species in a landscape; decisions on which species will take priority will require trade-offs
- Trade-offs between different social groups—traditionally more influential groups may have taken decisions, but primary stakeholders are those whose livelihoods are directly affected; in a truly representative process, trade-offs will need to happen across social groups and scales.
- Trade-offs among economic priorities, social welfare, and conservation.

The skills needed to assess and evaluate such trade-offs and support negotiations about them are often lacking amongst conservation organisations, although they are more likely to exist within aid or development bodies. Developing negotiating skills is one of the key priorities in developing the capacity to work at landscape level (see "Negotiations and Conflict Management").

2. Example: An Hypothetical Example for Negotiating the Restoration of a Landscape

There are as yet few examples where a truly negotiated discussion and trade-offs led to a restored landscape.

A theoretical process to achieve this was presented at a workshop in Madagascar.[94] Possible steps to reach a negotiated outcome for a restored landscape are as follows:

Each stakeholder group describes the landscape as it was 50 years ago, the steps that turned it into the current landscape and the main drivers of the changes.

A facilitated discussion takes place to negotiate the general state of the landscape and its possible future state(s) (characteristics, products, and services it could offer, etc.).

Each group develops a precise and detailed vision for the landscape 10 years from the present, identifying the most important characteristics (i.e., the nonnegotiables), categorising the possibly negotiable characteristics and the definitely negotiable characteristics.

The visions of different groups are then placed side by side, and a negotiation process begins that will culminate in a common vision for the future, restored landscape, that is acceptable to all.

Such a process most certainly takes a significant amount of time. It requires clear identification and representation of stakeholders, a genuine neutral broker (or group of brokers), and different tools and processes to allow each stakeholder group to understand the implications of different decisions.

3. Outline of Tools

Some of the tools available to allow the negotiation of trade-offs are as follows:

3.1. Focus Groups

Working in small groups builds confidence, especially amongst stakeholders who may be reluctant to air their views in large meetings or are not used to public speaking. It enables specific stakeholders to rehearse and deliberate in a safe structured environment, prior to larger meetings or workshops.

3.2. Surveys

Surveys can be valuable in generating baseline data and information to build believable scenarios or visions of the future and to illustrate management options. They are a means to learn about and approach different stakeholders. A particularly useful contribution is to feed back information generated from surveys to stakeholders as part of a social learning and triangulation process.

---

[94] Taken from a presentation by Tom Erdmann given at a workshop on Forest Landscape Restoration in Madagascar in March 2003.

3.3. Consensus Building Workshops

Different stakeholders may be brought together in workshops to negotiate trade-offs and agree on management strategies. A range of conflict resolution and consensus building techniques can be used, including visioning and scenarios, as well as ranking and voting on criteria and scenarios.

3.4. Multicriteria Analysis

Multicriteria analysis is a decision-support tool that can be used in a sophisticated and data intensive way or, in deliberative workshops, as a means to help stakeholders take a step back from concentrating on outcome to assess what criteria should guide decisions. Rather than discussing the outcomes of management, this forces people to look at *why* and *how* decisions should be made rather than on the impacts of the decisions. This aids a more consensus-based approach to negotiations.

3.5. Extended Cost-Benefit Analyses

A range of evaluation techniques can be used to draw attention to the nonmonetary and noneconomic impacts of different management options and to learn about how different stakeholders value the multiple functions of resources. Again it can help to validate and build confidence in stakeholders by recognising their priorities and values.

3.6. Scenario-Building

A useful way to discuss different options without them being directly linked to interests of specific stakeholders is to define scenarios or coherent, internally consistent, and plausible descriptions of the future. These must be believable and understandable to all stakeholders and must be linked to specific changes. Discussing and evaluating scenarios are a way of talking about management options without having to argue against one person's project or strategy, and therefore can be useful for building consensus.[95]

4. Future Needs

Evaluating and negotiating trade-offs is rarely part of conservation projects, let alone restoration ones. Much more practical experience is needed in negotiating trade-offs when looking at restoring forest functions in a landscape. This is particularly the case when considering limited resources and the urgency of some restoration needs. In other words, how does one balance a truly participatory trade-off analysis with urgent needs to restore habitat for a threatened species?

Capacity needs to be created among conservationists to engage stakeholders in constructive trade-off discussions and to deal with the outcomes of these.

References

Aldrich, M., Belokurov, A., Bowling, J., et al. 2003. Integrating Forest Protection, Management and Restoration at a Landscape Scale, WWF, Gland, Switzerland.

Brown, K., Tompkins, E., and Adger, W.N. 2002. Making Waves: Integrating Coastal Conservation and Development. Earthscan, London.

Brown, K. 2004. Trade-off Analysis for Integrated Conservation and Development. In: Mc Shane, T., and Wells, M.P., eds. Getting Biodiversity Projects to Work. Columbia University Press, New York.

Franks, P., and Blomley, T. 2004. Fitting ICD into a Project Framework: A CARE Perspective. In: Mc Shane, T., and Wells, M.P., eds. Getting Biodiversity Projects to Work. Columbia University Press, New York.

Mc Shane, T., and Wells, M.P. 2004. Getting Biodiversity Projects to Work. Colombia University Press, New York.

Sayer, J., Elliott, C., and Maginnis, S. 2003. Protect, manage and restore: conserving forests in multifunctional landscapes. Paper prepared for the World Forestry Congress, Quebec, Canada, September.

Sheng, F. (No date.) Wants, Needs and Rights: Economic Instruments and Biodiversity Conservation, a dialogue. WWF, Gland, Switzerland.

[95] Brown et al, 2002.

Part B
Key Preparatory Steps Toward Restoring Forests Within a Landscape Context

Section IV
Overview of the Planning Process

9
An Attempt to Develop a Framework for Restoration Planning

Daniel Vallauri, James Aronson, and Nigel Dudley

Key Points to Retain

While no two restoration experiences will follow the same pattern, indicative steps to planning a restoration initiative are important, particularly when dealing with large scales or landscapes.

Success depends on wise planning, balancing short-term with long-term goals, and allocating the funding available for the restoration programme as efficiently as possible.

Learning from past restoration programmes and their successes and failures is an important starting point to help plan better restoration actions in the future.

There are few tools dealing with planning restoration in large scales. A five-step logical planning process is being proposed.

1. Background and Explanation of the Issue

1.1. Why Planning?

Restoration of natural systems is a difficult, energy-consuming, and expensive undertaking. It is almost always a long-term, complex, and transdisciplinary process.[96] This is particularly true when dealing with highly degraded ecosystems and landscapes. Inevitably, conflicts of interest and other problems arise.

Ecologically speaking, the restoration of highly degraded forest usually requires initiating an embryonic ecosystem within a few years (usually less than 10 to 15 years after degradation), which will be only fully restored—very often after additional corrective or fine-tuning interventions—after a period of at least 50 years in the tropics, and of 100 years or more in the extratropical zones. However, forest policies and restoration programmes are generally financed only on a short- to medium-term basis. A 10- to 15-year project span, in most cases, is the longest possible perspective, both for political and financial reasons. Bearing this in mind, restorationists should (1) adapt short-term restoration goals and techniques to minimise the number of costly corrective actions; and (2) plan ahead to secure funds for carrying out monitoring and evaluation, corrective actions, or "aftercare" in the long term.

Also, forest restoration requires inputs and expertise from various academic and practitioner fields[97] like ecology, silviculture, economics, public policy, and the social sciences, which need to be combined in an efficient way.

Meanwhile, the relative lack of experience with broad-scale conservation means that filling the knowledge gaps through research programmes also takes time. Five to 10 years is the minimum period needed to investigate critical

---

[96] Pickett and Parker, 1994.

[97] Clewell and Rieger, 1997.

questions like natural dynamics, nursery and plantation techniques for native species, etc. However, very little money is available to finance pure research programmes unless they can be linked to real implementation and visible successes in the field. Bearing this in mind, restorationists should define short-term goals and activities that get restoration underway, along with long-term goals for how it can be sustained over the time period required. A critical, pragmatic aim is to achieve at least some rapid field results, for example on carefully selected pilot sites, to build support for longer term efforts.

Finally, forest landscape restoration, as developed in this book, requires a concerted approach among stakeholders and communities, to develop a shared and accepted vision and goals for the future of the landscape in question. This also takes time and should be planned for, but at the same time should lead rapidly to tangible changes or outcomes that really engage stakeholders and people living in the region in a lasting and meaningful manner.

Success in forest restoration depends on wise planning,[98,99] both in time and in space, balancing short-term goals with long-term goals, and allocating the funding available for the restoration programme as efficiently as possible. Accordingly, a clear step-by-step plan of action is needed for success. This was very often lacking in past restoration programmes, especially site-oriented ones, and has led to many failures or difficulties that often emerge only decades after the first restoration efforts were begun.

1.2. Restoring Step by Step

Where restoration is to be carried out as part of a wider conservation effort, at the landscape or ecoregional levels, we would propose that it be planned as an embedded element within an integrated programme that also involves protection of whatever is left of untouched nature, and the promotion of good ecosystem management, as guided by the principles of stewardship, sustainability, and sustained use. We have already outlined some possible elements in a protect–manage–restore programme in the introduction to this book. This approach includes identifying a series of conservation targets—in this context, what forest functions we wish to restore—and "reconciling" these with the needs, tastes, and expectations of other stakeholders, especially the indigenous populations.

Conceptualisation of the process of implementing restoration programmes is very new. We propose below an outline of a planning framework, following a five-step logical planning process. In the context of a broad-scale conservation strategy, then, the following steps help lead to the development and realisation of restoration achievements.

1.2.1. Step 1: Initiating a Restoration Programme and Partnerships

An essential first step of any forest landscape restoration programme is the identification of the problem being addressed and agreement on the solutions and the targets for restoration. Such targets should ideally contribute to wider ecological and socioeconomical objectives at a landscape scale. Very often, restorationists must start from zero to raise awareness on the state of degradation in the landscape, analyse the root causes, and then convince other stakeholders of both the need for and the feasibility of forest restoration. Depending on the context (the existing level of awareness, politics, funds available, etc.), this step could last for several years and require extensive effort.

Experience suggests that restoration usually only works in the long term if it has support from a significant proportion of local stakeholders. Finding out the needs and opinions of stakeholders is therefore important: What forest functions do they want to restore and are there potential clashes of interest? It should be recognised that the restorationists (conservation NGO or other) are themselves stakeholders with a particular interest (i.e., restoring biodiversity), which may need to be reconciled with other stakeholders' priorities.

---

[98] Aronson et al, 1993.
[99] Wyant et al, 1995a,b.

Outputs of this step are:

- recognition and common understanding of the degradation, root causes, and solutions;
- stakeholders' involvement and participation;
- partnership development for an efficient restoration programme (written key ideas of the programme and memorandum of understanding); and
- secured budget for the restoration programme for at least a first pluri-annual period (e.g., five years).

1.2.2. Step 2: Defining Restoration Needs, Linking Restoration to Large-Scale Conservation Vision

Here is a step that is not necessarily easy to "sell" to local stakeholders. The geographical scope can be much wider than many people are used to working with or even conceptualising (or want to work with, as it has some implications for development, too). Ideally, as mentioned above, a vision and strategy for restoration should be developed within an integrated "protect–manage–restore" approach, especially because the investment needed to restore has to be reinforced through synergy with management and protection activities.

Assessment is needed to determine how restoration targets might be achieved, including determining current or potential benefits from forests in the landscape (biodiversity, environmental services, and resources for subsistence or sale) and the potential for restoration through use of reference forests and other techniques. An important part of the process is deciding the realistic boundary of the area or areas that we wish to restore. Definition of key areas for protection, analysis of degradation, and the predictive anticipation of threats can all help to define priority landscapes where investment in restoration is most justified.

Outputs of this step are:

- definition of conservation targets at various pertinent scales (ecoregion, landscape);
- analysis of the broad consequences on the landscape of past degradation, active pressure, and potential threats;
- definition of the role of restoration along with identification of protection and management needs; and
- identification of the priority areas that require restoration and explanation of the reasons why: Which landscapes, landscape units, or landscape functions do we need to restore? Which species do we need to eradicate, control or reintroduce?

1.2.3. Step 3: Defining Restoration Strategy and Tactics, Including Land-Use Scenarios

Considering ecological characteristics, but also socioeconomical context or goals assigned to the restoration project, several trajectories and restoration options could be developed for the same project. Choosing among these options requires careful study and data gathering.

This will necessarily mean reconciling different points of view and opinions. Agreement can be a phased and continuing process; that is, it may be possible to agree to some specific and useful restoration interventions without reaching agreement about the whole future of the landscape. The way in which such agreements are reached will naturally depend on the political and social realities of particular countries or regions; the general principle that decisions should be as participatory as possible applies throughout.

Outputs of this step are:

- assessment of current/potential benefits from the landscape for people, and for biodiversity;
- assessment of the current, past, and reference landscape states;
- definition of what we can expect to restore;
- development of possible land-use scenarios in space (including maps);
- development of possible restoration trajectories to achieve short-term and long-term goals (including models, time frames, and maps);
- reconciliation of land-use options: how can we achieve specific goals while meeting or reconciling conflicting demands, tastes, and needs?;

- set of goals, strategies, and tactics for each zone and problem in the landscape;
- set of priorities in space and time;
- identification of restoration trajectories, technical options, steps, and phases, (especially remembering the monitoring and "fine-tuning" phases necessary to fully achieve long term restoration goals); and
- A written restoration plan, strategy, and set of tactics, with identified time frames, maps, allocated funds, and quantified targets.

1.2.4. Step 4: Implementing Restoration

This step is the most visible part of the work, and usually the most costly. Some projects start here, for example, by directly investing all the available funds to plant trees on an emblematic or strategic site. However, this ignores the previous planning steps recommended above and can easily end up wasting time and resources in restoration activities that either do not work or are in suboptimal locations. It is of course judicious to start small-scale actions, such as one or more pilot sites, for the sake of "learning by doing," to demonstrate the feasibility of key restoration goals and to test silvicultural techniques (for example planting, but also natural regeneration). But we would strongly recommend that larger scale activities also be undertaken in the context of careful planning and assessment as outlined in steps above.

Outputs of this step are:

- development of pilot sites;
- implementation of large-scale actions;
- lessons learned from first results, both successes and failures; and
- design and implementation of changes/adaptation in the restoration programme.

1.2.5. Step 5: Piloting Systems Toward Fully Restored Ecosystems

In practice, a few years or decades after starting implementation, even if restoration has hitherto been successful, unexpected results of previous work or changing circumstances (evolution of the socioeconomic context, for example) could alter the most preferable restoration trajectory. This could even lead in some cases to redefining overall project goals. Such modifications should not be considered as a failure of the overall programme, but rather as a normal step in the restoration of a complex set of ecosystems within a larger landscape matrix.

Thus, the restoration work is not "finished after planting." To sustain restoration success in the long run, and to anticipate potential problems, a simple monitoring and evaluation framework (see section "Monitoring and Evaluation") needs to be set up from the outset of the programme in order to facilitate adaptive management and corrective actions.

Outputs of this step are:

- regular evaluation (social, economical, ecological);
- restoration trajectory reappraisal; and
- design and implementation of corrective actions.

2. Examples

As yet, there are few full-scale forest landscape restoration programmes, although their numbers are rapidly increasing. The following examples show both the need for planning and broad-scale restoration planning in practice. These examples show not only how a planning framework can be implemented, but also how problems can arise by forgetting one step.

2.1. New Caledonia: From Awareness to Restoration of Tropical Dry Forests (Step 1)

It took 15 years from the first alarm signals by scientists to the first significant pilot plantings or protection of sites within a forest landscape restoration initiative in New Caledonia. Attention to the tropical dry forests of New Caledonia began to grow in the early 1990s. In 1998, WWF, the global conservation organisation, launched an effort to organise a consortium of research institutions, local government agencies, and NGOs (10 partners) to create a tropical dry forests programme. Underway since 2001, this programme has already carried out much of the preliminary reconnaissance and mapping in different tropical dry forest fragments, as well as ecological, silvicultural, and horticultural studies of great importance to restoration efforts slated to begin in the field in 2005. Two of the authors (Aronson and Vallauri), who have been involved in this restoration programme, consider that partners should work to prepare now as soon as possible a protect–manage–restore approach and restoration at broad scale in a large priority landscape, like the ecologically outstanding landscape of Gouaro Deva (see "Restoring Dry Tropical Forests").

2.2. Vietnam: Integrating Restoration into a Landscape Approach Across Seven Provinces (Step 2)

The Central Truong Son initiative, covering seven provinces in central Vietnam inland from Dalat, is developing an integrated approach to forest protection, management, and restoration. Comparatively large areas of natural forest remain standing, although often in poor or highly degraded condition. There are major plantation developments of varying success, and the government is committed to maintaining protected areas. The new Ho Chi Minh Highway is bringing rapid social and environmental changes, some of which directly threaten remaining natural forests. The Central Truong Son initiative has identified priority landscapes and used a gap analysis, coupled with a detailed study of forest quality, to pinpoint the most effective areas for restoring natural forest in terms of increasing forest connectivity and protecting biodiversity; these are currently around the buffer zone of Song Thanh nature reserve and in a so-called green corridor area linking several patches of natural forest. Elsewhere, more generally the project is seeking to increase the proportion of forest restoration funds used for natural regeneration (see case study "Monitoring Forest Landscape Restoration—Vietnam").

2.3. France: The Consequences of a Lack of Ecological Monitoring (Step 5)

In the early 1860s, an ambitious "Restoration of Mountain Lands" initiative was set up by the French forest administration in the southern Alps, primarily for the purpose of erosion control. A wide range of plant material was used, including native shrubs and grasses, but no particular preference was given to native trees for replanting. Over 60,000 hectares were thus planted between 1860 and 1914, using mainly *Pinus nigra* Arn. subsp. *nigra* Host. These efforts have proved effective at stopping the average erosion rate (of 0.7 mm per year) on black marls. Nevertheless, although rehabilitated in the sense that erosion has been halted and badlands forested, these ecosystems were not fully restored. No fine-tuning assistance and ecological evaluation was carried out until recently.[100] The forest soils were now better protected, as shown by the study of soil biological activity, especially earthworm communities. However, the rehabilitated ecosystems were facing two new ecological problems: lack of natural regeneration, and development of an infestation of the pine trees by mistletoe (*Viscum album*). Once management priorities have been revised, the goal for the future is to restore the diversity, structure, and functioning of a native forest ecosystem. The absence of long-term monitoring and evaluation for about 100 years did not allow a rapid adaptation of the restoration trajectory. After a necessary short pioneer stage with Austrian pine, the restoration strategy should have been pursued 30 years later by a phase of autogenic restoration of native biota [oak (*Quercus*), maple (*Acer*), mountain ash (*Sorbus*), and others].

3. Outline of Tools

There are still few specific planning tools designed specifically for restoration. However, many existing conservation planning tools could be adapted for or could include a restora-

---

[100] Vallauri et al, 2002.

tion component. For example, Conservation International has developed guidelines for corridors that include reference to restoration to fill gaps in existing forest cover, although with little detail.

The reader will find more details on the potential tools step by step in the following sections. They include among others:

Step 1. Initiating a restoration programme and partnerships
- Lobbying
- Participatory approaches
- Capacity building

Step 2. Defining restoration needs, linking restoration to large-scale conservation vision
- Ecoregional planning process (WWF)
- 5-S process and systematic conservation planning (The Nature Conservancy)
- Landscape planning

Step 3. Defining restoration strategy and tactics, including land-use scenarios
- Conceptual modelling
- Geographic information systems
- Ecological modelling
- "Restoration vision and strategy" meetings

Step 4. Implementing restoration
- Tools on plantation, natural regeneration, species' selection, etc., are covered in other sections of this book.

Step 5. Piloting systems toward fully restored ecosystems
- Restoration projects' databases: A lot could be learned from past restoration successes and failures. The analysis of databases of long-term restoration projects is very useful, like the world restoration database launched by UNEP-WCMC (http://www.unepwcmc.org/forest/restoration/database.htm) or the database of evaluated restoration programmes in the Mediterranean (http://www.ceam.es/reaction/)
- Criteria and indicators for monitoring (see section "Monitoring and Evaluation")

4. Future Needs

Restoration planning in landscapes or large scales is still in its infancy. Much further work is needed to refine and improve the planning process and define appropriate tools. Thus, specific work on restoration planning is highly needed in the coming years, both in theory and in practice. Learning from past restoration programmes and their successes and failures could prove an efficient starting point. In time, lessons might usefully be captured in a step-by-step guidebook or manual specifically on this subject and perhaps with associated software programmes if appropriate.

References

Aronson, J., Floret, C., Le Floc'h, E., Ovalle, C., and Pontanier, R. 1993. Restoration and rehabilitation of degraded ecosystems in arid and semi-arid lands. I. A view from the south. *Restoration Ecology* 1:8–17.

Clewell, A., and Rieger, J.P. 1997. What practitioners need from restoration ecologists. *Restoration Ecology* 5(4):350–354.

Pickett, S.T.A., and Parker, V.T. 1994. Avoiding old pitfalls: opportunities in a new discipline. Restoration *Ecology* 2(2):75–79.

Vallauri, D., Aronson, J., and Barbéro, M. 2002. An analysis of forest restoration 120 years after reforestation of badlands in the south-western Alps. *Restoration Ecology* 10(1):16–26.

Wyant, J.G., Meganck, R.A., and Ham, S.H. 1995a. A planning and decision-making framework for ecological restoration. *Environmental Management* 6:789–796.

Wyant, J.G., Meganck, R.A., and Ham, S.H. 1995b. The need for an environmental restoration decision framework. *Ecological Engineering* 5:417–420.

Section V
Identifying and Addressing Challenges/Constraints

10
Assessing and Addressing Threats in Restoration Programmes

Doreen Robinson

Key Points to Retain

Threats may be direct, indirect, or potential. Before undertaking a large-scale restoration effort, it is important to understand threats in all three categories.

A variety of tools for undertaking threat assessment and integrating the results into forest restoration programmes have been tested around the world. In most cases, tools will need to be used in conjunction with others or may need to be modified to fit local circumstances.

A key challenge for restoration programmes is to expand the breadth of expertise integrated into assessment and analysis through multidisciplinary teams.

1. Background and Explanation of the Issue

The key to any successful restoration programme lies in good project design that is based on sound science, a thorough understanding of threats and opportunities, and a strategic and pragmatic suite of interventions chosen to mitigate identified threats while capitalising on key opportunities. A comprehensive threat assessment goes beyond merely identifying the factors, behaviours, and practices that pose a challenge to forest restoration, but includes an analysis of the underlying social, economic, and political incentives that drive such behaviours.

1.1. Information Needed for Threat Assessment

For restoration programmes, a good threat assessment provides actionable information that can be used to define the scope of interventions. Information should be timely, verifiable, and collected in a cost- and time-effective manner. Restoration programmes are not immune to the all too common pitfall of investing considerable time and resources in collecting a tremendous amount of data that, while perhaps new and interesting, is not particularly relevant to making decisions about the best way to undertake restoration activities. To avoid this pitfall it is often useful to frame a threat assessment by exploring different types of threats—direct, indirect, and potential.

1.2. Types of Threats

Direct threats are those with immediate and clear causal links to the negative impact of forest degradation or loss. Indirect threats, often referred to as root causes,[101] are the underlying drivers behind direct threats. Potential threats are those threats that, while currently not posing a significant challenge to forest restoration, have the potential to under-

---
[101] Wood et al. 2000.

mine such investments in the future. Given that forest restoration is a necessarily long-term conservation intervention, it is important to include such a temporal component in threat analysis.

For restoration programmes around the world a number of common direct threats have been identified, including habitat fragmentation, unsustainable use, and overharvesting of forest resources, pollution, and invasive species—all contributing to the breakdown of ecological processes that are critical to the healthy functioning of natural forest systems.

Underlying drivers of such threats are often related to policies that favour rapid and unsustainable conversion of forests for short-term economic gains. Markets for forest products, including global markets for products like timber and palm oil or local markets for fuelwood, can drive forest degradation and loss, particularly when market dynamics externalise true costs.

Persistent conflict and civil unrest may force local dependence on forest resources to expand rapidly, given both a lack of alternatives to meet livelihood needs or an influx of migrants and displaced persons fleeing from conflict zones into forest areas. Moreover, in many cases, forest resources are the only resources readily available to generate the cash necessary to continue such conflicts. In such situations, the prospects for successful restoration are limited if underlying governance and conflict issues are not addressed.

Other common indirect threats to forest restoration include a lack of knowledge and skills regarding the science and research behind appropriate habitat restoration and a lack of technical capacity to implement activities on the ground. A lack of political will and broad stakeholder support for restoration activities plagues many restoration programmes worldwide. Such a lack of support is often tied to a perception of high transaction costs or limited benefits associated with undertaking restoration. Given the time frame required for restoration projects, both a lack of sustained financial resources and unsure resource and land tenure rights combined can create a strong disincentive for undertaking restoration activities.

2. Examples

2.1. Madagascar

In southern Madagascar the U.S. Agency for International Development is partnering with the Communes of Ampasy-Nahampoana and Mandromodromotra, the Department of Water and Forests (La Circonscription des Eaux et Forêts–CIREF) and QIT Madagascar Minerals (QMM) to undertake forest restoration activities in the Mandena Conservation Zone. The region's forests are highly fragmented as a result of extraction of forest resources to meet the rising fuelwood needs of a growing population and increasing slash-and-burn agriculture, among other threats. This is one of the poorest regions of Madagascar, and the reliance of local populations on the forests to meet livelihood needs is driving forest loss and degradation.

A thorough understanding of the threats and opportunities of this region identified by QMM in collaboration with the communes, community leaders, and regional government representatives produced a diverse set of innovative activities intended to mitigate direct threats of forest fragmentation and indirect threats associated with poverty. For example, in exchange for rights to mine ilmenite across the region—intended to stimulate economic growth and generate income within the region—QMM has agreed to invest in forest restoration in blocks adjacent to existing protected areas of primary forest harbouring significant biodiversity. The restoration will not only expand the area of contiguous forest, but also improve the health of the forest, protect critical water cycling processes, and is also tied to investment and development of ecotourism in the region. To mitigate deforestation of remaining intact areas driven by increasing local demand for fuelwood and charcoal, plantations of fast-growing species on already degraded or deforested land are also being supported.

Even with a solid understanding of threats, the ability to address forest restoration, biodiversity, and local development needs in southern Madagascar is certainly not without challenges. A lack of knowledge and capacity in local forest ecology made the identification of relevant native pioneer species a significant challenge, requiring over 8 years of research and a multimillion dollar investment to develop appropriate protocols for forest restoration. Perhaps the greatest challenges faced by partners now are how to scale up interventions beyond initial target restoration sites and to engage new collaborators in order to effectively address the true magnitude of threats driving forest degradation and loss across the entire region.

2.2. Atlantic Forest in Argentina

In the Andresito region of Misiones, Argentina, Fundación Vida Silvestre Argentina (FVSA) and WWF are helping to restore key areas of forest adjacent to the Green Corridor, the largest remaining area of contiguous Atlantic forest in the world. The area has been significantly deforested by rapidly growing human populations to support small-scale agriculture and meet human fuelwood needs.

To develop a detailed restoration strategy for the region, FVSA undertook a thorough analysis of threats and opportunities, combining on-the-ground surveys, economic analyses, and GIS tools. FVSA began by developing detailed land use maps for each parcel of land in the region based on the current tenure. Detailed land use maps were then overlaid with biological and socioeconomic data to identify key opportunities for creating forest restoration corridors that could meet overarching forest restoration goals. Research on biodiversity-friendly production practices for local forest and shade products was also undertaken with several universities in Argentina to assess potential economic gains from alternative conservation friendly enterprises. Pilot restoration plots using different species and production techniques were established to assess both ecological and economic costs and benefits (also see case study "Finding Economically Sustainable Means of Preserving and Restoring the Atlantic Forest in Argentina"). With poverty on the rise in the region, alternative income generation opportunities are a critical incentive for landowners to begin undertaking forest restoration.

Armed with these analyses and research results, FVSA continues to engage in a participatory process with individual private landowners, local cooperatives, government representatives, and others to develop appropriate long-term land use management options that include a mix of reforestation, timber harvesting, nontimber forest product production, and other uses. By including a spatially explicit component of such land use management plans, stakeholders are continuously able to see not only how restoration practices benefit them, but also how they are contributing to a broader sustainable vision for the entire region. Currently, the major challenge for this project also involves scaling up. FVSA is focussed on helping stakeholders expand the adoption of new production alternatives, sustainable resource use management practices, and developing carbon credit schemes to mitigate high restoration costs in order to achieve restoration goals over the long term.

2.3. Using a Three-Dimensional Model to Identify Threats in Vietnam

In the area surrounding the Song Thanh Nature Reserve in the Quang Nam Province of Vietnam, WWF and partners undertook a participatory landscape planning process with community members from nine villages.[102] A "papier-mâché" 1:10,000 model of the 30,000-hectare landscape surrounding the reserve was used to facilitate planning and decision making amongst villagers and forestry sector employees.

Using paints, pins, and yarn to depict land use, natural resource elements, threats, and relationships, animated discussions and debates helped inform an integrated management plan focussed on a suite of protection, management, and restoration activities. In particular, through the modelling process, threats from illegal gold mining activities were identified and hotly debated, and have been raised with relevant authorities. Elderly people, women, and children were all able to contribute to the model-

---

[102] Hardcastle et al, 2004.

ling exercise, facilitating broader community involvement in decision making and buy-in for the planning process. While the three-dimensional (3D) mapping of threats provided a good way to engage communities in restoration planning, solid facilitation and conflict resolution skills were critical in ensuring success. This relatively cost-effective activity is now being replicated in other areas in the region in order to develop an integrated land and resource management plan at a larger landscape scale.

3. Outline of Tools

A variety of tools for undertaking threat assessment and integrating such analysis into forest restoration programmes have been tested around the world. While no one tool is ideal for all situations, certain aspects are useful for programme implementers to consider when selecting and modifying existing tools to meet specific forest restoration goals, including stakeholder participation, flexibility/adaptability of analysis, costs (e.g., time, human resources, financial resources, etc.), iterative nature of information gathering and analyses, processes to include new and updated information, communicability of outputs to appropriate audiences, and ability to incorporate different types of data (i.e., qualitative vs. quantitative).

Research studies, literature reviews, ecological and socioeconomic surveys, focus groups, and key informant interviews are all techniques that are used to gather relevant information needed to undertake threat analyses. A number of tools can be used, singularly or in combination, to carry out the actual analysis.

Conceptual modelling[103] is commonly used to show linkages and complex relationships between threats and their impacts while providing a strategic framework for thinking about appropriate project interventions. Conceptual models explicitly identify the restoration factors that programmes are intended to influence while characterising both direct and indirect forces affecting these factors. Conceptual models are particularly good for teasing out root causes, integrating interdisciplinary perspectives and are generally supported by a mix of quantitative and qualitative background data. They can be quite participatory if multiple stakeholders are brought in as part of facilitated discussions. However, conceptual models can get very complex and make it challenging to identify and prioritise interventions.

Threat matrices are a useful way to link threat assessment to project goals and specific activities. Matrices can vary from relatively simple to complex logframes where forest restoration targets are explicitly stated, with relevant threats, activities, and potential indicators for monitoring change over time explicitly tied to these targets. Matrices are good for tying threat analysis to specific activities and strategic interventions and are easily updated as adaptive management is practised. The underlying assumptions linking threats to targets and activities can be obscure and should be explicitly stated and supported by both qualitative and quantitative analysis.

Threat mapping[104] can be used to assess threats for a forest restoration area—in the form of either a pictorial map or 3D models made out of clay, wood, or other materials (see above example in Vietnam). These maps are the basis for discussion of changes in forest habitat quantity or quality, often with community groups. The process involves facilitated discussion to ensure that different members of the community with differential knowledge of threats offer their insights. For example, elders may have knowledge of the historical extent of the forest, women and men may have very different perceptions of threats related to the different forest resources they use and manage, and so on. When used appropriately this is a highly participatory tool that effectively incorporates qualitative data and generates a product that multiple stakeholders can use. Threat mapping is often most effective when used in combination with some of the other more quantitatively oriented tools.

GIS-based tools offer more advanced threat mapping by reflecting quantitative data in

---

[103] Robinson, 2000; WCS, 2004.

[104] Biodiversity Support Programme, 1995.

sophisticated spatial maps. Direct threats, such as habitat fragmentation, can be represented in maps by showing changes in data over time. GIS-based threat assessment tools can range from simple maps that reflect data collected on the ground to complex decision-support systems incorporating threat data into programmes that model alternative scenarios and outcomes using criteria established by users. Visual products reflect alternative scenarios, and an appropriate and transparent criteria and value-setting process can help generate significant buy-in from stakeholders engaged in the process. These tools are heavily reliant on quantifiable data, and depending on the specific technology, their utility may suffer from limited or unreliable data. GIS-based threat assessment requires technical skills and equipment. These tools are particularly useful for generating baseline data sets and for monitoring change over time from restoration interventions.

4. Future Needs

A key challenge to forest restoration programmes is more effective integration of relevant threat analysis that is critical for making pragmatic and real decisions. Threat analysis has been seen as a discrete background research activity that, once completed, often gets put on a shelf, never to be revisited as part of strategic programme development and adaptive management. The gap between threat assessment, often seen as primarily scientific and academic investigations, and actual project implementation needs to be more effectively breached.

To improve the rigour and utility of threat assessments for forest restoration, approaches for undertaking integrated and multidisciplinary analyses also need to be refined. Biologists, social scientists, conservation practitioners, policy makers, economists, community leaders, and investors all bring a different lens to threat analysis. Through a combined view of the factors affecting restoration, more informed and pragmatic decisions can be made regarding trade-offs that inevitably must be made in the real world.

References

Biodiversity Support Programme. 1995. Indigenous peoples, mapping and biodiversity conservation: An analysis of current activities and opportunities for applying geomatics technologies. Washington, DC, 83 pp.

Hardcastle, J., Rambaldi, G., Long, B., Le Van Lanh, and Do Quoc Son. 2004. The use of participatory three-dimensional modelling in community-based planning in Quang Nam province, Vietnam. *PLA Notes* 49:70–76.

Robinson, D. 2000. Assessing Root Causes—A User's Guide. WWF Macroeconomics Programme Office, Washington, DC, 40 pp.

Wildlife Conservation Society (WCS). 2004. Creating conceptual models—a tool for thinking strategically. Living Landscapes Technical Manual 2, 8 pp.

Wood, A., Stedman-Edwards, P., and Mang, J. 2000. The Root Causes of Biodiversity Loss. WWF/Earthscan, 398 pp.

Additional Reading

Salafsky, N., and Margoluis, R. 1999. Threat reduction assessment to: a practical and cost-effective approach to evaluating Conservation and Development Projects. Conservation Biology 13(14): 830–841.

Verolme, H.J.H., and Moussa, J. 1999. Addressing the Underlying Causes of Deforestation and Forest Degradation—Case Studies, Analysis and Policy Recommendations. Biodiversity Action Network, Washington, DC, 141 pp.

Wildlife Conservation Society. 2004. Participatory spatial assessment of human activities—a tool for conservation planning. Living Landscapes Technical Manual 1, 12 pp.

WWF. 2000. A guide to socio-economic assessments for ecoregion conservation. Ecoregional Conservation Strategies Unit, 18 pp.

11
Perverse Policy Incentives

Kirsten Schuyt

Key Points to Retain

Many government incentive programmes in reforestation and afforestation suffer from poor design, lack of enforcement, and lack of monitoring, and are aimed at short-term tree-planting activities.

As a result, government support for such schemes acts as a perverse incentive that can sometimes undermine efforts at introducing more balanced or equitable forms of restoration.

Instead, incentives need to be redirected toward a wider more integrated approach. This allows broader benefits to society, the involvement of local partners and stakeholders, and effective monitoring and evaluation.

1. Background and Explanation of the Issue

In some countries, government incentives for particular kinds of restoration have distorted approaches to the conservation, restoration, and management of forests. Government incentives to the forest industry for restoring forest cover have traditionally been aimed mainly at supporting plantation development. In light of the financial costs of these incentive schemes, and criticism from some environment and social welfare groups, questions have been raised about the economic, environmental, and social benefits of these schemes. Although many public incentives in forestry have provided some employment and income opportunities, questions remain about the overall costs of such schemes and about who will bear these costs in the longer term. For example, some studies have pointed out social and equity concerns when subsidies are captured by a few actors, such as large companies and landowners. In Chile, 80 percent of public incentive payments for the establishment of plantations have gone to three companies.[105] Other poorly designed incentive schemes have resulted in increased conversion of natural forests and land degradation. The key question is: Are public funds for afforestation and reforestation directed toward projects that provide net benefits to society?

A case study review by Perrin[106] showed that government incentive programmes in reforestation and afforestation activities tend to suffer from poor design, a lack of enforcement mechanisms, and little or no monitoring. Public incentives are often applied for short-term tree planting activities that inadequately address sustainability, biodiversity, and livelihood concerns. Little emphasis is paid to ensuring that public incentives contribute to restoring forest functions and resources, and they seldom benefit from adequate stakeholder participa-

---

[105] Bazett and Associates, 2000.
[106] Perrin, 2003.

tion. There is also a general lack of adequate monitoring and enforcement mechanisms, meaning that incentives are easily misused.

The Convention on Biological Diversity[107] identifies three common types of perverse policy incentives:

- Environmentally perverse government subsidies: Many different definitions exist in the literature as to what a subsidy is. In general, they include direct subsidies (such as grants and payments to consumers or producers); tax policies (tax credits, exemptions, allowances, and so on); capital cost subsidies (preferential loans or debt forgiveness); public provision of public goods and services below cost; and policies that create transfers through the market mechanism (such as price regulations and quantity controls). Such subsidies may have a negative impact on biological resources by directly encouraging behaviour that leads to biodiversity loss. Another example of perverse effects of subsidies is that they may drain scarce public finances that could have been used to conserve biodiversity.
- Persistence of environmental externalities: Some governmental policies may contribute to the persistence of negative externalities. For example, government policies may weaken traditional property rights systems, where such rights reside within customary law or cultural traditions. This absence of well-defined property rights at private or communal level may lead to pollution and overexploitation of natural resources, resulting in negative externalities or costs to third parties.
- Laws and customary practices governing resource use: An example of formal law generating perverse incentives is beneficial use laws requiring land users to make productive use of water and forest resources to secure land entitlement. On the other hand, the clearing of land may be rooted in customary law to indicate a claim to an area, leading to perverse incentives.

Perverse incentive schemes, however, can be redirected to promote restoration practices that will offer benefits to conservation and to a wide range of stakeholders. In this respect, forest landscape restoration offers important tools for good practices in restoration, and the key lies in promoting these tools to redirect existing perverse incentive schemes toward restoration that benefits conservation and society. Some examples are provided below.

2. Examples[108]

2.1. Public Incentives for Plantation Development, Indonesia

Deforestation is a major problem in Indonesia. The Indonesian government began promoting the development of industrial tree plantations in the 1980s to boost industrial development in wood-based industries and the oil palm sector. Several government incentives were put in place to stimulate timber plantation development, including interest-free loans, allocation of state-owned land, absence of land taxes, and so on. Large sums of money could also be obtained through the Reforestation Fund. Another incentive came from the International Monetary Fund–backed restructuring of the corporate and banking sector in the late 1990s, which was poorly implemented and led to subsidies and financing being provided to badly managed and corrupt forest companies.

In an attempt to redirect some of these public incentives, WWF, the global conservation organisation, has collaborated with the Centre for International Forestry Research (CIFOR) to restructure debt agreements related to the forest and oil palm assets of the Indonesian Bank Restructuring Agency. This reform is to include a series of checks and balances among the state, private sector, and civil society to mitigate structural pressures on the economy and forests, which should help prevent the use of funding for unsustainable and sometimes illegal plantation development as has happened in the past.

[107] CBD, 2002.

[108] Perrin, 2003.

2.2. CAP and SAPARD Forestry-Related Incentives, European Union

Two key programmes of the European Commission (EC) that provide incentives for afforestation and reforestation are the Community Regulation Directive 2080/92 (later introduced as part of the Common Agricultural Policy, CAP), which promotes afforestation of agricultural land, and the Special Action for Pre-Accession Measures for Agriculture and Rural Development (SAPARD), which focusses on rural development in European Union (EU) accession countries and includes funding for afforestation. Both of these schemes have been widely criticised as perverse incentives (also see the case study that follows this chapter).

Under the CAP, detailed analysis in 1997 suggested that the decrease in utilised agricultural land was marginal and that the role of afforestation under CAP had been overestimated. Also, the application of the directive varied between member states, with six countries accounting for more than 90 percent of total area planted. Lastly, the analysis found examples where funds had been misspent—for instance, in Spain, where farmers frequently planted, cleared, and replanted the same plots, all with subsidised funds from the EU.

Under SAPARD, it has been noted that the procedures have proven to be a big burden for many countries. In addition, concerns have been raised about some of the damaging impacts of SAPARD, such as the use of chemical protection, fence building, and construction of new roads. Also, no requirements are given under SAPARD for a minimum percentage of native tree species to be planted or incentives to enhance environmentally sound management practices. Environmental measures related to forests are only marginally included in national plans.

WWF is working both in the context of CAP and the EU enlargement process to ensure that EC policies promote sustainable rural development. For example, in 2001 WWF undertook a comprehensive review study of SAPARD-related forestry measures, and it also took part in the midterm review of the CAP. Some of the main issues that emerged relate to improving monitoring and follow-up with different beneficiaries of afforestation subsidies.

2.3. Grain-for-Green Programme, China

The goal of China's Grain-for-Green programme, launched in 2000, was to convert steep cultivated land to forest and pasture. It was initiated as a result of severe flooding in China that was blamed on excessive logging and cultivation along the Yangtze and Yellow Rivers. The programme is expected to turn more than 340,000 hectares of farmland and 430,000 hectares of bare mountain back to forests. These activities are to be carried out by the communities and subsidised by the government. In return for afforestation and reforestation activities, communities receive grain, cash, and seedlings.

The positive effects of the incentive programme so far are that the incentives have contributed to afforestation and reforestation activities as well as natural forest protection. However, the long-term sustainability of the programme remains uncertain along with its ability to prevent soil erosion. Much restoration has involved planting orchards on steep slopes, which do little or nothing to stop soil erosion. An important weakness of the programme has been a lack of monitoring and virtually no evaluation of the policy implementation.

The Chinese government has been open to reviewing its scheme following preliminary recommendations by WWF. The Centre for International Forestry Research has also undertaken a thorough assessment of the lessons learned from this scheme (see "Local Participation, Livelihood Needs, and Institutional Arrangements") as well as other reforestation/rehabilitation efforts in China and provided a number of concrete recommendations.

3. Outline of Tools

Options to remove or mitigate public perverse incentives in the forestry sector are described here. Perrin[109] recommends redirecting public incentives within the context of the forest landscape restoration approach. This means governments and donor agencies need to (1) allocate resources to the development of alternative forms of afforestation and reforestation activities that provide broader benefits to the environment and society, (2) involve local partners and stakeholders in incentive schemes (mechanisms for consultation and participation need to be put in place), and (3) spend resources on regulating the application of incentive programmes for afforestation and reforestation activities and monitoring the impacts of such activities (including developing sets of indicators and criteria to assist monitoring). This needs to be accompanied by the necessary policy measures, institutional arrangements, and monitoring and compliance mechanisms. In this respect, the CBD[110] recommends three ideal phases:

- Identify policies or practices that generate perverse incentives. This includes: analysing underlying causes of biodiversity loss, identifying the nature and scope of perverse incentives, identifying costs and benefits to society from removing the perverse incentives, doing a strategic environmental assessment, and so on.
- Design and implement appropriate reform policies. Reforms can include the total removal of policies or practices, or their replacement with other policies with the same objectives but without perverse incentives, or with the introduction of additional policies, and so on.
- Monitor, enforce, and evaluate these reform policies. This includes institutional and administrative capacity building, development of sound indicators, stakeholder involvement, and transparency.

4. Future Needs

Despite the fact that numerous suggestions on how to address perverse policy incentives can be found (as described in the previous section), the reality is that many perverse policies still exist in the forestry sector. The key need is to start putting these new policies into practice, including the need for redirecting public incentives toward a forest landscape restoration approach at all levels in cases where policies have promoted habitat alteration or destruction and unsustainable use of natural resources. We also need to improve our understanding of the impacts caused by policies and practices on biodiversity. In this respect, the CBD[111] recommends undertaking further work on the use of valuation tools to assess the extent and scope of negative impacts of policies and practices on biodiversity.

References

Bazett, M., and Associates. 2000. *Public Incentives for Industrial Tree Plantations*. WWF, Gland, Switzerland, and IUCN, Gland, Switzerland.

Convention on Biological Diversity (CBD). 2002. *Proposals for the Application of Ways and Means to Remove or Mitigate Perverse Incentives*. Note by the Executive Secretary, Quebec, Canada.

Perrin, M. 2003. *Incentives for Forest Landscape Restoration: Maximizing Benefits for Forests and People*. WWF Discussion Paper, WWF, Gland, Switzerland.

Additional References

Myers, N., and Kent, J. 1998. *Perverse Subsidies—Tax $ Undercutting our Economies and Environments Alike*. International Institute for Sustainable Development, Winnipeg, Canada.

Sizer, N. 2000. *Perverse Habits, the G8 and Subsidies the Harm Forests and Economies*. World Resources Institute, Washington, DC.

---

[109] Perrin, 2003.
[110] CBD, 2002.
[111] CBD, 2002.

Case Study: The European Union's Afforestation Policies and Their Real Impact on Forest Restoration

Stephanie Mansourian and Pedro Regato

The European Commission has been promoting afforestation since 1992 under the Common Agriculture Policy (CAP) (Directive 2080/92) as a solution to reducing agricultural land and therefore, agricultural surpluses (which are currently supported financially through subsidies). More recently a sister scheme has been developed, the Special Action for Pre-accession Measures for Agriculture and Rural Development (SAPARD), which is applicable to European Union (EU) accession countries and covers the period 2000 to 2006, with a budget of over 333 million Euros.

Today, Directive 2080/92 is part of the Rural Development Regulation (RDR), which establishes a new framework for European Community support for sustainable rural development.

While the afforestation measures under the EU had spent four billion euros by 1999 and planted 900,000 hectares of trees, the results in terms of the original aims of the scheme, and also in terms of restoring forest cover and forest functionality remained disappointing.

Some of the key problems with the CAP afforestation directive include the following:

- Limited role in taking land out of agriculture: In most member states, only 1.3 to 1.4 percent of land has actually been set aside from agriculture following its application.
- Conflicting objectives: While the subsidy scheme was largely centred around taking land out of agriculture, many governments and companies used the scheme to establish timber plantations. In Ireland, for example, the subsidies were used to establish plantations with a high economic return (Sitka spruce, pines) in order to achieve the country's aim to double its forest area over the next 30 years.
- Unequal distribution of subsidies and "double dipping": Six countries accounted for more than 90 percent of the total area planted (Spain, the U.K., Portugal, Ireland, Italy, and France). In addition, individual examples show that funds were easily misspent. In Spain, the largest recipient of the EU afforestation funds, "double dipping" was discovered to be common, with farmers planting, clearing, and replanting the same plots all with subsidised funds from the EU.
- Unnecessary manipulation of natural processes: In many cases, subsidies were applied to reforest areas that were regenerating naturally. It is estimated that up to 62.5 percent of the area benefiting from the subsidy did not actually qualify as producing an oversupply of crops.
- Inappropriate methods and species: Over 65 percent of afforestation was carried out in areas believed at risk of fire under Council Regulation (EEC) No. 2158/92 on protection of the community's forests against fire. Planting was often done in an ad hoc fashion, without selecting optimal

areas to restore forest cover, nor were these properly integrated into land use plans.

References

Perrin, M. 2003. Incentives for forest landscape restoration: maximizing benefits for forests and people. WWF Discussion Paper, WWF, Gland, Switzerland.

Report to Parliament and the Council on the application of Regulation No. 2080/92 instituting a community aid scheme for forestry measures in agriculture, 1996.

12
Land Ownership and Forest Restoration

Gonzalo Oviedo

Key Points to Retain

Forest ownership regimes matter for forest restoration because the end result of restoration, the trees, are the centrepieces of the ecosystem, and their consequent, associated goods and ecological services are of direct value to people. The ownership regime determines how such goods and services are accessed and distributed, and therefore, is the basis for restoration incentives.

It is necessary to undertake further research on experiences (successful and unsuccessful) of forest restoration under different types of ownership, to better understand how ownership rights' systems impact on the results.

1. Background and Explanation of the Issue

1.1. Forest Ownership: An Overview

The reports "Who Owns the World's Forests"[112] and "Who Conserves the World's Forests?"[113] indicate that globally, 77 percent of forestlands are owned by governments, 7 percent by indigenous and local communities, and 12 percent by individual and corporate landowners, and that in the last 15 years the forest area owned and administered by indigenous and local communities has doubled, reaching nearly 400 million hectares. This reflects important changes in forest ownership worldwide.

This chapter discusses the relationships between forest ownership and restoration, more specifically, the implications of the various types and conditions of forest ownership for successful restoration of forestlands. The basic assumption in this chapter is that forest ownership regimes matter for restoration because the end result of forest restoration, trees, are the centrepieces of the ecosystem, and their consequent, associated goods and ecological services are of direct value to people. In other words, the basic nature of the link between forest ownership and forest restoration is the fact that forest owners (whatever their specific regime and bundles of rights) are driven to restore (or not) by the expectation of goods and services that restored forests offer.

1.2. Definitions

The literature often does not distinguish "tenure" from "property" or "ownership" of forests, although in a more general sense "tenure" could be linked to custom-defined bundles of rights that are socially acknowledged, and "property" would be identified as a status in which customary tenure becomes more "institutional" through legal and political procedures and means.

Ownership or *property* itself is in essence a bundle of rights which are defined according to the nature of the subject and the legal frame-

---
[112] White and Martin, 2002.
[113] Molnar et al, 2004.

work in a given situation. Such rights can be listed[114] as the rights to (1) possess and exclusively physically control, (2) use, (3) manage, (4) draw income, (5) transmit or destroy capital, (6) have protection from expropriation, (7) dispose of interest on death, (8) potentially hold property forever, (9) reversionary/residual interests arising on expiration, (10) liability to seizure for debts, and (11) prohibitions on harmful use. There are many differences in the way in which these various rights are defined and apply to forests in different countries and social and historical contexts; some of these specific rights appear to be particularly important when dealing with sustainable forest management and forest restoration, as will be discussed later.

The literature distinguishes four main types of property applicable to lands and forests: private (individual or corporative), state, common or communal, and open access. These systems have been studied extensively, and their advantages and disadvantages with regard to natural resource use are well documented (for a useful typology and comparative analysis, see GTZ, 1998).

In country regimes of the 20th and 21st centuries, the rule for forest ownership is typically a combination of these four types of property, with significant changes in the composition of property according to historical moments and with great differences among countries. Generally, however, the predominant pattern is for the majority of forest areas to be in the hands of government, and only a small proportion being communal forests. In modern times, legally speaking there is little if any open access in forestlands, as any forestlands without private owners are automatically converted by law to state lands. In practice, however, state-owned forest has in many cases meant open access, as governments, particularly in developing countries, have had little capacity to control access to their forests. In developing countries, however, the establishment of large state-owned forest areas was in most cases the result of the expropriation of forestlands from their traditional users, who until colonial times were owners of those lands (or parts of them) under customary tenure. In this sense, and in cases where traditional forest-owning communities still exist and inhabit their traditional lands, there is an overlap of state property and communal, customary tenure.

Partly due to the recognition of customary tenure as legal communal (or individual) property, forest ownership is undergoing a major change in the world, with the main trend being the transfer or "devolution" of ownership rights to the local level, and the consequent expansion of community-owned forests.

1.3. Degree of Dependence on Forests

From the perspective of goods and services that forests (standing or future) offer, there are roughly two types of owners: forest-dependent people and non–forest-dependent people (and institutions). This distinction is important because of the expectations of the end result of forest restoration and their implications. Forest-dependent communities basically expect from restored forests an array of goods and services of direct economic value. They may value other associated benefits, such as ecological services at a landscape scale—climate change mitigation, regulation of the hydrological cycle, watershed protection, etc.—but they will normally not place higher values on associated ecological services than on those related to direct forest produce.[115] In the cases of non–forest-dependent owners, such as the absentee forest owner and the state and public agencies, the scale and hierarchy of values may vary for some areas, and their expectations, therefore, may not directly be linked to the economic importance of forest produce, but to ecosystem protection and services, biodiversity conservation, aesthetic aspects (which in turn can become economic values for example from tourism), etc.

---

[114] Ziff, 1993, cited by Clogg, 1997.

[115] Some exceptions exist to the hierarchy of values of forest restoration from the perspective of forest-dependent owners, but they are exceptions that do not contradict the primary expectations on forest produce or alternative livelihoods. For example, this is the case of restoration of degraded forest areas with sacred or particular spiritual value to local communities.

1.4. Ownership of Land but Also of Forest Goods and Services

Forest ownership differs significantly from other types of land and resource tenure—agricultural land, for example. The differences rely basically on the wide array of goods and services of the forest, and more specifically on the fact that forest ownership consists of a complex mixture of three types of ownership rights: rights to the land, rights to the forest resources, and rights to the trees. Further, ownership rights in forestlands overlap frequently with, and are different from, user rights. As Neef and Schwarzmeier[116] illustrate for Southeast Asia, in some cases groups or individuals holding the property of the land recognise rights of other individuals or groups to use the trees existing on that land, as long as there are no competing uses over the trees. There could even be multiple layers of rights on a single plot of land; for example, when a group or individual has property on the land, another group has rights on nontimber forest products, and another group holds rights on timber exploitation.

1.5. Opportunity Cost and Intergenerational Equity

Tree growth takes place over long or relatively long periods, when the forest ecosystem under restoration can offer only limited services; therefore, we are dealing with situations where there is a high, or relatively high, opportunity cost in the use of the land for forest-dependent people. In these conditions, only significant incentives and economic alternatives can cover the opportunity cost of forest restoration. The nature of benefits and incentives from forest restoration in terms of the time horizon (especially in cases of slow-maturing tree species) adds a time perspective to tenure security. For forest owners and users, it is not sufficient to know that their rights to forests and trees are secure now; it is more important to know that they will be secure and enforceable after one generation or more. In this sense, changing ownership and rights policies are even worse than the absence of them, since they cause a great lack of confidence in restoration as something socially beneficial.

1.6. Stability of Forest Ownership

In the case of China, Liu Dachang[116a] finds no conclusive evidence that user rights on trees are the best option (e.g., compared to state regulations), but does find evidence that changing rights policies were the basis of ups and downs in forest cover, and especially that lack of stability of forest ownership policies was the main reason for decline in forest cover and tree planting in certain periods; in fact, over approximately 25 years of China's modern history (from 1956 to the early 1980s), there was a succession of at least five major forest ownership policy paradigms, thus an average of a major policy change every 5 years. In practice, a few years after villagers planted trees, a major policy change would affect dramatically their rights to those trees and forests. The results were simply lack of confidence in the system and lack of incentives for tree planting.

Generally, the evidence is that where tenure security was greatest, tree planting was most successful. Tenure security means basically three levels: land tenure security, forest ownership security, and also user rights security.

1.7. Communal Systems

Several researchers have pointed to the fact that communal forest tenure, especially in conditions of market economies, requires a "critical group size" to be effective, where enforcement of rights and regulations can be optimally implemented, and where economies of scale and diversification make opportunity costs affordable, particularly when the community has to invest in forest restoration or reforestation. In other words, in any particular situation of communal forest ownership, it seems that there is a certain size of the group where forest management works best; if it is too small or too big, management is inefficient.

In many places, forest communities have tended to solve this issue by establishing a dual community/user group system, where forest

---

[116] Neef and Schwarzmeier, 2001.

[116a] Dachang, 2001.

ownership remains at the community level, but user rights (especially for trees) are allocated to smaller groups that act as forest management units. For example, in Honduras group-based management has proven better than community-based management, but the experience also shows that links between both are critical at decision-making levels on broader issues such as natural resources linked to forests: "What is required, therefore, is an institutional arrangement that retains forest management under group control, but which also provides a protocol for liaison between group and community and possibly some form of profit-sharing"[117] i.e., an arrangement where land and forest ownership remains in the community, where decision making for the entire area or landscape lies, while user rights for trees and other products are allocated to forestry groups who act on behalf of the community.

The same logic applies to the duality community-households in many communal ownership regimes.

An effective articulation of forest ownership and use rights between small units (even individuals) and larger units (community) seems therefore a critical element for successful forest management and restoration (although not the only element, as already indicated). It is also a fundamental tool to deal with the very important elements of equity and social stratification or differentiation. It has been documented that as much in agricultural lands as in forestlands, the egalitarianism that dominated ideological paradigms of agrarian reform and forest estate reform in the 20th century produced large fragmentation of lands and forests as a result of the distribution of family plots. The intention of the reformers, who were probably aware of the need to address problems of stratification within rural communities, was to overcome community differentiation by allocating equal plots to all families.[118]

In areas where this type of reform took place, fragmentation often made forest management extremely inefficient, and restoration virtually impossible, as a "critical size" is required in plots of forestland to make restoration or reforestation viable; tree planting in these conditions is often reduced to small numbers of trees around houses and within agricultural plots—normally fruit trees.

1.8. Equity Issues

Stratification of local communities in relation to forest ownership is one of the equity issues that need to be addressed in community-owned forests. Experience shows that often the most forest-dependent groups have the least user rights, especially women,[119] a situation that creates obstacles to developing solid, long-term, rights-based incentives for forest restoration. As in the case of the relationship group/community, finding the appropriate articulation of forest ownership and use rights between specific groups of users, including individual users, and larger units (forests groups and communities), in a stable, long-term policy framework, is critical to forest rehabilitation success.

2. Examples

2.1. China: Restoration Benefits and Incentives

Liu Dachang[120] has extensively researched the experience of China on forest policies, and concludes that generally user rights on trees are of greater importance than forest ownership per se for sustainable management and particularly for tree planting, reforestation, and restoration. For example, Liu Dachang shows that despite clear tenure policies on forestlands in China, in periods of stringent protective regulations on trees there was no incentive for reforestation; strict market regulations, aimed at protecting forests by discouraging commercialisation of

---

[117] Markopoulos, 1999, p. 46.
[118] As an example, in China, under the Land Reform Campaign initiated in 1950, "all rural households in a given geographical area were given equal forest resources" (Liu Dachang, 2001, p. 241). Exceptions to this policy were Tibet and the ethnic minority areas in the South of Yunnan, where community forests were established.

[119] Neef and Schwarzmeier, 2001.
[120] Dachang, 2001, 2003.

timber, ended up discouraging tree planting and therefore slowing down or totally stopping reforestation of degraded lands owned by villagers. The conclusion here is that, at least in the case of China, regulations to protect forests by restricting tree owners' rights to trees and timber in fact removed incentives for tree planting and therefore for reforestation and restoration. Successful forest restoration depends on incentives for tree owners to use the trees when they are mature, and for forest owners to use also other forest products and services; it thus depends on the clarity, extent, and enforceability of user and owner rights over trees and forest products, where timber use seems to play a major role.

But, if forest ownership rights are insufficient or even ineffective for successful restoration when not combined with user rights on trees and products, total lack of regulations on the use of timber and forest products can create perverse market incentives, especially when the conditions of clarity and enforceability of rights are not present in other adjacent forest areas. In such conditions, perverse market incentives discourage owners and users from tree planting, as the pressures from unregulated markets where competition exists from unsustainably managed forest areas (for example, areas subject to illegal timber extraction) would make it impossible for forest owners to meet the opportunity costs of tree planting and forest restoration.

2.2. Forest Rights in Ethnic Groups of Thailand and Vietnam

"The concept of individual rights to planted trees on agricultural fields applies to virtually all ethnic minority groups in the uplands of northern Thailand and Vietnam,"[121] but there are considerable differences in gender-specific rights to plant trees due to distinct inheritance laws.

"In strictly patrilineal societies like the Hmong, women are not allowed to inherit land. Thus, tree planting by women is usually limited to the area around the houses.... In contrast to the Hmong, the Black Thai and Tay societies have strong matrilineal elements. Although land inheritance of women is not common, there are a few exceptions giving women fully individual use rights, including the rights to plant trees.... Marketing of forest products such as bamboo shoots, medicinal plants and fuelwood is mainly done by women. Despite the strong involvement of women in collection and marketing of products from the forests, they do not play a role in setting management rules."[121a]

2.3. Strengthening User Rights for Forest Restoration in Northeast Highlands of Ethiopia[122]

The Meket district in the North Wollo administrative zone of Ethiopia ranges in altitude from 2000 to 3400 m above sea level and has a mix of agroclimatic zones. Its inhabitants are almost wholly dependent on agriculture. As rising numbers of people have put more pressure on the land, fallow periods have shortened, and continuous ploughing has become commonplace. Local people say that within a generation, there has been dramatic deforestation, and the grazing has declined in both quantity and quality. Expanding cultivation and increasing demand for wood have left even the steepest slopes unprotected. Only about 8 percent of the total area remains under forest. Much of the rainfall is lost through runoff, causing severe soil erosion and floods. Indigenous trees are not commonly allowed to regenerate (except on some church lands), and efforts to plant trees have had little impact.

The Ethiopian people have had negative experiences of land reallocation over the last 20 years, and are hence unwilling to invest effort in reforestation or regeneration activities. Different types of forest ownership (individual, church, service cooperative, and community) can be found in the district, but none has reversed the natural resource depletion.

Weak land-tenure and user rights were clearly hindering effective community-led environmental conservation in Meket.

---

[121] Neef and Schwarzmeier, 2001, p. 22.

[121a] Neef and Schwarzmeier, 2001.
[122] International Institute of Rural Reconstruction, 2000.

In mid-1996, SOS Sahel, an international nongovernmental organisation (NGO), began working with local authorities and agriculture ministry staff to seek a way to work with communities and solve these problems. Central to these was the establishment of official user rights for villagers.

In the community reforestation project, communities were allowed to define their own objectives for their sites, but long-term plans (5 to 10 years, or more if indigenous trees were established) were required. Within communities, reforestation groups were established, and each group decided how to share the benefits among its members, and this had to be included in the management plan. Similarly, each village developed its own strategy for guarding the site.

The proposed plan was then presented for approval at the kebele (subdistrict) level by relevant bodies: community representatives, subdistrict officials, and church leaders. It was then submitted to district officials and the agriculture office. If the plan was approved, official user rights were given to the group for their site.

As a result of this approach, farmers' participation in reforestation efforts increased. At first, 14 villages received official user rights; 20 more communities have since become involved, directly benefiting more than 2000 households.

Natural regeneration of indigenous grass, shrub, and tree species has been dramatic. There are very clear differences when compared with unprotected sites.

Sufficient short-term benefits have been realised—such as improved forage and increased production of thatching grass—to motivate communities to strengthen and expand their enclosure sites.

More secure user rights have created confidence among the communities. They have expressed strong interest to plant indigenous species (e.g., *Hagenia abyssinica, Juniperus procera, Olea africana*) instead of eucalyptus.

Communities have started to expand their sites, and new communities want to establish their own enclosures. Some are seeking compensation from the subdistrict administration for individual farmers who are cultivating land within the future enclosures. Some villages have even begun a similar process without outside intervention or support.

Farmers seem to have accepted the introduction of cut-and-carry fodder systems. This may prove to be one of the most significant impacts for the Ethiopian highlands.

2.4. Limited Success in the Protection Forest Walomerah, Indonesia[123]

The province of East Nusa Tenggara consists of the main island of Flores, Sumba, the Western part of Timor and a number of smaller islands. In 1992 the population of the province totalled 3.3 million. With an average rainfall ranging from 2196 mm in Manggarai district to 805 mm in Alor district and not so fertile soils, the conditions for agriculture are not very favourable.

About 36 percent of the land area of the island of Flores has by ministerial decree been classified as forest land and one third of this forest land as Protection Forests. The largest part of this has in reality little or no tree cover and has for generations been tilled by the population living there.

The protection forest of the mountain Walomerah in Ngada district is one such area. As part of the Presidential Instruction Programme (INPRES) for the development of Indonesia, this particular protection forest was to be reforested. The project, which began in 1995, was to start with the reforestation of 500 hectares, including part of the village Wangka, which covers 9000 hectares. Almost all of the 2400 inhabitants secure their livelihoods from subsistence farming, as their ancestors have done for generations. They are totally dependent on the land. Their traditional rights to land had been recognised by the government, but all 9000 hectares of this village lie within the protection forest. According to the legislation applying to such areas, the villagers were not allowed to occupy this area on a permanent basis.

---

[123] Vochten and Mulyana, 1995.

The Forest Service decided it was necessary to consult with these villagers with the purpose of better understanding their living situation and see to what extent the reforestation project could be modified to accommodate their needs and aspirations. Several problems directly or indirectly connected with the proposed reforestation were identified by the villagers who took part in such consultations. The problem concerning the status of their land tenure rights surfaced as a key conflict. Even though they had been paying their land ownership taxes regularly, rights to use forest products could not be granted to them.

This key issue, land tenure rights, was not solved in this reforestation project. Some useful compromises were reached, and an attempt was made to balance the undisputed need for reforestation with the primary need of farmers—land. But clearly it was not possible to move ahead with enough confidence in the project's success without addressing further the issue of land and forest produce rights.

3. Outline of Tools

Tools useful to addressing ownership issues in forest restoration are basically the same that have proven useful in the case of examining land and resource tenure in different conditions.

1. Land and resource mapping: This can be done at any level, to learn about the environmental, economic, and social resources in the community. A variation of mapping is the technique of transects, which focusses on specific areas of a community's land, for learning about the community's natural resource base, land forms, and land use, location and size of farms or homesteads, and location and availability of infrastructure and services, and economic activities.

2. The International Tropical Timber Organisation (ITTO) restoration guidelines are a useful tool addressing ownership issues. To ensure secure land tenure, these guidelines recommend (recommended actions 13 to 16): "13) Clarify and legitimise equitable tenure, access, use and other customary rights in degraded and secondary forests among national and local stakeholders. 14) Strengthen the rights of forest dwellers and indigenous people. 15) Establish a transparent mechanism for conflict resolution where property and access rights are not clear. 16) Provide incentives for stabilizing colonists/farmers in agricultural frontier zones."

3. Participatory rural appraisal (PRA) or participatory rapid rural appraisal have been described many times in the literature.[124] A methodological illustration of a PRA exercise for forest restoration in Indonesia[125] is as follows:

> The PRA facilitator team included 14 people: from the government . . . , from local NGOs . . . , and the authors. . . . The main actors were the residents of two of the four hamlets of the village Wangka, which adjoined the proposed reforestation site. They collected the information, analysed the problems, considered the options, and drew up the final reforestation plan. The facilitators supported this by introducing certain techniques to structure the information. They also listened and learned. The entire PRA lasted only three days in the field, from October 12–14, 1993. It was preceded by a one day gathering of the facilitators to exchange information about the PRA techniques to be used and to inform themselves about the village of Wangka. At the start of the PRA, the facilitators introduced themselves and the purpose of their visit and then split into two groups each to cover one of the hamlets. On the first day a map of the village including the proposed reforestation site was made. Then a seasonal calendar, presenting the main events and activities of the community (agricultural, religious, festivals, etc.) was made. On the second day a transect of the respective hamlets and the proposed reforestation site was made. Later in the day a matrix ranking was done to learn about the preferred tree species. On the final day the results of the PRA exercise in both hamlets were combined and presented by the villagers who had been involved in the PRA at a village meeting. This was also attended by representatives from the other two hamlets, the village head (kepala desa), and the head of the Forestry Service of Ngada District. During this meeting, spiced with animated discussions, problems were reviewed

---

[124] Notably, Chambers, 1994; Chambers and Guijt, 1995.
[125] Vochten and Mulyana, 1995.

and compromises made. Finally a work programme for implementing the reforestation project was produced. To ensure its future implementation the facilitators met with representatives of the concerned government agencies and presented the proposal to them the next day.

4. FAO's Socio-economic and Gender Analysis (SEAGA): This is an approach to development based on an analysis of socioeconomic patterns and participatory identification of women's and men's priorities. The objective of the SEAGA approach is to close the gaps between what people need and what development delivers. It uses three toolkits: the Development Context Toolkit, for learning about the economic, environmental, social, and institutional patterns that pose supports or constraints for development; the Livelihood Analysis Toolkit, for learning about the flow of activities and resources through which different people make their living; and the Stakeholders' Priorities for Development Toolkit, for planning development activities based on women's and men's priorities.

5. Dachang approaches the analysis of drivers for forest restoration in South China through a logical procedure consisting of three stages: diagnosis, design, and delivery (Tri-D). This procedure is the result of an adaptation of farming system approaches and rapid rural appraisal (RRA) or PRA to the identification of problems and to the design and testing of forestry and agroforestry options. This procedure has been used commonly in community-based agroforestry research.

6. User rights/stakeholder analysis: A general long-term objective is to gain knowledge about the community, and to appreciate "how to approach and structure a collaboration process."[126] For WWF, stakeholder analysis "is the process by which the various stakeholders who might have an interest in a conservation initiative are identified. A stakeholder analysis generates information about stakeholders and their interests, the relationships between them, their motivations, and their ability to influence outcomes. There are numerous approaches to stakeholder analysis, ranging from the formal to informal, comprehensive to superficial." A frequent problem of these approaches, however, is a narrow understanding of stakes and differentiation within communities, associated with the absence of consideration of tenure rights. A second conceptual and methodological problem is that often conservation organisations define primary stakeholders as "those who, because of power, authority, responsibilities, or claims over the resources, are central to any conservation initiative," while in reality primary stakeholders are those with closer dependence and rights on the resources involved.

7. The German agency GTZ proposes four principles to assist decision makers in the process of drafting and enforcing property related legislation. The principles also serve as yardsticks for evaluating existing land tenure systems and reforms, and thus they can be used to assess the forest ownership situation in any given country, and monitoring progress in establishing clear tenure systems. The proposed principles are (1) certainty in law, (2) the rule of law and human rights, (3) political participation of the population in land issues, and (4) definition of property in market economies. Ideally, the development of forest restoration interventions should be preceded and accompanied by a process by which these principles guide an appraisal of the situation of forest ownership, and help identify the critical interventions to follow to ensure success of the initiatives in the long term.

8. The International Institute of Rural Reconstruction[127] offers advice as shown in Box 12.1 on addressing land tenure issues. This is largely applicable to situations where forest restoration is planned, and where forest ownership is an issue requiring specific actions.

4. Future Needs

The following areas require further development:

- Understanding better the complex issues of rights and how they interact with various factors, such as incentives and policy environ-

---

[126] WWF-US, 2000a,b.

[127] International Institute of Rural Reconstruction, 2000.

> Box 12.1. Do's and Don'ts from International Institute of Rural Reconstruction (2000)
>
> **Do's**
> Begin with a clear understanding of the local situation and policy context.
> Use a two-pronged approach for advocacy and lobbying work—at the top with policy makers, and on the ground to demonstrate impact.
> Start with a clear shared vision with partners at all levels.
> Have a clear understanding of policies and strategies.
> Prepare clear guidelines in the local language and share with all stakeholders.
> Actively share experiences and ideas.
> Be patient: be prepared to invest a lot of effort and time.
> Strive to build the technical and managerial capacity of communities.
> Full coordination with local government officials and line agencies is essential; they can play a key role in monitoring the entire process.
> Work toward establishing official legislation for user rights to greatly strengthen the process.
> Help communities understand that a short-term reduction in fuelwood availability will result from enclosure, and assist them to find ways to deal with this problem.
>
> **Don'ts**
> Don't start with sensitive issues (e.g., discussing the problems of the land-tenure situation).
> Don't allow conflicts to become too large. Try to resolve them as soon as possible.
> Don't impose plans.
> Don't monopolize the intervention. Partners should be key players in the process.

ments, is a task that needs to be undertaken on the basis of specific cases of forest restoration. It is therefore recommended that such initiatives include in their plans the ongoing accompaniment of the process by researchers equipped to understand the links between rights and incentives.

- Use experience to synthesise guidance in the form of option menus for dealing with tenure issues in different situations. For the moment, most of the experiences of forest restoration offer lessons of mostly local or national value on ownership matters, difficult to generalise and to apply to other situations. An analytical effort of learning more from those lessons and then systematising them for guidance would be valuable, always with the understanding that lesson-based guidance is indicative only, and any mechanistic application of experiences from one place to another needs to be avoided.
- Research further on experiences (successful and unsuccessful) of forest restoration under different types of ownership, to better understand how rights' systems (including from creation or granting of rights to law enforcement and judicial processes) impact on the results—in the short, medium, and long terms. In undertaking such research, it is fundamental to use a conceptual and methodological framework that is based on the understanding of the complexities of the bundle of forest ownership rights, avoiding for example an exclusive focus on land tenure.

References

Chambers, R. 1994a. The origins and practice of participatory rural appraisal. World Development 22(7):953–969.

Chambers, R. 1994b. Participatory rural appraisal (PRA): analysis of experience. World Development 22(9):1253–1268.

Chambers, R. 1994c. Participatory rural appraisal (PRA): challenges, potentials and paradigm. World Development 22(10):1437–1454.

Chambers, R., and Guijt, I. 1995. PRA—Five years later. Where are we now? Forests, Trees and People Newsletter 26/27:4–13.

Clogg, J. 1997. Tenure reform for ecologically and socially responsible forest use in British Columbia. A paper submitted to the Faculty of Environmental Studies in partial fulfilment of the requirements for the degree of Master in Environmental Studies, York University, North York, Ontario, Canada.

Dachang, L. 2001. Tenure and management of non-state forests in China since 1950: a historical review. Environmental History 6(2):239–263.

Dachang, L., ed. 2003. Rehabilitation of Degraded Forests to Improve Livelihoods of Poor Farmers in South China. CIFOR, Bogor, Indonesia.

International Institute of Rural Reconstruction. 2000. Sustainable Agriculture Extension Manual. IIRR, Silang, Cavite, Philippines.

Markopoulos, M.D. 1999. The Impacts of Certification on Campesino Forestry Groups in Northern Honduras. Oxford Forestry Institute (OFI), Oxford, UK.

Molnar, A., Scherr, S., and Khare, A. 2004. Who conserves the world's forests? Comunity-driven strategies to protect forests and respect rights. Forest Trends, Ecoagriculture Partners, Washington, DC.

Neef, A., and Schwarzmeier, R. 2001. Land Tenure Systems and Rights in Trees and Forests: Interdependencies, Dynamics and the Role of Development Cooperation, Case Studies from Mainland Southeast Asia. GTZ, Division 4500 Rural Development, Eschborn, Germany.

Vochten, P., and Mulyana, A. 1995. Reforestation, protection forest and people—finding compromises through PRA, Forests, Trees and People Newsletter, FAO, issues 26/27.

White, A., and Martin, A. 2002. Who Owns the World's Forests? Forest Tenure and Public Forests in Transition. Forest Trends, Washington, DC.

World Wildlife Fund USA. 2000a. A Guide to Socioeconomic Assessments for Ecoregion Conservation. WWF–US Ecoregional Conservation Strategies Unit, Washington, DC.

World Wildlife USA. 2000b. Stakeholder Collaboration: Building Bridges for Conservation. WWF–US Ecoregional Conservation Strategies Unit, Research and Development, Washington, DC.

Ziff, B. 1993. Principles of Property Law. Carswell. Scarborough, Canada.

Additional Reading

Agrawal, A., and Ostrom, E. 1999. Collective action, property rights, and devolution of forest and protected area management. Research paper. S/l.

Barton Bray, D., Merino-Perez, L., Negreros Castillo, P., Segura-Warnholtz, G., Torres, J.M., and Vester, H.F.M. 2003. Mexico's community-managed forests as a global model for sustainable landscapes. Conservation Biology 17(3):672–677.

Chambers, R. 1983. Rural Development: Putting the Last First, Longman, London.

Chambers, R. 1993. Challenging the Professions. Frontiers for Rural Development. Intermediate Technology Publications, London.

Chambers, R. 1996. Whose Reality Counts? Intermediate Technology Publications, London.

Chambers, R. 2002. Participatory Workshops: A Sourcebook, Institute of Development Studies, Brighton, UK.

Chambers, R., and Leach, M. 1990. Trees as Savings and Security for the Rural Poor. Unasylva 161(41):39–52.

Food and Agriculture Organisation of the United Nations, FAO. 2001. SEAGA—Socio-Economic and Gender Analysis Package. FAO Socio-Economic and Gender Analysis Programme. Gender and Population Division, Sustainable Development Department, Rome.

GTZ. 1998. Guiding Principles: Land Tenure in Development Cooperation. Deutsche Gesellschaft für Technische Zusammenarbeit, Abt. 45, Div. 45.

Jaramillo, C.F., and Kelly, T. 2000. La deforestación y los derechos de propiedad en América Latina. http://www.imacmexico.org/ev_es.php?ID=5587_203&ID2=DO_TOPIC.

Lamb, D., and Gilmour, D. 2003. Rehabilitation and Restoration of Degraded Forests. IUCN/WWF, Gland, Switzerland.

13
Challenges for Forest Landscape Restoration Based on WWF's Experience to Date

Stephanie Mansourian and Nigel Dudley

Key Points to Retain

Some of the most important challenges identified by WWF's forest landscape restoration programme in its first four years, include the following:

The need to better value forest goods and services

The need to increase capacity to deal with landscape restoration issues

The need to better monitor the return of forest functions at a landscape scale

1. Introduction

Since the start of its Forest Landscape Restoration programme in 2000, WWF, the global conservation organisation, has faced a number of challenges related to (1) the planning of restoration in large scales, (2) the integration of social and ecological dimensions, and (3) the implementation of restoration programmes on a large scale. A more detailed analysis of specific lessons learned from forest landscape restoration projects can be found in this book in the part entitled" Lessons Learned and the Way Forward." This chapter focusses instead on specific challenges anticipated for future programmes to restore forest functions in landscapes, based on experience in the first 4 years of WWF's restoration programme. While this draws on experience within one organisation, we hope that the brief summary of some of the tasks we have identified will also be useful to governments, nongovernment organisations, (NGOs) and others interested in developing restoration projects, large or small.

We started WWF's restoration initiative with some concepts (e.g., the need to integrate socioeconomics, the concept of trading off land uses within landscapes, the idea of working at a landscape scale), and also some principles (e.g., balancing ecological and social needs, adopting where possible a participatory approach). For the last 4 years, we have been testing out these theories in practice in field programmes around the world. One early result was recognition that there was a lack of succinct information for practitioners, which was the driving force behind this book. In light of WWF's experience to date, a number of future challenges and opportunities have been identified[128]:

1.1. Setting Realistic Goals for Restoration Within a Landscape

A failure of past restoration efforts can be traced back to having started with unrealistic goals or alternatively with very narrow goals that fail to take into account local and surrounding socioeconomic realities. For this reason it is important to set goals that are at once realistic but also consider the many dif-

[128] Mansourian, 2004.

ferent outputs required from most landscapes. In a landscape context, restoration goals for conservation organisations will often be closely linked to other activities relating to protected areas and sustainable forest management. Thus, restoration may seek to complement a protected area or relieve pressure on it. Equally, restoration can happen within and around the estate of a managed forest. Forest restoration goals within a landscape generally have to address both social and ecological needs; they may, for instance, relate to restoration of species' habitat in one location but also to the establishment of fuelwood plantations elsewhere. In all cases, the key will be to attempt to balance those goals to provide optimal benefits (also see "Goals and Targets of Forest Landscape Restoration," "Negotiations and Conflict Management," and "Addressing Trade-Offs in Forest Landscape Restoration").

1.2. Ensuring that Restoration Is Not Used as an Excuse for Uncontrolled Exploitation

One reason many conservationists still balk at restoration is that it can be seen to provide a justification for failing to address the problems of degradation. Given the cost, duration, and difficulty of restoration, we do not believe that this is a viable argument. However, the fact that conservation organisations encourage restoration should not be interpreted as licence for degradation, because in many circumstances restoration activities will not be able to recover all of the values that have been lost. There is a fine line between actively offering restoration as a solution to dwindling natural resources without undermining efforts at protection or good management of these resources.

1.3. Active or Passive Restoration?

In some cases it is clear that restoration is already urgently necessary. At this point the first question for a community, conservation organisation, or government becomes one of choice between passive and active restoration. Passive restoration, which means creating suitable conditions for restoration to happen through natural processes (e.g., by fencing an area against grazing or preventing artificial fire) is usually considered to be the most desirable solution, being simpler, cheaper, and more akin with natural processes. However, there comes a point (a status of degradation or particular set of ecological and social conditions) when active restoration is necessary, either because recovery needs to be speeded up to protect threatened biodiversity or because ecological conditions have changed so profoundly that natural processes need some assistance. The challenge for conservation planners is sometimes whether to wait for passive restoration, and risk further degradation and in the future a more expensive restoration process, or to jump straight into active restoration. Development of a more sophisticated set of criteria or tools for helping make these kinds of decisions will be one of the major needs in the future.

1.4. Promoting the Concept of Multifunctional Landscapes

If conservation organisations are to address the big emerging issues related to forestry and biodiversity, we will need to engage much more closely with social actors, an example is the emerging WWF-CARE partnership. An emphasis on "multifunctional landscapes," that is, landscapes that provide a mixture of environmental, social, and economic goods and services through a mosaic of sites managed with differing but harmonised objectives, can help to provide balanced approaches in landscapes that contain both environmental and social problems. One implication of this is that forest restoration in most cases will not be a viable activity unless it goes hand in hand with forest management and usually also with forest protection.

1.5. Sustainability of Restoration—Valuing Forest Goods, Services, and Processes to be Restored

Active restoration is an expensive process, and in most cases conservationists (both state government and NGOs) still opt to direct available conservation budgets toward protection instead. However, in many cases these decisions are not being taken in full knowledge of the long-term costs and benefits. For instance, it is often easier to build political support for setting aside a mountainous area of forest to protection because it appears to entail limited cost, or at least delayed costs, whereas the apparent cost of restoring a more accessible or economically valuable habitat such as a lowland forest appears immediately. But if the long-term value of a restored forest were properly estimated, then on balance the net costs might not appear to be as high. In some cases, it may make more sense to focus efforts on protection, in others more on restoration or a mixture of both. One future challenge is to increase skills and tools for valuation of the costs and benefits of various approaches so that more balanced judgements can be made.

1.6. Long Term Monitoring and Evaluating Impact of Restoration within Large Scales

Monitoring and evaluation are essential in any conservation programme, to help facilitate adaptive management, and have been identified as one of the most critical elements in success. They become particularly crucial in a large-scale restoration effort, which will span several decades and will involve many different actors. Mistakes need to be redressed and improvements need to be made. Proper monitoring tools that are adapted to a large scale need to be developed and then applied rigorously.

1.7. When Can We Claim Success? When Is a Landscape Restored?

There is no clear end point for restoration. A natural forest is itself not a fixed or static ecosystem but is generally in constant evolution and flux. In any case, many restoration projects will not be aiming to re-create an "original" forest. Agreeing and then finding ways of measuring an end point is therefore a challenge particularly for organisations such as WWF, which work in time-limited programmes and to targets that are often agreed to between NGOs and donors. In practice, targets need to be set at the level of a specific landscape. For instance, is the ultimate aim of a forest landscape restoration programme to return a certain endangered animal species to a viable population? Or is it to improve water quality? Or is it to reverse the decline in forest quality? Many restoration projects have multiple aims, such as restoring habitat for species but also increasing nontimber forest products for local communities. By setting goals, conservation organisations should be able to establish meaningful programmes, whilst recognising that forest landscape restoration is never a short-term project with a clear beginning and end. Efforts should be longer term, and specific measures of success will necessarily be steps along a trajectory toward a healthier and more sustainable forest landscape.

1.8. Resources

Forest restoration at the scale of large landscapes can be enormously costly. In addition, the longer we wait before undertaking restoration, the more degraded the landscape is likely to have become (for instance, seeds of original species may no longer be present, soil conditions will have changed) and therefore the higher the costs of restoration are likely to be. Many restoration efforts have failed through lack of resources. Ideally, systems that integrate the cost of restoration within landscape-level activities via taxes (for instance on ecotourists) or via payment for environmental services (for instance, for the provision of clean water, also see "Payment for Environmental Services and Restoration") should provide long-term and sustainable financing for restoration activities. However, this assumes both that costs and benefits can be measured accurately, which is still often a challenge, and that there is sufficient political support for restoration that such payments can be levied. Establishing means for long-term funding that go beyond donor project cycles remains a key challenge for the future.

1.9. Capacity

A restoration programme carried out over large areas is likely to require many different skills, for instance negotiating skills, lobbying skills, monitoring skills, small enterprise development skills, plantation skills, nursery development skills, etc. It is important to ensure that local capacity to support the long-term restoration effort exists. In many cases this requires training as well as the partnering of different institutions to share their respective knowledge and expertise.

2. Examples

These examples demonstrate some of the practical challenges that have been encountered. They may not all be as fundamental as those listed above, but are interesting to highlight as they demonstrate the full range of challenges that may emerge from real experiences.

2.1. Vietnam: The Challenge of Dealing with Pressures on Remaining Forests

The government of Vietnam is well aware of the importance of its forests, for instance to ensure water quality, and has taken significant forest areas out of production. But pressures remain because local people face serious land shortages, and restoration efforts have until now mainly been aimed at intensive plantations that supply only a small proportion of the potential goods and services. Restoration efforts in Vietnam therefore need to embrace demonstration projects both to show what is possible and to work with government authorities to modify current restoration policies (see case study "Monitoring Forest Landscape Restoration in Vietnam").

2.2. Madagascar: The Challenge of Choosing a Priority Landscape for Restoration

In a country like Madagascar that has lost over 90 percent of its forest, it would seem straightforward to decide where to restore. Nonetheless, given scarce resources and given a difficult socioeconomic context (Madagascar is one of the poorest countries on the planet, and poor people survive largely from slash and burn agriculture), it is necessary to select priority area(s) to begin a large-scale restoration programme. In 2003 WWF brought together a number of stakeholders from government, civil society, and the private sector to define together what might be criteria for choosing a priority landscape in which to restore forest functions.

The group identified the following categories of criteria:

1. Sociocultural
2. Economic
3. Ecological/biophysical
4. Political

Within these categories, some of the 24 criteria were, for example:

- Type of land tenure
- Values attributed to forests by local people
- Proximity of fragments to a large forest plot
- Level of diversification of revenue sources
- Presence of management entity for the landscape
- Numbers of species used by local communities that have been lost
- Level of involvement of communities in local environmental actions

Members of the national working group on forest landscape restoration then visited a short-listed selection of landscapes and rated each against the 24 criteria. The outcome was a prioritised list of landscapes that need to be restored based on criteria that were developed locally and that were very specific to local conditions.[129]

2.3. New Caledonia: The Challenge of Dealing with Multiple Partners

It took 2 years to develop an agreed to partnership, strategy, and plan, and to engage eight other partners in the dry forest restoration

---

[129] Allnutt et al, 2004.

programme for New Caledonia. While this may seem a long time to invest in building a partnership, the fruits of such an effort are now being felt as the programme is taking off. The programme carries much more weight in the eyes of all stakeholders because of the partnership.

2.4. Malaysia: The Challenge of Identifying Priority Species for Restoration

While restoration along the Kinabatangan river was identified as a priority in order to reconnect patches of forest for biodiversity, the selection of appropriate species was not clearly done. For this reason a demonstration site has been set up where different species and techniques (from simply fencing to weeding or active planting) are being tested and monitored in order to identify the approach that is best suited to local conditions and which can then be propagated along the corridor.

References

Allnutt, T., Mansourian, S., and Erdmann, T. 2004. Setting preliminary biological and ecological restoration targets for the landscape of Fandriana-Marolambo in Madagascar's moist forest ecoregion. WWF internal paper. WWF, Gland, Switzerland.

Mansourian, S. 2004. Challenges and opportunities for WWF's Forest Landscape Restoration programme. WWF internal paper. WWF, Gland, Switzerland.

Section VI
A Suite of Planning Tools

14
Goals and Targets of Forest Landscape Restoration

Jeffrey Sayer

The most fundamental (question) relates to the definition of the goals and targets for restoration projects. It would seem that definition would be simple, but it is often complex and involves difficult decisions and compromises. Ideally, restoration reproduces the entire system in question, complete in all its aspects—genetics, populations, ecosystems, and landscapes. This means not merely replicating the system's composition, structure and functions, but also its dynamics—even allowing for evolutionary as well as ecological change (Meffe and Carroll, 1994).

Key Points to Retain

Outside experts cannot alone set goals and targets because they are never self-evident.

Careful multi-stakeholder processes are needed to set goals and targets that will be broadly accepted.

Goals and targets will change with time and need to be adapted.

Pristine "pre-intervention" nature is only one of many possible goals.

1. Background and Explanation of the Issue

A broadly shared understanding and acceptance by all stakeholders is fundamental to the success of any restoration project. There are countless examples of attempts at restoration failing because one person's "restoration" is often another person's degradation. Here are some examples:

- Attempts by the Indonesian Ministry of Forestry to "restore" Imperata grasslands by planting trees failed because local people had no use for the trees (they belonged to the foresters) but they made extensive uses of the grasslands. The grasslands provided fodder for their cattle and grass for roofing.
- Attempts to plant spruce forests to restore the degraded moorlands of northern England and Scotland were opposed by amenity and conservation groups because the moorland scenery had come to be accepted as "natural" and "beautiful" and it was the habitat of rare birds.
- Government attempts to restore tree cover on the uplands of Vietnam were opposed by local people because the types of trees planted by the government were not the ones that local people needed or could use.
- Government-sponsored tree planting schemes in China have denied local people access to medicinal plants and have damaged the habitats of rare plants and animals in the dry mountainous areas of South Western and Western China.
- Attempts to restore pristine nature in degraded areas in the United States are opposed by some conservationists who consider that such artificially restored areas can never have the value of a pristine landscape.

Pretending that restoration is possible is seen as a ploy by commercial interests to justify activities that degrade nature.

The basic problem is that what is perceived as "degraded" by one interest group may be perceived as desirable by another group. Foresters consider land degraded if it does not support a crop of commercially valuable trees. Ecologists consider a forest degraded if it does not have multiple layers of vegetation and a reasonable number of dead or decaying trees as habitat for birds and invertebrate. Amenity groups do not like dense forests; they want mosaics of woodland and open land with extensive views. The list is endless. The basic lesson is that there can never be a single vision of an "end point" for restoration that will automatically meet with the approval of all interested parties.

2. Steps to Success

The first task in any broad-scale restoration initiative, therefore, is to find out what everyone would ideally like to see as an outcome and then to negotiate compromises between what will inevitably be a collection of different viewpoints and attempt to come up with a scenario that is acceptable to all.

It is unwise to assume that once an end point has been negotiated that the "visioning thing" is done. As landscapes change so the perceptions and needs of interest groups will evolve. Restoration is often a moving target. Markets, recreational needs, conservation priorities, etc. all change with time, and what people want today will not necessarily be what they will want tomorrow.

Dunwiddie[130] has argued that objectives for restoration projects should be defined as "motion pictures" rather than "snapshots." The problem is that objects such as species are much easier to specify and monitor in projects than are processes such as ecosystem function and community dynamics.

The following concepts and approaches can be used as tools to ensure that forest landscape restoration projects are moving in the right direction:

2.1. Answer the Questions: Restoring What, for Whom and Why

These are the most important questions yet they are frequently not properly addressed in restoration projects.

These questions should be answered by real stakeholders—local people, conservation organisations, etc.—those who will do the work or incur the costs and benefits.

Avoid programmes that are "expert driven" and ensure that development assistance agencies stay honest, that they are explicit about their real objectives and recognise that they also are interested parties.

2.2. Work with Scenarios, Visions, and Stakeholder Processes

There is an abundant literature on methods for involving stakeholders in the development of scenarios and visions. Care has to be taken to ensure that the interests of less powerful groups are addressed. Achieving genuine public participation is not just common sense—it requires professional skills. Neutral professional facilitation is almost always necessary. The Centre for International Forestry Research (CIFOR) and the International Institute for Environment and Development (IIED) Web sites provide access to the literature on these approaches.

Simple modelling tools exist for exploring options and making assumptions explicit. STELLA, VENSIM, and SIMILE are widely used. These models are the best tools for developing scenarios, understanding the drivers of change in a system, making stakeholder assumptions and understanding explicit, and then tracking progress toward goals that are identified as desirable.

The concept of *getting into the system*[131] is fundamental. This means engaging for the long-term, becoming a stakeholder, and making one's interest explicit. In the case of WWF, as

---

[130] Dunwiddie, 1992.

[131] Sayer and Campbell, 2004.

with other conservation organisations, this interest is principally biodiversity, and we have to make commitments for what we are prepared to contribute in cash or other contributions to support the achievement of our biodiversity goals.

2.3. Understand Development Trajectories

What would happen if we did not intervene? What is the underlying development trajectory? What are the principal *drivers of change*? It is vital to get the correct answers to these questions. Modelling can help. Normally only a small number of drivers of change are significant at any one time. We have to know which ones they are and how they can be influenced.[132]

We must also understand the underlying processes of ecological succession.[133] The factors that influence restoration at a single location are not necessarily confined to that place. A variety of extrasectoral influences such as economic and trade policies and levels of public understanding of issues will have a continuing and variable influence on restoration processes.

2.4. Use Monitoring and Evaluation as a Management Tool

Monitoring and evaluation have to be linked to the desired outcomes of interventions. Negotiating these outcomes is the first and most important activity in any programme. Indicators of the desired outcomes have to be agreed to or negotiated at the beginning, and they then become the tools for adaptive management.[134] The book by Sayer and Campbell has a chapter on this issue that gives further references to the monitoring and evaluation literature.[134a]

2.5. Find and Protect Reference Landscapes

Whether or not the objective of forest landscape restoration is to restore the "original" vegetation cover, it will always be useful to have reference areas that are as near as possible to the natural conditions of the area (see "Identifying and Using Reference Landscapes for Restoration"). These are useful as benchmarks, for understanding ecological processes, for education, and as sources of plants and animals to be used in assisted restoration.

Much has been written about attempts to restore a pristine, climax, "natural" land cover. There are lots of problems with this approach, not least of which is the difficulty of knowing what the preintervention situation was. It is also important to avoid falling into the trap of assuming that natural systems reach a climax condition and are then constant—this is rarely the case. Even in the remotest and least disturbed parts of the Congo Basin or the Amazon the species' composition of the forests today is not the same as it was 100, 500, or 5000 years ago. Natural landscapes are highly dynamic, and decisions to restore to "natural" conditions will always be arbitrary and open to multiple interpretations. Reference landscapes, or plots, with minimal intervention remain valuable in helping us to understand landscape processes and can be useful components of any large-scale restoration programme. They can be valuable as examples to look at during negotiation processes.

Normally restoring "natural conditions" is just one of a range of possible objectives, and in most situations what one restores will be defined by more precise production and environmental objectives.

2.6. Be Realistic About Designer Landscapes

Once a comprehensive stakeholder participation process is engaged, it will gradually become possible to begin to talk about desirable outcomes. Eventually a vision of a "designer landscape" may begin to emerge. Different approaches and tools are useful to

---

[132] See the Web site of the Resilience Alliance and publication by Berkes et al, 2003.
[133] Walker and del Moral, 2003.
[134] CIFOR's work on Adaptive Collaborative Management provides guidance.
[134a] Sayer and Campbell, 2004.

explore what the landscape should look like in order to respond to the needs and wishes of different interest groups.

3. Outline of Tools

Stakeholders may decide that a certain landscape configuration and condition is ideal for their objectives. But usually different stakeholders have different ideals. To fine-tune a landscape vision, some specific approaches can be used depending on the restoration goal:

- *Biodiversity*: Modelling tools developed by the United Nations Environment Programme-World Conservation Monitoring Centre (UNEP-WCMC) are useful.[135] Some assumptions about corridors and connectivity have to be treated with caution.[136] One should not always assume that protected areas should be as big as possible. There are often significant opportunity costs that protected areas create for local people. Protected areas should be of an optimal size, not necessarily as big as possible.[136] The importance of seral stages in vegetation development is often underestimated. Many wildlife species require early successional vegetation for their survival.
- *Poverty mapping and assessment*: The World Agroforestry Centre has a lot to offer on this topic (see "Agroforestry as a Tool for Forest Landscape Restoration").
- *Land care*: The Landcare programme in Australia and now expanding elsewhere is an interesting model for participatory multi-stakeholder restoration programmes.
- *Water*: Lots of common assumptions about the value of land cover for water quality and quantity are not borne out by empirical evidence. Forest cover may consume more water than it conserves; it all depends on the type of trees, the frequency and intensity of rainfall, and the nature of the underlying substrate. Expert advice should be sought on the hydrological implications of restoration programmes (also see "Restoring Water Quality and Quantity").
- *Amenity*: The Netherlands, the United Kingdom, and the United States have restoration programmes with a heavy emphasis on amenity. This is the realm of landscape architecture.[138]
- *Avalanche control*: This is an important issue in temperate and boreal countries and there is an abundant literature.
- *Timber*: Timber is the real objective of much so-called restoration. Caution is needed because narrow timber production objectives are rarely consistent with the broader objectives of local people and the environment.
- *Tree crops*: Tree crops include oil palm, coffee, cacao, rubber etc. More can be found on this topic in the chapter on agroforestry, cited above, but also in publications on extractive reserves and jungle rubber.

4. Future Needs

4.1. Improved Economic Analysis

Restoring landscapes is expensive, but can and should yield economic benefits. The valuation of environmental goods and services is still an imprecise science. The valuation of the subsistence products used by poor subsistence farmers is also a challenge. But all large-scale restoration initiatives have to be rooted in economic realism. The cost-benefit ratios are essential in determining what is possible and desirable. There are countless examples of forest restoration programmes that have cost a lot of money and yielded few real benefits.

It is especially important to remember that investments in restoration carry opportunity costs—the same money could be invested in employment creation, establishing protected areas, etc. Even though complete economic valuation will only rarely be possible or necessary, it is always important to thoroughly examine options from an economic perspective.

---

[135] UNEP-WCMC, 2003.
[136] Simberloff et al, 1992.
[137] Zuidema et al, 1997.

[138] Liu and Taylor, 2002.

4.2. A Capacity for Learning by Doing

The above consideration may suggest a need for heavy planning processes, but this should be avoided at all costs. It is much better to start immediately with a few experimental restoration activities on the basis of outcomes of the initial discussions amongst stakeholders. These trials will establish the credibility of outside stakeholders and will permit learning. They will greatly enrich ongoing stakeholder negotiations that should continue throughout the programme. The initial objective should be to build a community of interest groups that can experiment and learn together.

A sense of community or "social capital" can really enhance efforts to restore landscapes. Voluntary groups have accomplished some remarkable restoration achievements. People can work together and develop a shared passion for restoring the habitat of a rare animal or the beauty of a disfigured landscape. Such communities will fine-tune their objectives and adapt their programmes as they advance. They will provide an excellent mechanism for setting and updating goals and end points.

To get real "buy-in" from diverse interest groups, it is important to start small, provide outside inputs as drip-feeding, not as big cash injections, avoid setting up bureaucracies, and learn and adapt as you progress.

4.3. Tracking Tools for "Landscapes"

As restoration programmes unfold it is essential to have feedback mechanisms so that success can be assessed, stakeholders consulted, and activities adapted to reflect changed perspectives. Such tracking tools (or monitoring and evaluation) need to be negotiated at the beginning of the process to ensure that they genuinely track the attributes of the site that people value. Since landscapes are complex and stakeholders' views often divergent, such tracking tools will inevitably be complicated.[139]

---

[139] See penultimate chapter in Sayer and Cambell, 2004.

References

Berkes, F., Colding, J., and Folke, C. 2003. Navigating Social-Ecological Systems. Cambridge University Press, Cambridge, UK.

Dunwiddie, P.W. 1992. On setting goals: from snapshots to movies and beyond. Restoration Management Notes 10(2):116–119.

Liu, J., and Taylor, W.W. 2002. Integrating Landscape Ecology into Natural Resource Management. Cambridge University Press, Cambridge, UK.

Meffe, G.K., and Carroll, C.R. 1994. Ecological Restoration. In: Principles of Conservation Biology, pp. 409–438. Sinamer Associates, Inc., Sunderland, MA.

Sayer, J.A., and Campbell, B. 2004. The Science of Sustainable Development. Cambridge University Press, Cambridge, UK.

Simberloff, D., Farr, J.A., Cox, J., and Mehlman, D.W. 1992. Movement corridors: conservation bargains or poor investments? Conservation Biology 6: 493–504.

UNEP-WCMC. 2003. Spatial analysis as a decision support tool for forest landscape restoration. Report to WWF.

Walker, L.R., and del Moral, R. 2003. Primary Succession and Ecosystem Rehabilitation. Cambridge University Press, Cambridge, UK.

Zuidema, P.A., Sayer, J.A., and Dijkman, W. 1997. Forest fragmentation and biodiversity: the case for intermediate-sized conservation areas. Environmental Conservation 23:290–297.

Additional Reading

Aide, T.M., Zimmerman, J.K., et al. 2000. Forest regeneration in a chronosequence of tropical abandoned pastures: implications for restoration ecology. Restoration Ecology 8(4): 328–338.

Ashton, M.S., Gunatilleke, C.V.S. et al. 2001. Restoration pathways for rainforest in Southwest Sri Lanka: a review of concepts and models. Forest Ecology and Management 154:409–430.

Bradshaw, A.D., and Chadwick M.J. 1980. The Restoration of Land: The ecology and reclamation of derelict and degraded land. Blackwell Scientific Publications, Oxford, UK.

Buckley, G.P., ed. 1989. Biological Habitat Reconstruction. Belhaven Press, London.

Cairns, J., Jr., ed. 1988. Rehabilitating Damaged Ecosystems, vols. 1 and 2. CRC Press, Boca Raton, Florida.

Gobster, P.H., and Hull, R.B., eds. 1999. Restoring Nature: Perspectives from the Social Sciences and Humanities. Island Press, Washington, D.C.

Holl, K.D., Loik, M.E., et al. 2000. Tropical montane forest restoration in Costa Rica: overcoming barriers to dispersal and establishment. Restoration Ecology 8(4):339–349.

IUFRO. 2003. Occasional paper no. 15. Part 1: Science and technology—building the future of the world's forests. Part II: Planted forests and biodiversity. ISSN 1024-1414X. IUFRO, Vienna, pp 1–50.

Jordan, W.R. III, Gilpin, M.E., and Abers, J.D., eds. 1987. Restoration Ecology: A Synthetic Approach to Ecological Research. Cambridge University Press, Cambridge, UK.

Lamb, D. 1998. Large scale ecological restoration of degraded tropical forest lands: the potential role of timber plantations. Restoration Ecology 6(3):271–279.

Luken, J.O. 1990. Directing Ecological Succession. Chapman and Hall, London.

Nilsen, R., ed. 1991. Helping Nature Heal: An Introduction to Environmental Restoration. A Whole Earth Catalogue, Ten Speed Press, Berkeley, California (Deals with restoration in a U.S. context.)

Perrow, M.R., and Davy, A.J. 2002. Handbook or Ecological Restoration, vols. 1 and 2. Cambridge University Press, Cambridge, UK.

Reiners, W.A., and Driese, K.L. 2003. Propagation of Ecological Influence Through Environmental Space. Cambridge University Press, Cambridge, UK.

Smout, T.C. 2000. Nature Contested; Environmental History in Scotland and Northern England Since 1600. Edinburgh University Press, Edinburgh, UK.

Whisenant, S.G. 1999. Repairing Damaged Wildlands A Process-Oriented, Landscape-Scale Approach. Cambridge University Press, Cambridge, UK.

Case Study: Madagascar: Developing a Forest Landscape Restoration Initiative in a Landscape in the Moist Forest

Stephanie Mansourian and Gérard Rambeloarisoa

Starting in March 2003, WWF, the global conservation organisation, and its partners began developing a Forest Landscape Restoration programme in the moist forest ecoregion of Madagascar. This case study highlights the different steps in the process.

Only about 10 percent of Madagascar's forests are left, and much of this is in poor condition. For this reason forest landscape restoration was identified as a useful approach to tackle conservation and development concerns in the country. In March 2003, when WWF began its restoration programme, a moist forest ecoregion process was already underway to develop a comprehensive conservation programme for the whole area (i.e., data were being gathered, maps developed highlighting key habitats, the range of different species were being surveyed, etc.) which helped to feed crucial data into the development of the restoration initiative.

The key steps in the development of the restoration programme are as follows:

1. Short-listing priority landscapes (March 2003): In a national workshop with participants representing civil society, researchers, government, and the private sector, a number of potential landscapes were selected for restoration based on coarse criteria developed together in the workshop.

2. Reconnaissance to focus on one landscape (June–August 2003): The criteria were then further refined by a national working group set up at the workshop. Using the selected criteria (which included both ecological and social issues, for instance, distance from large forest patch, literacy rate, presence or absence of land tenure conflict), the members of the national working group visited the five short-listed landscapes and rated each according to the criteria in order to select one priority one.

3. Proposal development and funds raised (August 2003–June 2004): A proposal was developed, submitted, and approved for the priority landscape.

4. Beginning the process for selecting biological and ecological targets (June 2004): To begin identifying the biological and ecological priorities for the landscape, data from the ecoregion process was used to define what might be priority areas for restoration within the landscape and with which biological/ecological objective (e.g., restoring the habitat for a specific lemur, buffering a protected area, etc.).

5. Socioeconomic analysis (September–December 2004): Before taking the biological data further, it was felt that a better understanding of the social and economic situation inside the landscape was needed, leading to the commissioning of a socioeconomic analysis.

Next Steps

Some of the key next steps that have been already identified include the following:

- Setting common targets in landscape: Using a merge of the ecological and the socioeconomic data, it will be possible to identify "compromise targets" for the landscape in consultation with stakeholders.
- Partnerships: Key partnerships with stakeholders will be important to the process, from a point of view of both political support and technical complementarity.
- Setting up a monitoring system at the landscape level: To measure progress against those targets, a monitoring system will need to be set up.
- Beginning small-scale activities: Small-scale activities need to start rapidly to identify the most suitable techniques, species, species' mix, training needs, and alternative economic activities that the population can engage in.
- Extracting lessons learned from the process and revisiting the work plan: On an annual basis, it is necessary to revise work plans and review data to determine whether the process is progressing according to plan or if adjustments are necessary.

15
Identifying and Using Reference Landscapes for Restoration

Nigel Dudley

Key Points to Retain

Reference forests are carefully preserved natural or near-natural forests that can provide information about natural species' mix and ecology, that can be used in planning and measuring the success of restoration.

Formal and informal networks of reference forests are building up around the world.

Use of reference forests often needs to be supplemented with other data such as historical records, old maps, identification of past vegetation through pollen mapping from peat cores, etc.

- Restoration of deforested land with a staged process leading to a more natural forest over time, e.g., as in Guanacaste, Costa Rica, where exotic species are used as nurse crops for natural forest[141]
- Restoration of forest with specific social values, e.g., *tembawang* fruit gardens of western Borneo, which are planted for their nontimber forest products but are also high repositories of biodiversity
- Restoration of specific values within managed forests by specific interventions, such as re-creation of dead wood components in southern Swedish and Finnish forests
- Restoration as a centuries-long process, where initial intervention is then augmented by natural changes and aging, as in the previously deforested *Agathis* forests of northern New Zealand

1. Background and Explanation of the Issue

Because forest restoration is a process, a good restoration programme starts with a fairly clear idea of what type of forest is being created, that is, the *target* for restoration and the associated activities. This can only be approximate, because ecosystems change and evolve, but can help set the approach and time scale.[140] There can be many different aims and end points, for instance:

Although it is often assumed that restoration aims to re-create a "natural" forest, this is not always the case. Many efforts aim instead at culturally important forests, as in parts of the Mediterranean, or even seek to limit the spread of trees to maintain game animals, as in many of the eastern African savannahs. Whatever the aims, good restoration needs to be planned and monitored against some framework, usually a similar forest type that identifies a template for the type of forest being restored.

Reference forests provide a model to follow. The best reference forests are those that have

---

[140] Peterken, 1996.

[141] Janzen, 2002.

been identified, protected, and monitored over time, so that they have an associated body of understanding about their ecology. They will often, although not invariably, be old forests, although younger forests can provide valuable reference for successional stages. Even quite newly identified reference forests can provide valuable information if their history is known and it will often be necessary to find a reference forest or reference landscape as part of the planning for forest restoration at a landscape scale. Sometimes reference forests need to be re-created theoretically from historical records and pollen diagrams. Although most valuable in relating to forest types in the same ecosystem, reference forests also provide information of value to forests far away. It is important to understand the relationship between the historical reference forest and the future forest being re-created or modified; the reference forest is not necessarily the same as the target forest being restored. Sometimes it will be possible, over time, for the latter to become very similar to its reference, while in other cases this will be impossible either because of other pressures on and needs from the forest or because conditions have changed and certain elements of the original forest are irrecoverable. A clear understanding of this relationship is important when setting targets for restoration.

Reference landscapes provide information on different aspects of ecology, particularly composition, ecological processes and functioning, and, crucially but often the most difficult to pinpoint, cyclical changes over time. Locating forests undisturbed enough to exhibit natural changes either through a gradual process of aging and renewal or from evidence of natural catastrophic events is now increasingly difficult in many areas, yet an understanding of how forests renew themselves is important in re-creating near-to-natural forests and in understanding likely pressures on managed forests.

Other elements to consider in defining targets for restoration include long-term human interaction with forests and the evolution of cultural landscapes (many forests have never existed without the presence of humans so that the idea of a pristine, human-free ecosystem is often little more than a myth). The probability of future climate change and other forms of environmental disturbance means that targets should be tailored with this in mind, also suggesting the limitation of following reference landscapes too closely, when they may be undergoing change themselves. More generally, targets for restoration should be developed with an understanding of likely changes. The idea that vegetation evolves to some climax type and then stays the same is now largely disproved, at least at the level of a particular stand, where flux is expected and is likely to be constant. In the end, choices usually need to be made about levels of biodiversity, naturalness, and livelihood values contained in particular restored forests, and reference forests can only provide information to help with these more political choices.

2. Examples

The presence of reference forests has played a fundamental role in understanding forest ecology and in developing responses to forest loss and degradation. Some reference forests are outlined below.

2.1. Oregon, United States

The H.J. Andrews Experimental Forest was protected by the U.S. Forest Service in 1948 as part of a network of forests intended to serve as living laboratories for studies by the service's scientific research branch. The forest is administered cooperatively by the U.S. Department of Agriculture (USDA) Forest Service Pacific Northwest Research Station, Oregon State University, and the Willamette National Forest, with funding from the National Science Foundation, U.S. Forest Service, Oregon State University, and others. Long-term field experiments have focussed on climate dynamics, stream flow, water quality, and vegetation succession. Currently, researchers are working to develop concepts and tools needed to predict effects of natural disturbance, land use, and climate change on ecosystem structure, function, and species' composition. Over 3000 scientific publications have used data from the forest. The research has been used in developing ways of restoring old-growth characteristics within managed forests in the Pacific Northwest through "new forestry," including retention of standing dead wood and coarse woody debris in streams.[142]

2.2. Centre for Tropical Forest Science (CTFS), Smithsonian Institute, Washington, DC

The CTFS has developed an international network of standardised forest dynamics plots. Within each plot, every tree over 1 cm in diameter is marked, measured, plotted on a map, and identified according to species. The typical forest dynamics plot is 50 hectares, containing up to 360,000 individual trees. An initial tree census and periodic follow-up censuses yield long-term information on species' growth, mortality, regeneration, distribution, and productivity, which currently provides an almost unique information source for developing restoration strategies within managed tropical forests. Utilising the data from the standardised, intensive forest dynamics plots throughout the tropics, CTFS researchers are exploring tropical forest species' diversity and dynamics at a global scale. Plots currently exist in Panama, Puerto Rico, Ecuador, Colombia, Cameroon, Democratic Republic of Congo, Malaysia, Thailand, Sri Lanka, India (see below), the Philippines, Singapore, and Taiwan.

2.3. India

The Mudumalai Wildlife Sanctuary and Bandipur National Park are part of the wildlife-rich protected areas within the Nilgiri Biosphere in the Western Ghat Mountains of southern India. These reserves are sites of long-term ecological research by the Centre for Ecological Sciences. A 50-hectare permanent plot in Mudumalai, where the dynamics of a tropical dry forest is investigated in relation to fire and herbivory by large mammals, is part of the international network of large-scale plots coordinated by the CTFS (see above).

2.4. Europe

Under the auspices of the European Cooperation in the Field of Scientific and Technical Research (COST) programme of the European Commission, a network has been established to help coordinate research taking place in strict forest reserves in 19 European countries. The process established protocols for data collection both in a core area and over the whole reserve, primarily to develop repeatable methods of describing the stand structure and ground vegetation. A Web-based forest reserves databank is helping to coordinate information. Natural forests are perhaps more critically threatened in Europe than in any other region, and the information will be used to help identify and manage protected areas and increase component of naturalness in managed forests.[143]

2.5. Mediterranean Europe

In some cases, changes have progressed so far that fully natural or near to natural reference forests have been lost. The origin of many of the fruit trees commonly found in Mediterranean forests is often only very generally known for example. Here the most useful references are often old cultural forests that contain many elements of biodiversity, and restoration programmes often aim to re-create these.[144]

Changes in access to reference forests can dramatically increase our level of understanding of forest dynamics and therefore management options. For example, when Finnish forest ecologists gained access to more natural forests in the Russian Federation at the end of the 1980s, they revised their understanding about disturbance patterns, recognising that snow damage was a proportionately larger agent of change than had been suspected. However, reference forests seldom provide *all* necessary information, particularly when changes have been so profound that no natural forest remains. Living reference forests are therefore a useful tool but by no means the only method

[142] Luoma, 1999.
[143] Broekmeyer et al, 1993.
[144] Moussouris and Regato, 1999.

for determining targets. Some of the other tools that may be used as surrogates for living reference forests are outlined below.

3. Outline of Tools

In most cases, reference landscapes are developed using a suite of different tools, the main ones of which follow:

- Reference forests: As described above, these are probably the most valuable single source of information.
- Comparison with other ecologically similar forests: Even if no nearby forests exist to act as a reference, use of cumulative data around the world can help to build our understanding about a forest's ecology. For example, knowledge about breeding patterns and population in many birds of prey allows ornithologists to make reasonably good predictions about stable reproduction rates for species based on body weight. Understanding about forest fire ecology can, with caution, be transferred from one ecosystem to another, at least to develop working hypotheses. Other elements, such as old growth characteristics, have been found to translate rather poorly from one forest ecosystem to another.
- Comparison with "original" forest types: Although it is often impossible to find a wholly unaltered forest ecosystem, numerous well-thought-out attempts have been made to describe ancient or natural forests: some examples are given in Table 15.1.
- Historical records: Written records can tell us a great deal and sometimes stretch back for hundreds or even thousands of years. The oldest known written records of forest management are 2000 years old and refer to forests maintained to supply timber for Shinto temples in Japan. Records from written histories, religious scriptures, sagas, and trade accounts can all provide valuable, albeit usually fragmentary, information about forests. Many supposedly "natural" forests in the U.K. can be traced back to recorded planting (often with the names of the people who planted them). More recent travellers' accounts are frequently used to provide information on past vegetation patterns, such as the records kept by Italian travellers in Eritrea a century ago that

TABLE 15.1. Definitions of original forests.

Definition	Explanation
Ancient woodland	Woodland that has been in existence for many centuries: precise time varies but in the U.K., 400 years is commonly used[1]
Frontier forest	"Relatively undisturbed and big enough to maintain all their biodiversity, including viable populations of the wide-ranging species associated with each forest type"; criteria include primarily forested; natural structure, composition, and heterogeneity; dominated by indigenous tree species[2]
Native forests	Meaning is variable: often forests consisting of species originally found in the area—may be young or old, established or naturally occurring, although in Australia often used as if it were primary woodland[3]
Old-growth in the Pacific Northwest, United States	"A forest stand usually at least 180–220 years old with moderate to high canopy cover; a multi-layered multi-species canopy dominated by large over-storey trees"[4]
Primary woodland	"Land that has been wooded continuously since the original-natural woodlands were fragmented. The character of the woodland varies according to how it has been treated."[5]
Wildwood	"Wholly natural woodland unaffected by Neolithic or later civilisation"[6]

[1] Bunce, 1989.
[2] Bryant et al, 1997.
[3] Clark, 1992.
[4] Johnson et al, 1991.
[5] Peterken, 2002.
[6] Rackham, 1976.

now provide information for restoration activities.
- Forest fragments: Even quite unnatural forest fragments or remnant microhabitats can with care and caution, be used as partial surrogates in areas where full reference forests no longer exist. For instance, park land and hedgerows both contain important elements of natural forests in Western Europe and can help set targets for restoration. Similarly sacred sites, preserved for religious reasons, can contain species that have disappeared from the surrounding area, as in forest gardens and sacred groves in, for instance, Indonesia, Laos, China, Kenya, and Malawi.
- Pollen analysis and soil microcarbon analysis: Analysis of pollen in peat cores, lake beds, or soil profiles can identify plants from thousands of years ago, as pollen is highly resistant to decay, particularly in the anaerobic conditions found in peat, and can often be identified to the level of individual species. Analysis along a core can show how vegetation changed over time, the presence and frequency of fires, and sometimes information about pollution. Such analysis is often the only sure way of building a picture of past vegetation where changes have been dramatic and living reference landscapes have disappeared.
- Gap analysis using enduring features: This approach consists of a coarse-filter conservation assessment of protected areas based on a landscape approach using "enduring features" (essentially land forms or physical habitats) as geographic units that reflect biological diversity. The gap analysis involves three main stages. First, natural regional frameworks are reviewed to ensure that natural region boundaries reflect broad physiographic and climatic gradients. Next, within each natural region maps are used to identify enduring features. An enduring feature is a land form or landscape element or unit within a natural region characterised by relatively uniform origin of parent material, texture of parent material, and topography-relief. Finally, the relationship of biodiversity to enduring features of the landscape is derived from more detailed tertiary sources.[145]

4. Future Needs

Although a lot of the tools are in place, there is still little experience in combining them to develop realistic targets for restoration exercises. Gaps go right back to the philosophical roots of restoration and at what is being aimed for—for example, original vegetation or just a workable ecosystem at the present time. Much better understanding of the likely process of forest restoration itself is needed, along with more accurate methods of measuring progress.

References

Broekmeyer, M.E.A., Vos, W., and Koop, H., eds. 1993. European Forest Reserves. Pudic Scientific Publishers, Wageningen, The Netherlands.

Bryant, D., Nielsen, D., and Tangley, L. 1997. The Last Frontier Forests: Ecosystems and economies on the edge. World Resources Institute, Washington, DC.

Bunce, R.G.H. 1989. A Field Key for Classifying British Woodland Vegetation. Institute of Terrestrial Ecology and HMSO, London.

Clark, J. 1992. The future for native logging in Australia. Centre for Resource and Environmental Studies Working Paper 1992/1. The Australia National University, Canberra.

Iacobelli, T., Kavanagh, K., and Rowe, S. 1994. A Protected Areas Gap Analysis Methodology: Planning for the Conservation of Biodiversity. World Wildlife Fund Canada, Toronto.

Janzen, D.H. 2002. Tropical dry forest: Area de Conservación Guanacaste, northwestern Costa Rica. In: Perrow, M.R., Davy, A.J., eds. Handbook of Ecological Restoration, vol. 2, Restoration in Practice. Cambridge University Press, Cambridge, UK, pp. 559–583.

Johnson, K.N., Franklin, J.F., Thomas, J.W., and Gordon, J. 1991. Alternatives to Late-Successional Forests of the Pacific Northwest. A Report to the US House of Representatives, Washington, DC.

Luoma, J.R. 1999. The Hidden Forest: The Biography of an Ecosystem. Owl Books, New York.

---

[145] Iacobelli et al, 1994.

Moussouris, Y., and Regato, P. 1999. Forest harvest: Mediterranean woodlands and the importance of non-timber forest products to forest conservation. Arborvitae supplement, WWF and IUCN, Gland, Switzerland.

Peterken, G.F. 1996. Natural Woodland: Ecology and Conservation in Northern Temperate Regions. Cambridge University Press, Cambridge, UK.

Peterken G. 2002. Reversing the Habitat Fragmentation of British Woodlands. WWF UK, Goldalming, UK.

Rackham, O. 1976. Trees and Woodland in the British Landscape. Weidenfeld and Nicholson, London.

16
Mapping and Modelling as Tools to Set Targets, Identify Opportunities, and Measure Progress

Thomas F. Allnutt

> **Key Points to Retain**
>
> Forest landscape restoration can benefit from mapping and use of geographical information systems (GIS) in several key ways, but in particular by measuring and monitoring progress toward meeting biological and socioeconomic targets via restoration.
>
> Many potential methods exist to utilise maps and GIS for landscape-scale restoration, from the simple to the highly customised and experimental.

1. Background and Explanation of the Issue

Successfully planning, implementing, and monitoring projects that aim to restore forest landscapes involves the management and analysis of spatial information, that is, quantitative and qualitative two-dimensional data covering the area of interest. For example, understanding how a potential restoration site may or may not meet a biodiversity goal such as "increase overall habitat connectivity from x to y to maintain the viability of species z" requires maps and basic statistics (size, isolation, etc.) for all forest patches that occur across the landscape. Many other spatial variables influence the suitability and likely success of a given area for restoration. Therefore, map-based technologies, such as satellite remote sensing, aerial photography, and geographic information systems (GIS) have and will continue to provide many benefits to forest landscape restoration.

There are many ways GIS and other spatial technologies can assist forest landscape restoration projects. At one end of the spectrum, simple maps of forest cover, elevation, rivers, communities, and roads are inherently useful for understanding the ecological and human context of the landscape. At the other extreme, sophisticated and custom spatial models may be constructed to simulate, for example, the hydrological effects of forest restoration on downstream watersheds. Here we focus on the use of spatial data to develop spatial scenarios that meet biological and socioeconomic targets. Known as "suitability modelling" or "multicriteria evaluation," this approach is one type of GIS-based modelling utilising readily available commercial GIS packages.

Specifically, in this chapter we provide (1) examples of the types of spatial data and some common map-based measures useful for planning and monitoring restoration of forest landscapes, (2) examples of spatial tools and technologies for deriving this information, and (3) reviews of several recent applications of spatial technologies to restoration.

1.1. Mapping Areas to Meet or Set Targets

The targets and goals of the project determine the types of spatial data to collect and spatial analyses to conduct. There are two main types of targets, biological and socioeconomic. Although not all targets are spatial in nature (e.g., "prevent the extinction of species x"), many are. Some examples of spatial targets include "Protect x hectares of habitat y" or "Establish x hectares of community forest reserves." Planning for and evaluating progress toward a target such as the latter type requires appropriate spatial data.

1.1.1. Biological Targets

Often, biological targets are derived directly from existing large-scale conservation planning processes such as ecoregion conservation (ERC).[146] An initial product of an ERC vision is a set of priority landscapes designed to meet specific biological objectives, such as the conservation of an endangered primate. Where this is the case, these targets can be used directly to prioritise and implement restoration areas, for example, preferentially conduct restoration adjacent to known populations of the target primate.

In other cases, no such information may exist. Here, participants may rely on basic principles of biological conservation to guide what targets to select, and thus what spatial data sets are needed. In general, space-based biological targets involve individual species (e.g., cheetah),[147] habitat, or vegetation types (e.g., wetlands), or ecological and evolutionary processes (e.g., migration, hydrology).[148] Targets for these features are typically expressed as quantitative areas or percentages of the total distribution of the biological element in question (e.g., 1000 hectares of oak-savannah).

Once biological targets are established, several classes of spatial data are necessary to map where they may be achieved on the ground. In many cases, existing map sources may be used; in others, maps will have to be created using modelling or technologies such as remote sensing.

To evaluate species-based targets, one first needs to know the current distribution of all target species within the landscape at the finest level of detail possible. Range maps are one potential surrogate for this information and they are increasingly available for a number of taxa worldwide.[149] In other cases, modelling may be used to predict species' distributions from field collections coupled with environmental data.[150] Often, and particularly at fine scales, field-based inventories will be required to assess the presence or absence of certain key species.

Another common type of biological target involves particular habitat and/or vegetation types. Several sources of data are available to evaluate this type of target. Existing maps and classifications are often used, from national or regional inventories, for example. In other cases, new maps may be created from raw photographs or the processing of photographs or digital images. The most widespread source is remote sensing—typically photographs or digital imagery from airplanes or satellite-borne sensors. New, high-resolution imagery (submetre) provides a good source for mapping natural habitats as well as human land uses, though cost can be a significant constraint.

In areas of high species and habitat heterogeneity, optical remote-sensing may not be able to distinguish biological differences to a necessary degree. Forest that is indistinguishable spectrally—from the perspective of a camera or satellite—is often very diverse biologically. Here, habitat modelling can be used to map areas where one expects species to differ significantly. A range of approaches are available, from the quick and approximate, to more formal statistical methods.[151] Elevation, for example, is often used as a proxy for species' distributions, and can be used to quickly divide a continuously mapped forest type into several or more forest habitats (lowland, sub-montane, montane, etc.).

---

[146] Dinerstein et al, 2000.
[147] Lambeck, 1997.
[148] Pressey et al, 2003.
[149] Ridgely et al, 2003.
[150] Boitani et al, 1999.
[151] Ferrier et al, 2002.

The spatial configuration of the restoration landscape is of critical importance for biodiversity conservation for several reasons. One, the long-term survival of many species often depends directly on the size and connectivity of available habitat. The reasons for this are generally (a) individuals and populations require sufficient outbreeding opportunities that are only available in habitat blocks of a particular size, and (b) the species in question has ecological requirements (e.g., seasonal migration) that require large connected blocks of habitat. In both cases, research may be necessary to assess the habitat configuration necessary for the target species. Two, many environmental and ecological processes will not be maintained once habitat fragments drop below a particular threshold of isolation or fragmentation. The maintenance of natural hydrological flows in watersheds, for example, can depend on the size and connectivity of intact forest blocks.

1.1.2. Socioeconomic Targets

The second major class of targets are socioeconomic. In some cases, socioeconomic targets will have been specified when the landscape was identified within a priority setting exercise (e.g., the visioning process in ecoregion conservation), though this is less often the case than with biological targets. Socioeconomic targets that require spatial data generally specify target amounts of land uses within the landscape. This may involve zoning one portion of the landscape for a particular land use. For example, participants may wish to have one third of the landscape devoted to community forestry. In other cases, the entire landscape (apart from those areas reserved for biodiversity conservation) may be zoned for particular land uses, akin to a traditional land-use plan or zoning map.

Mapping areas to meet socioeconomic targets requires a detailed and up-to-date land-cover map. This map shows the current distribution of natural and human-oriented areas in as much detail and at as fine a scale as possible and it can be derived from existing land-use/land-cover maps for the area, or may be created from aerial and remote sensing sources coupled with ground truth. The map of current land uses serves as the starting point; a map of future land uses shows those areas where changes in land uses will be necessary to meet socioeconomic targets.

1.1.3. Land Tenure and Land Value

The legal status and ownership of land (land tenure) within the landscape, and the economic value of that land are also important for planning forest landscape restoration. Sometimes this information can be derived from existing maps available from local or national government organisations, particularly in the case of land tenure. In other cases, ground surveys will need to be conducted to establish tenure and land value of unknown areas. Spatial economic modelling has also been used to estimate land value. Rules are constructed that allow one to estimate the value of every parcel of land within the area of interest, based on variables such as market access, for example.

1.3. Mapping Opportunities: Integrating Biological and Socioeconomic Data to Meet Targets and Map Opportunities

Some areas are more suitable than others for particular uses. Analysis of spatial data has the potential to efficiently allocate areas to one use or another. This idea is formalised in land-use plans or more formally via suitability modelling otherwise known as multicriteria evaluation (MCE).[152]

Suitability modelling or MCE using GIS can be used to systematically combine spatial, biological, physical and socioeconomic data detailed above in order to meet biological and socioeconomic objectives via restoration. Here are two generic examples:

1. Map suitability for a single biological or socioeconomic target. As an example, imagine

---

[152] Eastman et al, 1993.

one biological target for the landscape is to maintain a viable population of a primate. It is estimated that the target primate requires 25,000 hectares of habitat between 1000 and 3000 m in elevation, in a single, connected block of forest. There are currently only 15,000 hectares of suitable forest within the landscape, in two disconnected blocks. Therefore, the challenge is to map at least 10,000 hectares to restore based on the habitat criteria required for the species: elevation, size, and connectivity. Three maps are created. One shows all areas in the target range of 1000 to 3000 m, one ranks areas according to their potential to rejoin the disconnected blocks, and one ranks areas by their proximity to existing good habitat for the primate. These three maps are standardised to a common numeric range, and then combined by means of a weighted average, to produce a continuous map of suitability. The most suitable areas are those that are close to existing intact habitat, connect the two blocks, and are the right elevation. The highest scoring areas (those that come close to meeting all three criteria) are selected until the target of 10,000 hectares is met. These form the priority restoration areas for this biological target. The same process may be used to map suitable areas for socioeconomic targets.

2. Incorporating socioeconomic data as a constraint on suitable areas for biological targets. Just as physical and biological criteria may be combined to identify suitable restoration areas to meet biological targets, socioeconomic criteria, such as land use or land value, can also be incorporated in the process. For example, imagine two parcels of land that, when restored, would be equal in every way for meeting the above biological target. They are equivalent in elevation, in proximity to existing forest, and in terms of connecting the two forest blocks. One parcel is currently actively used for agricultural production, whereas the other has been abandoned for several years. For several reasons, it would likely be easier to restore the abandoned parcel. Thus, including socioeconomic data in the MCE process can help to efficiently identify restoration priorities when there are choices of areas to meet biological targets.

1.4. Monitoring

A key benefit of using quantitative spatial data and targets for both biological and socioeconomic variables throughout the planning and implementation process is that it facilitates long-term monitoring as the project proceeds. Remote sensing in particular provides a relatively quick and inexpensive, synoptic, repeatable view of large-scale changes to land uses and land cover over time within the landscape. Clearly this will have to be paired with reviews of progress toward those biological and socioeconomic targets that cannot be measured remotely. A current disadvantage is the lack of long-term large-scale attempts at systematic monitoring of conservation programmes, though efforts are currently underway at a number of places and institutions.

2. Examples

Examples abound of the use of maps and GIS in the fields of planning and conservation.[153] Generally speaking, however, there are few examples of its application to forest restoration planning. One exception is the recent work of J. Halperin, in which GIS was used for participatory, community-based, large-scale restoration planning in Uganda.[154]

The WWF network has only recently begun to apply GIS to its restoration initiatives. The United Nations' Environment Programme-World Conservation Monitoring Centre (UNEP-WCMC) used GIS to prioritise areas for WWF-based restoration projects in North Africa.[155] Biological attributes such as species' richness, forest integrity, and patch size were balanced against human pressures including road density, grazing pressure, and resource use. As of early 2004, there are two additional projects underway. In one, in the Andresito landscape (Argentina) of the Atlantic Forest, there are plans to use suitability modelling with IDRISI to identify key restoration corridors in

---

[153] see e.g., Eghenter, 2000; Herrman and Osinski, 1999.
[154] Halperin et al, 2004.
[155] UNEP-WCMC, 2003.

conjunction with a set of stakeholders from the region. Similarly, GIS is being used in Madagascar to map and prioritise suitable areas for restoration within a large landscape that needs to be restored. Here, biological targets are being established for six IUCN red-listed vertebrates. Criteria are being established to map suitable habitat for each species in order to evaluate current status within the landscape. Where current habitat is insufficient for long-term viability of each population, areas will be prioritised for restoration based on connectivity, proximity to known populations, and habitat characteristics. Socioeconomic data will be used as a constraint where options exist to meet biological targets. This work is in its initial stages and is expected to continue through 2005.

3. Outline of Tools

Standard vector-based GIS software—ESRI (ArcMap, ArcView, Arcinfo)—is the standard GIS virtually worldwide. It is available at low cost to conservation organisations, and it performs all types of GIS functions, from basic mapping to advanced analyses, especially when customised or linked to other programmes (e.g., statistical software, etc.).

Standard raster-based GIS—IDRISI, ESRI (Spatial Analyst, GRID for Arcview, ArcMap, and Arcinfo), ERDAS. The IDRISI and ESRI products are low cost (for educational or non-profit companies) GISs capable of doing raster-based analyses (e.g., most analyses involving remotely sensed imagery). IDRISI includes functions for easily stepping through suitability models and MCE as part of its decision support package. ERDAS is a much more expensive software designed primarily to analyse satellite imagery and other remotely sensed data.

4. Future Needs

A key need is for participatory GIS-based decision-support tools designed specifically for restoration in a biodiversity conservation context. Similarly, research is needed into tools to strengthen linkages between site-based restoration research and spatial decision making with GIS. Recently, several new GIS models are in use that have been used extensively for spatial planning in conservation, notably C-Plan[156] and SITES/Marxan.[157] These particular applications are currently, generally speaking, spatial optimisation tools designed to meet representation targets in conservation plans. There is tremendous potential, however, especially with the simulated-annealing algorithm used by Marxan (and now SPOT among other tools) to optimise any given set of objectives (such as restoration) in a spatial model. Research is urgently needed to expand these tools to meet other objectives beyond simple reservation and representation.

References

Boitani, L. (coordinator), Corsi, F., De Biase, A., et al. 1999. A databank for the conservation and management of African Mammals. Institute of Applied Ecology, Rome, Italy.

Dinerstein, E., Powell, G., Olson, D. et al. 2000. A Workbook for Conducting Biological Assessments and Developing Biodiversity Visions for Ecoregion-Based Conservation. Conservation Science Programme, World Wildlife Fund, Washington, DC.

Eastman, J.R., Kyem, P.A.K., Toledano, J., and Jin, W. 1993. GIS and Decision Making, UNITAR. Explorations in GIS Technology, Vol. 4. UNITAR, Geneva.

Eghenter, C. 2000. Mapping People's Forests: The Role of Mapping in Planning Community-Based Management of Conservation Areas in Indonesia. Biodiversity Support Programme, Washington, DC.

Ferrier, S. 2002. Mapping spatial pattern in biodiversity for regional conservation planning: where to from here? Systematic Biology 51:331–363.

Halperin, J.J., Shear, T.H., Munishi, P.K.T., and Wentworth, T.R. 2004. Multiple-objective forestry planning in biodiversity hotspots of east Africa. In preparation.

Herrman, S., and Osinski, E. 1999. Planning sustainable land use in rural areas at different spatial levels using GIS and modelling tools. Landscape and Urban Planning 46:93–101.

[156] Pressey et al, 1995
[157] Leslie et al, 2003; McDonnell et al, 2002.

Lambeck, R.J. 1997. Focal species: a multi-species umbrella for nature conservation. Conservation Biology 11:849–856.

Leslie, H., Ruckelshaus, R., Ball, I.R., Andelman, S., and Possingham, H.P. 2003. Using siting algorithms in the design of marine reserve networks. Ecological Applications 13:S185–S198.

McDonnell, M.D., Possingham, H.P., Ball, I.R., and Cousins, E.A. 2002. Mathematical methods for spatially cohesive reserve design. Environmental Modelling and Assessment 7:107–114.

Pressey, R.L., Cowling, R.M., and Rouget, M. 2003. Formulating conservation targets for biodiversity pattern and process in the Cape Floristic Region, South Africa. Biological Conservation 112:99–127.

Pressey, R.L., Ferrier, S., Hutchinson, C.D., Sivertsen, D.P., and Manion, G. 1995. Planning for negotiation: using an interactive geographic information system to explore alternative protected area networks. In: Saunders, D.A., Craig, J.L., Mattiske, E.M., eds. Nature Conservation: The Role of Networks. Surrey Beatty and Sons, Sydney, pp. 23–33.

Ridgely, R.S., Allnutt, T.F. Brooks, T., et al. 2003. Digital Distribution Maps of the Birds of the Western Hemisphere. Version 1.0. CD-ROM. NatureServe, Arlington, Virginia.

UNEP-WCMC. 2003. Spatial analysis as a decision support tool for forest landscape restoration. Report to WWF.

Additional Reading

George, T.L., and Zack, S. 2001. Spatial and temporal considerations in restoring habitat for wildlife. Restoration Ecology 9:272.

Huxel, G.R., and Hastings, A. 2001. Habitat loss, fragmentation, and restoration. Restoration Ecology 7:309.

Jankowski, P., and Nyerges, T. 2001. Geographic Information Systems for Group Decision Making. Taylor and Francis, New York.

Loiselle, B.A., Howell, C.A. Graham, C.H., et al. 2003. Avoiding pitfalls of using species distribution models in conservation planning. Conservation Biology 6:1591–1600.

Wickam, J.D., Jones, B.K., Riiters, K.H., Wade, T.G., and O'Neill, R.V. 1999. Transitions in forest fragmentation: implications for restoration opportunities at regional scales. Landscape Ecology 14:137–145.

17
Policy Interventions for Forest Landscape Restoration

Nigel Dudley

> **Key Points to Retain**
>
> Changing policy toward restoration or land use is often the most effective way of stimulating large-scale restoration.
>
> Such policy changes can be addressed, in different ways, at a local scale (e.g., changing grazing patterns), a national scale (e.g., modifying forestry laws), or a global scale (e.g., ensuring that international conventions favour high-quality restoration).
>
> Key tools in policy interventions include good analysis, especially economic analysis, case studies, and advocacy.

1. Background and Explanation of the Issue

Localised and site-based interventions to restore habitat can be very useful, and much of what we have learned about ecological restoration comes from small-scale initiatives, primarily carried out by nongovernmental organisations (NGOs) and local communities but also to an increasing extent by forward-looking companies and government departments. We also describe further in this book (see "Practical Interventions that Will Support Restoration in Broad-Scale Conservation Based on WWF Experiences") how strategic use of such initiatives can have wider benefits, for example by linking patches of existing habitat, by providing fuelwood to places that are otherwise without energy sources, or by preventing erosion. However, small-scale initiatives are inevitably limited in what they can achieve on their own and are usually expensive, stretching the resources of the organisations or communities that carry them out. Accordingly, it is often more effective to spend effort in changing policies at local, provincial, national, regional or even global level to encourage restoration at a broader scale. Many NGOs undertake restoration initiatives to use them as a lever to change policies, by, for example, showing that different approaches can be more effective or cost less money. But although working examples can be powerful tools in stimulating change, they usually need to be accompanied by effective advocacy and a thorough understanding of the policy climate.

Policy change can operate at many different levels. At the most local level, it can include changing policies within a single community[158] or landscape to stimulate forest restoration. Examples include:

- Agreed changes in grazing regimes to allow natural regeneration, perhaps agreeing to protect different zones at different times
- Voluntary controls on collection of nontimber forest products to ensure that these are not degraded

---

[158] Sithole, 2000.

- Collective investment in tree planting, for instance to establish fuelwood plantations

Whilst such interventions are already a regular feature of many large conservation or conservation and development projects, they are again quite limited in scope. A far more significant change can be affected if national policies are changed in favour of more sympathetic restoration, for example:

- Modification of national forestry laws to allow old-growth forest to remain, facilitate retention of deadwood, or remove perverse incentives that discourage restoration
- Changing national forest restoration or afforestation programmes to increase the range of goods and services that they provide (for example, reducing the proportion of intensive plantations and increasing assisted natural regeneration)

There are also increasingly opportunities to change policies that transcend national borders,[159] thus potentially having an impact on a global or a regional scale. Along with intergovernmental bodies, such transnational policy can also involve companies that operate in many countries or bilateral and multilateral donors, including the following:

- Introduction of pro-restoration clauses within international treaties or incentives, such as using carbon offsets for forest restoration under the U.N. Framework Convention on Climate Change, or specific policy recommendations of global forest initiatives such as the U.N. Forum on Forests
- Integration of restoration into funding opportunities or legislative requirements from regional agreements such as those of the European Community
- Development of company policies for restoration after mineral extraction, infrastructure developments, etc.
- Modification of projects funded by bilateral or multilateral donor agencies

2. Examples

2.1. Altai Sayan, Russia

Russia's first woodland area to be certified under the Forest Stewardship Council is still managed collectively and includes large areas of woodland on sandy soils dominated by birch—used for specialist products sold by the Body Shop chain. The certification process included agreement by farming cooperatives on changes in sheep grazing to leave some areas untouched for long enough to foster regeneration of birch woods.[159a]

2.2. Latvia

Latvian forestry inherited legislation crafted by the Soviet Union, which included the use of large clearcuts and a requirement to manage forests including removal of deadwood. As a result, dead standing and lying timber is in short supply in many woodlands, leading to a decline in many saproxylic (deadwood living) species.[160] This is particularly serious at a European scale because Latvia's forests contain some of the richest biodiversity in the continent. WWF in Latvia has worked with the government to change the forestry regulations to allow retention of deadwood in managed forests, thus opening the opportunity of increasing this threatened microhabitat.

2.3. Vietnam

The government's five million hectare reforestation programme aims to restore forest cover but in practice hampers local flexibility. Although large plantations have been established, it seems likely that in several provinces much money has been wasted in places where forest cover remains high. In theory funding can be used to support natural regeneration, for example in the buffer zones of protected areas, as is already happening around Song Thanh Nature Reserve. The WWF Indochina Programme is working with the government to

---

[159] Tarasofsky, 1999.
[159a] Information drawn from site visit as part of certification team, 1998.
[160] Rotbergs, 1994.

modify the way in which funds are used, both to increase natural forest restoration and to ensure that established forests are retained and gain higher value (see detailed case study "Monitoring Forest Landscape Restoration in Vietnam").

2.4. European Community

Throughout the European Union (EU) region, restoration of natural woodlands is hampered in areas of sheep or goat grazing because farmers receive hectare-based payments depending on the area capable of being grazed.[161] To obtain maximum funds, woodlands are opened to grazing, which means that young seedlings fail to establish, resulting in gradually aging forest. In some cases, woodlands that have been fenced with EU funds to encourage regeneration are now being opened up again. It is recognised that the key to facilitating regeneration in many areas is not further grants for tree planting but a removal of perverse incentives (see "Perverse Policy Incentives" and case study "The European Union's Afforestation Policies and their Real Impact on Forest Restoration") by changing incentives' schemes within the Common Agricultural Policy to reduce the reasons for allowing sheep grazing in woodlands.

2.5. Central America

The Kyoto protocol of the U.N. Framework Convention on Climate Change allows for governments to offset some of their carbon emissions, or trade other countries' emissions, through tree planting. Initial proposals focussed largely on the establishment of intensive plantations of exotic species, but research suggests that the long-term carbon sequestration benefits of such plantations are very limited, as they are used mainly for short-term products such as paper and cardboard that are quickly abandoned and break down. Central American governments have been amongst those most active in lobbying for modification of the Kyoto protocol to allow different kinds of forest management including natural regeneration and increase of retention of deadwood and humus components. Research suggests that innovative use of carbon markets has aided forest regeneration, with the side benefit of also increasing tourism in these areas.[162]

2.6. Lafarge—Quarry Restoration in Kenya

Lafarge, based in France, is now the largest quarrying company in the world. The development of its policy toward forest landscape restoration is an example of how small-scale interventions can lead to larger restoration policy initiatives.

Lafarge's forest restoration work started with a series of site-based interventions. The former quarry of the Bamburi cement plant near Mombasa in Kenya was mined for 20 years. In the early 1970s, a rehabilitation programme was started to restore the site as a nature reserve. After a phase of soil formation using the leaf litter of introduced pioneer trees, a large number of tree and other plant species typical of the indigenous coastal forests were also planted. The success of these was observed over time in order to select those species that proved suitable for planting on a larger scale to replace the pioneer trees. In addition to trees of potential economic value (such as Iroko and other indigenous hardwood, which is valuable for local crafts such as carving), endangered species and those that provide habitat or food for indigenous wildlife have also been planted: to date, 422 indigenous plant species have been introduced into the newly created ecosystems of forests, wetlands, and grasslands in Bamburi's former quarries. Of these 364 have survived, including 30 that are on the IUCN Red List of Threatened Species for Kenya.

Lafarge also started working with WWF on policy issues, including supporting the organisation's forest landscape restoration initiative. In April 2002, Bamburi signed a partnership agreement with WWF East Africa, and identified forest landscape restoration as one of the priority partnership activities, including the

---

[161] Joint Nature Conservation Committee, 2002.

[162] Miranda et al, 2004.

need to establish a biodiversity monitoring system in partnership with WWF, in order to define guidelines for ecological quarry rehabilitation.

In 2001 Lafarge adopted a formal quarry rehabilitation policy with the participation of WWF to spread best practice in terms of quarrying work and relations with local stakeholders. The most important elements of this policy are to plan restoration from the outset and coordinate restoration with quarrying activities. In addition to biodiversity issues, land planning considerations are also taken into account when defining a rehabilitation project in order both to preserve the environment and to generate income for the local communities. In this framework quarry rehabilitation often leads to the creation of wetlands and natural reserves or leisure areas.

3. Outline of Tools

Stimulating policy changes requires hard and convincing analysis, including economic analysis, a clear message, and sometimes some targeted and effective advocacy. In cases where financial support is being changed around in favour of more balanced forms of restoration, it may also include economic incentives. Some key tools are as follows:

Economic analysis is useful to make the case for restoration or for different kinds of restoration. Examples might include demonstrating that retention of deadwood within managed forests does not entail excessive cost, or showing that natural regeneration is cheaper than replanting. For example, a WWF/World Bank economic analysis convinced the government of Bulgaria to change plans for establishing intensive poplar plantations on islands in the Danube with natural regeneration,[163] and an analysis for Forestry Commission economists in Wales, U.K., persuaded the government agency to use natural regeneration in an area of forest because it proved cheaper than replanting.

Economic incentives encourage individuals and groups to make space for restoration, including both official incentive schemes and incentives through the market, such as certification. Targeted incentives have been used very successfully to encourage restoration, for instance through conservation easements to take land out of production, as has occurred widely in the U.S., through direct support for tree planting as successfully implemented on a large scale in parts of Pakistan, or through tax incentives as in several Latin American countries.[164]

Case studies show that restoration can work and pay for itself. The case of the restored quarry near Mombasa showed that restoration was not an impossibly expensive task and helped to encourage Lafarge, the company concerned, to introduce a wider policy. Case studies only work, however, if they are carefully prepared and include all the relevant information needed to make policy decisions, and if they reach the attention of the right policy makers.

Advocacy entails campaigns or lobbying to encourage change.[165] Targeted lobbying has been successful, for example, in changing some conditions in the Kyoto Protocol to allow greater latitude for natural regeneration.

Codes of practice are developed by working with other stakeholders (e.g., industry) to agree and implement them voluntarily and to encourage restoration. The International Tropical Timber Organisation recently completed detailed guidelines for natural regeneration, in association with IUCN and WWF, which provide an example of this approach.[166] As with case studies, however, such codes are only worth the investment in developing them if they are implemented in practice.

4. Future Needs

Many of these ideas remain in their infancy. We still require far better understanding of the economic and other benefits of environmental goods and services from restoration in order to make the case, for example, for natural regen-

---

[163] Ecott, 2002.
[164] Piskulich, 2001.
[165] Byers, 2000.
[166] ITTO, 2002.

eration rather than other land uses or for changes in major funding initiatives such as those under the European Common Agricultural Policy. More generally, major changes are still needed in global trade policy to remove the perverse incentives that currently act against restoration in many areas.

References

Byers, B. 2000. Understanding and Influencing Behaviour. Biodiversity Support Programme, Washington DC

Ecott, T. 2002. Forest Landscape Restoration: Working Examples from Five Ecoregions. WWF, Gland, Switzerland.

International Tropical Timber Organisation. 2002. ITTO Guidelines for the Restoration, Management and Rehabilitation of Degraded and Secondary Tropical Forests. ITTO, Yokohama, Japan

Joint Nature Conservation Committee. 2002. Environmental effects of the Common Agricultural Policy and possible mitigation measures. Report to the Department of Environment, Food and Rural Affairs, Peterborough, UK.

Miranda, M., Moreno, M.L., and Porras, I.T. 2004. The social impacts of carbon markets in Costa Rica: the case of the Huetar Norte region. International Institute of Environment and Development, London.

Piskulich, Z. 2001. Incentives for the Conservation of Private Lands in Latin America. Biodiversity Support Programme. The Nature Conservancy and USAID, Arlington, Virginia.

Rotbergs, U. 1994. Forests and forestry in Latvia. In: Paulenka, J., and Paule, L., eds. Conservation of Forests in Central Europe. Arbora Publishers, Zvolen, Slovakia.

Sithole, B. 2000. Where the Power Lies: Multiple Stakeholder Politics Over Natural Resources—A Participatory Methods Guide. Center for International Forestry Research, Bogor, Indonesia.

Tarasofsky, R. 1999. Assessing the International Forest Regime. IUCN Environmental Law Centre, Bonn, Germany.

18
Negotiations and Conflict Management

Scott Jones and Nigel Dudley

Key Points to Retain

Forest landscape restoration relies on achieving broad consensus among a variety of stakeholders.

However, stakeholders may have very different perceptions of what forest landscapes should provide.

This will require a certain amount of negotiation and possible conflict resolution.

1. Background and Explanation of the Issue

Forest landscape restoration approaches use the restoration of forest functions as an entry point to identify and build a diversity of social, ecological, and economic benefits at a landscape scale. As such they rely on achieving broad consensus on a range of restoration interventions from a variety of stakeholders, who may have very different perceptions of what forest landscapes should provide. This requires effective negotiation among stakeholders whose negotiation skills, interests, needs, and power are often markedly different. However, the success of forest landscape restoration approaches often hinges on how successfully such negotiations are conducted. The principles of forest landscape restoration, therefore, aim at restoring forests to provide multiple social and environmental benefits through processes that involve stakeholder participation. The achievement of these ambitious goals relies on finding a successful passage through an array of practical challenges. These include the implications of current and future land tenure, competing land uses, and reaching a balance between different management regimes. Success depends on the ability of those initiating or guiding a forest landscape restoration project to manage the tensions and conflicts that will arise on the way. This, in turn, implies a certain amount of knowledge about how to identify, analyse, and manage conflict, retaining the varied, useful perspectives that are helpfully expressed through conflict, while resolving or mitigating those aspects of conflict that are dangerous or prevent project success.

1.1. Types of Conflict

There are two aspects that characterise conflicts: their openness and the type of conflict.

Conflict can be concealed or open[167]; either can cause problems in developing successful landscape-scale approaches to restoration:

- Open conflicts: everyone can see them and knows about them.
- Hidden conflicts: some people can see them and know about them, but hide them from others (particularly outsiders), perhaps because of cultural or social reasons (e.g.,

[167] DFID, 2002a; Fisher et al, 2000.

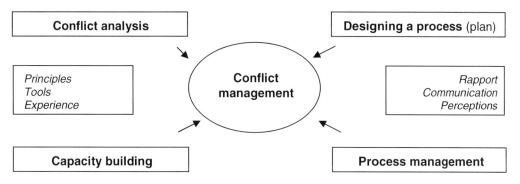

FIGURE 18.1. Building blocks in the conflict management process: elements in a conflict situation.

many gender-related conflicts) or because disputes may be embarrassing to the community (e.g., disagreements between young people and elders).
- Latent conflicts: these come to the surface when something changes the status quo. For example, if a restoration project brings benefits (money, power, influence, equipment), their distribution can create conflicts that were not there before the project arrived.

There are also different types of conflict. It is important to understand which type of conflict one is facing since each needs addressing in a different way.

- Interpersonal conflicts: between two or more people relating to personality differences
- Conflicts of interest: someone wants something that another has (e.g., money, power, land, influence, inheritance)
- Conflicts about process: how different people, groups, and organisations solve problems (e.g., legal, customary, institutional)
- Structural conflicts: the most deep-seated type relating to major differences that are hard to address (e.g., unequal social structures, unfair legal systems, economic power biased toward certain stakeholders, or differences in deep-seated values, such as cultural or religious)

Sometimes one type of conflict, perhaps unthinkingly, is disguised as another, for instance a personality clash may be presented as an issue of process.

1.2. Elements in a Conflict Situation

Managing conflict is not a straightforward process. Rather, there are a number of key building blocks in a conflict management process that interrelate and must often be undertaken in parallel (Figure 18.1[168]):

- Conflict analysis is about understanding who the different stakeholders are, what are their strengths, fears, needs, and interests, and how they perceive or understand the conflict(s).
- Capacity-building is about helping people to manage conflict. It may be required at any time. For example, it may take place prior to negotiations because some stakeholders need to develop negotiation skills. It may take place before agreements are signed because different groups like to have agreements in different forms; it is important that all groups have the capacity to understand each other's approaches to problem solving and reaching agreements. Capacity-building often takes the form of training (e.g., in negotiations or "people" skills), but sometimes other resources are needed.
- Designing a process is about planning who to bring together, where, when, and how. The most effective conflict management processes are usually flexible, iterative, and capable of keeping stakeholders on board as events, issues, and even the attitudes of the conflicting parties change.

[168] Modified from Warner and Jones, 1998.

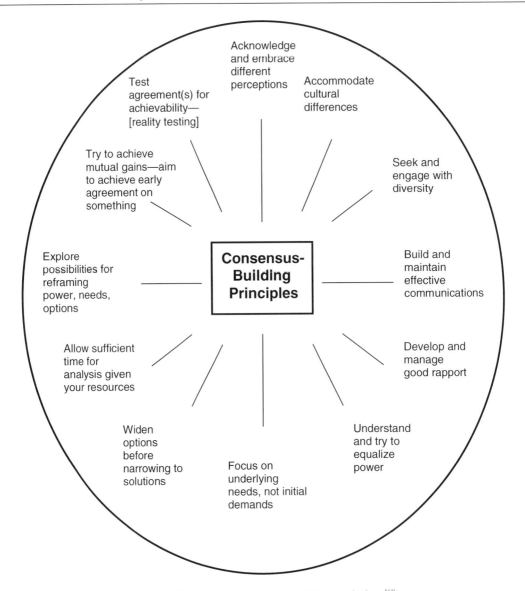

FIGURE 18.2. Principles for successful negotiation.[168a]

- Process management is about how to build and maintain effective ways of working with the parties, to retain flexibility and patience, while still keeping focussed on outcomes and working toward success on the criteria that stakeholders have agreed to, for example, how to convene an effective meeting with clear goals, or how to monitor an agreement.

Achieving these things requires adhering to certain *principles* (e.g., mutual respect, being accountable, recognising the potential and limits of your influence, see Figure 18.2), using certain *tools* (e.g., stakeholder and gender analysis), and applying key *experience* (e.g., with similar projects or with these people in other projects). They also require key people *skills*, among the most important of which are maintaining good *rapport* and effective *communications*, and effectively engaging with the multiple *perspectives*.[169]

[168a] Modified from Warner, 2001.
[169] Jones, 1998.

> Box 18.1. Examples of Best Alternative to a Negotiated Agreement (BATNA) in the Context of Forest Landscape Restoration
>
> *The loggers simply don't want to negotiate at all. They are going to go ahead and cut those trees.*
> BATNA—What about going to the newspapers? Let the media know that this biodiversity hotspot is threatened and local people are suffering.
> *The donor is not able to give you another grant to add an extra component to this work.*
> BATNA—Perhaps write a report that helps to bring the donor's expectations in line with your capacity to deliver.
> *The people in the community feel powerless to enter face-to-face negotiations with the government and the large Geneva-based and Washington, DC–based agencies.*
> BATNA—Possibly see if a mediator can be found who would be acceptable to both sides.
> *The negotiations went well and trust is high, but the government was unable to agree involvement of their officials due to government rules.*
> BATNA—Perhaps work with another NGO with relevant expertise that can complement you but has no government restrictions over committing official staff.

1.3. BATNA (Best Alternative to a Negotiated Agreement)[169a]

Negotiations are a voluntary process. But what if the other person is completely inflexible, breaks the ground rules you agreed to, and only wants his or her own way. In short, what if the other person does not want to negotiate? Similarly, what if the other person is negotiating in good faith, you have excellent communications, and trust each other, but it is simply not possible (in his or her view) to meet even your "bottom line" needs? Under these circumstances, you need an alternative to negotiation. There may be several alternatives. What you really need is the best one.

So what would be your best alternative to a negotiated agreement? In the (unfortunate) language of conflict management, this has become known as a BATNA (best alternative to a negotiated agreement). Box 18.1 illustrates some examples of where a BATNA may be appropriate.

1.4. Project and Process Management

Any approach to forest landscape restoration requires time and resources to identify, to agree to, and to manage the process. Different agencies have different approaches to project and process management, developed perhaps from commercial approaches or international development models. Clearly, in the world of logical frameworks, multi-stakeholder partnerships, and collaborative management schemes, the management process itself is a subject for negotiation that requires the full range of skills and principles discussed above.

Conflicts over one form of management indicate an opportunity to search for other approaches that can helpfully deal with the legal, financial, political, and operational issues that any complex project or programme involves. It follows that successful forest landscape design will be able to identify and engage with different management approaches and use the negotiation process to build ownership while deciding roles and responsibilities. Sometimes one agency or another will desperately seek management control, and the task is to negotiate shared understandings and responsi-

---

[169a] Fisher and Ertel, 1995.

bilities. At other times, it is a hard task to identify any agency that feels able to take management responsibility. Again, this is an opportunity to explore why, and to undertake a collective search for a solution that supports stakeholders who are willing to put their names forward.

1.5. Negotiation Health Warning

Finally, it is important to note that like other aspects of conflict management, negotiation is a culturally bound process. Different societies, groups, agencies, and organisations all have different cultures and approaches to managing conflict. While much of the literature on negotiations is Western and business-oriented, there needs to be a high degree of cultural sensitivity and contextually located understanding to proceed with negotiations, especially where many different cultures are involved in multi-stakeholder negotiations.

2. Examples

There is very limited experience in applying conflict resolution and negotiation skills to landscape initiatives in forest restoration. We highlight here just a few examples from other chapters in this book that have shown some successful or interesting outcomes through negotiations.

- In Vietnam, a three-dimensional paper and cardboard model was used to bring stakeholders together around "their" landscape to identify specific elements within it. The process was aimed at reconciling different views of the landscape and what it could look like in the future. It provided those around the model with the opportunity to express their views on the importance of different elements in the landscape (more information on this example can be found in "Assessing and Addressing Threats in Restoration Programmes").
- In Malaysia, an ongoing negotiation process with oil palm plantation companies is gradually ensuring a change in the companies' policies related to restoration. Whereas initially the companies converted their entire estates to oil palm, they are now gradually allocating part of their land for natural regeneration and plantation of local species (for more on this example see "Restoring Quality in Existing Native Forest Landscapes").
- In Jordan, negotiation between goat herders and park authorities ensured a reduction in grazing, thus allowing for more natural regeneration (for more on this example see "Restoration of Protected Area Values").

3. Outline of Tools

Learning and applying the tools and skills for successful conflict management cannot come from reading books or attending courses alone, but also involves long periods of trial and error, and observation—"learning by doing." Many participatory techniques described elsewhere in this book are relevant. Tools and skill sets for conflict management that are particularly relevant include those relating to analysis, capacity building, communications, creative thinking, negotiation, and project and process management.

3.1. Negotiation Process

Negotiating involves meeting to discuss ways of reaching a mutual agreement or arrangement. A negotiation is a voluntary process in which each person or group (often called a party) has a position that is not fixed, but that does have its limits. A successful negotiation can create a sense of ownership and commitment to shared solutions and shared follow-up actions. This sense of ownership and commitment makes negotiated solutions often more desirable, for example, than legal solutions, where one party may feel it lost out. In a conflict, some things cannot be negotiated, and some things can. Usually it turns out that many more things can be negotiated than people first thought. This is another reason why negotiated agreements are a valuable way, though not the only way, of trying to manage conflicts in forest landscape restoration. It follows that a first step in negotiation is reaching agreement on what is negotiable. Successful negotiations follow certain important principles (see Box 18.2) and require

Box 18.2. Some Principles and Skills Involved in Negotiating Forest Landscape Restoration (See also Figure 18.2)

- Be clear on what everyone means by the issue and the problems, opportunities, and people/agencies involved
- Adopt a positive attitude, for example, being clear that conflicts are not just problems but also opportunities
- Have in mind some kind of a route map, some idea about ways in which key stakeholders wish to proceed
- Address role, responsibility, and legitimacy issues, including the limitations (boundaries) to your negotiating authority
- Build and maintain effective rapport and relationships
- Active listening
- Identify high-quality, relevant questions
- Embrace multiple perspectives and perceptions
- Build on what is already there (including cultural aspects of conflict management and problem solving)
- Consider process (law, custom, institutional) as well as structural conflicts and conflicts of interest
- Keep in mind options for withdrawing or not getting involved further
- Keep an eye on capacity building for self-development and organisational development
- Separate and focus on the problem and not the personalities
- Separate and focus on underlying needs and motivations, not initial positions
- Know what you would do if the negotiations did not work, perhaps because the other party broke the ground rules or tried to use unacceptable force (this is also called knowing your BATNA: best alternative to a negotiated agreement; see Box 18.1)
- Seek, explore, and emphasise common ground
- Put your case in terms of their needs, not just why *you* want something
- The more you know about the other's position, the better able you are to find consensus-based solutions; do some homework to find out their situation
- Maintain a creative, positive approach
- Use paraphrasing and other communication skills to understand and describe the other's points
- Create a positive environment for the negotiation (think about the physical setting, the comfort and acceptability of the place, the time, and the way you manage yourself)
- Look for an early, small successes (reach agreement on something early, even if that is just the venue, then emphasise that agreement; common ground—start small)
- Make sure your preparations are as complete and accurate as possible. Write down what you have done to prepare. Check with a colleague. Check with another colleague. Seek constructive feedback.

Keep in mind:

1. The process and conflict management style
2. Your goals and boundaries (your limit or bottom line)
3. Opportunities to address power inequalities
4. Your colleagues' needs, expectations, and ability to act as resources
5. Your personal values and principles
6. Time and space for reframing issues
7. Capacity building needs that may emerge
8. The needs for more analysis that may emerge

Multiple perspectives and perceptions can be useful. A diversity of opinion helps us shed light on the issue from different directions. Treat difference and diversity not as an emotional trigger to fight against, but as a moment of opportunity to engage with.

knowledge, skills, and a *positive attitude*. It is helpful to look at each of these things in relation to three phases in negotiations:

- Preparation—what we need to do before the negotiation
- Negotiation itself—could take place in one meeting or over several meetings
- Follow-up—what we need to do after the negotiation is over and agreement has been reached

A negotiation can happen at any time. Entering a community or a government official's office may require a negotiation. The gatekeeper may want to know some details before people just walk in, including when a group or agency will arrive, how long it will stay, under whose authority, with what level of formality, and to do what.

Having agreed to who are the stakeholders who need to be involved, a process of negotiations in forest landscape restoration will probably look something like this:

1. Each group works to understand the other group's initial positions relating to the landscape.
2. Each group then asks high-quality questions and uses listening skills to try to understand underlying needs, fears, and motivations in identifying restoration interventions.
3. The parties try to deploy creative thinking and other skills to generate a wide range of options that could address these needs, fears, and motivations.
4. This range of options is prioritised and brought together in ways that allow everyone to gain as much as possible.
5. An agreement is sought, to which everyone can commit.
6. That agreement is tested against the real world to make sure it is achievable.
7. The parties agree on the next steps, on how to manage the restoration interventions and the resources that are needed, and on ways of monitoring the agreements and commitments they have made.

3.2. Analytical Tools

A large number of analytical tools and skills that are used in participatory forest management, project management, and development can be brought to bear in conflict management. Examples include *participatory appraisal*,[170] a variety of approaches for measuring and *analysing sustainability*,[171] and more general tools that help to frame and guide further analysis, such as *STEEP, SWOT, problem trees, and forcefield analyses*.[172] The key is to use those that are relevant for different stakeholders and that help to bring understanding and wider perspectives on the issues. Key analytical tools, though, include the following:

- Stakeholder analysis[173]
- Conflict mapping and situation analysis[174]
- Tools that address power relations, culture, and gender[175]

A variety of analytical tools can feed into a summary conflict analysis. Conflict analysis can be done in the office (alone or in a group) or in the field (for example, in participatory exercises) or in combination. Successful analyses are clear about who undertook the analysis, when, and why, and make it clear how different groups were involved in verifying and agreeing to analysis summaries from different stakeholder perspectives. Of course, as events change and time moves on, analyses need to be revisited. This is especially important when new stakeholders enter the picture or established stakeholders leave, and when critical events change key stakeholders' circumstances.

Analysis helps to identify the domain of conflict (e.g., domestic, social, cultural, economic, or political) and whether conflict is nested within several domains. Conflict mapping with key individuals or stakeholder groups, can help to summarise information and show up major differences and possible ways forward. One example is given as a matrix (Fig. 18.3). However, flow charts, Venn diagrams, and other visually powerful mapping tools can help

---

[170] Jackson and Ingles, 1998; www.fao.org/participation.
[171] Bell and Morse, 2003; Dalal-Clayton and Bass, 2002.
[172] Pretty et al, 1995.
[173] DFID, 2002b, section 2; Ramirez, 1999; Richards et al, 2003.
[174] DFID, 2002b, section 3; Fisher et al, 2000; Wehr, 1998.
[175] Fisher et al, 2000.

Name of person or party	A	B	C
Position or stance in relation to the conflict			
Needs			
Concerns, anxieties, or fears			
Attitudes toward the others			
Assumptions about the others			
Values and beliefs			
Historical issues (e.g., past misunderstandings)			
Types of power (e.g., moral, financial, political)			

FIGURE 18.3. Matrix to help analyse conflict.

communicate the outcomes from an analysis. It is important to remember, though, that the process of analysis itself is a part of managing conflict. Done well, the process itself can help foster trust and mutual understanding. An early agreement on the individual and collective concerns and opportunities can help establish the stage for positive negotiation of emerging issues.

3.3. Capacity Building

Undertaking a process of analysis often requires capacity building. Some stakeholders will be familiar with negotiating from a business perspective. Others will see negotiations as embedded within their own culture and society—the way they negotiate and problem solve will be different. Others may use legal frameworks or a scientific approach to analysis. Again, addressing the process of analysis is itself a part of the overall approach to managing conflict. Capacity building skills and tools may need to be deployed at an early stage.

Identifying and responding to gaps in conflict management skills or to gaps in resources requires a sophisticated approach to capacity building backed up by appropriate levels of resourcing (e.g., for training and stakeholder support). Building capacity is best seen as an ongoing activity rather than a linear one. High-quality capacity building forms part of addressing inequalities in power relations. Strengths and needs analysis and some form of training needs analysis are important first steps in capacity building.[176] Capacity building actions also need to be linked with reflection, so that interventions can be monitored and evaluated on an ongoing basis. This process, too, helps to build confidence and trust, when people appreciate the fact that someone somewhere is taking responsibility for empowering key stakeholders to participate effectively.

3.4. Effective Communications

Building and maintaining effective communications are key aspects of conflict management and multi-stakeholder partnerships in forest landscape restoration. Providing, managing, using, and facilitating access to information is part of any communication strategy.[177] What is additionally important in conflict management is ensuring that these things translate into meaningful understanding. Indeed, effective communications are vital to generating and disseminating the high levels of understanding of different stakeholders' perspectives and needs that good conflict management requires. Some aspects of effective communications relate to general communications strategies: the frameworks and mechanisms for enabling stakeholders to engage with one another on relevant matters. This includes documents, meetings, the use of different media, and an overall information, communication, and monitoring management system, such as a logical framework or

---

[176] Bartram and Gibson, 1997.
[177] Dalal-Clayton and Bass, 2002, Ch. 8.

> **Box 18.3. Barriers to Good Listening**
>
> "On-off listening"—drifting off into personal affairs while someone is talking
>
> "Switch off" listening—words that irritate us so that we stop listening
>
> "Open ears–closed mind" listening—we decide the speaker is boring and think that we can predict what he or she will say, so we stop listening
>
> "Glassy eyed" listening
>
> "Too deep for me" listening—when ideas are complex or complicated there is a danger we will switch off
>
> "Matter over mind" listening—when a speaker says something that clashes with what we think and believe strongly, we may stop listening
>
> Being "subject-centred" instead of "speaker-centred"—details and facts about an incident become more important than what people are saying themselves
>
> "Fact" listening—we try to remember facts but the speaker has gone on to new facts and we become lost
>
> "Pencil" listening—trying to put down on paper everything the speaker says usually means we are bound to lose some of it and eye contact is also lost
>
> "Hubbub" listening—there are many distractions that we listen to instead
>
> "I've got something to contribute" listening—something the speaker says triggers something in our own mind and we are so eager to contribute that we stop listening
>
> An awareness of the above barriers to listening can be a first step in avoiding them.

Adapted from training materials, Centre for International Development and Training, University of Wolverhampton, UK.

action plan. Other aspects relate more to interpersonal communications, such as getting the balance right between telling and asking, or become a good listener (Box 18.3).

In dealing with conflict, one important distinction is between *telling* and *asking*. Giving free information is an important part of building communications. However, if one is usually "telling" people, this can be perceived as aggressive and dominating (e.g., "I'm going to tell you what the law says—and that is the end of the story"). Asking relevant questions in an involving, open way can communicate a sense of concern and interest, that someone has bothered to identify questions that may help mutual understanding. Of course, a balance between the two is needed.

3.5. Creative Thinking

People and agencies tend to think and react in the ways that they always have done. The way we think is constrained by many things, including our experience, worldview, education, and degree of comfort with new ideas. Creative thinking is about breaking these patterns to look at situations in new ways—thinking "outside the box." Creative thinking is an important asset to conflict management at all stages, not just analysis. Often, a breakthrough can come when creative thinking allows the situation to be reframed—changing the way we construct and represent the conflict.[178] Reaching agreement requires strong skills in synthesis—thinking creatively about how to develop an agreement and monitoring process that everyone can live with can be challenging. A number of tools exist that can help enhance people's creative thinking skills. One-on-one and in small groups, good facilitators and trainers can help to build creative thinking skills. Where things get trickier is moving through organisations' management and decision-making structures to translate the creative, useful thoughts into actions that are helpful. Creative thinking is culturally embedded. Indeed, culture plays a major part in resisting

---

[178] Lewicki et al, 2003.

and improving creative thinking skills, in organisations as well as other groups.[179]

4. Future Needs

Most conservation organisations, forestry departments, and companies have only very limited knowledge about conflict resolution. Capacity building for conflict management and negotiation within conservation and forestry organisations is a critical need in terms of building the ability to work across broad scales and mainstream conservation. Most of the tools and expertise are known but have been applied in only a very limited way within the field of natural resource management.

References

Bartram, S., and Gibson, B. 1997. Training Needs Analysis. Gower Publishing, London.
Bell, S., and Morse, S. 2003. Measuring Sustainability. Earthscan, London.
Dalal-Clayton, B., and Bass, S. 2002. Sustainable Development Strategies. OECD, Earthscan and UNDP. Earthscan Publications, London.
Department for International Development (DFID). 2002a. Conducting conflict assessments: guidance notes, DFID. Government of the United Kingdom, http://www.dfid.gov.uk/pubs/files/conflictassessmentguidance.pdf.
Department for International Development (DFID). 2002b. Tools for development. DFID, Government of the United Kingdom. http://www.dfid.gov.uk/pubs/files/toolsfordevelopment.pdf.

[179] Hofstede, 1994.

FAO, 2002.
Fisher, S., et al. 2000. Working with Conflict. Zed Books, London.
Fisher, R., and Ertel, D. 1995. Getting Reading to Negotiate, Penguin Books, London.
Hofstede, G. 1994. Cultures and Organisations: Software of the Mind—The Successful Strategist Series. Harper Collins, London.
Jackson, W.J., and Ingles, A.W. 1998. Participatory Techniques for Community Forestry. World Wide Fund for Nature, IUCN-World Conservation Union and Australian Agency for International Development, Gland, Switzerland.
Jones, P.S. 1998. Conflicts about Natural Resources. Footsteps No. 36 (September). Tearfund, Teddington, London.
Lewicki, R.J., Gray, B., and Elliott, M. 2003. Making Sense of Intractable Environmental Conflicts: Concepts and Cases. Island Press, Covelo and Washington, DC.
Pretty, J.N., Gujit, I., Thompson, J., and Scoones, I. 1995. Participatory Learning and Action: A Trainer's Guide. International Institute for Environment and Development, London.
Ramirez, R. 1999. Stakeholder analysis and conflict management. In: Buckles, D. ed. Cultivating Peace—Conflict and Collaboration in Natural Resources Management. World Bank, Washington, DC.
Richards, M., Davies, J., and Yaron, G. 2003. Stakeholder Incentives in Participatory Forest Management. ITDG Publishing, London.
Warner, M., and Jones, P.S. 1998. Conflict resolution in community based natural resources management. Overseas Development Institute Policy Paper (No. 35), August.
Warner, M. 2001. Complex Problems, Negotiated Solutions. ITDG Publishing, London.
Wehr, P. 1998. International on-line training programme on intractable conflict. http://www.colorado.edu/conflict/peace/problem/cemerge.htm.

19
Practical Interventions that Will Support Restoration in Broad-Scale Conservation Based on WWF Experiences

Stephanie Mansourian

Key Points to Retain

Urgent conservation or livelihood problems may necessitate short-term, strategic interventions even in the absence of a longer-term programme.

A series of 10 different tactical interventions are suggested, ranging from threat removal to positive economic incentives.

1. Background and Explanation of the Issue

In the face of increased threat of massive species' extinction, with estimates that more than half of the world's threatened species live on less than 1.4 percent of the earth,[180] it may be important to consider a range of practical and tactical interventions to begin to reverse this rapid degradation, particularly in highly threatened areas that are extremely rich in biodiversity.

There are still surprisingly few examples of successful forest restoration from a conservation perspective, particularly at a large scale.[181] Elsewhere, we have discussed the importance of carrying out restoration as a component of larger conservation and development programmes, but in some cases there may also be opportunities to carry out useful restoration more opportunistically. This chapter is intended to highlight some tactical interventions that could be undertaken if framed within a forest landscape restoration process or approach.

Planning at a landscape or ecoregional scale is difficult enough, but actually intervening at that scale is generally harder still. In a forest landscape restoration context, activities such as planning, engagement, priority setting, negotiation, trade-offs, modelling, etc. are usually all best carried out at a landscape scale. However, with the exception of some policy interventions, most of the *practical* restoration actions will take place at sites within the landscape or ecoregion. Although planning processes are often lengthy, some actions can often start in anticipation of the overall long-term strategy to restore forest landscapes; generally some responses will be clear and uncontroversial and these can often be initiated even whilst more difficult issues remain unresolved.

This chapter discusses the types of specific and punctual interventions related to restoration that a field programme may consider undertaking. Some of these would be expected to arise within a longer term strategy to restore ecological and social forest functions but may also come in advance of such a strategy due to lack of funds for the overall process, lack of buy-in from stakeholders, and other issues relating to expediency or urgency. When a species is facing immediate threats of extinc-

---

[180] Brooks et al, 2002.
[181] TNC, 2002.

tion, for instance, short-term measures may be needed even while long-term planning is still in process. None of the proposed interventions below replace larger scale efforts, nor are they meant to be implemented in isolation from a broad-scale planning process. Rather, they are to be seen as elements of the larger process and as possible entry points; success at a small scale is one of the most effective ways of gaining support for larger-scale programmes.

When selecting one of the proposed entry points listed below (see Outline of Tools), it is important to think of the desired impact of this tactical intervention:

- Is it to influence a specific group of stakeholders? Which one and what is the desired effect?
- Is it to understand better the dynamics (biological or social) in the landscape?
- Is it to change sociopolitical conditions in the landscape before engaging in restoration within the landscape? Which conditions? And what is the most cost-effective way to change them?
- What are the resources (human and financial) and time involved? Can we afford them?
- What are the priority issues that need addressing soonest?

2. Examples

2.1. Research into Different Restoration Methods in Malaysia

Some palm oil companies along the Kinabatangan River in Sabah, Borneo, have agreed to set aside land for restoration. Initial trials showed limited success. Starting in 2004, in an effort to identify the most successful techniques for restoration, tests began using different methods on a small plot of land. These are the methods proposed (during a field visit by the author):

- Natural regeneration with no intervention (including a smaller study area fenced against browsing animals)
- Assisted natural regeneration (mainly some land preparation and weeding around regenerating species)
- Planting with native species (using species adapted to local conditions and including if possible both commercially valuable dipterocarp trees and fruit trees)
- Planting an exotic species as a nurse crop to foster natural regeneration

Each approach is to be monitored on a regular basis in order to determine which one yields the highest survival rates. The long-term aim of this research is to disseminate the most suitable restoration methods in all the areas set aside for restoration along this important biodiversity corridor.

2.2. Changing the Forest Policy in Bulgaria Thanks to a Cost-Benefit Analysis[182]

Bulgaria's 75 islands on the Danube river are rich in biodiversity, and are an important stopover site for migratory birds. Yet, over the last 40 years, the government has systematically converted natural floodplain forest to hybrid poplar plantations to supply the local timber industry. Until the year 2000, the government had plans to continue conversion of this ecosystem, leaving only 7 percent of the original forest. Thanks to a comprehensive cost-benefit analysis, sponsored by the World Bank and WWF, it was shown that financial losses from suspending timber production on certain islands could be offset by intensifying production in areas already converted to poplar plantations. Additional benefits that were highlighted by the analysis included the potential use of original forest for recreational purposes, improved fishing (by creating more spawning grounds), the harvest of nontimber forest products, and possible ecotourism development. In 2001 the government, therefore, changed its policy, adopting one that called for the immediate halt of all logging and conversion of floodplain forests to poplar plantations on the Danube islands, restoration of native species

---

[182] Ecott, 2002.

in selected sites, as well as strengthening of the protected areas network on the islands. Although a longer term forest landscape restoration programme for the Danube is underway, this tactical intervention helped to maintain a unique habitat that might well have disappeared before the more detailed programme was implemented.

3. Outline of Tools

3.1. Focussing on Removing or Reducing the Identified Threats

Sometimes it will be sufficient to remove, reduce, or mitigate a particular threat or pressure on forests in a landscape to set them on a positive path toward regeneration. Because threats often originate from political or economic decisions, changing them may require significant lobbying, backed up by negotiations, research, and building of strategic partnerships. If these threats can be reduced or removed, natural regeneration can often be significant (if there are no other biophysical constraining factors).

Examples of threats that are common as an impediment to natural forest regeneration include the following:

- Alien invasive species (e.g., electric ants, *Wasmannia auropunctata*, in New Caledonia)
- Government incentives that foster forest conversion (e.g., Chile's subsidies for plantations)
- Infrastructure projects (e.g., the construction of the Ho Chi Minh highway in Vietnam)
- Demand for cash crops (e.g., valuable soya expansion in Paraguay causing forest conversion)
- Unsustainable agricultural practices (e.g., Slash and burn agriculture in Madagascar)
- Illegal logging (e.g., in Indonesia)
- Uncontrolled and "unnatural" fires (e.g., in India)

Concentrating first on removal of threats is appropriate when it is clear that addressing the identified threat can lead to natural regeneration or restoration with only limited interventions. This is also a necessary choice in cases when a field project cannot start until the threat has been addressed.

Depending on the social and economic context, some threats may be much easier to address than others. For instance, illegal logging is in itself a very complex issue, which may well be beyond the remit of a restoration project. However, knowledge of key areas affected can help determine where (or even whether) and how to establish a restoration programme. It is important to recognise threats that cannot be addressed, or resources may be pumped into a hopeless situation.

3.2. Changing Government Policies

Often, a change in government policy may provide the right conditions to promote restoration (also see "Policy Interventions for Forest Landscape Restoration"). In some cases it may be necessary to lobby for more supportive policies, while in others, it may be necessary to remove destructive ones. The European Union's (EU's) Common Agriculture Policy (CAP) has for instance invested significantly in afforestation with limited social and ecological results (see case study "The European Union's afforestation Policies and their Real Impact on Forest Restoration"). WWF and other local partners are trying to address this in many EU countries (particularly in southern Europe) by demonstrating alternative, more socially and environmentally appropriate forms of restoration that could be financed by the same CAP subsidies. It will be important and relevant to focus efforts on government policies when these have been identified as a key factor in causing the loss and degradation of forests (e.g., perverse incentives) or when there is a clear opportunity to engage the government in supportive policies (e.g., a new forest plan being developed). In some countries, like Vietnam or China, there are huge government programmes promoting investments in reforestation/afforestation. Because of the scale of these programmes, it is often wiser (and economically more efficient) to engage in these processes than to invest efforts in a separate project.

3.3. Using Advocacy Levers

Some advocacy, lobbying, and economic tools can be used to encourage change that supports forest restoration or that removes or reduces the pressure on forests.

- Market pressure: The market may be used to promote the use of products from well-managed forests or forests that are being restored. For example, WWF has worked on the palm oil markets in Switzerland to promote better practices in Malaysia where the oil palm plantations have significantly damaged natural forest cover and where restoration of natural forest is now having to take place. This signifies engaging in research on market routes and raising awareness at the consumer end, as well as promoting solutions for better practices at the production end.
- Pressure using multilateral donors: Multilateral donors may be used as a lever for change either through their own projects or through imposing conditionality on loans. For example, agencies such as the Asian Development Bank (ADB) have active projects related to forest policy, but they also finance plantation projects. In Vietnam, for instance, the ADB is one of the main donors to the government's Five Million Hectares Reforestation Programme. Working together with such institutions may be a way of improving practices within their projects and also encouraging change in those projects that they finance.
- Communications/media tools such as Gifts to the Earth: WWF developed the Gifts to the Earth tool, a public relations mechanism, to pay tribute to major acts that favour the environment. This is one of many creative tools that may be used as an incentive for a government or other decision maker to change current policies or adopt new ones that would be more beneficial to or supportive of restoration.
- Campaigning: mobilising many stakeholders to put pressure on the relevant decision makers (governments, multilateral agencies, the private sector) is an effective means of ensuring change. It does need to be used carefully, however, and must be founded on good data.

3.4. Changing Companies' Practices

Traditionally, conservation organisations have not worked much with the private sector. Yet given that the largest companies are larger financial players than most governments and that they often determine future land-use options (e.g., mining companies, plantation companies, infrastructure companies), it is important to work with them in any large-scale restoration effort in order to ensure that restoration is well integrated in their plans.

This is, for instance, an effective way of encouraging companies to adopt best (or at least "better") practices. Many companies are happy to work with civil society organisations especially if improvement in their standards means some form of certification, media opportunities, and even in some cases the additional bonus of more efficient (cheaper) production. The sorts of sectors that may be influential include the infrastructure sector, the mining sector, and the forestry sector. WWF is currently engaging with large plantation companies such as Stora Enso to not only promote better management of their estates but also assist them to restore areas of the land that they manage.

3.5. Valuing Forests

Governments sometimes neglect or mismanage forests because the goods and services that they produce have not been properly valued. By obtaining recognition of the value of forests from either the government (if it is the major cause of concern) or local communities, restoration of those values can be promoted.

This can be done a number of ways:

- Through a traditional cost-benefit analysis that would provide a good argument for restoration for governments (see the Bulgaria example, above)
- Through research and surveys with local communities, particularly elders, to identify what values have been lost and what values

they would like to see restored. For example, in Vietnam WWF has engaged with communities and the provincial government in the central Annamites to identify the forest values that have been lost as a starting point for setting future restoration objectives.

While recognising the value of forests is one important step, it is but the first step. Governments and other decision makers then need to take necessary measures to ensure that those values are protected and where relevant restored.[183]

3.6. Specific Research

Often a large-scale programme to restore a range of forest functions cannot start until a number of specifications of the landscape are better understood. Initial research can be carried out with limited funds as a way to start a larger-scale programme.

This research may be related to any of the following, for example:

- Restoration techniques: While a number of restoration techniques have been tried and tested, it is not always easy to know which one will work best under local conditions. A small-scale trial plot can help identify those (see example on Borneo, above).
- Species' mix: Often exotic species have been used because they are better understood than local ones. Research money may be well spent on identifying the growth rate of and necessary conditions for specific local species as well as on the optimal mix of species.
- Removal of invasive species: Invasive species can often be the single most important impediment to natural regeneration or maintenance of forest quality within existing forests. Applied research can help test different techniques to remove the invasive species while promoting indigenous ones.
- Communities and stakeholders: Socioeconomic research may be necessary to understand better the profiles of stakeholders in the landscape and their motivations, pressures, livelihood conditions, and aspirations.

- Market research: Market research may be helpful when seeking to promote alternative income generating activities.
- Upstream versus downstream: In a landscape context, it may be important to identify the types of activities upstream and their impact downstream. For example, deforestation upstream may be causing sedimentation problems downstream. To encourage restoration within the landscape context, such cause and effect will need to be clearly demonstrated to stakeholders and substantiated by suitable research.

The above represent but a few of the numerous research topics. There are many others that are specific to different conditions.

3.7. Awareness Raising

If there is no identified need from the local population for restoration, then attempts at restoration are likely to fail. It is important to ensure that relevant stakeholders understand the linkages between restoration and the things that matter to them (availability of useful plants, soil protection, provision of forest products, etc.), and this may necessitate an awareness-raising campaign. For example, in New Caledonia, WWF is one of nine partners engaging in the protection and restoration of the dry forest. The project has a number of components, including active engagement of stakeholders (particularly land owners), and it has spent considerable time and resources working with local landowners to mobilise their support for restoration and to help them understand the implications of restoring the dry forest (benefits and costs).

There are a number of different forms of publicity (different media, workshops) and part of the skill in successful advocacy is in identifying the one that will reach the target audience (e.g., radio is often a good way of reaching rural populations in poorer countries).

3.8. Training and Capacity Building

One tactical intervention may consist of offering training in relevant restoration techniques. For instance in Morocco, WWF has been

---

[183] Sheng, 1993.

invited to help redesign the university's forestry curriculum to include specific restoration elements.

The sorts of training that can be provided include the following:

- Nursery design and development: Training can be provided to farmers and other community members on managing tree nurseries. This may also include elements of seed recognition and collection.
- Agroforestry techniques: When agricultural practices are an issue, training farmers in techniques such as agroforestry that are more compatible with some form of natural forest cover can be a useful approach within a forest landscape restoration initiative.
- Training can be provided in alternative income-generating activities (see below) to reduce the impact people are having on forests while offering them a realistic livelihood alternative.
- Improved grazing practices may sometimes be a simple way of returning areas of land to natural forest.
- In relevant cases, training may involve better fire management practices (to remove fire risks, to control them, or to undertake prescribed burns).

3.9. Forest-Friendly Economic Activities (Microenterprise Development)

In many countries the pressure on forests, the conversion of forests, or the hindering of natural regeneration is driven by the poorest people, who rely on forests for their immediate needs but are under too much short-term pressure to invest in long-term restoration strategies. One way of addressing this may be by providing training in improved practices that will help both sustain their own resource base and reduce forest degradation, or, on the other hand, by offering new economic activities that reduce their detrimental impact on forests. For a conservation organisation, this will generally require partnering with development organisations with expertise in, for example, microenterprise development.

For example, in Madagascar, the main threat to forests is slash-and-burn agriculture with short fallow periods. In a country with such high poverty levels, the only way to reduce this pressure on forests is to provide alternative livelihood options for those local communities. A number of successful microenterprise development programmes have been attempted by entities such as USAID (US Agency for International Development),[184] the U.N., and CARE. These programmes may not have been explicitly intended to reduce pressure on forests, but in partnering with conservation organisations two objectives could be reached: improving livelihoods while ensuring that forests are protected and, where appropriate, restored. When promoting such alternative livelihood options, it is important to undertake suitable feasibility and market studies, and not engage people, for instance, in honey production if there is no market for it.

3.10. Paying Communities for Better Practices

It may sometimes be necessary or appropriate to use project money to compensate communities for the loss they suffer by accepting restoration on land they own or use. This could be a first activity before developing alternative livelihood options. It can also be a way of engaging communities that may not otherwise be very receptive to the project. One risk with this approach is that of getting communities accustomed to compensation and expecting it over the long term. This clearly needs to be a short-term activity with a clear plan to move into other activities.

4. Future Needs

In an ideal world, a comprehensive restoration programme would be well thought out, would address a range of stakeholders' priorities, would be implemented at various scales (national, local, regional), and would be given the necessary resources and time to succeed.

[184] ARD-RAISE Consortium, 2002.

Unfortunately, this is often not the case, and therefore punctual interventions like those listed above may become necessary first actions. All of the actions listed above would benefit from being integrated into large programmes that aim to restore forest functions within landscapes for the benefit of people and biodiversity. One future need, therefore, is for decision makers and donors to allocate sufficient resources to allow for the implementation of the large-scale programmes that are required to achieve the restoration of forest functions in many regions of the world. Another need is for more creative partnerships between public, private, and civil society organisations, as well as between development and conservation organisations to achieve the ambitious aims of restoring forest functions in landscapes.

References

ARD-RAISE Consortium. 2002. Agribusiness and forest industry assessment. Report submitted to USAID–Madagascar, November 18.

Brooks, T.M., Mittermeier, R.A., Mittermeier, C.G., et al. 2002. Habitat loss and extinction in the hotspots of biodiversity. Conservation Biology 16(4):909–923.

Ecott, T. 2002. Forest Landscape Restoration: Working Examples from Five Ecoregions. WWF, Gland, Switzerland.

Sheng, F. 1993. Integrating Economic Development with Conservation. WWF International, Gland, Switzerland.

The Nature Conservancy (TNC). 2002. Geography of Hope Update: When and Where to Consider Restoration. The Nature Conservancy, Arlington, Virginia.

Additional Reading

Lamb, D., and Gilmour, D. 2003. Rehabilitation and Restoration of Degraded Forests. IUCN and WWF, Gland, Switzerland.

Mansourian, S., Davison, G., and Sayer, J. 2002. Bringing back the forests: by whom and for whom? In: Sim, H.C., Appanah, S., and Durst, P.B., eds. Bringing Back the Forests: Policies and Practices for Degraded Lands and Forests. Proceedings of an International Conference, 7–10 October 2002. FAO, Thailand, 2003.

Ormerod, S.J. 2003. Restoration in applied ecology: editor's introduction. Journal of Applied Ecology 40:44–50.

Sayer, J., Elliott, C., and Maginnis, S. 2003. Protect, manage and restore: conserving forests in multifunctional landscapes. Paper prepared for the World Forestry Congress, Quebec, Canada.

Section VII
Monitoring and Evaluation

20
Monitoring Forest Restoration Projects in the Context of an Adaptive Management Cycle

Sheila O'Connor, Nick Salafsky, and Daniel W. Salzer

Key Points to Retain

Monitoring is a process of periodically collecting and using data to inform management decisions.

Monitoring is best done not as a separate activity at the end of a project, but as an integral part of an adaptive management cycle.

A complete monitoring plan outlines information needs, specifies the least number of indicators to meet these needs, the methods for collecting the indicator data and who is responsible, and when the data are collected.

The amount of resources spent on monitoring should vary inversely to the degree of certainty that project activities will be effective.

There are tools and guidance available for doing monitoring in the context of adaptive management, but not enough has been done specifically for long-term multiparty forest restoration projects.

1. Background and Explanation of the Issue

Monitoring is the process of periodically collecting and using data to inform management decisions. Monitoring is important for projects of all sizes and for all areas of conservation, including forest restoration, to demonstrate impact and to help improve project effectiveness. Monitoring becomes particularly vital when projects become complex and include many different types of goals and a variety of stakeholders, as is often the case with forest restoration projects.[185]

Although there are many different approaches for monitoring conservation projects, over the last decade there has been an increasing convergence on doing monitoring in the context of an adaptive management approach.[186] The key to this approach is that monitoring cannot be tacked on at the end of a project.[187] Instead, it must be integrated into the overall project cycle[188] (Fig. 20.1).

The first step in any type of restoration project is to carefully define the site and issues, and to identify what elements of biodiversity and other values that you want to focus on. This should be followed by a thorough situation analysis that establishes the causal chains that link your restoration targets (features) to the threats (pressures) and root causes that affect these targets. The third step is to identify where along these causal chains you think you can intervene with your actions (responses) and to develop specific objectives for how you need to change the system to improve the chances of success. Once you have done this basic work, it should now be readily apparent as to what key

[185] Ecological Restoration Institute and USDA-CFRP, 2004.
[186] Stem et al, 2005.
[187] Ralph and Poole, 2002.
[188] CMP, 2004; Salafsky and Margoluis, 1998; TNC, 2000.

WWF Programme Management Cycle

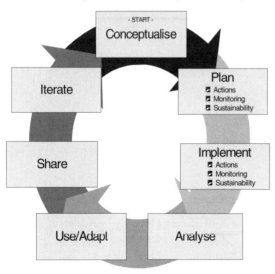

FIGURE 20.1. A project/programme management cycle adapted for WWF use. (Adapted from the Conservation Measures Partnership (CMP), 2004.)

indicators you need to track in order to determine how the targets are doing and whether your restoration actions are having their intended results. A complete monitoring plan clearly outlines your information needs, specifies the *least* number of indicators needed to meet these needs, details methods for collecting the indicator data, and describes who has this responsibility and when these data are collected. In addition, the monitoring plan identifies what analysis is undertaken by whom, and to whom information is circulated and when.[189]

The amount of project resources that you invest in monitoring should generally vary depending on the situation you are facing.[190] If you are in the rare situation where you are highly confident that forest conditions will restore themselves passively, then you would likely spend only a limited amount of resources on monitoring the situation and making sure that no new threats emerge. If the restoration effort warrants the use of straightforward restoration techniques that have a proven record of success, then you would likely invest most of your resources in taking action and only limited amounts on monitoring the results. And if there are restoration needs, but you are unsure how to effectively address them, you may have to experiment with different actions and spend relatively more resources to monitor and analyse the results. In general, the percentage of project resources spent on monitoring should vary inversely with your degree of certainty that your activities will be effective.

2. Examples

We present a case study showing how monitoring and adaptive management were used to improve forest restoration efforts and a fictitious case study illustrating some of the traps that monitoring efforts commonly fall into.

2.1. Case 1: Using Monitoring to Improve the Effectiveness of Restoration Actions in an Adaptive Management Cycle[191]

Problem: Deciding which strategies and activities to undertake in a major restoration effort of Longleaf Pine Ecosystems in the southeastern U.S., and how to monitor the effectiveness of these actions so that effective adaptive management can take place.

Solution: The goal of the project was to identify which management techniques most effectively reduced hardwood density and moved the ecosystem toward predetermined values found in natural high-quality sand hills. The project established a reference condition (or a set of targets related to the biodiversity values—these included composition, structure, and function). They also determined a set of metrics that would possibly be useful as indicators of both management success (effective actions) and the state of the sand hill ecosystem. To help determine the strategic management actions, a conceptual model was developed that looked at both the degradation

---

[189] Earl et al, 2001; Hartanto et al, 2002; Margoluis and Salafsky, 1998.
[190] Earl et al, 2001; Hartanto et al, 2002; Margoluis and Salafsky, 1998.
[191] Provencher et al, 2001.

of the sand hill ecosystem as well as its restoration. Through the experimental implementation of actions, they monitored the impact of the actions themselves (did it meet the assumptions made in the conceptual model?) as well as looked at the overarching improvement to the values defined for the ecosystem. This allowed for a complete and iterative process to achieve the objective of the project as well as make progress toward the long-term goal, which was restoring a functional diverse sand hill system and restoring a habitat for the endangered red-cockaded woodpecker and other long leaf pine–associated species of special concern.

2.2. Case 2: Common Mistakes in Monitoring[192]

Problem: Deciding what to monitor as part of the implementation of a large forest restoration project.

Solution: For the first 2 years of the project, the team does no monitoring whatsoever; it states that it is so busy taking important restoration actions that it has no money or staff resources to devote to monitoring.

The project team members first begin to consider monitoring at the start of the third year because they realise that they need to report on their results to their financial donors. The project managers convene a meeting in which they consider the indicators that they will assess. One biologist on the team, who studied deer for her graduate dissertation, recommends doing an intensive and expensive long-term study of the forest deer population. Another researcher discusses the need to start setting up forest plots and belt transects in various types of the forest to assess plant species' abundance. A third team member goes on the Internet and pulls down a long list of indicators collected by other forest projects including identifying animal and plant species, surveying bird populations, tagging trees, counting hunting parties, sampling water quality, and tracking resource extraction permit applications, and recommends that the project team members consider which of these they should use. Overwhelmed and frustrated, the project manager is about to give up on monitoring altogether.

Finally, the team decides to put its monitoring work in the context of an adaptive management approach. The team takes the time to develop a conceptual model of its situation and realises that the major assumption behind its work is that working with local communities to reduce hunting pressure on key seed dispersers will lead to enhanced forest regeneration. To this end, the team members develop a series of simple indicators to assess whether the community members are responding to their efforts to reduce hunting and to measure whether seedling regeneration is occurring. When they implement this work, they realise that although they are being successful with stopping the hunting, the seedlings are not coming back as expected, especially in large gaps. This forces the team members to focus in more detail on studying why seedlings are not coming back in the gaps and leads to changing their focus to actively planting seeds in large gap areas.

3. Outline of Tools

Different conservation groups have developed more or less similar project management systems for helping practitioners to design, manage, and monitor their conservation work. An overview of some of these systems can be found in the "Rosetta Stone of Conservation Practice" that has been developed by the Conservation Measures Partnership (CMP).[193] Likewise, the Partnership's "Open Standards for the Practice of Conservation" provides a generic listing of the steps in this process.[194]

One specific system that can be useful to practitioners is the Nature Conservancy's (TNC) "Enhanced 5-S Project Management Process,"[195] which can help identify the integrity of biodiversity targets (critical in forest restoration work), as well as help evaluate and prioritise critical threats and other factors from the

---

[192] Adapted from Salzer and Salafsky, in press.

[193] Conservation Measures Partnership (CMP), 2004a.
[194] CMP, 2004b.
[195] The Nature Conservancy (TNC), 2004.

situation analysis, develop objectives, and identify critical indicators. This system is based on an Excel workbook tool that walks practitioners through the steps in the process. A simpler version of this process can be found in *Measures of Success*,[196] which uses visual conceptual models to help show the causal chains linking key factors in your situation analysis as a basis for setting objectives and selecting indicators.

In addition, many government agencies that work on forest management and restoration also have guidance and tools available to help in the design of monitoring plans and the selection of specific indicators and methods (for example, in the United States there is extensive literature on the subject from the Ecological Restoration Institute or USDA's Collaborative Forest Restoration Programme). One example of this type of effort is offered by the Forest Biodiversity Indicators Project.[197] They have developed an online Forest Biodiversity Indicators Selection Web Tool (www.manometmaine.org/indicators/) that provides for rapid searching and comparison of different forest biodiversity monitoring indicators. Indicator search criteria include spatial scale, forest type, forest organisational level, indicator type, category of information need, regional context, and ecological values measured by the indicator. Indicators are rated based on their practicality, relevance, utility, scientific merit, and ecological breadth.

4. Future Needs

To date, most of the adaptive management based monitoring approaches being developed by conservation organisations have not been rigorously tested with forest restoration projects. In addition, almost all forest restoration work involves multiparties, yet there is still no volume of best practices on how to design, implement, and learn from multiple stakeholder monitoring work. Some early examples are cited elsewhere.[198]

Ideally, forest restoration practitioners could come together and begin to agree on a common way of designing, managing, and monitoring such that it is inclusive yet functional. In particular, it would be useful to develop common assumptions, indicators, and methods as well as metrics of long-term success.

References

Conservation Measures Partnership (CMP). 2004a. Rosetta Stone of Conservation Practice. www.conservationmeasures.org.

Conservation Measures Partnership (CMP). 2004b. Open Standards for the Practice of Conservation. www.conservationmeasures.org.

Earl, S., Carden, F., and Smutylo, T. 2001. Outcome Mapping. IDRC, Ottawa, Canada.

Ecological Restoration Institute and the U.S.D.A. Collaborative Forest Restoration Programme. 2004. Handbook FIVE. Monitoring Social and Economic Effects of Forest Restoration. USDA, Washington, DC, and Ecological Restoration Institute, Flagstaff, Arizona.

Hagan, J.M., and Whitman, A.A. 2004. A primer on selecting biodiversity indicators for forest sustainability: simplifying complexity. Forest Conservation Programme of Manomet Center for Conservation Science. FMSN-2004-1. www.manometmaine.org/indicators/.

Hartanto, H., Lorenzo, M.C.B., and Frio, A.L. 2002. Collective action and learning in developing a local monitoring system. International Forestry Review 4(3): 184–195.

Margolius, R., and Salafsky, N. 1998. Measures of Success: Designing, Managing and Monitoring Conservation and Development Projects. Island Press, Washington, DC.

Provencher, L., Litt, A.R., Galley, K.E.M., et al. 2001. Restoration of fire-suppressed long leaf pine sandhills at Eglin Air Force Base, Florida. Final report to the Natural Resources Management Division, Eglin Air Force Base, Niceville, Florida. The Science Division, The Nature Conservancy, Gainesville, Florida.

Ralph, S.C., and Poole, G.C. 2002. Putting monitoring first: designing accountable ecosystem restoration and management plans. In: Montgomery, D.R., Bolton, S., Booth, D.B., and Wall, L. eds. Restoration of Puget Sound Rivers. UW Press, Seattle, WA, pp. 222–242.

Salzer, D., and Salafsky, N. (In press). Allocating resources between taking action, assessing status,

---

[196] Margoluis and Salafsky, 1998.
[197] Hagan and Whitman, 2004.
[198] Ecological Restoration Institute/USDA CRFP, 2004.

and measuring effectiveness of conservation actions. Natural Areas Journal.

Stem, C., Margoluis, R., Salafsky, N., and Brown, M. 2005. Monitoring and evaluation in conservation: A review of trends and approaches. Conservation Biology, 19(2): 1–15.

The Nature Conservancy (TNC). 2004. The Enhanced 5-S Project Management Process. Links to guidance and the Excel Workbook are available at http://www.conserveonline.org/2004/03/a/Enhanced_5S_Resources.

Additional Reading

Brown, R.J., Agee, J.K., and Franklin, J. 2004. Forest restoration and fire: principles in the context of place. Conservation Biology 18(4): 903–912.

Carey, A.B., Thysell, D.R., and Brodie, A.W. 1999. The forest ecosystem study: background, rationale, implementation, baseline conditions and silvicultural assessment. USDA General Technical Report. PNW-6TR-451.

Groves, C. 2003. Drafting a Conservation Blueprint. Island Press, Washington, DC.

Johnson, K.N., Holthausen, R., Shannon, M.A., and Sedell, J., 1999. Case study. In: Johnson, K.N., Swanson, F., Herring, M., and Greene, S., eds. Bioregional Assessments. Island Press, Covelo, California. pp. 87–116.

Kaufmann, J.B., Beschta, R.L., Otting, N., and Lytjen, D. 1997. An ecological perspective of riparian and stream restoration in the western United States. Fisheries 22(5): 12–24.

Lamb. D., Parotta, J., Keenan, R., and Tucker, N. 1997. Rejoining habitat remnants: restoring degraded rainforest lands. In: Laurance, W.F., and Bierregaard, R.O., Jr., eds. Tropical Forest Remnants. pp. 366–385.

Simberloff, D.J., 1999. Regional and continental restoration. In: Soulé, M.E., and Terborgh, J., eds. Continental Conservation. Island Press, Washington, DC, pp. 65–98.

21
Monitoring and Evaluating Forest Restoration Success

Daniel Vallauri, James Aronson, Nigel Dudley, and Ramon Vallejo

Key Points to Retain

An effective monitoring and evaluation system is recognised as an essential part of a successful restoration project, allowing measurement of progress and more importantly helping to identify corrective actions and modifications that will inevitably be needed in such a long-term process.

We propose that in addition to measuring obvious indicators such as area of forest, such monitoring and evaluation systems will usually need to cover issues relating to naturalness of the forest being created at a landscape scale (not necessarily at an individual site), environmental benefits, and livelihood issues.

Some useful indicators are starting to emerge, although much work is still needed on monitoring and evaluation in broad-scale restoration.

1. Background and Explanation of the Issue

1.1. Why Evaluate and Monitor?

Worldwide, monitoring and evaluation have become in the past decade a major issue[199] with

[199] Sheil et al, 2004.

strong repercussions in national forest policies both for conservation (e.g., efficiency of protected areas, status of endangered species) and management (e.g., sustainability standards, impact assessment, ecocertification, and market driven demand). At various scales (from local to international), issues like the design of the best framework for evaluation and monitoring, the choice of an efficient—but not too expensive—set of criteria and indicators, has led to intense debates between major stakeholders in forest management, including nongovernmental organisations (NGOs).

Forest restoration, as defined in this book, is a difficult, energy-consuming, and expensive undertaking. It is almost always a long-term, complex, and multidisciplinary process. On the one hand, forest restoration requires recreating within a few years (usually less than 10 to 15 years) an embryo ecosystem that will only be fully developed after several decades. On the other hand, forest restoration requires inputs and expertise from fields like ecology, economics, public policy, and social sciences, further complicating monitoring and assessment.

For a long time, some forest restoration issues have been the subject of considerable raised tensions and interest, especially, for instance, when comparing the economical benefits of some large afforestation programmes, with their ecological and social disadvantages. How can we be sure that the choices made when starting restoration projects will succeed in reaching the defined goals in the long run? Forest restoration successes are seldom complete or easy to evaluate, and the type of global indicators used by foresters (such as planted trees' height or diameter growth, or plantation cover) give very little information to help assessment in the modern sense of restoration in large-scale conservation.

Thus, monitoring and periodic evaluation of advances in the restoration process is not an optional extra, but a critical and essential part of restoration, that restorationists need to consider mainly in order to do the following:

- Confirm the hypotheses used to develop the restoration programme and ensure that defined goals are reached and the time frame respected. For example, from an ecological perspective, it is important to restore damaged components of forest ecosystems and reintegrate them within the landscape.
- Proceed to fine-tuning management actions that correct problems encountered during restoration (e.g., lower or higher survival of seedlings than expected) or incorrect choices.
- Adapt restoration actions to changes along a restoration trajectory, which will inevitably last several decades, especially with respect to aspects that go far beyond what those initiating the project could forecast (e.g., social issues such as demand for land, awareness of environmental issues; economic issues such as wood prices or demand for nontimber forest products (NTFPs); and ecological issues such as climate change).
- Prove to stakeholders that the investments (not only financial) in the restoration programme are worthwhile.

1.2. What to Monitor and Evaluate?

First of all, the scope of restoration evaluation should fit the goals of the programme or help to redirect them. Nowadays, for forest landscape restoration as defined in this book, the framework for monitoring restoration success should analyse the following issues[200]:

- *Naturalness/ecological integrity*: Under forest landscape restoration, some sites may—if appropriate and in a first stage—be dedicated to highly unnatural tree cover if these fulfil legitimate social and economic needs. However, restoration should have a net increase in naturalness and integrity (biodiversity and ecosystem functioning) within the landscape.
- *Environmental benefits*: Forest management that results in environmental damage—such as soil erosion, fertiliser run-off, pesticide spray drift, or downstream hydrological effects—is incompatible with the wider aims of forest landscape restoration.
- *Livelihoods and well-being*: Forest landscape restoration may not improve social well-being at every site, but should improve it on a landscape scale. The involvement of key stakeholders in decision-making processes should help to ensure that issues relating to human well-being are fully addressed.

Not all projects will have such a broad range of objectives: the framework outlined above is one for restoration projects that seek to balance social and environmental benefits. We believe that this should become the norm.

1.3. How to Evaluate? The Difficult Selection of Criteria and Indicators

A set of pertinent indicators should be agreed upon and tested to reflect the restoration advances for each issue. They should reveal current conditions, and reflect on what has been done in the past by foresters and other forest managers. They should capture information on ecosystem health (i.e., relative absence of disease or pests of epidemic proportions) as well as diversity and productivity at plot and landscape scales. They should also reveal to what extent the explicitly restoration-oriented project has improved the delivery of ecosystem services.

To be effective, each indicator should be SM(a)RRT. That is:

- *S*imple (e.g., vegetation cover [percent], number of tree species present)

[200] WWF, 2003.

- *M*easurable (e.g., percent of "badlands" in a given landscape or watershed, biodiversity indices, and indices of productivity for timber and nontimber products, and money flow for restoration and monitoring)
- *R*eliable (e.g., ecological function demonstrated, indicators of structure and composition)
- *R*elevant: It should be linked, if possible, to critical stage(s) of ecosystem change in response to restoration or other management (the notion of ecological thresholds; e.g., criteria expressing or reflecting biodiversity, flows and functions, structure, and contingency)
- *T*imely: Indicators should be chosen to take into account the contingency factors imposed by past uses and degradation, and the restoration process. The framework for monitoring should be ideally developed starting with an initial evaluation before the beginning of the project and thereafter be reappraised regularly. The periodicity of the evaluation needs to be in accordance with the planned process of restoration, taking into account goals, phases, and stages.

Ideally, indicators should also be sensitive to small changes in a system's trajectory, as expressed in structure, composition, and functioning, and broadly able to be generalised to other systems and situations across a range of ecological and socioeconomic conditions.[201]

1.4. Setting a Framework for Monitoring and Evaluation

A large number of descriptors and indicators are possible, and many have been described in the technical literature. How to choose among them? In line with the above-mentioned criteria, and in light of the specific objectives and budgetary constraints (data collecting is costly), it should be possible to collectively set priorities.

It should be noted that in attempting the diagnosis, evaluation, and monitoring of something as complex as a forest ecosystem, landscape, or, to use a newly emerging term, socioecosystem, a degree of subjectivity can never be excluded. To increase objectivity and fairness, two strategies pertain:

- A complementary portfolio of several attributes should be selected, covering at least two different hierarchical levels (Table 21.1). In a forest landscape restoration initiative, the evaluation at landscape level is compulsory. It is both the most critical and the most difficult to evaluate of the four included in Table 21.1.
- All such evaluations ideally should be considered as relative. Thus, the exercise can benefit greatly if comparisons are carried out between comparable sites within a landscape, or among landscapes.

2. Examples

2.1. Evaluating Ecological Components of Badlands Restoration in Southwestern Alps (Saignon, France), 130 Years After Planting

In the Saignon case study,[202] a pioneer stage dominated by exotics (Austrian black pine) planted in 1870 was evaluated only from the perspective of erosion and forest production. Fine-tuning and corrective actions were limited until the site faced problems 110 years after planting: mainly lack of regeneration and specific infestation of the stands by mistletoe (*Viscum album* L.). Regeneration potential and sanitary conditions and opportunities for the dissemination of native broad-leaved species should have been monitored earlier to avoid problems and to speed up the ecological restoration process—an error not to be repeated! In the 1990s a full set of indicators was identified and evaluated, aiming to highlight the functions that have recovered and to identify the main constraints and trade-offs currently affecting ongoing restoration of native broad-leaved forest. Indicators captured information

---

[201] Aronson and Le Floc'h, 1996; Aronson et al, 1993a,b.

[202] Vallauri et al, 2002.

TABLE 21.1. Partial list of vital attributes, classified by hierarchical organisation level and according to relation to the diversity, flows and functioning, structure, and contingencies of the ecological system.

Hierarchical level	System components			
	Diversity	Flows and functions	Structural factors	Contingency factors
Population	Genotypic and phenotypic diversity	**Gene flow:** pollination, seed production **Matter and energy:** food and energy available **Functions:** intraspecific interaction	Age structure, sexual ratio Height, productivity	**Human impact:** present and past uses **Environment:** chorology, autecology, distance to seed sources
Community	Diversity of species and functional groups among plants, animals, and microorganisms Keystone species	**Gene flow:** hybridation **Matter and energy:** water efficiency, cations exchange capacity, cycling indices **Functions:** productivity, interactions among populations	Tree species richness, life form spectrum Total vegetation cover, vertical heterogeneity Age, above-ground and below biomass, productivity	**Human impact:** present and past uses **Environment:** ecological niche
Ecosystem	Diversity of species, habitat, and functional groups Keystone communities	**Gene flow:** vector of seed dissemination and pollination, seed stocking, predation **Matter and energy:** soil cycles indices **Functions:** regeneration, productivity, soil biological activity, seed distribution, host population control	Total land cover, soil surface conditions Microbial biomass Number of dead trees	**Human impact:** present and past uses **Environment:** type of sites
Landscape	Ecodiversity, diversity of functional groups Keystone ecosystems	**Gene flow:** patterns of dissemination **Matter and energy:** cycling indices, fluxes among ecosystems **Functions:** disturbance regime, connectivity	Land forms and units, ecotones, corridors Organisms regularly crossing ecotones	**Human impact:** present and past land-use **Environment:** ecosystem zonation

For further discussion see Aronson and Le Floc'h (1996).
Note: This list of attributes, which could be analysed to evaluate the restoration success, must be complemented by socioeconomic attributes indicating the socioeconomic success of the restoration programmes.

on a wide range of issues like diversity (of trees and birds at community level), structure (of the soil, of the Austrian pine population), functions (dissemination of tree seeds at the landscape level, soil biological activity) and contingency factors (land use at site and landscape levels).

2.2. Vietnam: A Participatory Monitoring System Covering Biological and Socioeconomic Elements of Restoration

A monitoring and evaluation system for the Central Truong Son in Vietnam has been developed by the Forest Protection Department and WWF. It aims to measure environmental and social trends, communicate achievements, and identify threats and opportunities. Over 60 meetings took place with stakeholders at the national, provincial, district, and commune level to identify 20 core indicators to measure progress on four fronts: forest condition and biodiversity, forest ecosystem services, livelihoods, and capacity for good natural resource management. Many of the indicators come from existing government statistics, sometimes with extra analysis, and some additional indicators will be monitored by other stakeholders. Indicators include natural forest; private and public plantations; legal and illegal timber production; non-timber forest products; measures of sustainable forest management; proportion of reforestation budget for natural regeneration; number of restoration projects; areas needing restoration; forest fires; statistics relating to the wildlife trade and protected areas; catchment protection and irrigation; social indicators including, amongst others, life expectancy, health centres, and education; government training; ratio of arrests for illegal hunting and wildlife trade to successful prosecutions; and specific targets of the initiative. It is notable that only a proportion of indicators relate directly to biodiversity restoration; many are there to give context and to measure other aspects of the broader project, which aims to restore a range of forest functions for people as well as biodiversity (see case study "Monitoring Forest Landscape Restoration in Vietnam').

2.3. A Framework and Database to Evaluate Restoration Programmes in the Mediterranean Region

Forest restoration experience in the Mediterranean region is long-standing, both in the north and in the south. During the last two centuries, a large number of restoration initiatives have been implemented at the site or landscape level, although several distinct phases can be identified with very different approaches, aims, and techniques. A first phase started in the mid-19th century, considering restoration of specific forest functions (like erosion control for example) by slope engineering and planting and seeding of trees, grasses, and shrubs. A second phase, since the 1950s, has been considering afforestation for wood production in the context of reducing fire damage. The latest phase is currently considering ecological restoration in the modern sense, both at the site level and at wider scales. To take advantage and learn from this long experience, a knowledge project (funded by the EU-Directorate General V) was set up and conducted by the CEAM (Centro de Estudios Ambientales Mediterráneos) Foundation (Valencia, Spain) and partners from five Mediterranean countries (Spain, Greece, Italy, Portugal, and France). Named REACTION, (Restoration Actions to Combat Desertification in the Northern Mediterranean), this programme aims at establishing a database of land restoration in the northern Mediterranean by collecting well-documented restoration projects; selecting and applying the most appropriate methodology to evaluate the results of restoration projects; facilitating access to high-quality information for forest managers, policy makers, and other stakeholders; and providing restoration guidelines in light of a critical analysis of contrasted past and innovative techniques. Although it is still underway at the time of writing, this programme already provides online access to a wide range of evaluated restoration programmes in various ecological, historical, and socioeconomical contexts (http://www.ceam.es/reaction).

3. Outline of Tools

Monitoring and evaluation of broad-scale restoration is still in the early stages of development, but some tools are already available for use:

- Ecological attributes: A list of vital attributes at various hierarchical levels (population, ecosystem, and landscape attributes for biodiversity, naturalness, functions, etc.) has been provided and tested by several authors. Table 21.1 presents an attempt at a formulation for monitoring.

- Restoration plan, including monitoring and evaluation definition: Unlike forest management plans, relatively few restoration plans have been fully conceptualised and written in a form that allows comparison. Furthermore, monitoring and evaluation is very often absent at the beginning of the programme. A list of indicators and monitoring protocols such as the periodicity of monitoring (which may be variable along the restoration trajectory) should be defined before inclusion in the restoration plan.
- Restoration databases (learning from past projects): A lot could be learned from past restoration successes and failures. The analysis of databases of long-term restoration projects is very useful, like the world database launched by UNEP-WCMC (http://www.unep-wcmc.org/forest/restoration/database.htm) or the database of evaluated restoration programmes in the Mediterranean region (http://www.ceam.es/reaction).
- Photographs, mapping, experimental design and statistics,[203] and field notes are important tools for understanding the restoration process.
- Criteria and indicators: Although poorly developed for restoration, there is already considerable experience in the development and use of criteria and indicators for sustainable forest management, and some of these could easily be adapted for restoration projects, particularly when they are capable of measuring trends in forest quality over time.

4. Future Needs

The needs for further development are important here. They include the following:

- Improvement in methodologies for monitoring and evaluating human well-being in the context of restoration: Although lists of attributes, indicators, and methodologies exist in the literature, very few have been adapted to forest restoration. Adapting and field testing them will be necessary in the coming years.
- A unified procedure for monitoring restoration programmes: Attempts to develop a common form and approach to monitoring and evaluating large-scale restoration efforts, such as the REACTION programme described above, are essential, although they pose considerable challenges. Development of these programmes are needed in other geographical regions, coupled with field tests and modifications.
- Economic tools to secure funds for assistance in long-term monitoring and fine-tuning: Sustainable financing remains a key problem to restore forest ecosystems in the longer term. Designating a specific part of a state's forest service to be responsible for forest restoration, and subsequently integrating restoration into normal management procedures (through the management plan) could be part of the solution.
- Finally, field testing and learning from years of experience are still essential to build up a database of knowledge.

References

Aronson, J., Floret, C., Le Floc'h, E., Ovalle, C., and Pontanier, R. 1993a. Restoration and rehabilitation of degraded ecosystems in arid and semi-arid lands. I. A view from the south. Restoration Ecology 1:8–17.

Aronson, J., Floret, C., Le Floc'h, E., Ovalle, C., and Pontanier, R. 1993b. Restoration and rehabilitation of degraded ecosystems in arid and semi-arid lands. II. Case studies in southern Tunisia, central Chile and northern Cameroon. Restoration Ecology 3:168–187.

Aronson, J., and Le Floc'h, E. 1996. Vital landscape attributes: missing tools for restoration ecology. Restoration Ecology 4:377–387.

Michener, W.K. 1997. Quantitatively evaluating restoration experiments: research design, statistical analysis and data management considerations. Restoration Ecology 5:324–337.

Sheil, D., Nasi, R., and Johnson, B. 2004. Ecological criteria and indicators for tropical forest landscapes: challenges in search of progress. Ecology and Society 9(1):7 (online). URL:http//www.ecologyandsociety.org/vol9/Iss1/art7.

---

[203] Michener, 1997.

Vallauri, D., Aronson, J., and Barbéro, M. 2002. An analysis of forest restoration 120 years after reforestation of badlands in the southwestern Alps. Restoration Ecology 10:16–26.

WWF. 2003. Indicators for measuring progress towards forest landscape restoration: a draft framework for WWF's Forests for Life Programme. Unpublished report, Gland, Switzerland.

Case Study: Monitoring Forest Landscape Restoration in Vietnam

Nigel Dudley and Nguyen Thi Dao

The challenge: the government of Vietnam is committed to forest restoration and protection and has major reforestation grants available. But although these can in theory support both natural regeneration and plantations, virtually all funds have been used for exotic plantations, particularly of *Acacia mangium*. The structure of the Five Million Hectare Reforestation Programme hampers flexibility, and although large plantations have been established, it seems likely that in several provinces a lot of money has been wasted. In some areas planting is rumoured to cover the same land repeatedly, with seedlings quickly being cut and sold as firewood and the land used for swidden agriculture before being planted again. Because the job security of many Forest Protection Department officials is tied to the programme, they are under pressure to maintain the status quo even when this makes little environmental or economic sense. Restoration is needed both in terms of tree cover and in particular forest quality, especially in protected area buffer zones and along the route of the Ho Chi Minh highway. Successful restoration will depend on the support of local communities and the political will to take into account the importance of indigenous plant species, yet there is little experience of stakeholder involvement or participatory approaches in Vietnam.

The Opportunity

The multidonor Forest Sector Support Programme is funding forest management developments in Vietnam and provides an opportunity to look at the programme afresh, to find ways of realigning it to maximise environmental and social gains. The government has been working with various stakeholders, with facilitation from WWF, in developing a conservation strategy for the Central Truong Son (Annamites) Landscape across seven provinces in the middle of the country, which aims to use a mixture of protection, good forest management, and restoration to create a landscape that will support both biodiversity and local livelihoods.[204] There have already been some good, local-level forest restoration projects (including some run by the German technical development organisation GTZ and WWF's MOSAIC (Management of Strategic Areas for Integrated Conservation) project), which provide lessons that can be applied more widely.[205]

Interventions

A monitoring and evaluation system was developed to measure progress on forest landscape restoration in the Central Truong Son Landscape Biodiversity Conservation Initiative's Action Plan by WWF working in cooperation with the Government of Vietnam's Forest Protection Department.[206] Over 60 stakeholder meetings took place at the national, provincial, district, and commune level to identify around 30 core indicators.

---

[204] Baltzer et al, 2001.
[205] Hardcastle et al, 2004.
[206] Dudley et al, 2003.

Indicators measure progress on four fronts: forest condition and biodiversity, forest ecosystem services, livelihoods, and capacity for good natural resource management. Many indicators come from existing government statistics, sometimes with extra analysis, and some additional indicators will be monitored by WWF and other stakeholders, augmented by information from research reports and surveys so that as complete a picture as possible is developed. Simple benchmarks have also been agreed upon for the different indicators, for instance "an increasing area of natural forests" and "life expectancy reaching a regional average," which help to set measurable targets for the programme. The indicators include measuring the impact and success of restoration, including the proportion of the Five Million Hectare Programme budget used for natural regeneration. By talking to different interest groups, and getting agreement from the government, the monitoring system also serves as a way of negotiating policy; for instance, by agreeing to measure trends in use of funds for natural regeneration as opposed to just large-scale plantation, stakeholders including the government are recognising this as a target, making it easier to plan restoration interventions. Since the initial work, the importance of monitoring and evaluation is increasingly being recognised. The Forest Sector Support Programme is developing a monitoring and evaluation system based on the one in the Central Truong Son, and other long-term restoration projects are also recognising the need for good monitoring and evaluation.

Lessons Learned

A well-designed monitoring and evaluation system has been identified as a critical step in a successful integrated conservation and development project.[207] The experience in Vietnam bears this out but also shows that a shared monitoring system applied at a landscape scale, which integrates different projects, actors, and stakeholders towards a larger goal, can play a key role in scaling restoration and conservation efforts up to a landscape.

References

Baltzer, M., Dao, N.T., and Shore, R. 2001. Towards a Vision for Biodiversity Conservation in the Forests of the Lower Mekong Ecoregion Complex. WWF Indochina Programme, Hanoi.

Dudley, N., Cu, N., and Manh, V.T. 2003. A Monitoring and Evaluation System for Forest Landscape Restoration in the Central Truong Son Landscape. WWF Indochina Programme and Government of Vietnam, Hanoi.

Hardcastle, J., Rambaldi, G., Long, B., Lanh, L.V., and Son, D.Q. 2004. The use of participatory three-dimensional modelling in community-based planning in Quang Nam province, Vietnam. PLA Notes 49:70–76.

McShane, T.O., and Wells, M.P. 2004. Getting Biodiversity Projects to Work: Towards More Effective Conservation and Development. Columbia University Press, New York.

---

[207] McShane and Well, 2004.

Section VIII
Financing and Promoting Forest Landscape Restoration

22
Opportunities for Long-Term Financing of Forest Restoration in Landscapes

Kirsten Schuyt

Key Points to Retain

The key to tapping into private and public sector funding opportunities for forest landscape restoration lies in making it financially and economically attractive. This requires estimating and recognising the economic values of forests and the role restoration can play in increasing this economic value. It also requires proper pricing of forest goods and services and setting up mechanisms where money is transferred to pay these prices, such as payments for environmental services (PES).

In light of economic liberalisation, private sector funding, including PES, provides a lucrative opportunity for financing restoration activities.

In terms of public funding, it will be increasingly important to mainstream forest landscape restoration in other programmes, including poverty reduction programmes.

1. Background and Explanation of the Issue

The economic, social, and biodiversity values of forests are increasingly being recognised, and many countries have understood the need to better manage their forest resources. At the same time, in 1997 the Intergovernmental Panel on Forests (IPF) found that domestic financial resources were insufficient to achieve sustainable management, development, or conservation of forests. With the threat of worsening forest depletion in many parts of the world leading to further degradation of forest goods and services, it is recognised that there is a critical need to explore new and innovative ways of financing improved forest management and conservation, including the restoration of forest resources.

Forest landscape restoration is a long-term process and will generally require sustained sources of funding. All too often, overreliance on grants means that funds can only be obtained for short-term projects, and a long term-effort such as the restoration of forests suffers. Grants, however, are not the only source of funding, and a number of options for long-term financing of forest landscape restoration are highlighted below (see Outline of Tools).

Traditional financing sources for forestry in developing countries have been domestic public and private, foreign public and private, and international organisations, including NGOs. Depending on the objective of the forestry activities (environmental conservation, subsistence needs for local people, commercial purposes), different financing sources have been sought. However, global financing trends in general are changing, and a wave of economic liberalisation is providing impetus for increased private sector participation.[208] These trends allow for new financing opportunities from the private sector

[208] Joshi, 1998, p. 6.

for restoration activities. In light of declining external public funding and weak prospects for new and additional public funding of overseas development assistance (ODA) in forestry, private capital flows represent potential opportunities for restoration initiatives.

The key to financing opportunities from both private and public funding sources for landscape-scale forest restoration lies in recognising its full economic and financial value. This requires estimating and recognising the economic values of forests and therefore recognising the benefits provided by restoring these forest values. The restoration or loss of these values can then be more realistically weighted against other possible uses of the land. In a landscape context, it then becomes possible to better select areas within the landscape for different uses, allowing a potentially more complete range of values and benefits to be offered. This also requires proper pricing of forest goods and services and setting up mechanisms where money is transferred to pay these prices. One way to do this is by selling environmental services of forests, such as carbon sequestration, watershed protection, and biodiversity, to finance restoration—a mechanism called payments for environmental services (PES) (see "Payment for Environmental Services and Restoration"). The PES mechanisms ensure that those who supply environmental services are paid by those who use these services. These range from public payments to self-organised private deals. For example, private companies such as downstream bottling companies pay upstream communities for sustainably managing the forests in the watershed that provide services such as watershed protection on which the bottling companies depend. At the basis of sustainable watershed management should be restoration, where the key is convincing investors that such activities will ensure sustainable environmental services as sustainable "production inputs," thereby making landscape scale restoration financially and economically attractive. Another example of PES is paying for carbon sequestration; energy companies could invest money in restoration projects to increase the carbon sequestration service of forests for the purpose of meeting their carbon offsets, as is allowed under the Kyoto protocol.

2. Examples

Notwithstanding the need for continued public investment in restoration, the two examples below illustrate private sector involvement in forest restoration activities. Both examples illustrate how restoration can be made economically interesting to attract new investors—the private sector—to mobilise innovative sources of financing.

2.1. Private For-Profit Sources: Outgrower Schemes, South Africa[209]

In an outgrower scheme, a company provides marketing and production services to farmers to grow trees on their land under specific agreements. In 2002, 12,000 smallholder tree growers were involved in these schemes in South Africa on about 27,000 hectares of land. Although the outgrower timber provides only a small percentage of a larger company's pulp mill output and is more expensive per tonne than wood from other sources, it provides important fibre that would otherwise be unavailable due to land tenure constraints. It also provides companies with a better image at a time when the distribution of land rights in South Africa is being discussed. Community motivations are mostly for cash income at harvest, while trees are also seen as a form of savings. The two schemes with the largest membership are Sappi and Mondi, where smallholders grow eucalyptus trees with seedlings, credit, fertiliser, and extension advice from the companies. The companies in return expect to buy all the harvest at the end of the growing cycle.

2.2. Payments for Forest Services: Pimampiro Payment for Watershed Services Scheme, Ecuador[210]

The Paluarco river is used for irrigation and drinking but is of poor quality due to agri-

---

[209] Taken from Gutman, 2003.
[210] Taken from Gutman, 2003.

cultural discharge upstream. Under a pioneering project for Ecuador, landowners in the Paluarco river sub-watershed are being paid to manage the forest in the watershed in order to protect water sources. In 2001 the municipality approved an ordinance that established the Water Regulation for the Payment of Environmental Services from Forest and Paramo Conservation. A fund was created to channel payments from beneficiaries (mostly domestic water users) to those providing good quality of water through maintenance of forest cover upstream.

3. Outline of Tools[211]

As outlined in section 1, new opportunities for financing large-scale restoration are arising from the private sector. Opportunities, however, still exist in public funding sources. This section discusses how specific financing sources, including private and public sources as well as international organisations, can be mobilised for forest landscape restoration activities.

3.1. Financing from Domestic Public Sources

General strategies to increase public sources for large-scale restoration involve activities like improving expenditure policies on forestry, reforming macroeconomic policies (including taxes and subsidies), and putting in place new incentives, subsidies, and technical and institutional changes to support restoration that provides wider benefits (also see "Perverse Policy Incentives"). It is, however, also important to improve the administrative capacity of forestry agencies themselves to increase their efficiency to collect revenue and to use the resources efficiently for restoration. Other ways to increase forest revenues from public funding are to ensure the proper pricing of forest goods and services (through charges, policies that demand full-cost pricing, permits, licensing, etc.) or setting up special forest trust funds with earmarked taxes to finance specific restoration activities. It is also possible to use tax measures that tax downstream beneficiaries to fund restoration upstream.

3.2. Multilateral and Bilateral Donors

Given the declining trend in ODA, efforts must be directed at maintaining current funds from multi- and bilateral aid. In general, however, environment is no longer a top priority of development and cooperation agencies, and it has now been mainstreamed in all development activities under the new sector approach embraced by many donor agencies. Therefore, successful proposals for forest landscape restoration from multilateral and bilateral donors increasingly need to explain how forest landscape restoration activities will address poverty alleviation. Furthermore, it is also useful to use ODA to leverage private funding for restoration. The World Bank's Sustainable Forest Market Transformation Initiative (SFMTI) is a good example, which promotes private sector participation in forest management. Another example is USAID's (US Agency for International Development) Biodiversity Conservation Network, which provides seed money to promote the participation of the private sector in biodiversity-based business.

3.3. Private Not-for-Profit Sources

Private not-for-profit sources include financing channelled from local communities, international foundations, and NGOs for forest landscape restoration activities. International NGOs have become important for providing new financing mechanisms, of which environment trust funds or foundations are particularly interesting for providing financing to natural resource management in general. Trust funds are not philanthropic foundations. Rather, they raise money to carry out their own programmes and have specific missions and interests and sometimes geographical focusses. The main purpose of setting up a trust fund has traditionally been to provide long-term stable funding for national parks and other protected

---

[211] Based on Joshi, 1998; Gutman, 2003; and the Conservation Finance Alliance online guide, 2002.

areas or small grants to local NGOs and community groups for projects aimed at conserving biodiversity and using natural resources more sustainably. Such trust funds could be set up to support the restoration of forest values over the long term.

3.4. Private for-Profit Sources

Private for-profit sources range from mobilising households to invest in restoration to investments from large international corporations. Household investments will have an effect only if the projects offer short-term benefits with an acceptable level of risk. These benefits can be an increased income for households or indirect payments in, for example, alternative livelihoods, roads, schools, and so on. On the other hand, a more grant-type of financing from large private companies like dam, oil, plantation, and mining companies can be mobilised to pay for forest restoration as compensation for environmental disruption they may cause. This motivation may also come from business ethics and thus be part of a company's public relations campaign. An example is where environmental NGOs are invited by a plantation company to restore part of their land according to standards compatible with forest landscape restoration. Lastly, engaging conventional capital markets by channelling capital toward forest management and restoration has potential. For example, Xylem Investment Inc. is an international timber investment company based on equity investments in plantation forests in developing countries that attracts U.S. pension funds, insurance companies, and others that prefer safer and steadier-growth investments. This company manages forest assets worth $235 million. Another example is Precious Woods, an international timber company that focusses on sustainably produced timber in Latin America. Funding from these sources could also be mobilised for forest landscape restoration.

3.5. Payments for Forest Goods and Services

Market-based financing has both potentials and limitations but it does provide real opportunities for mobilising funds for forest landscape restoration. A good example of payments for environmental goods is the certification body, the Forest Stewardship Council (FSC), which developed a market for sustainably produced wood and wood products that come with a seal of approval or certificate. In terms of payments for environmental services, a good example is the increase in projects that create payment mechanisms where downstream beneficiaries pay for the sustainable management of forests upstream. Such systems provide significant opportunities for innovative funding for forest landscape restoration.

3.6. International Systems of Payments for the Environmental Commons

There has been some progress at international level to pay for the global commons. The best known is the Global Environmental Facility (GEF), which provides partial grant funding to eligible countries for projects that address threats to the environment in four areas: biodiversity loss, climate change, ozone depletion, and degradation of international waters. Under its biodiversity programme, the GEF can support conservation and sustainable use of significant biodiversity, including forest ecosystems. Funding from GEF for forest landscape restoration could be mobilised under this area.

In a landscape context, it will be possible to initiate a restoration activity with public funding in order to address immediate livelihood needs (e.g., provision of traditional medicines, reduction in people's vulnerability). In the longer term, and still within the context of landscapes and the restoration of many forest benefits, it may become possible to ensure sustained funding by the private sector in order to meet additional benefits (such as certified non-timber forest products, for instance).

4. Future Needs

The key need for further development across all funding opportunities is to become more innovative in finding funding in an increasingly competitive market. Whether this means creating partnerships with organisations that were previously unheard of, making forest landscape restoration financially lucrative for actors with funding to become involved in such projects, mainstreaming restoration into other types of projects such as development projects, or mobilising funding from other nonenvironmental sources toward forest landscape restoration, there is a real need to think "outside the box" and search for innovative funding opportunities. In light of economic liberalisation, private sector funding, including PES, might provide a lucrative opportunity for financing broad-scale restoration. Establishing clearer links with livelihood concerns is also a clear need, whether it be poverty reduction, disease control and prevention, postconflict resolution, etc.

References

Gutman, P., ed. 2003. From Good-Will to Payments for Environmental Services—A Survey of Financing Natural Resource Management in Developing Countries. WWF-MPO, Economic Change, Poverty and Environment Project, DANIDA, Copenhagen, Denmark and WWF, Washington, DC.

Joshi, M. 1998. Innovative Financing for Sustainable Forest Management. UNDP, PROFOR, New York.

Additional Reading

Chandrasekharan, C. 1996. Status of financing for sustainable forestry. Proceedings of the UNDP/Denmark/South Africa Workshop on Financial Mechanisms and Sources of Finance for Sustainable Forestry, Pretoria, South Africa, 4–7 June.

Conservation Finance Alliance. 2002. Mobilizing funding for biodiversity conservation—a user-friendly training guide for understanding, selecting and implementing conservation finance mechanisms. http://guide.conservationfinance.org.

EFTRN News. 2001/2002. Innovative finance mechanisms for conservation and sustainable forest management. European Tropical Forest Research Network, No. 35.

Lapham, N.P., and Livermore, R.J. 2003. Ensuring Conservation's Place on the International Biodiversity Assistance Agenda. Conservation International, Washington, DC.

WWF-MPO. 2000. Wants, Needs and Rights—Economic Instruments and Biodiversity Conservation: A Dialogue. WWF, Washington, DC.

23
Payment for Environmental Services and Restoration

Kirsten Schuyt

> **Key Points to Retain**
>
> Payments for environmental services provide real opportunities for innovative conservation financing.
>
> Payments for environmental services can work effectively in landscape-level restoration projects where large scales are involved as well as many different stakeholders.
>
> Payments for environmental services are still relatively new, and opportunities for regrouping services ("bundling" them) seem to offer an interesting way forward.

1. Background and Explanation of the Issue

Forests provide a wide variety of benefits. They provide goods such as fuel wood, construction materials, and nontimber products, as well as services including watershed protection, carbon sequestration, reduction of sedimentation, water purification, and biodiversity. Despite these benefits, forests are severely threatened in many parts of the world. Deforestation is taking place at alarming rates, accompanied by a loss in forest goods and services.

The causes of deforestation are complex, and include market and institutional failures. Many forest benefits lack well-defined property rights. This is especially the case with forest services. For example, cutting down trees upstream can increase the amount of sedimentation and flooding downstream. Since the costs associated with sedimentation and flooding are not borne by the upstream communities that cut down the trees, these costs will not be incorporated in their decisions. The value of the forest to these upstream communities is perceived to be much less than their full value, and the result is the cutting of more trees than is optimal.[212]

Payments for environmental services (PES) (also known as payments for ecosystem services) are instruments that arose as a response to remedy market failure; PES implies that those who use the ecosystem service pay those who provide the service, and can include a wide range of mechanisms for financing conservation, such as the following[213]:

- Self-organised private deals: direct, closed transactions with little government involvement, involving private entities who are usually offsite beneficiaries of forest services
- Public payments: government payments for the protection of specific ecosystem services through better land and forest management
- Open trading: a government regulation creates demand for a particular environmental service by setting a cap on the damage to an ecosystem service or establishing a floor

---
[212] Pagiola et al. 2002.
[213] Inbar and Scherr, 2004.

Ecolabelling: certifying forest and farm products that were produced in ways consistent with biodiversity conservation

Many examples of PES systems exist, where the most common forest services that have been addressed by PES are carbon sequestration, watershed protection, landscape beauty, and biodiversity conservation. Since payment mechanisms are very different across these four services but also across countries, it is difficult to generalise about how PES works. However, there are certain elements of success.[214] First, as with any market, there needs to be supply and demand.

There needs to be a product: supply. There needs to be a product (the forest service) to sell, such as watershed protection, carbon sequestration, biodiversity conservation, and landscape beauty. Also, many services do not come alone. Is it possible to regroup or "bundle" the services? It is very important to clearly document the relationship between the provision of the service and the economic benefits: for example, what is the relationship between upstream watershed protection and downstream land use?

There need to be buyers: demand. There needs to be a demand for the forest services. Just because a forest provides a service does not mean that there is a market for it. This demand may be local, national, or global. For example, the demand for watershed protection arises mostly from local or national buyers, while the demand for carbon sequestration may come from anywhere in the world. The type of demand determines the type of system to establish—water markets are very site-specific, depending on the institutional context, while carbon markets can actually learn from each other and even compete.

In addition to supply and demand, other elements must be in place to ensure success:

Mechanisms to capture willingness to pay: These mechanisms must capture part or even all of the benefits provided by the forest services and transform them into actual payments to encourage forest conservation or restoration. The key is to establish a mechanism with low transaction costs, where the costs of capturing the benefits (including the opportunity costs—the lost benefits associated with other land uses) are lower than the benefits. For watershed protection, for example, benefits are easiest to capture and at a lower cost when users are already organised (municipal water supply, irrigation systems, etc.) and when some form of payment mechanism is already in place, such as a domestic water fee. Payments for watershed protection can then be added to this fee.

Identification of key actors: A key step is to identify who the key actors are that supply the forest services. Different actors can be involved—NGOs, commercial companies, private landowners, farmers, governments, donors, community groups, and so on. Each of these stakeholders may be able to play a crucial role in the PES system, which must be identified. It is also important to understand their motivations, for example for logging, and what is required for them to conserve or restore.

Developing the institutional structure: It is necessary to develop the market infrastructure: access to information on values and quantity, negotiation, monitoring and enforcement mechanisms, and so forth. A key institution is property rights, which define who owns the carbon sequestered in the forest or the trees that protect the watershed. Without clear ownership or usufruct of the services, they cannot be bought or sold.

A more detailed discussion on these elements can be found elsewhere.[215]

The opportunities from PES for forest landscape restoration are potentially enormous. Because of the dramatic loss in forest cover worldwide, and the consequent loss in forest goods and services, there is great potential to incorporate payments for environmental services into a broad-scale approach to restoration. The sorts of goods and services that restored forests can provide and that can be quantified

---

[214] Pagiola et al, 2002.

[215] Pagiola et al, 2002.

include payments for the carbon sequestered by forests, watershed protection of forests, and biodiversity conservation of forests.

Concerns have been raised as to how PES will affect the environment and the poor. Does it help conservation and do the poor benefit or is it a mere "silver bullet"?

The next section gives three examples of PESs in relation to forest conservation that provide opportunities for forest landscape restoration.

2. Examples

2.1. Payments for Watershed Protection: The Case of Costa Rica[216]

The hydrological impact of widespread deforestation has been a major concern throughout Central America, followed by a strong interest to tackle deforestation. Within this context, Costa Rica pioneered a PES approach in which land users were directly compensated for the environmental services they generated. Costa Rica has had one of the highest rates of deforestation in the world, mostly driven by conversion to agriculture and pasture. As a result of deforestation, water services deteriorated, but responses, mostly regulation, to deal with deforestation had largely failed.

In the beginning of 1997, Costa Rica developed an elaborate system of PES to deal with deforestation called Pago por Servicios Ambientales (PSA). In this system, land users are compensated directly for the environmental services they provide, which enables them to include the services in their decisions. When the PSA was created, however, Costa Rica already had a payments system (essentially through tax incentives) for reforestation and forest management in place. Most importantly, the institutional structure to contract landowners and pay them for specific activities already existed. As part of the PES process, a forestry law was enacted that built on these institutions. The law specifically recognised four environmental services supplied by forests and provided the regulatory basis for the government to contract landowners for the services provided by their lands. It established a financing mechanism for this called FONAFIFO (Fonda Nacional de Financiamiento Forestal). The two key differences between the PSA and past incentives are (1) that financing through the PSA focusses on the services provided by forests rather than on the timber, and (2) that the financing comes from users of those services rather than public funds.

Under the PSA, all participants must have a sustainable forest management plan that is certified by a licensed forester. Once the plans have been approved, land users begin implementing the different activities and receive payments over 5 years. FONAFIFO in cooperation with other institutions contracts the service providers and collects and manages the payments from service beneficiaries. The PSA programme is overseen by a governing board that consists of representatives of the public sector and the private sector. Most of the financing comes from a system that allocates one third of the revenues from a fossil-fuel sales tax to FONAFIFO. Other financial supporters of the PSA programme have been the World Bank and the Global Environment Facility (GEF). The idea is that eventually all beneficiaries of water services (irrigators, domestic users, power plants, and so on) would pay for the services they receive.

2.2. Payments for Carbon Sequestration: The Case of British Columbia

A valuable service provided by forests is that forests sequester carbon. The Kyoto protocol has expanded opportunities for markets for carbon, in which income from traditional forest products can be supplemented with the sale of carbon sequestration services provided by forests. British Columbia in Canada has started developing a market for carbon and this section discusses these developments.[217]

---

[216] Pagiola et al, 2002.

[217] Bull et al, 2002, cited in Pagiola et al, 2002, pp. 201–221.

Creating markets for carbon is a complex process that requires efforts from scientists, forest companies, energy companies, and government. A necessary first step is to understand and quantify forest carbon dynamics and carbon budgets. At the national level in Canada, forest carbon budgets have been measured using remote sensing and a carbon budget model developed by the Canadian forest sector. At the provincial level, carbon budget calculations are also underway to forecast carbon budgets into the future. Other models have been developed to calculate the contribution of British Columbia's forest carbon to the global carbon cycle as well as models to estimate the amount of carbon in carbon pools above ground and in roots, soils, litter, and deadwood.

Another important step is creating the necessary institutional arrangements in government policies and markets. In this respect, several initiatives have been carried out in British Columbia, including the establishment of an emissions' trading platform at the national level called the Greenhouse Emissions Reduction Trading (GERT) pilot. This was launched in 1998 and has allowed Canadian business to gain experience in emissions' trading. Private initiatives to establish an emissions' trading platform have also emerged. It has also been necessary to create a national registry to document sequestration, emission, and buying and selling of carbon, which was established in 1994 (called Voluntary Challenge and Registry, VCR). Other necessary institutions are incentive-based policies in which the Canadian government recognises the role of forests in global warming and recognises the need to better understand the role carbon sinks can play to mitigate global warming. These policies are currently still under review.

The key to a carbon market is buyers and suppliers. In British Columbia, there is considerable caution on behalf of potential buyers, resulting in insufficient incentive on behalf of forest growers to supply forest carbon. Uncertainty over the role of forest carbon in the Kyoto protocol also adds to this. The result is that the market for forest carbon in British Columbia is still in its infancy, despite strong expressions of interest from both buyers and sellers.

2.3. Payments for Biodiversity Conservation: RISEMP in Colombia, Costa Rica and Nicaragua[218]

The Regional Integrated Silvopastoral Ecosystem Management Project (RISEMP) is a GEF-funded project implemented by the World Bank in Colombia, Costa Rica, and Nicaragua. The project pays farmers directly for the provision of biodiversity services. Silvopastoral systems combine trees with pasture. They provide a range of benefits to farmers: (1) additional production from trees; (2) maintaining and/or improving pasture productivity; and (3) contributing to the overall farming system. Furthermore, the trees provide shade that may enhance livestock productivity, especially milk production. In terms of biodiversity, silvopastoral systems support much higher species' diversity than traditional pastures. They also help to connect protected areas. Other benefits include carbon sequestration (additional carbon is sequestered by the trees found in silvopastoral systems) and watershed services.

So why, despite such benefits, do farmers not adopt silvopastoral systems more often? The main reason is the limited profitability for the individual farmer. First, it requires high initial costs and there is a long time lag before the system actually becomes productive. Second, biodiversity carbon and watershed services are externalities from the farmer's perspective; other parties benefit from these services. Therefore, farmers will not take these benefits into account when making decisions. To deal with this issue of externalities, RISEMP was initiated, its goal being to encourage silvopastoral systems in degraded areas (micro-watersheds) in Central and South America. Farmers enter into contracts under which they receive annual payments for the environmental services they generate. Annual payment levels are based on the opportunity cost to farmers of the main alternative land use, and the payment for carbon is set at around the current world price of U.S. $2 per tonne of CO_2 equivalent. It will

[218] Pagiola et al, 2004.

be some time before the effectiveness of this project can be determined, but the intensive monitoring of this project will allow a detailed analysis of its effectiveness.

3. Outline of Tools

As has been illustrated by the three case studies, the creation and development of PES is complex and requires a wide variety of skills and tools. It is impossible to list all these tools, but one that is common to many PESs in one form or another is the economic valuation of the goods and services forests provide. The key is recognising and understanding economic values of forest services in decision-making processes related to forests in addition to their biological and sociocultural values. Economic valuation tools exist that quantify these economic values in monetary units, which allows them to then be recognised and weighed against other values. Examples are the contingent valuation method, which estimates people's willingness to pay for an environmental service or people's willingness to accept compensation if that service is lost. Another tool is the replacement cost method, which uses the costs of replacing an environmental service as an indication of its value. Yet another example is the travel cost method, where the costs people are investing to travel to a forest area can be used as an indication of the value those people attach to the area.[219]

4. Future Needs

Although PES systems are rapidly becoming more common, many are still in their infancy and much remains to be learned. For example, there is a need for a better understanding of what mechanisms need to be in place for PES to work. It is also necessary to better understand the impacts of PES schemes on poor people and how the poor can really benefit from PES. Lastly, it is increasingly being suggested that there is a need to sell "bundles" of environmental services as an incentive for sustainable forest management—jointly selling the forest services of carbon sequestration, watershed protection, biodiversity, and landscape beauty as a package. There is, however, a need to further develop possibilities of linking forest services successfully.

References

Bull, G., Harkin, Z., and Wong, A. 2002. Developing a market for forest carbon in British Columbia. In: Pagiola, S., Bishop, J., Landell-Mills, N., eds. 2002. Selling Forest Environmental Services: Market-Based Mechanisms for Conservation and Development, Sterling: Earthscan Publications, London.

Campbell, B.M., and Luckert, M.K., eds. 2002. Uncovering the Hidden Harvest: Valuation Methods for Woodland and Forest Resources. Sterling: Earthscan, London.

Inbar, M., and Scherr, S. 2004. Getting Started: A Guide to Designing Payments for Ecosystem Services (draft). Forest Trends, Washington, DC.

Pagiola, S., Agostini, P., Gobbi, J., et al. 2004. Paying for Biodiversity Services in Agricultural Landscapes. World Bank, Washington, DC.

Pagiola, S., Bishop, J., and Landell-Mills, N., eds. 2002. Selling Forest Environmental Services: Market-Based Mechanisms for Conservation and Development. Sterling: Earthscan Publications, London.

Additional Reading

Forest Trends. 2004. Learning More About Payments for Environmental Services—Case Studies and Suggested Resources (draft). Forest Trends, Washington, DC.

Landell-Mills, N., and Porras, I. 2002. Silver Bullet or Fools' Gold? A Global Review of Markets for Forest Environmental Services and Their Impact on the Poor. International Institute for Environment and Development, London.

---

[219] See Campbell and Luckert, 2002, for an overview of economic valuation tools for forest resources.

24
Carbon Knowledge Projects and Forest Landscape Restoration

Jessica Orrego

Key Points to Retain

The biggest carbon reductions should be achieved through a reduction in emissions rather than an expansion of "sinks."

The carbon market is still in its infancy.

The potential value of forests as carbon sinks is important. With agreements such as the Kyoto protocol as well as voluntary carbon markets, it is possible to finance "carbon knowledge" projects that test out, monitor, and improve knowledge on forest restoration and carbon.

An approach that integrates, among others, a carbon sink target can improve the current afforestation approach and help to address the traditional social and ecological weaknesses.

1. Background and Explanation of the Issue

There is still limited knowledge concerning the long-term impact of climate change, and the real role that trees can play in absorbing carbon and in the costs and benefits involved in using restoration as a mechanism to offset carbon emissions. For these reasons, carbon knowledge projects are proposed as a way of testing these parameters in the context of landscape-based forest restoration activities.

The atmospheric concentration of carbon dioxide (CO_2) has increased by over one third since the Industrial Revolution. This increase is primarily attributed to fossil fuel combustion and also significantly to land use cover changes (e.g., conversion of forests to agriculture). There is broad consensus among scientists that CO_2 is linked to climate change and global warming. Of course, reducing human dependence on fossil fuels and imposing legally binding targets for reduced CO_2 emissions is essential to curb atmospheric CO_2 concentrations and must be the central focus of any policy programme. However, to stabilise atmospheric CO_2 concentrations, the international community must also slow the destruction of natural ecosystems that are important stocks and sinks of carbon. In addition to slowing the rate of land conversion, increasing land coverage of carbon-absorbing vegetation (or carbon sinks) has been considered a mitigation tool to stabilise the burgeoning concentration of CO_2 in the atmosphere. The concept of carbon sinks is based on the natural ability of trees and other plants to take up CO_2 from the atmosphere and store the carbon in wood, roots, leaves, and the soil. The theory behind land-based carbon trading is that governments or institutions that wish to, or that are required to, reduce their fossil fuel emissions can offset some of these emissions by investing in afforestation and reforestation activities, where trees sequester carbon. Indeed, in some cases private companies are voluntarily electing to offset some of their fossil fuel CO_2 emissions through the purchase of carbon credits from land-based carbon sequestration projects.

The concept of carbon trading, and the subsequent carbon market that has emerged out of it, is rooted in the U.N. Framework Convention on Climate Change, which resulted from the Rio Earth Summit in 1992, and the subsequent 1997 Kyoto protocol. The Kyoto protocol sets forth legally binding reductions in greenhouse gas emissions for governments in developed countries (so-called "Annex I countries") to be accomplished during 5-year commitment periods, with the first commitment period set for 2008–2012. On average, Annex I countries would be subject to a 5 percent reduction below their 1990 emissions' levels.

The Clean Development Mechanism (CDM), article 12 of the Kyoto protocol, provides a flexible mechanism through which Annex I parties can meet their emissions' reduction targets by purchasing carbon that is sequestered through afforestation and reforestation (and energy) activities being implemented in Annex II countries (developing countries). Since its creation, CDM procedures and modalities have evolved significantly in response to strong criticism and debate.

A concern of environmentalists is whether carbon stored in sinks projects will be sequestered permanently. Clearly forests are subject to natural death and also to a variety of disturbances that result in the release of CO_2 back into the atmosphere. This was addressed at the 9th Conference of the Parties (COP 9) of the Kyoto protocol signatories in Milan in December 2003. It was decided that temporary credits must be reissued or recertified every 5 years and then replaced by another credit.

There are also ways to make forestry activities last for the long term by, for example, introducing land-use systems that are beneficial to local communities, incorporating fire management activities into the project, and retaining a risk buffer from all carbon finance to cover the costs of reestablishment in case of losses.

Other issues surrounding the sinks' debate include the risk of leakage, whereby afforestation or reforestation project activities in one area displace forest felling or destruction to another area. Leakage can be avoided if the needs of local communities and local market trends are analysed and incorporated into project design. It is also necessary that carbon sequestered in sinks projects would not have been stored even in the absence of the project, thereby proving their additionality.

Critics also fear that the CDM reduces pressure from governments to take real action toward reducing fossil fuel emissions at their sources. Parties will be able to use the CDM to meet 1 percent of their below-1990 emissions target, which equates to approximately 20 percent of a country's target.

Furthermore, opponents of the CDM are concerned that efforts to sequester carbon will result in large-scale monoculture plantations that have no socioeconomic or ecological benefits.

Some of the key policy issues being discussed today relate to which types of forest and land-use projects should be undertaken under the umbrella of climate change mitigation and to what extent these types of projects should be integrated with mainstream carbon markets. The following examples illustrate two contrasting types of projects that are part of this debate.

2. Examples

2.1. Plantar in Brazil

One example of this type of project that is being promoted as potentially CDM eligible is the Plantar project in Minas Gerais, Brazil. The project consists of 23,100 hectares of eucalyptus plantations that are used to make charcoal for pig iron production. Plantar plans to claim CDM emissions' reductions from both the sequestration by the eucalyptus trees and from the avoided use of coal. This project has attracted numerous criticisms because of its scale and manner of implementation. There have been allegations that the local Geraiszeiro inhabitants were forcibly evicted when the plantations were first established and that run-off from the plantations has polluted local water supplies affecting the livelihoods of local farmers and fisher folk. However, viewed within the context of recent industrial history of Brazil, which has seen many factories move from Minas Gerais to the Amazon region, sourcing energy from trees cut from virgin rainforest, such efforts may not be wholly negative. In a landscape context, the choice of trees and their location would play a significant role as well to not only minimise social and ecological impacts, but also seek to enhance the wider benefits.

2.2. Scolel Té in Mexico

In contrast with the industrial plantation approach of Plantar, the Scolel Té project for rural livelihoods and carbon management aims to demonstrate how carbon finance can allow low-income rural farmers to invest in forest conservation, sustainable land-use systems, and livelihood improvements that would otherwise be inaccessible to them.

Operating since 1996, the project works in over 25 communities, among seven different indigenous Mayan and mestizo groups of Chiapas and Oaxaca, Mexico (Fig. 24.1). The project engages rural farmers in a fully participatory manner. All potential participants attend a training workshop prior to making a decision to enter into the project. Participants then work with technical experts from the project to design land use activities that will suit their own needs and that are ecologically viable. Technical specifications are produced for each land-use system, and these provide information about the area of land, tree species and planting density, the intended management regime, and local ecological conditions. From this information a credible carbon sequestration estimate can be made. Subsequently, an evidence-based monitoring protocol is used to verify carbon stocks using easy-to-measure indicators. Farmers engage in forestry activities, including integrated community restoration of forests, afforestation of degraded and fallow land, and shade coffee. Carbon payments allow participating farmers to invest in these land use systems and also in other livelihood improvements such as livestock, cooking stoves, and fire and erosion prevention.

Since 1997 the project has attracted a variety of carbon buyers, including the Fédération Internationale de l'Automobile (FIA), which committed to an ongoing purchase of approximately 20,000 tonnes of CO_2 offsets per year to compensate for greenhouse gas emissions associated with the Formula 1 and World Rally Championships and others.

FIGURE 24.1. Farmers in Chiapas, Mexico, learning how to monitor carbon stocks in above-ground biomass (Photo © Jessica Orrego).

These purchases have been made through the voluntary carbon market. Companies that wish to offset their carbon emissions for corporate social responsibility or "good practice" reasons are more compelled by projects that have added social and environmental co-benefits associated. Indeed, there is a growing trend in the private sector to take voluntary actions to offset CO_2 emissions, and projects that contribute to both sustainable development and conservation are the most appealing for this.

The Clean Development Market will also provide an additional market for land-based carbon credits, although the size of this market during the first commitment period (2008–2012) is uncertain, as "sinks" credits are not permissible under the European Union Emissions Trading Scheme for this period. This does not mean that individual countries will not be enticed by sinks projects, especially those that provide strong social and environmental benefits. Furthermore, the World Bank's BioCarbon Fund will provide carbon finance for CDM-eligible projects that sequester carbon in forests and other landscapes in developing countries. However, it is likely that the bulk of the Kyoto carbon market will focus on emissions' trading and energy projects, and less on sinks projects.

3. Outline of Tools

Carbon management can provide an excellent vehicle for channelling funds into sustainable development and forest conservation and restoration activities while playing a key role in mitigating climate change. Stringent standards must be set for both compliant (e.g., Kyoto protocol) and voluntary markets to weed out projects with negative impacts, such as the Plantar project described above. In addition to providing socioeconomic and environmental benefits, projects must be promoted that can demonstrate transparent and credible baseline assessments and carbon verification systems. Organisations such as Winrock International, El Colegio de la Frontera Sur (ECOSUR), and the Edinburgh Centre for Carbon Management (ECCM) have made strides in developing methods for determining regional baselines for forest conservation carbon management. These methods are promising; however, currently there is no standard methodology that is used across projects. Furthermore, carbon monitoring protocols and frequency can vary between projects; therefore, standardisation of these procedures across projects is necessary.

Several models exist for estimating carbon sequestration potential. CO_2Fix, for example, offers a relatively easy-to-use method for estimating carbon sequestration (the model can be downloaded for free on the Internet). Subsequent and ongoing monitoring and forest measurement to verify carbon estimates is necessary. Remote sensing methods for estimating carbon stocks are in place and are undergoing further enhancements and validation via land-based studies.

A consistent set of standards and procedures is necessary to ensure the overall credibility of carbon sequestration projects and the carbon credits sold through them, whether in the voluntary or compliant market. The Plan Vivo system (www.planvivo.org) used in the Scolel Té project (mentioned above) and in similar projects in Africa (Fig. 24.2) and India provides a rigorous set of standards and procedures to ensure a high level of community participation, sustainable land use practices, and verifiable carbon credits. Plan Vivo projects are now among the most credible and widely recognised form of carbon offsets available in the voluntary sector.

The Climate, Community, and Biodiversity (CCB) standards,[220] resulting from a partnership among research institutions, corporations, and environmental groups, are a rigorous set of criteria that aim to combine climate, biodiversity, and sustainable-development benefits.

The IPCC Good Practice Guidelines for Land Use, Land-Use Change and Forestry (LULUCF)[221] provides useful guidance about methods for estimating, measuring, and monitoring carbon stocks as well as a wealth of default figures. If designed properly, such land-based carbon sequestration projects can benefit

---

[220] Climate, Community and Biodiversity Alliance (CCBA), 2004.
[221] IPCC, 2003.

FIGURE 24.2. Nursery workers in Sofala, Mozambique, preparing seedlings for carbon agroforestry activities (Photo © Jessica Orrego).

rural communities, slow destruction, and increase the restoration of vital forest ecosystems, while contributing to a combination of activities that will help slow increases in atmospheric concentrations of greenhouse gases.

4. Future Needs

The greatest limiting factor in carbon projects is the carbon market. As the carbon market is developed and expanded, so too will small-scale carbon management projects. As more carbon finance is channelled into these projects, the carbon models and baselines will be refined and more sophisticated methods will be developed. It is also important for accurate information to replace speculation when it comes to the importance of the carbon market, as well as its real value in mitigating climate change.

References

Climate, Community and Biodiversity Alliance (CCBA). 2004. Climate, Community and Biodiversity Project design standards (Draft 1.0). CCBA, Washington, DC: www.climate-standards.org.

IPCC. 2003. Good practice guidance for Land Use, Land-Use Change and Forestry National Greenhouse Gas Inventories Programme Technical Support Unit. Kanagawa, Japan. http://www.ipcc-ggip.iges.or.jp/public/gpglulucf/gpglulucf.htm.

Additional Reading

Bass, S., Dubois, O., Moura-Costa, P., Pinard, M., Tipper, R., and Wilson, C. 2000. Rural livelihoods and carbon management. IIED Natural Resource Issue Paper No. 1. International Institute for Environment and Development, London.

Landell-Mills, N., and Porras, I.T. 2002. Silver bullet or fools' gold? A global review of markets for forest environmental services and their impact on the poor. Instruments of Sustainable Private Sector Forestry Series. IIED, London.

Smith, J., and Scherr, S.J. 2002. Forest Carbon and Local Livelihoods: Assessment of Opportunities and Policy Recommendations. Center for International Forestry Research, Jakarta, Indonesia.

WB Carbon prototype Fund. http://carbonfinance.org/pcf/Home_Main.cfm.

25
Marketing and Communications Opportunities: How to Promote and Market Forest Landscape Restoration

Soh Koon Chng

Good communication cuts through the clutter, it doesn't add to it. It does this by getting the right message, in the right medium, delivered by the right messengers, to the right audience.
—Now Hear This, Fenton Communications

Key Points to Retain

Forest landscape restoration needs to be clearly communicated and different target audiences will require different channels and media.

Communicating the issue can be planned to respond quickly and strategically to news items that emerge and where restoration can create a positive message.

Marketing complex restoration programmes is equally important and it is essential to clearly understand what are the key triggers that might make the chosen audience engage in a forest landscape restoration programme.

1. Background and Explanation of the Issue

Communications is about moving people from awareness to action. If done well, it can help achieve conservation goals. Communicating about forest landscape restoration (FLR) can be done either proactively/strategically or opportunistically.

1.1. Communicating Forest Landscape Restoration

Because of its complexity, communicating forest landscape restoration is challenging. Messages should ideally cover the following:

- What are we restoring?
- Why restore?
- Who is going to benefit from the restoration?
- How can the target audience help?

Messages have to be relevant to each target audience. For example, for landowners in New Caledonia who are not at all enthusiastic about nature conservation, telling them that the island has only 1 percent of dry forests left may not be motivating or inspiring enough to make them take any action to prevent its further decline. What could grab their attention may be the economic value of these forests—their land—and therefore the need to return good quality forest. Table 25.1. lists some examples of key messages for various target audiences to whom we may want to reach out to help us restore forest landscapes. The messages are examples, and a more targeted approach will be needed for specific audiences.

1.2. Marketing

Marketing or the "selling" of projects to potential funders or donors requires good communications and research. Just as you need to understand your target audience when communicating, so it is with marketing. You need to

TABLE 25.1. Different messages for different audiences.

Target audience	Possible key message
Governments	Current reforestation practices are costing a lot of money and not providing much environmental or social benefit. FLR achieves a balance between socioeconomic and environmental benefits. Let us show you how.
Technical experts	FLR is an approach that requires an integrated effort. Join us and be a part of an initiative, working with others to share expertise and know-how.
Development organisations working locally	FLR aims to restore forest goods and services for both people and nature. It takes an integrated approach. Work with us in this initiative so that together we can meet our collective goals.
Conservation organisations and programmes already implementing FLR	Let's share lessons so we can help one another in implementing and advancing FLR initiatives in our respective countries/areas.
Conservation organisations and programmes not yet involved in FLR	Protection and management of forests are no longer enough in achieving forest conservation in the face of increasing forest loss and degradation. We need also to work on forest restoration. More and more forest conservation projects are integrating a landscape-level forest restoration approach. Don't get left behind. Jump on board.

FLR, forest landscape restoration.

know what makes the funders "tick," what are their pet interests, goals, history of giving, etc. Such information is useful in helping us draw up approaches that are appropriate to the donor, and also in developing good funding proposals. Remember, marketing is about creating a win-win situation—matching your objectives and those of the donor. Proper background research is essential.

To nonpractitioners, landscape-scale forest restoration may be a complex concept. Don't pass on the complexity to potential donors. Even if they are versed in the technicalities of the concept, their supervisors may not be. Simplicity and speaking in the donors' language are important. It is also important to do in-depth research to better know and understand the donors and their priorities, in order to address them. Above all, remember that you are talking to people; even if they are working in government aid agencies or multinationals, they are just like us—they have feelings and emotions too.

So you got the funds. Well done! But the marketing job's not over. Most businesses know it is important to keep their customer base. Likewise, we need to keep our pool of donors. Never, as we say, take the money and run! Donor engagement throughout the project is all important. Most donors appreciate being involved, and it could be as simple as receiving regular updates on how the project is progressing. In many ways, donor engagement is like making and keeping friends. So invite them home: invite donors to see how the project is progressing and to understand your challenges. It's also more fun than just reading progress reports, and they would certainly love seeing how their funds are being spent. Like investors, donors like seeing how their investments are doing. Finally, don't forget to acknowledge and thank the donor.

2. Examples

2.1. Responding to a Crisis—The Big Storm of 1999

A third of France's forests were damaged when the country was hit by one of the biggest storms ever in December 1999. Damage was extensive, shocking foresters and the public. The news made headlines and for the first time in France's forest history, forest problems and the links between forest and society were hotly debated by the media for at least 6 months.

During the weeks immediately following the big catastrophe, WWF, the global conservation organisation, "surfed" the wave, taking advantage of the media and public interest to reach out to a broad audience, developing its arguments on the need for improving forest management and the problems and threats to biodiversity. In the months that followed, WWF communicated the need for renewing forestry practices that take nature into account as well as promoting ecological restoration. Television publicity and print advertisement cajoled people into "making a wish for forest restoration." In late 2000 a press conference was called to present WWF and other NGOs' proposals for improving forest management and restoration.

The first anniversary of the storm was well covered by the mainstream media. It was an opportunity to repeat the messages while interest was still high. Subsequent anniversaries, however, did not generate as much media interest—the topic became "cold", covered only by those journalists on the forest/environment beats. In 2003, for example, there was little media interest in a WWF-released study on the implementation of forest restoration, including criticisms of the use of subsidies and bad practices in the management of habitat of key endangered species.

As a result of its communications' efforts on this issue, WWF was identified as a major actor in forest management in France—something that was not obvious before. It was successful in setting up partnerships with companies to implement restoration programmes.

WWF France's Daniel Vallauri noted, "An important lesson learnt for us in communicating during the storm crisis was the need for rapid response, coupled with a specific strategy to communicate at least for the first six months after a big storm."

2.2. Prestige Oil Spill—Responding Rapidly

While this example is not about forest restoration, it shows how quick mobilisation of a multidisciplinary team helped to deal efficiently and effectively with communications in the aftermath of an environmental disaster.

In November 2002 one of the worst oil spills in history occurred in Spain's Galicia province. It was the eighth marine environmental disaster in Galicia in the last three decades, and involved a tanker called *Prestige*.

Immediately after receiving news of the crisis, WWF Spain formed a multidisciplinary crisis group, led by its CEO, to deal with the issue. Within an hour it had alerted both the national and international media. The group designed and planned an integrated rapid-response strategy covering conservation, policy, and communication. It also developed action plans for fund raising and a membership drive.

At the same time, there was strong coordination with WWF International's Communications Department and the Endangered Seas Programme, and national offices, on policy and communication. A Web site was created to provide daily updates from the field, strengthen WWF's demands on marine security, and attend to international media queries.

A very rapid response, clear key messages, rigorous and factual information, presence on the ground, and coordination with the WWF Network ensured that WWF was the media's main reference point. This in turn ensured that WWF was mentioned in almost all media coverage with its calls for urgent action by those concerned. Most importantly, the fast and integrated response enabled WWF Spain to obtain strong conservation results, including significant policies on improving marine security adopted by the European Union (EU).

As WWF Spain summed it up, it is unfortunate but true that "An environmental crisis is a great opportunity for an NGO in terms of communications and achieving policy goals." It also has the following tips to share with offices that may have to embark on rapid response communications:

- Respond very rapidly.
- Send clear, sound, and single messages.
- Use strong visuals.
- Use integrated strategy (conservation, communications, and fund raising).
- Have a presence on the ground.
- Provide scientific and factual information.
- Use the WWF Network for expertise.

3. Outline of Tools

3.1. Communicating Proactively

Proactive communications means having a concerted and long-term plan that supports the restoration strategy. The plan involves knowing the following:

- Why we are communicating (the communications objective)—In some cases communications may be for fund-raising purposes, in others to mobilise public opinion, and yet in others, to share knowledge.
- Who we need to communicate with (the target audience)—These could be NGOs, decision makers, students, farmers, etc. Unfortunately, they are rarely a homogeneous group. The key to the communications plan is in knowing your audience. Find out as much as you can about them, particularly what inspires and motivates them. Such information is vital as it helps answer the "when" and "how" questions and also in crafting the appropriate messages.
- What should we be saying to the target audience (the message)—It is important to be clear when communicating. In some cases there may be a clear message and call to action (e.g., lobby for a change in policy) and we can even measure success of that message. In others, when disseminating knowledge or experiences for instance, there is no explicit "call for action."
- How to reach the target audience (the tools or approach)—Once the audience is identified (given that it is not always a single group but often a mixture), it is necessary to identify the best tool to reach them (see the note below about the Web, for example).

3.2. Opportunistic Communications

Opportunistic or rapid-response communication entails communicating in response to an event, for example, a sudden policy change, or a sudden natural event such as fires or storms damaging large forest areas. Because restoration is often considered as necessary once a disaster strikes, and because all too often short-term "quick-fix" solutions are offered to satisfy political and media needs, it is extremely important to be prepared with a suitable response that presents a broader-based forest landscape restoration approach as the solution (if indeed it is the right one under the circumstances). In most places, one can anticipate likely events, and therefore it is possible to prepare a "rapid response" package with the necessary recommendations for appropriate restoration and for mitigating future damage. Rapid response communications can help in reinforcing messages on forest landscape restoration and getting those results that are hard or take twice as long to achieve. Although opportunistic, this kind of communications still requires some degree of preparedness. In this regard, communications materials such as background information, including facts and figures and actions to be taken when disaster strikes, are useful to have ready.

For example, WWF has developed an information sheet with responses and recommendations on how to deal with storm damage in Europe. This proved useful after the significant storms that swept across much of France in 1999, destroying large areas of forests. What is important is that while this communications is responsive in many cases, we can anticipate a recurring natural disaster and therefore be suitably prepared for it. While a standard message or response may need some slight tailoring to the situation, the overarching message can be more carefully crafted in advance. This is particularly true of restoration, which is in itself about responding to a crisis. Remember, in an ideal world, restoration would not be needed.

The case study above on the Prestige oil spill provides an example that, while not related to restoration, demonstrates how an effective rapid response was organised.

New positive policy announcements also present good opportunities for communicating forest landscape restoration goals and objectives. For example, when former Indonesian President Megawati announced in early 2004 her government's support to implement restoration initiatives, this presented an opportunity to not only applaud the initiative but also offer support and help in ensuring that past errors are not repeated.

3.3. A Word About the Web

The explosion of Web sites makes it tempting to jump onto the bandwagon. But be aware that while nice to have, a Web site requires long-term investment in resources in maintenance as well as marketing to draw in visitors. Also, a Web site is not always the panacea for all communications. For example, in many countries, target audiences will not have access to a computer. Another common error is the failure to regularly update a Web site, which can quickly become obsolete.

4. Future Needs

A number of rapid-response messages and packages still need to be developed for anticipated crises. These are important because they allow for quick dissemination of the importance of restoration, when the audience is receptive. In some cases, such as for the linkage between floods and tree cover, more research is needed on the real linkages and cause-and-effect relationship in order to substantiate communications' claims.

References

Now hear this—the nine laws of successful advocacy communications. http://www.fenton.com/. Concise report by Fenton Communications detailing their approach to advocacy communication campaigns.

Part C
Implementing Forest Restoration

Section IX
Restoring Ecological Functions

26
Restoring Quality in Existing Native Forest Landscapes

Nigel Dudley

Key Points to Retain

In many countries the most pressing restoration need from a conservation perspective is not for new forests but for higher quality in existing forests.

Restoring ecological quality requires a proper understanding of the components of a natural forest: composition, pattern, functioning, process of renewal, resilience, and continuity in time and space.

Approaches to restoring quality include active management to restore missing microhabitats and steps to influence both process and the way in which the forest renews itself.

1. Background and Explanation of the Issue

Forest management has changed the composition and ecology of the remaining forests in many parts of the world. Intensive management of native temperate forests in Europe, North America, and parts of Asia has resulted in forests that are species-poor, artificially young, lacking many of the expected microhabitats and with radical changes to ecology and disturbance patterns. Logging in many tropical forests has removed the largest trees, fragmented habitats through the construction of logging roads and skid trails, and often opened forests up to exploitation by settlers and poachers. Although these forests still exist, their ability to support biodiversity or to supply goods and services for local human communities may have been radically reduced. Or more precisely, their structure has been altered to supply one particular good—timber products—at the expense of other goods and services. Changing priorities mean that there is now increasing interest in managing forests for biodiversity, environmental services, recreation, and cultural and social benefits, as well as for timber production. In places where there are large areas of intensively managed or logged-over forest, the primary focus of restoration activities may well be on restoring forest quality in existing stands of trees rather than extending the area under trees; in effect, this usually means returning the forest to a more natural composition and ecology. Six major components are important in defining the naturalness of a forest ecosystem:

1. The composition of tree species and other forest-living plant and animal species, where changes can include both loss of native species and problems from the occurrence of nonnative invasive species

2. The pattern of intraspecific variation, as shown in trees by canopy and stand structure, age-class, under-storey, with changes in managed forests commonly being toward younger, more uniform forest stands

3. The ecological functioning of plant and animal species in the forest as manifest in food

webs, competition, symbiosis, parasitism, and the presence of important microhabitats such as dead wood and leaf litter

4. The process by which the forest changes and regenerates itself over time, as demonstrated by disturbance patterns, forest succession, and the occurrence of periodic major disturbances from storms, fire, or heavy snowfall

5. The resilience of the forest in terms of tree health, ecosystem health, and the ability to withstand environmental stress, which is of increasing importance during a period of rapid climate change

6. The continuity of the forest particularly with respect to total size, but also the existence of natural forest edges (often lost in managed habitats), connectivity of forest patches and the impact of fragmentation[222]

Restoration of quality can sometimes be achieved just by withdrawing management or other pressures, allowing natural ecological functioning to reassert itself gradually. However, in other cases, where, for instance, species have been lost from a locality, or where remaining pressures are undermining natural disturbance patterns, more active restoration efforts may be needed. Over the past two decades, limited experience has built up in restoration of forest quality, although there is still a great deal to be learned.

2. Examples

Most of the experience in restoration of forest quality currently exists in temperate and boreal forests, as shown by the examples below, although the importance of restoring forest quality is also increasingly being recognised in the tropics.

2.1. Wales—Restoring a Native Forest Composition by Removing Invasive Species

The Ynyshir bird reserve on the Dyfi estuary contains some of the oldest native oak woodland in Wales, within the core of a projected UNESCO biosphere reserve. The wood is variable-aged with a natural ecology but has been substantially altered by invasive species, mainly sycamore (*Acer pseudoplatanus*) and rhododendron (*Rhododendron ponticum*). To restore a natural composition, sycamore has been progressively removed by ring-barking mature trees and cutting out saplings. Rhododendron has been cut and burned during the winter and stumps spot-painted with a short-life herbicide to prevent regeneration (information from reserve staff).

2.2. Sweden—Re-Creating Dead Wood Microhabitats in Managed Forest

Artificial high stumps were created as potential hosts for saproxylic (deadwood-living) beetles in managed forests in Fagerön, Uppland, and stumps and logs were also left as substrates for saprophytic fungi. The results showed that hundreds of beetle species, including many red-listed species, utilise high stumps, and two thirds of them favour stumps in semi- or fully sun-exposed conditions, showing that high stumps in logging areas and other open sites are potentially very valuable tools for conservation of saproxylic beetles. Cut wood, especially large-diameter logs, also hosted numerous species of saprophytic fungi. Thus, cut logs may support fungal diversity, both in managed forest landscapes and in forest protected areas (see "Restoration of Deadwood as a Critical Microhabitat in Forest Landscapes").[223]

2.3. Finland—Restoring Natural Fire Disturbance Patterns by Prescribed Burning

Controlled burning is used to restore forests where fire suppression has resulted in the decline of species that need fire for germination or to remove competitors. Finland's Natural Heritage Services' department uses prescribed burning in protected areas, particularly in the south of the country, and to date almost 4000

---

[222] Dudley, 1996.

[223] Lindhe, 2004.

hectares have been restored in this way. Burning has to be carried out with extreme care when weather conditions are suitable, that is, when the forest is not too wet to burn but not so dry as to create uncontrollable fire.

2.4. Sabah—Reconnecting Forest Fragments

Forest along the banks of the Kinabatangan River in the Malaysian state of Sabah, in Borneo, creates an important corridor between coastal mangrove and secondary forests in the highlands. Much of the remaining lowland forest has been converted into oil palm plantations. Substantial parts of the riparian corridor are now protected but these areas have become fragmented and oil palm reaches right to the river bank in places, cutting migration corridors for elephants and other species. The WWF Partners for Wetlands project has been liaising with villagers to promote targeted tree planting to reconnect the patches of remaining forest to form a larger and ecologically coherent whole. More importantly, WWF has been working with oil palm companies to find ways in which selected areas can be returned to forest (personal observation and discussions with field staff).

2.5. Lebanon—Building Capacity for Better Forest Management and Restoration

The Al Shouf Cedar Reserve in Lebanon covers 550 km², around 5 percent of the country, and contains around a quarter of Lebanon's remaining cedar (*Cedrus libani*) forest. The core of the reserve is strictly protected and is in mountainous territory of little economic value. The Shouf Forest Resource Centre was opened in 1998 to help improve forest quality particularly in the buffer zone of the park, through management of forest biodiversity and silvopastoral systems, forest fire prevention, production and commercialisation of nontimber forest products, tree nurseries and eco-forestry techniques, and environmental education (information from WWF Mediterranean Programme).

3. Outline of Tools

In many cases restoration of quality is best served by simply giving a forest time to recover its natural dynamic, although some additional help may be required to achieve this as the previous examples show.

- Assessment: The first step in restoring quality of forests is to determine what is missing. Many different definitions of naturalness exist at a site level, although most of these do not identify the different components involved (see "Identifying and Using Reference Landscapes for Restoration"). A simple site-level scorecard (Table 26.1) for assessing levels of authenticity in forest ecosystems[224] can be used to provide a quick reference to elements of authenticity that are either present or absent as an aid to planning restoration programmes.

- Influencing rate of change: Most aspects of quality restoration can be achieved by removing the pressures that are currently reducing quality, such as overgrazing, changes in fire regime (either unnaturally high or low incidence of fire), poaching, and overcollection. The simplest and cheapest tools available are agreements with stakeholders, for example, ensuring that shepherds keep sheep or goat flocks away from certain forests or reducing nontimber forest product collection. More expensive options include fencing against grazing animals, antipoaching patrols, and fire watching.

- Active management to restore natural dynamics: Where particular natural elements are missing from the forest ecosystem, or unnatural elements (e.g., invasive species) are present, more active intervention may be required. Many invasive species only become established when there are gaps in the canopy so that removal for a period can lead to their virtual elimination, in other cases more long-term control strategies may be needed (particularly in the cases of invasive animals). Re-creation of missing microhabitats, such as dead wood (see "Restoration of Deadwood

[224] Dudley et al, in press.

TABLE 26.1. Data card for stand-level assessment of forest authenticity (Dudley et al, in press).

Indicator	Elements			
Assessors should fill in as much of the table as possible. Space is left for further observations				
Composition	How natural is composition of tree species?	Fully	Partly	Exotic
	How natural is composition of other species?	Fully	Partly	Exotic
	Are significant alien species present?	Yes	No	
	Is the ecosystem functioning naturally?	Yes	No	
Notes on composition:				
Pattern	What is the tree age distribution?	Mixed—old	Mixed—young	Mono
	Is the canopy natural or artificial?	Natural	Artificial	
	Is the forest mosaic natural or artificial?	Natural	Artificial	
Notes on pattern:				
Functioning	Are viable populations of most species present?	Yes	No	
	Does a natural food web exist?	Yes	No	
	What are the soil characteristics?	Stable	Seriously eroding	
	What are hydrological characteristics?	Healthy	Problems	
	What is the age of the forest?	Old growth	Mature	Young
	What is the period of continual forest cover?			
Notes on functioning:				
Process	Does a natural disturbance regime exist?	Yes	No	
	Does an unnatural disturbance regime exist?	Yes	No	
	Is a significant amount of deadwood present?	Snags	Down logs	
Notes on process:				
Continuity	Size (in hectares):			
	Age (approximate length of continuous forest cover)			
	Are the forest edges natural or artificial?	Natural	Artificial	
	Is the forest connected to other similar habitat?	Yes	No	
	Is the forest fragmented?	Yes	No	
Notes on continuity:				
Resilience	What is the tree health?	Good	Average	Poor
	Are there important introduced pests, diseases, and invasive species?	Yes	No	
	What are the pollution levels?	High	Medium	Low
Notes on resilience:				

as a Critical Microhabitat"), riparian forest strips, or particular species, may also be necessary in cases where there is either some urgency or where these are unlikely to reappear naturally.

- Influencing disturbance patterns: Various techniques for reintroducing or mimicking natural disturbance patterns exist or are being developed. Most aim to "manage" disturbance mainly by controlling it so that it influences smaller areas (for example, because the forest is already fragmented or because land tenure agreements mean that only limited areas can be disturbed). Techniques such as prescribed burning, artificial creation of standing deadwood, and mimicking storm damage are all now available.

4. Future Needs

Much more information is needed about the ability of different forest ecosystems to recover quality over time and particularly about the likely speed of recovery; this information is important in making decisions about whether or not to undertake more active (and expensive) forms of restoration. Methods for control of invasive species are in some cases still also poorly developed as is management of artificial disturbance. Codes of practice and perhaps principles for artificial disturbance remain to be developed.

References

Dudley, N. 1996. Authenticity as a means of measuring forest quality. Biodiversity Letters 3:6–9.

Dudley, N., Schlaepfer, R., Jeanrenaud, J.-P., and Jackson, W.J. In press. Manual on Forest Quality.

Lindhe, A. 2004. Conservation Through Management. Doctoral dissertation, Department of Entomology, Swedish University of Agricultural Sciences (SLU). Acta Universitatis Agriculturae Suecia, Silvestria, vol. 300.

Case Study: Restoring a Natural Wetland and Woodland Landscape from a Spruce Plantation in Wales, UK

Nigel Dudley and Martin Ashby

The Challenge

Sixteen hectares of salt marsh on the Dyfi Estuary in Wales had been planted with a dense stand of Sitka spruce (*Picea sitchensis*), a North American conifer. The estuary is a Ramsar Site (i.e., listed as an area of outstanding wetland) and UNESCO Man and the Biosphere reserve, and the plantation adjoins a strictly protected area. Most natural vegetation had been shaded out and the peat underneath the canopy supported few plant species except for flushes of fungi in the autumn. Several plants (e.g., the heathers *Erica* spp.) appeared after the spruce was removed, having presumably been dormant in the peat. The natural water table had also been altered as a result of constructing an embankment for the railway and through subsequent drainage, and the soil layer disturbed by deep ploughing when the plantation was established.

The Opportunity

Sell-off of a proportion of state forest land meant that the area was available for purchase. As the plantation never produced a commercially viable crop, the sale price was fixed only slightly higher than the value of standing timber, creating the chance of a cheap net land purchase. The plantation was bought by a private individual, who has leased this to the local wildlife trust as a nature reserve. Agreeing the purchase involved lengthy negotiation because under United Kingdom law any trees that are felled must be replaced, whereas the conservation opportunity here was *not* to replace but instead to see what emerged through natural regeneration (with an assumption that a proportion of the area would be naturally treeless). There was considerable uncertainty about how the site would regenerate, although support from the Countryside Council for Wales eventually helped to encourage a change of policy to allow natural regeneration rather than replanting.

Interventions

The spruce trees were felled and cleared, along with most remaining brash (branches, etc). Early ideas of replacing ploughed soil were abandoned because of the scale and costs of the operation. Some drainage ditches were blocked on an experimental basis, raising the water table and in addition, with National Heritage Lottery funding, several new ponds were dug. Subsequently, a boardwalk circuit has been established for visitors, which is being upgraded to allow wheelchair access to part of the protected area, and a simple bird hide is being constructed from living willow at the edge of one of the new ponds.

Results

The area has changed, over 6 years, from a place almost devoid of natural life to a rich woodland and fen habitat, with emerging birch (*Betula pendula*) and willow (*Salix* species) in places, along with large areas of wetland plants including stands of reedmace (*Typha latifolia*). Up to three nesting pairs of the nationally endangered nightjar (*Caprimulgus europaeus*) have successfully raised young and other wetland birds such as the common snipe (*Gallinago gallinago*), sedge warbler (*Acrocephalus schoenobaenus*), and grasshopper warbler (*Locustella naevia*) have bred, and the area has become a hunting ground for the rare barn owl (*Tyto alba*) and for otters (*Lutra lutra*). It is hoped that a locally rare moth, the rosy marsh moth, might reestablish, and a survey is planned.

Future Issues

Under natural circumstances, the area would be mainly salt marsh with some freshwater inflow and emergent trees. Much of the challenge, therefore, has been to replace a habitat of nonnative trees with a smaller mosaic of native species. This habitat has almost disappeared in parts of the U.K. and is therefore a particular focus for restoration. As yet it remains uncertain as to whether the increasing water level will serve to restrict colonisation by birch and willow or whether the reserve manager will have to arrange periodic clearance to maintain the forest mosaic that would exist under completely natural circumstances. The extent to which Sitka spruce will regenerate is not clear, although some clearance may be needed, ideally before young trees become mature enough to reproduce themselves. Grazing would help keep scrub regeneration under control and leave open areas for nightjars, although it may prove difficult to find native species able to live successfully in the wet conditions of the site.

Lessons Learned

Despite the warnings about the irreversibility of changes associated with plantation establishment, reversal has been rapid and so far highly successful. The fact that the soil humus layer was badly damaged by ploughing has apparently made little difference to recovery: it makes access more difficult for people, but paradoxically this may be an advantage in a nature reserve.

27
Restoring Soil and Ecosystem Processes

Lawrence R. Walker

Key Points to Retain

Ecosystem processes, especially those directing successions, are the working parts of a successfully restored habitat.

Below-ground processes are the first key to many harshly degraded situations restorationists have to face, and thus require specific attention.

Reestablishment of biodiversity implies a fully functioning ecosystem.

1. Background and Explanation of the Issue

It is necessary to link human restoration efforts with the reestablishment of ecosystem processes in order to maximise biodiversity and ecosystem services (e.g., clean water, stable soils) while minimising additional human inputs. Simply planting local vegetation and adding agricultural levels of fertiliser is not necessarily sufficient. Restoration activities focussed solely on maximising substrate stability or primary productivity frequently result in arrested succession and require further effort to encourage successional change. Critical ecosystem processes are the working parts of a successfully restored habitat.[225] Without them, restoration is incomplete.

An ecosystem is defined as a series of interactions among a particular set of organisms and between those organisms and their physical environment. Restoration addresses inputs, outputs, and internal dynamics of the flow of energy and matter. Typical measures of inputs include sunlight, water, nutrients, and organisms. Typical outputs include water, eroded soil, and organisms. Internal fluxes include nutrient cycling, primary productivity, and decomposition. Additional ecosystem processes concern the interaction of the biota to disturbance (resistance, resilience, succession, invasion) and the development of structure and biodiversity. Successful restoration complements the natural recovery process of succession, following removal of constraints such as unstable, toxic, or infertile substrates or the lack of adequate soils. Successful restoration also allows succession to proceed and leaves an ecosystem both resistant and resilient to disturbance. Because we are able to predict successional trajectories only in the broadest sense (of functional groups, biomass, and nutrient accumulation), restoration that incorporates successional dynamics is often experimental. At best, unsuccessful restoration efforts help elucidate successional principles, as successional theories, in turn, guide restoration.[226]

---

[225] Ehrenfeld and Toth, 1997.

[226] Walker and del Moral, 2003.

FIGURE 27.1. Extreme soil erosion (left) in Iceland can be slowed by fencing to exclude sheep and horses (right) (Photo © Lawrence R. Walker.)

2. Examples

2.1. Substrate Stability in Iceland

Erosion is a major disturbance on over 40 percent of the terrestrial surface of the earth. Site stabilisation is essential for restoration, but care must be taken in how it is done. Iceland has the temperate world's worst soil erosion due to 1000 years of overgrazing of sensitive soils (Fig. 27.1). It used to have 2- to 3-m-tall birch forests (*Betula pubescens*), and Icelanders want them back (Fig. 27.2). The use of native ground cover to stabilise the erosive forces of wind, water, and ice heaving, combined with fences to keep out sheep and horses, leads (after about 50 years) to the return of native forests.[227] No success has been achieved by planting native trees without first stabilising the surface or without fencing (Fig. 27.3).

2.2. Substrate Stability in Puerto Rico

Reforestation of landslides in Puerto Rico requires slope stabilisation, best provided by native climbing fern (*Gleichenia bifida, Dicranopteris pectinata*) thickets that then delay forest growth for several decades.[228] Direct tree planting is rarely successful on erosive surfaces, even with fertiliser or organic soil amendments,

FIGURE 27.2. A 45-year-old Betula pubescens forest in Iceland, restored by protecting it from grazing by sheep and horses. (Photo © Lawrence R. Walker.)

due to continued erosion in this high rainfall habitat. Gabions, mats, and other human efforts to stabilise the slopes rarely function as well as the ferns that have extensive below-ground

---

[227] Aradottir and Eysteinsson, 2004.
[228] Walker et al, 1996.

FIGURE 27.3. A rofabard, or erosion remnant in Iceland where severe soil losses have removed several metres of soil, leaving only gravel barrens. (Photo © Lawrence R. Walker.)

rhizomes and add copious above-ground litter. Even though they take longer, natural successional processes thus appear to be most robust in achieving restoration goals.

2.3. Substrate Fertility: Iceland and Alaska

Adding fertiliser does not immediately establish critical nutrient cycles, and too much fertiliser may result in dominance by densely growing grasses or herbs that inhibit tree establishment through competition for nitrogen, phosphorus, water, or light. Appropriate levels of fertiliser, combined with species that are short-lived or grow less densely, can act as nurse plants for seedlings of later successional plants and facilitate succession. Legumes introduced to increase soil nitrogen may benefit tree growth if their densities are kept low. Attempts to accelerate reforestation in Iceland with commercial fertilisers or by planting tree seedlings into stands of the nonnative, nitrogen-fixing lupine (*Lupinus nootkatensis*) have shown some promise for nonnative trees such as Sitka spruce (*Picea sitchensis*). However, overfertilisation or overreliance on lupine may lead to dominance by nonnative herbs or conifers in some parts of Iceland. Fertilisation of nonnative grasses on the Alaska pipeline corridor delayed recolonisation of native tundra species by several decades. Low-fertility sites where competition is reduced and where all key species are introduced initially have the greatest chance of restoration success.[229]

2.4. Amelioration of Toxic Conditions in Mines in South Africa

Reforestation often involves addressing toxic site conditions. Landfills can have toxic liquids and gases; mine tailings can have extreme pH values or toxic levels of metals in addition to surface compaction or erosion problems. Reforestation of dunes mined for various ores in South Africa involved topsoil replacement, windbreaks, and sowing of various grasses that provided a nurse crop for slower-germinating native *Acacia karoo* trees from the seedbank. The acacia trees, in turn, promoted soil development through nitrogen fixation and were gradually replaced by larger native trees. In this case, normal successional processes replaced early intensive manipulations.[230] However, it is often difficult to restore some semblance of predisturbance vegetation due to alterations in drainage, fertility, and even topography. Forests may remain stunted if they do colonise toxic sites, and reclamation goals are often more modest than in less toxic situations.

3. Outline of Tools

Stabilising soil substrate: Substrate stability is essential before restoration can proceed. For example, the following actions treat successively more serious erosion conditions on

[229] Walker and del Moral, 2003.

[230] Cooke, 1999.

Puerto Rican landslides: mulch, fertiliser, transplants, silt fences, contouring, jute cloth covers, rock-filled gabions, redirecting water flow, and lining alternative drainage channels.

Adding organic matter: Soil processes are key to successful restoration. Beginning with severely disturbed substrates, organic matter additions are the fastest way to incorporate critical soil microbes. Earthworm additions, inoculations of mycorrhizae, and additions of limiting nutrients (with the caveats noted above) all potentially accelerate soil development and facilitate woody plant invasions or plantings, especially in severely disturbed habitats. However, mycorrhizae can act as parasites when nutrient limitations are severe. Minimal additions of topsoil or other sources of nutrients and soil biota can reduce the risk of overfertilisation and dominance by early successional species that preclude tree establishment. Additions of nitrogen-fixing plants can often benefit (but see Substrate Fertility, above).

Reducing soil nutrients: Restoration can also involve reducing soil nutrients (via carbon-rich straw, sawdust, or sugar, or additions of lignin-rich plant litter that immobilise nutrients) if the goal is a naturally infertile site. For example, native ohia (*Metrosideros polymorpha*) forests in Hawaii are out-competed by the introduced nitrogen fixing tree *Myrica faya*. In fact, the whole successional pathway on volcanic surfaces is altered to favour plants adapted to higher nutrients, particularly nitrogen.[231] Restoration of native Hawaiian communities and successional processes will most likely require nutrient-reduction treatments.

Reducing toxic conditions: Toxic conditions can be ameliorated by bioremediation, or the use of plants, mycorrhizae, and microbes. Once toxins are reduced, restoration of native communities can begin. Additions of topsoil from late successional communities, sometimes combined with sludge, composted yard wastes, or other concentrated organic matter source, often accelerate succession. Arrested succession can be avoided by dense plantings of native species, particularly ones that attract vertebrate dispersers.

Biodiversity is a key goal to restoration, and its reestablishment implies a fully functioning ecosystem. If a diverse biological community resembling the reference ecosystem is self-sustaining, then landscape and successional dynamics have likely been incorporated. In addition, adequate substrate stability, drainage, depth, and fertility have been achieved. However, restoration generally requires ongoing monitoring and strategic alterations.

4. Future Needs

We need to better understand the role that individual species have in the restoration of ecosystem processes. We have tended to focus on nitrogen fixers used in agricultural settings and neglected vascular species that concentrate nitrogen and phosphorus from infertile soils. We have also neglected the nature and specificity of plant mycorrhizal associations and their role in restoration. Species that have similar functional attributes (fix nitrogen, grow early and fast in succession, host key pollinators or dispersers, have deep roots that break through compacted soils, etc.) may offer insights into better approaches to restoration. Similarly, keystone species (ones with ecosystem and community impacts disproportional to their biomass) could be important to restoration efforts.

Invasive species are becoming ubiquitous and restorationists need to address the impact of such species on ecosystem processes. Do they alter nutrient dynamics, soil stability, soil salinity, fire frequency, or primary productivity? If so, restoration efforts must not ignore these new influences.

Restoration is essentially the manipulation of succession, yet we understand little about how ecosystem processes vary through succession. Temporal replacement of vascular plant species reflects and influences a complex of ecosystem processes, including, generally, a reduction in light availability and an increase in nutrient availability. How can restorationists maximise their manipulations of these trends to favour

---

[231] Vitousek and Walker, 1989.

desirable outcomes? Finally, much emphasis is placed on above-ground and visually obvious criteria for measuring restoration success. When below-ground processes are ignored or only treated in a crude way (through fertilisation or stabilisation, for example), restoration suffers. The interplay of soil organisms with soil stability, fertility, and/or toxicity and with animals and vascular plants is perhaps the ultimate key to successful restoration.[232]

References

Aradottir, A.L., and Eysteinsson, T. 2004. Restoration of birch woodlands in Iceland. In: Stanturf, J., and Madsen, P., eds. Restoration of Boreal and Temperate Forests, pp. 195–209. CRC/Lewis Press, Boca Raton, Florida.

Cooke, J.A. 1999. Mining. In: Walker, L.R., ed. Ecosystems of Disturbed Ground, vol. 16, Ecosystems of the World, pp. 365–384 Elsevier, Amsterdam.

Ehrenfeld, J.G., and Toth, L.A. 1997. Restoration ecology and the ecosystem perspective. Restoration Ecology 5:307–317.

Vitousek, P.M., and Walker, L.R. 1989. Biological invasion by *Myrica faya* in Hawaii: plant demography, nitrogen fixation and ecosystem effects. Ecological Monographs 59:247–265.

Walker, L.R., and del Moral, R. 2003. Primary Succession and Ecosystem Rehabilitation. Cambridge University Press, Cambridge, UK.

Walker, L.R., Zarin, D.J., Fetcher, N., Myster, R.W., and Johnson, A.H. 1996. Ecosystem development and plant succession on landslides in the Caribbean. Biotropica 28:566–576.

Wardle, D.A. 2002. Communities and Ecosystems: Linking the Aboveground and Belowground Components. Princeton University Press, Princeton, New Jersey.

Additional References

Palmer, M.A., Ambrose, R.F., and Poff, N.L. 1997. Ecological theory and community restoration ecology. Restoration Ecology 5:291–300.

Temperton, V.M., Hobbs, R.J., Nuttle, T., and Halle, S., eds. 2004. Assembly Rules and Restoration Ecology: Bridging the Gap Between Theory and Practice. Island Press, Washington, DC.

Walker, L.R., and Smith, S.D. 1997. Impacts of invasive plants on community and ecosystem properties. In: Luken, J.O., and Thieret, J.W., eds. Assessment and Management of Plant Invasions, pp. 69–86. Springer, New York.

---

[232] Wardle, 2002.

28
Active Restoration of Boreal Forest Habitats for Target Species

Harri Karjalainen

Key Points to Retain

The last natural habitats still hosting original species' composition are often small fragments, and successful conservation of these often requires the re-creation of new, larger, and better connected forest habitats.

Target species are the objective of restoration efforts for two reasons: either because the particular species has declined for a specific reason and therefore needs special attention, or because the target is used as an indicator of a wider biodiversity grouping that has also declined.

Target species (in particular endangered species) are often useful in assessing the results of certain restoration activities in the ecosystem.

1. Background and Explanation of the Issue

Loss of original forest cover to other land uses, increased degradation of remaining forests, and decreasing areas of authentic forest habitats have had a deep impact on biodiversity in many forest vegetation zones. Authentic forest habitats have become fragmented, and distances between suitable habitats hindered the spreading of specialised species. Indeed, small fragments of authentic forest habitats cannot maintain viable populations of many forest specialists.

Loss of authentic forest habitats below critical thresholds has resulted in a decline of many original forest species. In Europe the number of threatened taxa is alarmingly high: among mammals, typically 20 to 50 percent, and among birds 15 to 40 percent, of the forest dwelling species are categorised as threatened according to IUCN's red-data book classification. The situation is almost as bad even for lichens, mosses, and vascular plants.[233]

The last natural habitats still hosting original species' composition are often small fragments situated inside protected areas, or located within larger, degraded forests. Successful conservation of habitats of endangered species in these forest landscapes requires the re-creation of new, larger, and better connected forest habitats by the means of ecological forest restoration. Active ecological forest restoration is urgently required when natural forest recovery is too slow, or it is uncertain whether natural forest recovery could maintain or improve critical habitat qualities for the target species.[234]

At site levels, one short-term objective of ecological forest restoration is to enhance the populations of certain target species.

Target species fall into a number of categories:

- Species that are chosen as a focus of attention because they are representative of many

[233] Karjalainen et al, 2001.
[234] Rassi et al, 2003.

other species within the ecosystem, and therefore their recovery signals that other species are likely to be recovering. Such species may also be called "umbrella species," which means these species' habitat requirements are relatively wide (comprehensive) and hence conservation of umbrella species may protect many other important species with similar or less demanding habitat requirements.
- Species that influence significantly the viability of other species' populations, or play a key role for ecosystem functionality or structure. These are known as "keystone species."
- Species that are of particular importance within a conservation plan because they are, for example, endangered, endemic, culturally important, economically valuable, etc.
- Species that act as surrogates for certain habitat and/or landscape qualities that are considered important for maintaining biodiversity.

In the long term, ecological forest restoration objectives are to create self-sustaining forest landscapes, where natural succession dynamics prevail and forests form natural mosaics that are able to maintain viable populations of all naturally occurring species.

1.1. Importance of Restoration for Target Species

Target species are the objective of restoration efforts for two reasons: either because the particular species has declined for a specific reason and therefore needs special attention, or because the target is used as an indicator of a wider biodiversity grouping that has also declined.

In the second case, recovery of the target implies also recovery of other species. This is more often claimed than substantiated: target species are often relatively large, charismatic species and therefore also relatively adaptable. For instance, the recovery of a woodpecker species implies that the volume of its prey species have also recovered (probably due to deadwood retention) but not necessarily the diversity of its prey: it may be feeding on large numbers of a small group of saproxylic beetles.

So while target species are politically and practically useful in helping to stimulate restoration activity, they need to be treated with caution if they are also to be used as a surrogate for a whole cross section of biodiversity. This may imply, for instance, broader monitoring to check the wider implications of target recovery (refer to the Section "Monitoring and Evaluation").

Ideally, all restoration activities shall be based on in-depth knowledge of the structure and function of the forest ecosystem and target species in question.

1.2. Where to Start Restoration for Target Species

Target species' populations may have decreased, but may still be surviving in a degraded forest area. Priority should be set for the restoration of the habitats of the target species, as well as for the enhancement of the viability of the target population. Even those species surviving for now in forest fragments may not be viable in the longer term, and hence there is urgency for restoration. This argument provides another reason for intervention rather than relying on natural processes.

In the case where target species have become extinct to the region, it is necessary to know habitat requirements of the target species and possibilities for colonisation: species' capacity to disperse, location of the source population, distance to the restored habitats, and in the case of plants, the existence of the seed bank.

1.3. Target Species as Indicators of Successful Restoration

Target species (in particular endangered species) often play an important role in assessing the results of certain restoration activities in the ecosystem. The achievements may be measured structurally (e.g., by the abundance or number of target species or species' composition) or functionally (e.g., interaction of species, trophic structure, side effects).[235] However, the

---
[235] Palmer et al, 1997.

presence of certain target species does not necessarily mean that restoration activities have been successful. From the population biology viewpoint, only populations that are capable to reproduce, grow, disperse, and develop can be viable in the long term. This implies that successful restoration of target populations requires they become functionally connected with regional metapopulations in the long term.[236]

If restored target species' populations are too small, there is a risk for too narrow genetic variation that may become a limiting factor for successful restoration.[237] Narrow genetic variation may cause, for example, lower evolutionary adaptability and lower genetic population size. Small populations are also more vulnerable to sporadic factors.

2. Examples

2.1. Restoring Habitats for Species Requiring Deadwood

Old, dying, and decaying trees are important element in natural forests, providing habitats for numerous specialised species. For example, scientists estimate 20 to 25 percent (or some 4000 to 5000 species) of all forest-dwelling species are dependent on deadwood in Finnish boreal forests.[238] Forestry practises have made forests tidier and the amount of deadwood has fallen to critically low levels, resulting in a high numbers of those species relying on deadwood becoming endangered.

Therefore, one of the most common goals of ecological forest restoration is to re-create a proper environment for the species using decaying wood. Typical species are different beetle species and saprophytic biota, both of which are good indicators of the general deadwood conditions in the forests for other species' groups. Dead and dying wood can be created by damaging and felling trees and by triggering and starting the succession dynamics with self-thinning and natural disturbance. The key factor in restoration is to evaluate the restoration validity of the site compared to the naturalness of the forest structure, species' immigration, probability, and possibility of species' recovery.

New research in boreal forests in Finland suggests that at least $20\,m^3$ of deadwood per hectare on stand-level would probably meet, and at least $50\,m^3$/hectare would give a high probability to meet the ecological minimum requirements of many endangered forest species specialised in deadwood.[239] However, the quality of the deadwood is essential and it is important to offer deadwood that varies in quality to suit different specialised species. There should be a whole variety of natural tree species, as well as a variety of different decomposition classes (see also "Restoration of Deadwood as a Critical Microhabitat in Forest Landscapes").

2.2. Forest Fires Specialist Species

Many endangered specialist species are highly dependent on forest fires and burned wood. These species typically populate the burned area immediately after the fire, and revert some 5 years after the fire. Some endangered fire-dependent beetle species utilise certain fungi species, which only occur in recently burned wood. Most of the fire-dependent specialist species are capable of spreading long distances, which is necessary because forest fires have occurred randomly in the forest landscape. These species often have certain physiological and morphological adaptations, such as infrared sensors, which helps species to find suitable habitats from a distance.

Other groups of species are not as closely linked to fires, but clearly favour them. These species are typically the same that occur in other large-scale natural disturbances such as large-scale wind falls, flooded forests or even clear-felled forests. These species populate forest fire areas typically 5 to 25 years after the actual fire.

---

[236] Montalvo et al, 1997.
[237] Montalvo et al, 1997.
[238] Siitonen et al, 2001.
[239] Penttilä et al, 2004; Siitonen et al, 2001.

2.3. Restoring Habitats for Forest Bird Species

Many declining bird species are dependent on deadwood, and forest restoration activities may rapidly create new suitable habitats that these species can populate. For example in Finland it has been observed that the critically endangered white-backed woodpecker (*Dendrocopos leucotos*) utilises artificially created snags and deadwood as a source of insect nutriment. *Dendrocopos minor* and *Picoides tridactylus* have also benefited from an increase in deadwood availability in restored forest areas. It also appears that the higher numbers of nest holes created by woodpeckers also benefit other hole-nesting species that have declined due to critically low amounts of natural nest holes available in intensively managed commercial forests.

3. Outline of Tools

3.1. Planning of the Target Species' Restoration

Ecological forest restoration should be planned carefully at different levels. The first level should be ecoregional or country-based strategies where objectives for target species or species' groups are defined by major forest types. Such a plan should take into account the current occurrence of target species' populations, and present a strategy on how target species may colonise existing habitats, and how they may migrate into new, restored forest habitats.

All restoration activities should be based on the precautionary principle. Activities should include careful planning by ecological experts in the species' groups in question. If there is insufficient knowledge of the target species' ecology, it is advisable to leave the habitat to restore through natural succession, although even natural succession may in some cases require active management (e.g., fencing against livestock, changes in management interventions, etc.).

Restoration activities targeted at endangered species should be directed so that populations and local occurrences shall be maintained long-term in the ecoregion or country. Actual restoration activities shall be located in the vicinity of known and demarcated habitats of endangered species, not in the actual habitat of the target species.[240] The aim of these activities is to restore neighbouring low-quality forests and in that way re-create new potential habitat for the species. Results of scientific studies, simulations, and mathematical models[241] support the theory that restoration activities are most effective when located in the vicinity of existing source populations of target species.

In terms of landscape-level planning, restoration should aim first to maintain target species populations (endangered species) and abundance of crucial forest habitats. Restoration activities should be concentrated to re-create larger, unified ecological core units. Landscape-level restoration plans should aim to re-create forests that provide sufficient variety of all natural habitats in terms of quality and quantity.

Forest stand-level restoration activities should aim at strengthening the existing core area by re-creation of buffer zones, ecological connections, and minimising fragmentation. Restoration should be planned so that forest areas will become naturally connected to other ecosystems such as watercourses, open mires, or mountain areas. At its best, restored forest ecosystems form large, united ecologically self-sustaining units and cover natural drainage basins.

In certain extreme cases, target species (endangered species) could be transferred into restored forests that meet species' critical habitat requirements. There is, however, quite limited experience and scientific research on species' transfers. In Finland species' transfers have yielded both negative and positive results. For example, the endangered butterfly *Pseudophilotes baton* was transferred into its former restored forest habitat in southern Finland, but the butterfly population withered away. On the other hand, some endangered vas-

---

[240] Rassi et al, 2003.
[241] Hanski, 2000; Huxel and Hastings, 1999; Tilman et al, 1997.

cular plants (*Primula stricta, Pilosella peleteriana, Moehringia lateriflora, Elymus mutabilis*) have been successfully transferred into a test area, and there are plans to transfer species into nature, on restored river banks.

3.2. Stand-Level Restoration Methods

The type of ecological restoration aimed at specific plant and animal species tends to be aimed at changing certain elements of the forest (reintroducing microhabitats, changing successional stages, etc.) rather than at the whole forest ecosystem, at least in the first instance. More information on stand-level restoration methods can be found in Section XI. The ultimate aim of restoration for a target species is the immigration of lost species and populations back to previously suitable, though today only potential, sites. The objectives set for the restoration of target species determine the methods to be used. Usually there are several alternative methods that can be used, and some examples that have been used in the restoration of boreal forests are described below.

3.2.1. Restoring Homogeneous Monocultures

Typical planted forest may consist of tree species native to the site, but the spacing is not natural (trees are planted in rows), age structure is unnatural (even-aged, one canopy layer), and the forest is lacking the mixture of other natural tree species (planted for one species, and thinnings eliminated other tree species).

By felling tree groups, small openings can be created inside the homogeneous stands. Openings mimic natural small gap dynamics, for example created naturally by wind falls. Trees felled form deadwood, whilst open areas regenerate naturally (or by planting) to native pioneering tree species.[242]

3.2.2. Mimicking Natural Forest Fires

Forest fires have been an important ecological disturbance factor in many forest types, and many species have become endangered due to the elimination of natural forest fires. Mimicking forest fires is therefore often a key restoration activity. Since many fire-specialised species can only live some years in the burned forest, it is recommended that burning will be repeated in the region two to three times per decade.

Forest fires should be planned and controlled so that fire does not spread to other areas important for conservation, such as fire refugias or old-growth forests. Recommended size for the burning is 3 to 10 hectares, designed by using natural barriers such as wet open mires, lakes, and rivers. In the absence of natural barriers, unwanted spreading of fire must be eliminated by open channels where all forest and top soil shall be cleared.

Before burning, the target area shall be prepared for the operation: some trees should be felled and piled to feed the fire, and this should be done some months earlier so trees dry and burn well. The burning should ideally affect the forest in a versatile fashion: some trees should be entirely burned, some damaged but still languish alive, and some of the trees should be slightly affected and stay alive. Fire intensity should also be variable for the other ecosystem layers: bushes, surface vegetation, and ground layer.[243]

3.2.3. Creating Deadwood by Damaging Trees

If the forest that is subject to restoration consists of tree species native to the site, but is lacking deadwood, the easiest method to increase deadwood is to fell living trees or to damage the living trees mechanically. This can be done by peeling the bark from around the tree base by chain saw, axe, or billhook. Deadwood can also be created artificially by damaging living trees with small explosive charges or by artificially introducing fungal mycelia into otherwise healthy trees.

---

[242] Tukia et al, 2001.

[243] Tukia et al, 2001.

When creating deadwood, it is important to select large trees, and produce different qualities of decaying wood, for example, by directing the falling of trees so that they lie in varied microhabitats: some in moist soil, in the shadow, and some in dry, open, scorching hot sunny places. If a forest harvester or forest tractor can be used, some trees could be pushed down with their roots, thus creating disturbances to the soil conditions, mimicking natural damages such as storms.[244]

4. Future Needs

Our knowledge of the ecology and likely population trajectories of even quite common species is still very inadequate for many forest types. Particular needs include the following:

- Better methods for assessing the restoration of ecological integrity over time for a variety of forest ecosystems
- Understanding of population levels at which long-term decline and extirpation or extinction become likely, which should serve as a trigger for active restoration efforts (especially the impact of forest continuity in time and space on metapopulations of forest-dwelling species)
- Better knowledge on the precise relationship between habitat requirements of species or functional groups and the dynamics of key habitats that can be managed and monitored with greater facility than the 5000 species living in a small temperate forest. This should be particularly done through the development of long-term research investment in some of the best existing forest laboratories (i.e., remaining old-growth forests).

- Basic taxonomic knowledge, rapid sampling and monitoring techniques for groups that represent the highest species' richness of the temperate or boreal forest, such as fungi, lichens, and invertebrates (or their habitat)

References

Hanski, I. 2000. Extinction debt and species credit in boreal forests: modelling the consequences of different approaches to biodiversity conservation. Annales Zoologi Fennici 37:271–280.

Huxel, G., and Hastings, A. 1999. Habitat loss, fragmentation and restoration. Restoration Ecology 7:309–315.

Karjalainen, H., Halkka, A., and Lappalainen, I. (ed.) 2001. Insight into Europe´s Forest Protection. WWF International Report.

Montalvo, A., Williams, S., Rice, K., et al. 1997. Restoration biology: a population biology perspective. Restoration Ecology 5:277–290.

Palmer, M., Ambrose, R., and Poff, N. 1997. Ecological theory and community restoration ecology. Restoration Ecology 5:292–300.

Penttilä, R., Siitonen, J., and Kuusinen, M. 2004. Polypore diversity in mature managed and old-growth boreal *Picea abies* forests in Southern Finland. Biological Conservation, 117(3):271–283.

Rassi, P. et al. 2003. Committee Report on Forest Restoration in Finland. Ministry of the Environment, Finland.

Siitonen, J., Kaila, L., Kuusinen, M., 2001. Vanhojen talousmetsien ja luonnonmetsien rakenteen ja lajiston erot Etelä-Suomessa. Metsäntutkimuslaitoksen Tiedonantoja 812:25–53.

Tilman, D., Lehman, C., and Kareiva, P. 1997. Population Dynamics in Spatial Habitats. Spatial Ecology, pp. 3–20. Princeton University Press, Princeton, New Jersey.

Tukia, H., Hokkanen, M., Jaakkola, S., 2001. The Handbook of Ecological Forest Restoration (in Finnish). Metsähallitus and Finnish Environmental Institute, 87 pages.

---

[244] Tukia et al, 2001.

29
Restoration of Deadwood as a Critical Microhabitat in Forest Landscapes

Nigel Dudley and Daniel Vallauri

> **Key Points to Retain**
>
> Deadwood is one of the most critically threatened microhabitats in many temperate forests and supports up to 25 percent of forest biodiversity.
>
> Deadwood can best be re-created through policy changes that allow retention of veteran, dying, and dead timber, but in a few specific cases where biodiversity loss is likely because of the short-term nature of the lack of deadwood, management to create deadwood is sometimes justifiable.

1. Background and Explanation of the Issue

Managed forests often lack critical microhabitats, because these have been deliberately or inadvertently removed. Without them much of the naturally occurring biodiversity disappears and restoration of forest quality often involves re-creation of microhabitats. Perhaps the most important of all forest microhabitats are ancient trees and deadwood. These help to:

- maintain forest productivity by providing organic matter, moisture, nutrients, and regeneration sites for conifer trees—some tree species germinate preferentially on logs;
- provide habitat for creatures that live, feed, or nest in cavities in dead and dying timber, and for aquatic species that live in pools created by fallen logs and branches;
- supply a food source for specialised feeders such as beetles and for fungi and bacteria, which in turn help maintain the food web by their own role as food for predators;
- stabilise the forest by helping to preserve slope and surface stability and preventing soil erosion; and
- store carbon in the long term, which could help mitigate some of the impacts of climate change.[245]

A newly dead tree attracts specialised organisms, principally fungi, able to break down the tough lignin layer. In Sweden 2500 fungi species rely on deadwood.[246] Next come cellulose feeders including many beetles. Research in Czech floodplain forest found 14 saproxylic (deadwood loving) ant species and 389 saproxylic beetle species.[247] Specialised birds feed on these; the great spotted woodpecker (*Dendrocopus major*) relies on insects in deadwood for 97 percent of its winter food.[248] At least 10 European owls use tree holes for nesting along with many other birds and bats, while mammals like bears shelter in hollows in dead trees.[249] Over a quarter of mammals in European forests are associated with deadwood and cav-

[245] Humphrey et al, 2002; Maser et al, 1988.
[246] Sandström, 2003.
[247] Schlaghamersky, 2000.
[248] Royal Society for the Protection of Birds, undated.
[249] Mullarney et al, 1999.

TABLE 29.1. Habitats provided by deadwood.

Living old trees	Very old trees with large canopy and cavities
	Dead wood on live trees
Standing dead trees	Newly dead standing trees with branches and twigs
	Standing trunks (snags) of different ages
	Snags with major cavities
	Young dead trees
Lying timber	Recently fallen logs
	Down logs largely intact, wood starting to soften internally
	Down logs without bark, wood softening
	Down logs well decayed, wood largely soft and discoloured
	Down logs almost completely decayed, wood powdery
	Uprooted trees
Litter to soil	Large woody debris
	Fragments of woody debris
	Coarse woody debris in rivers and streams

ities.[250] Accumulation of coarse woody debris in streams slows downstream flow, creating fish habitat and providing substrate for algae. Research in the western United States found that pools created by logs and branches provide over half the salmonid spawning and rearing habitats in small streams. Deadwood creates a variety of habitats, as shown in Table 29.1, depending on the tree species, age at death, and stage of decay; its role as food and habitat varies depending on whether deadwood is part of an otherwise living tree, a standing tree or trunk or a down log in various stages of decay.[251]

In unmanaged European broad-leaved forest, deadwood comprises 5 to 30 percent of timber, normally 40 to 200 m^3 per hectare with average volumes for beech forest of 136 m^3/hectare.[252] Yet current national averages are often only a few cubic metres per hectare, and species associated with deadwood are often at risk. In Sweden, for instance, one of the most densely forested countries in Europe, 805 species dependent on deadwood are on the national Red List because forest management does not support suitable habitat.[253]

2. Examples

The following examples illustrate the importance of deadwood in both temperate and tropical forests.

2.1. Poland: Białowieza Forest

The Bialowieza forest is one of the most natural forests in Europe, between Poland and Belarus, protected as a hunting reserve since the 1300s. On the Polish side 17 percent of the forest (10,500 hectares) is a national park, of which half has been strictly protected for over 80 years. Deadwood (mainly logs and other lying material) contribute about a quarter of the total above-ground wood biomass in the reserve, ranging from 87 to 160 m^3/hectare.[254]

2.2. France: Fontainebleau

Fontainebleau is a 136-hectare forest reserve last cut over in 1372, protected since 1853 and consisting mainly of beech with oak, hornbeam, and lime. Volumes of deadwood are 142 to 256 m^3/hectare, with higher volume following a severe storm. Volume is linked to decay time, with higher volumes but shorter retention time in the case of stands being suddenly knocked down by storms and lower, more constant volumes when trees fall naturally with age. This contrasts markedly with the current national average of deadwood for France of 2.2 m^3/hectare; most forests have as little as 1 to 2 percent of the naturally occurring deadwood densities.[255]

2.3. Finland: Southern Region

An active restoration policy has been developed under METSO (forest biodiversity pro-

---

[250] Travé et al, 1999.
[251] Dudley and Vallauri, 2004.
[252] Christensen and Katrine, 2003.
[253] Sandström, 2003.
[254] Bobiec et al, 2000.
[255] Mountford, 2002.

gramme for southern Finland) with goals of restoring 33,000 hectares, including prescribed burning on 960 hectares, an increase in dead and decaying trees on 10,500 hectares, and creating small gaps in stands on 5200 hectares and peatland restoration on 16,000 hectares. So far, 56 operational restoration plans have been prepared and some have already been implemented.[256]

2.4. Canada: Pacific Northwest

Research in Canada shows that 69 vertebrate species commonly use cavities, and 47 species respond positively to the presence of down wood. Cavity users typically represent 25 to 30 percent of the terrestrial vertebrate fauna in these forests. Around two to three large snags [over 40 cm in diameter at breast height (dbh)] per hectare and 10 to 20 smaller (20 cm dbh) snags per hectare are required for cavity nesting birds.[257]

2.5. Australia: Southern Forests of Tasmania

In Tasmania around 350 beetle species have been collected from *Eucalyptus obliqua* logs in wet eucalypt forests along with many flies, earthworms, velvet worms, and molluscs. Fungi and lichens are also heavily dependent on deadwood, and 165 bryophyte species have been recorded from logs at the same habitat.[258]

2.6. U.S.: Hawaii

Many of the woody species in Hawaii's tropical montane cloud forest germinate on down logs, particularly those with a substantial moss covering. Research found that natural coarse woody debris volume varied between 136 and 428 m^3/hectare. The presence of logs is thought to be a critical factor in ensuring regeneration in these closed canopy tropical forests.[259]

3. Outline of Tools

Today, the most threatened species in many forests are often those associated with deadwood and very old forest stands, and as a result the retention and restoration of deadwood components is seen as one of the most important challenges facing forest managers interested in creating forests that are good for both people and wildlife.[260] Forest managers have a number of tools available to help in the assessment, planning, and restoration of natural deadwood components in forests:

- *Assessment:* Assessment systems are now available to give guidance in recording and classifying deadwood components in a range of forest types, and a few governments are starting to include such assessments as a standard part of their forest inventory. The Ministerial Conference for the Protection of Forests in Europe has identified deadwood as a necessary indicator for member states, and such survey techniques are likely to increase in the future. Most surveys rely on transects or random sampling plots and use standardised recording systems to classify deadwood components with respect to size, location, and stage of decay. Assessment and an understanding of the ecology of target species' ecology are the first stages in determining restoration needs.
- *Identifying and protecting key sites:* The richness of remaining natural forest fragments is increasingly being recognised, yet many are currently being threatened or degraded. Use of initiatives such as the Natura 2000 network in Europe and additions to national protected area networks can help to maintain essential reference forests and "arks" for deadwood species. Some reserves still practice forest management, particularly in Europe, for instance, through maintaining ancient coppicing systems, and these may need to be adjusted to increase deadwood and veteran trees; a greater number of strict nature reserves are also required in many regions.

---

[256] Väisänen, personal communication.
[257] Boyland and Bunnell, 2002.
[258] Grove et al, 2002.
[259] Santiago, 2000.
[260] Vallauri et al, 2003.

- *Zoning:* In forest landscapes the proportion of deadwood desired in any one place is likely to vary according to management needs, from a fully natural deadwood component in protected areas to inclusion of deadwood components in managed secondary forest, and perhaps very little deadwood retained in intensively managed artificial plantations. Landscape-scale zoning can be a useful tool to agree necessary and desired levels of deadwood in order to support biodiversity.
- *Forest management policies:* Forest management policies should include the retention of trees and wood components likely to support saproxylic species within managed forests. Guidelines are available for what size and shape of deadwood to leave; in general, it is the larger components of deadwood (logs and standing trunks) that are likely to be missing, although in intensively managed areas even branches and twigs may have been routinely cleared. Likely components include:
  - existing large, old, dying or dead trees, pollarding senescent trees if necessary to prolong the existence of this particular habitat if it is in short supply;
  - a proportion of middle-aged trees to ensure a balanced supply of deadwood in the future;
  - key habitat areas within managed forests where stands are allowed to mature in a natural manner; and
  - fallen deadwood, including brash from thinnings (possibly a mixture of cleared and uncleared areas) and, even more importantly, large logs.
- *Using other management interventions:* Other management interventions can be considered if these are likely to help support saproxylic species, either in designated areas or more generally, including:
  - prescribed burning in boreal and some other forest habitats (there is also a need to balance deadwood retention with management of fire risk);
  - after a storm, before grant-supporting expensive salvage logging, balance the ecological and economical benefit of leaving a large amount of deadwood on the ground (without perverse subsidies, economic factors will often create a near-to-nature form of management); and
  - creation of artificial snags by leaving a proportion of some trunks standing after felling.
- *Artificial restoration of deadwood and bridging substitutes:* In a crisis, where deadwood is in such short supply that dependent species face extirpation or even extinction, short-term restoration methods may be justifiable, whereby deadwood is created through artificial disturbance. However, these are costly and only partially successful in helping to protect a proportion of the expected species and are at best an interim measure. Several strategies have been tested, including:
  - deliberate creation of standing or fallen snags, uprooted trees, leaning dead trees, and standing dead trees;
  - hastening senescence and creating habitat trees;
  - drilling, for example, nest holes of different sizes so that species using secondary nest holes have instantly created habitat; and
  - creation of habitat surrogates such as nest boxes and bat boxes: the recovery of the pied flycatcher (*Ficedula hypoleuca*) in the U.K. has been ascribed to use of nest boxes.

4. Future Needs

Perhaps the most urgent need is for a better understanding of the dynamics and importance of deadwood to the biodiversity and ecology of forests, particularly in the tropics and in Mediterranean habitats, where research has generally been more limited to date. More information is also needed about the possible costs of deadwood retention policies, including the economic costs for commercial management and more about links between deadwood and the spread of pests and diseases. (Current research suggests that this should not be a major problem, but more detailed studies are required.) Simple-to-use assessment techniques are still needed for many forest types, and a better understanding of national or regional deadwood averages. In addition, national Red Lists generally contain scant information about deadwood species such as fungi and beetles, and this gap needs to be addressed. In addition, knowledge about the role of deadwood in tropical forests is far less complete, and much research is needed on its role and conservation.

References

Bobiec, A., van der Burgt, H., Zuyderduyn, C., Haga, J., and Vlaanderen, B. 2000. Rich deciduous forests in Bialowieza as a dynamic mosaic of developmental phases: premises for nature conservation and restoration management. Forest Ecology and Management 130(1–3):159–175.

Boyland, M., and Bunnell, F.L. 2002. Vertebrate use of dead wood in the Pacific Northwest, University of British Columbia, British Columbia.

Christensen, M., and Katrine, H., compilers. 2003. A Study of Dead Wood in European Beech Forest Reserves. Nature-Based Management of Beech in Europe Project.

Dudley, N., and Vallauri, D. 2004. Deadwood, living forests. The importance of veteran trees and deadwood to biodiversity. WWF brochure, Gland, Switzerland, 16 pages.

Grove, S., Meggs, J., and Goodwin, A. 2002. A Review of Biodiversity Conservation Issues Relating to Coarse Woody Debris Management in the Wet Eucalypt Production Forests of Tasmania. Forestry Tasmania, Hobart.

Humphrey, J., Stevenson, A., Whitfield, P., and Swailes, J. 2002. Life in the deadwood: a guide to managing deadwood in Forestry Commission forests. Forest Enterprise, Edinburgh.

Maser, C., Tarrant, R.F., Trappe, J.M., Franklin, J.F., eds. 1988. From the forest to the sea: a story of fallen trees. General Technical Report PNW-GTR-229. US Forest Service, Pacific Northwest Research Station, Oregon, 153 pages.

Mountford, E.P. 2002. Fallen dead wood levels in the near-natural beech forest at La Tillaie reserve, Fontainebleau, France. Forestry: Research Note 75(2):203–208.

Mullarney, K., Svensson, L., Zetterström, D., and Grant, P.J. 1999. Bird Guide. Harper Collins, London.

Royal Society for the Protection of Birds. Undated. Leaflet. Sandy, Bedfordshire.

Sandström, E. 2003. Dead wood: objectives, results and life-projects in Swedish forestry. In: Mason, F., Nardi, G., and Tisato, M., eds. Dead Wood: A Key to Biodiversity. Proceedings of the international symposium, May 29–31, Mantova, Italy. Sherwood 95 (suppl 2), Mantova.

Santiago, L.B. 2000. Use of coarse woody debris by a plant community of a Hawaiian montane cloud forest. Biotropica 32(4a):633–641.

Schlaghamersky, J. 2000. The saproxylic beetles (Coleoptera) and ants (Formicidae) of Central European Floodplain Forests. Published by the author.

Travé, J., Duran, F., and Garrigue, J. 1999. Biodiversité, richesse spécifique, naturalité. L'exemple de la Réserve Naturelle de la Massane. Travaux scientifiques de la Réserve Naturelle de la Massane 50:1–30.

Väisänen, R. Personal communication from the director of Metsähallitus Natural Heritage Services, Vantaa, Finland.

Vallauri, D., André, J., and Blondel, J. 2003. Le bois mort: une lacune des forêts gérées. Revue Forestière Française 2:3–16.

Additional Reading

Mason, F., Nardi, G., and Tisato, M., eds. 2003. Legno Morto: Una Chiave per la Biodiversita / Dead Wood: A Key to Biodiversity. Proceedings of the International Symposium May 29–31, Sherwood 95 (suppl 2), Mantova, Italy.

30
Restoration of Protected Area Values

Nigel Dudley

Key Points to Retain

Restoration is required even within many protected areas, either because they have previously been degraded or because of overexploitation since protection, often through illegal use.

A key element in promoting restoration is the careful zoning of protected areas, particularly if these permit some level of use, to include strictly off-limit areas to allow natural dynamics; sometimes these can be temporary exclusion zones.

Careful use of the IUCN protected area categories can help determine and describe management options in protected areas.

1. Background and Explanation of the Issue

Protected area networks are based on the assumption that designated areas will be protected in perpetuity and that their values (biodiversity, environmental services, cultural importance, etc.) will survive. Unfortunately, many protected areas are under threat or are actually losing habitat and biodiversity. Current threats to forest protected areas include illegal logging, overcollection of nontimber forest products (especially poaching and bush meat hunting), and encroachment. Other protected areas have been set up in areas where forests have previously been managed and otherwise altered, degraded, or destroyed. Many forest protected areas have become isolated from other forest habitat, creating long-term problems of viability.[261] Restoration, therefore, may be required to reestablish natural habitat or to re-create or improve corridors between forest protected areas and thus build a strong protected area network.

In all these cases some form of management may be needed to restore forests or more specifically to restore and maintain specific protected-area values. Sometimes restoration will simply require protecting forests to encourage natural regeneration, but in other cases more active intervention may be needed. Where species are under immediate threat, the time and expense involved in active restoration may be justified in order to speed up the process of reestablishing suitable habitat. In large protected areas, restoration itself needs to be focussed on the most important places and approaches, such as the identification of high conservation value forest.

Restoration in protected areas can take two forms. It is often a time-limited process to restore specific areas of forest or forest types that have been degraded or destroyed (i.e., planned interventions to increase forest quality from the perspective of natural plant and animal species). However, where loss of quality comes from more intractable problems such as

---

[261] Carey et al, 2000.

persistent invasive species, or where forests have been managed for so long that they have become cultural landscapes with their own associated biodiversity, restoration may be a longer-term process that requires constant intervention both to re-create and then to maintain desired habitat.

Decisions about the extent and type of restoration should be addressed within protected area management plans, based on overall management objectives, which themselves relate to the IUCN category assigned to the area (see below).

In some parts of the world, for instance Western Europe, the eastern United States, and Southeast Asia, virtually all protected areas have been altered and could thus be candidates for restoration. However, there is also a growing movement for re-creation of "wilderness," and this creates tension with restoration activities and sometimes a backlash against management interventions within protected areas.[262] There is an inherent contradiction between intervening to increase forest quality and reducing interventions to increase naturalness and wilderness. Promoting "passive restoration" (for example, by removing the threats and pressures that are altering forests) can sometimes achieve both ends. Sometimes forests are actively suppressed to enhance biodiversity values, such as in the various savannah habitats of national parks in East Africa where regular burning is used to prevent trees from encroaching. The extent to which it is possible to re-create wilderness values is still not well tested.

Restoration can be and is practised in all types of protected areas, from the most strictly protected to cultural landscape areas with relatively large resident human communities. In addition, IUCN has defined one type of protected area—category IV: habitat/species management area—as protected areas managed mainly for conservation through management interventions, which often include a large element of restoration.

1.1. Recognition of the Need for Restoration in Protected Areas

The international community has long recognised the importance of restoration within protected areas. For example in 1972 the original wording of the World Heritage Convention (Article 5-d) included this requirement: "To take appropriate legal, scientific, technical, administrative and financial measures necessary for the identification, protection, conservation, presentation *and rehabilitation* of this heritage" (our emphasis).

In February 2004, the Seventh Conference of the Parties of the Convention on Biological Diversity met in Kuala Lumpur to look specifically at protected areas. Its draft Programme of Work on Protected Areas includes the following in suggested activities for parties (1.2.5): "Rehabilitate and restore habitats and degraded ecosystems, as appropriate, as a contribution to building ecological networks and/or buffer zones."

2. Examples

Many restoration activities simply involve reducing pressures by force or by agreement; in other cases more active measures are also needed on the ground. The following examples show some of the ways in which restoration is being attempted within protected areas:

2.1. Jordan: Restoration Can Sometimes Simply Involve Removing Immediate Pressures

In Dana Nature Reserve in central Jordan, agreements between local Bedouin and park authorities have halved the number of goats grazing within the reserve to 9000, leading to large-scale forest regeneration in what had previously been almost a desert landscape. Here the efforts at restoration were more in negotiating agreements than in management interventions and have been accompanied by efforts to provide alternative livelihoods for local people through ecotourism, agriculture, and selling herbs (information from discussion with park guards, September 2000).

---

[262] Landres et al, 2000.

2.2. Finland: Active Restoration Is Used to Accelerate the Achievement of a Natural State in Areas Previously Utilised Commercially

In this example in Finland, the longer term aim is the creation of ecologically coherent, self-sustaining areas of woodland where natural dynamics are the driving forces behind change. Such interventions are used particularly in protected areas in the south, where long-term management has altered forest composition and structure. The main measures used are helping deciduous saplings to establish by making small clearings, deliberate creation of deadwood by damaging trees to hasten the restoration of natural decay patterns, and use of artificial forest fires.[263]

2.3. Costa Rica: Where Deforestation Has Been Severe, Active Planting May Be Needed to Restore Forest Cover

In the Guanacaste National Park in Costa Rica, severe forest loss necessitated artificial reforestation, including the use of *Gmelina* plantations to provide a nurse crop for natural forest regeneration.[264]

2.4. France: Even in Relatively Pristine Forests, Invasive Species Can Create Arguments for Restoration

In Fontainebleau Forest strict reserve, near Paris, native woodland has been left to regain natural structure and functioning, but the area has been invaded by Japanese knotweed (*Fallopia japonica*) where tree fall creates gaps in the canopy. There is a debate about whether nonintervention can work in situations where the natural ecology has already been radically altered.[265]

---

[263] Metsähallitus Forest and Park Service, 2000.
[264] Janzen, 2000.
[265] Dudley, 1996.

2.5. U.S.: Restoration of Wilderness Values Requires Particular Management Steps

Many officially designated wilderness areas have been settled in the past and are now being managed to restore values of naturalness and wilderness. For example, the Coronado National Forest in Arizona contains many wilderness areas that have previously been subject to gold mining, settlement, logging, and ranching. All logging has now been banned from these areas, and relics of human activity are left to decay over time. Current visitation is managed, with, for instance, camping permitted in only a few designated areas. These management actions reflect a desire to increase wilderness values in what is already a fairly natural forest from the perspective of biodiversity, although gold mining would still be legal in the area (information collected on site visit).

3. Outline of Tools

Protected area managers can choose from a range of assessment, planning, and management tools to re-create or restore natural forests in their reserves. Once needs have been identified, many restoration approaches described elsewhere in this book may be appropriate.

Assessment frameworks for wilderness and naturalness: A key element in developing restoration strategies is determining an end point for restoration. Fortunately, many definitions of natural forest exist and some have associated assessment methodologies. While these provide some useful assessment tools, most have been developed for temperate forests and do not translate well to tropical conditions, or necessarily between forest types even in temperate countries. More generalised tools for assessing naturalness and wilderness still need to be developed. In general, we would propose that protected area managers concentrate on re-creating the values and conditions that they are trying to manage for, rather than aiming to reproduce an (often largely hypothetical) "original" forest.

Management: Plans and zoning of use: most protected areas do not exist as single management entities, but instead are zoned into areas with different management approaches, and different regulations regarding use and level of protection. IUCN divides protected areas into six categories[266]:
- Category Ia: Strict nature reserve/wilderness protection area managed mainly for science or wilderness protection
- Category Ib: Wilderness area: protected area managed mainly for wilderness protection
- Category II: National park: protected area managed mainly for ecosystem protection and recreation
- Category III: Natural monument: protected area managed mainly for conservation of specific natural features
- Category IV: Habitat/species management area: protected area managed mainly for conservation through management intervention
- Category V: Protected landscape/seascape: protected area managed mainly for landscape/seascape conservation or recreation
- Category VI: Managed resource protected area: protected area managed mainly for the sustainable use of natural resources.

Although these categories describe the main purpose of the reserve (and should apply to at least two thirds of its area) other forms of management are possible in the remainder to meet the needs of local communities, visitors, or, for instance, because active restoration is needed in an otherwise strictly protected area.[267] Identification of the need, extent of, and timing for restoration should be a key part of management plans in those forest protected areas where restoration is needed, including the identification of specific targets, approaches, and timetables.

Access controls to allow regeneration: Protected areas in which one management authority controls the whole site can use zoning, including temporary zoning such as exclusion zones for visitors or for herbivores, to facilitate natural regeneration or to increase the speed and success of regeneration planting. A variety of different approaches exist:
- More or less permanent exclusion zones to allow long-term recovery of forest types that have lost old-growth characteristics. For example, it will take hundreds of years to recover fully old-growth characteristics in the recovering kauri (*Agathis*) forests of New Zealand, which were almost totally destroyed by miners but are now gradually regrowing in a series of national parks and reserves where grazing and felling are both controlled (information from reserve staff in 1991).
- Temporary exclusion zones to allow recovering forest to get a head start without trampling from visitors, once seedlings have established the exclusion zone can be removed. For example, such exclusion zones are established on Stradbroke Island off the coast of Queensland, Australia, in reserves established on former sand quarry sites where poor soils make tree establishment relatively difficult (information from a site visit, 2000)
- Agreements with landowners: protected areas under the control of multiple landowners, for instance many category V reserves, or with multiple stakeholders, need to rely instead on voluntary agreements with landowners, with or without compensation payments, to facilitate restoration.[268] Such agreements might be to exclude grazing stock from particular areas or for more active regeneration activities. If possible, agreements should be developed in such a way as to create benefits for all parties, for instance, a community agreement to restore a forest that

---

[266] IUCN, 1994; Phillips, 2003.
[267] IUCN, 1994.
[268] Phillips, 2003.

would be both a form of erosion control and a wildlife habitat.
- Active restoration activities: lastly, protected area managers will also have to resort to the kinds of active interventions that are described elsewhere in this book. Particular needs in the case of protected areas might relate to:
  - tourist impact (e.g., trampling, damage at camping grounds, trails, etc.)[269];
  - areas being reclaimed following past activity such as mining, quarrying, etc. (see "Opencast Mining Reclamation"); and
  - areas being restored through eradication of exotic invasive species (see "Managing the Risk of Invasive Alien Species in Restoration").

4. Future Needs

Key needs for the future include more systematic integration of restoration into protected area networks (for example, through buffer zones, corridors, etc.) and greater investment for restoration in protected areas, which is still generally approached as a minor part of protected area management.

[269] Eagles et al, 2002.

References

Carey, C., Dudley, N., and Stolton, S. 2000. Squandering Paradise: The Importance and Vulnerability of the World's Protected Areas. WWF International, Gland, Switzerland.

Dudley, N. 1996. Why research in natural forest reserves? A discussion paper for COST Action E4, Fontainebleau, September 12–14, 7 pp.

Eagles, P.F.J., McCool, S.F., and Haynes, C.D. 2002. Sustainable Tourism in Protected Areas: Guidelines for Planning and Management. Cardiff University and IUCN, Cardiff and Gland, Switzerland.

IUCN. 1994. Guidelines for Protected Area Management Categories. IUCN, Gland, Switzerland.

Janzen, D.H. 2000. Costa Rica's Area de Conservación Guanacaste: a long march to survival through nondamaging biodevelopment. Biodiversity 1(2):7–20.

Landres, P.B., Brunson, M.W., Merigliano, L., Sydoriak, C., and Morton, S. 2000. Naturalness and Wildness: The Dilemma and Irony of Managing Wilderness. Proceedings RMRS, Wilderness Science in a Time of Change Conference, Missoula, Montana, May 23–27, 1999, USDA Forest Service, pp. 377–381.

Metsähallitus Forest and Park Service. 2000. The Principles of Protected Area Management in Finland: Guidelines on the Aims, Function and Management of State-Owned Protected Areas. Natural Heritage Services, Vantaa, Finland.

Phillips, A. 2003. Management Guidelines for IUCN Category V Protected Areas: Protected Landscapes and Seascapes. Cardiff University and IUCN, Cambridge, UK.

Section X
Restoring Socioeconomic Values

31
Using Nontimber Forest Products for Restoring Environmental, Social, and Economic Functions

Pedro Regato and Nora Berrahmouni

Key Points to Retain

The economic and social significance of nontimber forest products (NTFPs) to sustain people's livelihoods and local, national, and international markets justify the need to invest resources in harvesting, growing, and planting a wide range of native plant species.

Applying and adapting the existing ecological restoration techniques to NTFPs can help secure focal species' habitat requirements and diversify natural resource production on which sustainable forest management is based.

Well-defined tenure and access rights and funding mechanisms can provide adequate incentives for creating community-based NTFP income-generating restoration initiatives.

1. Background and Explanation of the Issue

Nontimber forest products (NTFPs) are defined as biological resources of plant and animal origin, derived from natural forests, managed forests, plantations, wooded land, and trees outside forests. What distinguishes NTFPs from agricultural products is their origin: they come from species of flora and fauna native to the forest systems, and the wild or semidomesticated mode of production.[270]

An indication of the socioeconomic importance of NTFPs is the fact that 80 percent of the population from the developing world meets a proportion of its health and nutritional needs through NTFPs.[271] Several million households worldwide depend on NTFPs for subsistence consumption or income. Global attention to NTFPs has recently increased, mainly due to two factors:

- Their compatibility with environmental objectives, including the conservation of biological diversity
- Their contribution, not only to household economies and food security, but also to national economies

There are at least 150 NTFPs that contribute substantially to international trade, including honey, gum arabic, rattan and bamboo shoots, cork, forest nuts and mushrooms, oleoresins, essential oils, and plant or animal parts for pharmaceutical products.

The NTFPs' availability in forest landscapes is related to the maintenance of high plant diversity rates, and the existence of a rich mosaic of habitat types and well-structured forests.

---

[270] Moussouris and Regato, 1999.
[271] FAO, 1997.

1.1. The Multifunctional Forest Concept

Historically, in many forest areas, rural communities have developed forest management systems that meet multiple functions or purposes, in which their economies are based on the harvesting and production of a wide range of NTFPs channelled through local, national, or international markets. Under these circumstances, forest landscapes have been to a certain extent human-shaped, characterised by a rich mosaic-like structure integrating natural forests, several wooded, shrub and grassland formations, and seminatural agroforestry land areas, including extensive agricultural land.

Unfortunately, many traditional multipurpose forestry systems have been lost or collapsed in numerous forest areas due to sociopolitical instability or macroeconomic drivers. The result has been the intensification of one single forest use—the conversion of forest land into agriculture or nonnative tree plantations—and significant biodiversity loss and land degradation.

1.2. Forest Landscapes and Habitat Diversity: The Environmental Values of NTFPs

The production of NTFPs can be expected to produce less severe environmental impacts to forest ecosystems than timber extraction. Valuing and supporting new economic opportunities based on NTFPs as part of multipurpose forest systems can contribute to both improving the environmental benefits of forest landscapes and to sustaining and improving livelihoods, especially in less favoured rural areas.

1.3. Traditional Sustainable Management Systems: The Economic and Social Significance of NTFPs

Considering people's high dependence on NTFPs for their livelihoods, there is a significant economic incentive for many countries to develop the NTFP production potential of their forests and to generate positive socioeconomic benefits for rural populations while ensuring that these are compatible with conservation values. However, to deliver this potential there is a need to modify current economic notions that govern forest management, notably by enlarging and improving market opportunities, and securing payment mechanisms and incentives for land owners/users to restore forest resources and the goods and services that they provide.

The NTFP markets are also important at the regional and international levels as they provide revenues for the actors directly involved and for the government. At the international level, it is estimated that the trade in NTFPs amounts to $11 billion. The European Community (EC), the United States, and Japan account for 60 percent of world imports of NTFPs, and the general direction of trade is from developing to developed countries.[272]

1.4. NTFPs As a Response to Poverty and As a Safety Net for the Poorest Members of Society

Forest biodiversity, via NTFPs (harvested or hunted biological products from wild or cultivated sources), plays an important role in addressing poverty for marginalised, forest-dependent communities. The NTFPs contribute to livelihood needs, including food security, health and well-being, and income.[273] In many parts of the world these resources are critical for the poorest members of society who are often the main actors in NTFP extraction and may provide them with their only source of income. Ninety percent of people who earn less than one dollar a day depend on forests for their livelihoods, according to the World Bank (see Box 31.1).

---

[272] Ndoye and Ruiz-Perez, 1997.
[273] Pagiola et al, 2002.

> **Box 31.1. NTFPs in figures**
>
> It is estimated that 1.5 million people in the Brazilian Amazon derive their income from extractive products.
>
> In the forest zone of Southern Ghana, it is estimated that 258,000 people or 20 percent of the economically active population derive income from NTFPs.[1]
>
> In Nigeria, it is estimated that 78,880 tons of *Irvingia gabonensis* are marketed per year.[2]
>
> In the Mediterranean region, the production of NTFPs is well below its potential. For instance, the current cork production (3.7 million tonnes/year), game production (1.2 million tonnes/year), and medicinal/aromatic plants (4.5 million tonnes/year) represent in all around one third of their potential.[3]
>
> Source: Shanley et al, 2002.
> [1] Townson, 1995.
> [2] Shanley et al, 2002.
> [3] Moussouris and Regato, 2002.

2. Examples

2.1. NTFPs in the Mediterranean Region: Restoring the Ecological, Social, and Economic Functions of Cork Oak Forest Landscapes[274]

Cork oak (*Quercus suber*) characterises mosaic-like forest landscapes in the siliceous lowland and mid-mountain areas of the western Mediterranean region. Even though cork represents the main economic interest (270,000 tonnes/year, which represents $100 million) the environmental, social, and economic sustainability of cork oak forest systems depends on a diversified production of several NTFPs (i.e., edible nuts, fruits and acorns, honey, medicinal and aromatic plants, mushrooms, game, resins, spirits, basketry, pastures) from which farmers get their annual revenue (for example, a diverse NTFPs production of more than 10 products in cork oak and holm oak sylvopastoral systems represented a total amount of 433 million euros in 1986 in Spain).

Bad management practices, overexploitation of a few resources (i.e., firewood and grazing), land conversion, and climate change have all contrived to greatly threaten remaining cork oak forest areas.

More than 240,000 hectares of cork oak trees have been planted in Portugal and Spain since 1993, funded by the European Commission's agriculture subsidies. Nevertheless, the simple action of planting cork oaks may be neither environmentally sufficient nor seen as economically interesting for land owners who will not be prepared to wait 20 to 30 years to make a profit. On the other hand, by applying ecological restoration principles and emphasising multifunctionality in the landscape, land owners/users may benefit economically after 5 to 10 years. By restoring the forest ecosystem as a whole through planting a wide number of native trees, shrubs, and herbal plants—for example, strawberry tree (*Arbutus unedo*), harvesting for the production of spirits, aromatic shrubs' harvesting for distillation, game, honey, etc.—they can benefit from the harvesting of these various NTFPs well before the planted cork oak trees become productive. Appropriate incentives focussing on such multipurpose restoration practices may change people's attitudes from short-term choices to longlasting sustainable management systems.

---

[274] This example has been extracted from Moussouris and Regato, 1999; Moussouris and Regato, 2002; and Oliveira and Palma, 2003.

The restoration of cork oak forests implies a set of management options, among which we may highlight the following:

- Production of native trees and shrubs in tree nurseries for (1) developing mixed plantations—alternating oaks with faster growing small fruit trees and aromatic shrubs in degraded land; (2) diversifying the species' composition of high shrubs and forest stands; (3) increasing tree density and understorey species' composition in open woodlands; (4) creating vegetation lines along river networks and ravines
- Improving natural regeneration of oak species through pruning and rotating livestock systems
- Diversifying native species' composition in grasslands through seedlings
- Simulating natural fire breaks by creating a mosaic of forest gaps in sensitive areas with grasslands, small shrubs plantation lines, and scattered oak trees
- Specific management plans for controlling the dispersion of pioneer monospecific *Cistus* spp. formations through harvesting for *Cistus* distillation, and diversifying them through plantation of fruit and honey shrub species

2.2. NTFP Restoration in Southeast Asia: The Case of Rattan Species' Production[275]

Rattans are light-demanding climbing palms exploited for supplying cane for furniture, matting, and basketry markets. Moreover, rattans play an important role in the subsistence strategies of many rural populations in Southeast Asia (e.g., edible fruits and palm heart, medicines, and dyes). During the last 20 years, the rapid expansion of the international and domestic trade in rattan ($6.5 billion/year) has led to substantial overexploitation of the wild resources. In addition, the lack of adequate resource tenure contributes to their irrational exploitation in many forest areas.

Current attempts at long-term in situ management of rattan in the wild have demonstrated the value of developing a range of restoration options, which include the following:

- Specific management plans for creating "extractive reserves" in community forests and low-level protected areas, where local people harvest rattan population within carrying capacity margins, which secures its natural regeneration
- Enrichment planting and canopy manipulation (opening "artificial" gaps) in selectively logged natural forests, as a way to enhance rattan natural regeneration. This is perhaps the most beneficial form of cultivation, both in terms of productivity and maintenance of ecological integrity.
- Rattan cultivation as part of agroforestry systems, by rotating 7- to 15-year cycles of rattan with plant food crops
- Planting rattan within tree-based fast growing plantations, such as rubber (*Hevea brasiliensis*)

To improve harvesting techniques and avoid any impacts on potential sustainability, the younger stems of clustering species should be left to regenerate future sources of cane, and harvesting intensity should be based on long-term assessments of growth rate and recruitment.

2.3. NTFP Restoration in Latin America: The Dragon's Blood Case in Western Amazonia[276]

Dragon's blood is the generic name of neotropical trees of the genus *Croton*, used to treat a wide range of health problems. *Croton* species are all pioneer, light-demanding species, commonly associated with nonflooded riparian habitats, as well as low- and mid-elevation secondary forests in human-disturbed areas and forest gaps in mature forests. For many years, Dragon's blood has been used by rural inhabi-

---

[275] The case study text has been extracted from Shanley et al, 2002; Sunderland and Dransfield, 2002.

[276] The case study text has been extracted from Alexiades, 2002.

tants and urban dwellers within and beyond the tropical forests, and commercialised by an extensive and largely informal network. During the last decades, *Croton* latex has become an international commodity, reaching over 26 tonnes in 1998. Commercial harvesting is having a clear ecological impact on *Croton*, especially in the most accessible areas, affecting its distribution and demographics, which has been a source of concern for nongovernmental organisations (NGOs) and government agencies.

Management regimes for *Croton* propagation and reforestation have been adopted in Amazonian agroforestry systems, accompanied by a concomitant "professionalisation" of all concerned actors. *Croton*'s role as a pioneer species and its association with secondary forests make it an ideal candidate for increasing economic returns from fallow management. Abandoned crops and pastures are ideal environments for the establishment of mixed forest stands, including *Croton* seedlings together with other timber species. Restoration programmes with *Croton* in Peru have combined Dragon's blood trees with medicinal plants, several timber trees, including *Cedrela* and *Swietenia*, and crop species such as coffee, cacao, *naranjilla*, and manioc. The central government of Peru has established an official goal of planting two million *Croton* trees.

3. Outline of Tools

3.1. Valuing NTFPs in Rural Development

Quantifying in economic terms the value of NTFPs and the income they can provide rural families is an important step forward for understanding the prevalent role of forest resources in rural subsistence. If NTFPs were appropriately valued, this could provide a powerful argument to governments and the private sector to alter or reverse wrong spatial planning decisions in forest landscapes of outstanding biodiversity. When planning the conversion of forests into agricultural land for subsistence reasons, it is necessary to estimate the real economic value of these forest resources in order to make an informed decision. Economically oriented projects involving the use of native plant species should be subjected to a thorough cost-benefit analysis before being implemented. Generally speaking, there is a growing need to argue and reaffirm the fact that NTFPs significantly contribute to many local and national economies, and have an unknown potential that needs to be further researched.

There are a number of processes for evaluating what has been called the "hidden forest harvest"[277]: (1) understand and assess the role of forests in rural livelihoods, (2) assess the economic value of resources for rural households, (3) value the local and regional markets for forest products, (4) measure nonmarket values, and (5) develop economic decision-making frameworks. These methods are based on a set of general principles: (1) data collection must be done at the most appropriate social organisational unit—family, gender, or other major relationship; (2) collection of data on income, consumption, and expenditures should include as much as possible on uses of NTFPs; and (3) data must be quantitative for statistical analysis and must be harmonised to make sure there is coherence between different surveys. Participatory rural appraisal methods help understand the social context and help design the most appropriate survey form. Data are gathered through periodical interviews (i.e., semester interviews) in order to obtain fresh information about the yearly cycle of NTFP use.

3.2. Harvesting, Growing, and Planting NTFPs

There are a number of ecological guidelines and techniques applicable for restoring NTFP source species in degraded forest land, described in several chapters of this book. In all cases, specific research and field testing is needed to get the necessary know-how on harvesting, growing, and planting the wide range of trees, shrubs, and herbs native to each forest ecosystem, as well as to facilitate natural regen-

---

[277] Campbell and Luckert, 2002.

eration and habitat improvement techniques. Standardised protocols for seed collection, mycorrhization of nursery plants, nursery and field techniques for reduction of transplant shock, need to be developed through pilot experiences.

3.3. Establishing Community-Based Income-Generating Associative Systems Based on NTFPs

Well-defined tenure and access rights can provide an incentive for local communities to manage their natural resources sustainably.[278] Replacement of communal tenure systems with government management regimes and private property has reduced people's access to NTFPs, which have traditionally been an important part of their livelihoods. This fact has had detrimental consequences, by increasing both uncontrolled overexploitation of forest resources and biodiversity loss.

A number of treaties covering indigenous people's rights to tenure, resource access, benefit sharing, and intellectual property rights have been recently drafted and legally adopted in several countries.

For instance, in the last decade, the Tunisian government has established a legal framework to provide local communities with access to NTFPs in the state-owned forests and organisational means for people living in forest land to manage them. WWF, the global conservation organisation has assisted local communities to build pilot local associations of common interest in the cork forest land through education, institutional development, and training programmes for implementing forest management plans and NTFPs' harvesting.[279]

A pilot forest plan (Plan Piloto Forestal) was conducted in Quintana Roo State, on the Yucatan Peninsula of Mexico, with the aim to increase empowerment and control of forest extraction activities to communities. This programme was built with political and technical support, following a "bottom-up" approach, which emphasised local decision making and negotiation.[280]

The success of the Tunisian and Quintana Roo pilot experiences have gained domestic and international recognition, and these projects are now seen by governments, intergovernmental organisations, and NGOs as models for similar initiatives in both countries.[281]

4. Future Needs

4.1. NTFPs and Forest Certification

Certification is a policy tool that attempts to foster responsible resource stewardship through the labelling of consumer products. Even if forest certification has tended to focus on timber products, opportunities exist to promote sound ecological and social practices in NTFPs' management to support restoration in degraded forest landscapes of outstanding biodiversity and increase local communities' revenues and trade opportunities through this market tool.

The certification systems that are relevant for NTFPs include sustainable forestry, organic agriculture, and fair trade. The Forest Stewardship Council (FSC) promotes well-managed forests through the application of criteria that address ecological, social, and economic issues.[282] The International Federation of Organic Agriculture Movements (IFOAM) has criteria for wild-harvested products as well as specific criteria for some NTFPs like maple syrup and honey. The Fair-Trade Labelling Organisation (FLO), developed out of the alternative trade movement, currently certifies a limited number of agroforestry products, although its product range is increasing. The integration of the three certification schemes will appeal to a broader consumer market as it may address in a more cost-effective and harmonised manner environmental, harvesting, processing, sanitation, benefit sharing, social and worker welfare, and chain-of-custody criteria.

---

[278] Shanley et al, 2002.
[279] WWF, 2003.
[280] Shanley et al, 2002.
[281] WWF, 2003.
[282] Mallet, 2001.

Certification specific to NTFPs is very recent, and principles and processes are still being worked out. Two certification bodies have played a major role in NTFP certification: the Rainforest Alliance's Smartwood Programme is certifying and labelling NTFPs through FSC, and the Soil Association's Woodmark Programme offers a joint FSC/IFOAM certification. In 2002 seven FSC certificates were issued that permitted commercial NTFP harvesting (Chicle latex in Mexico; maple syrup in the U.S.; Acai juice and palm hearts in Brazil; 30 cosmetic plants in Brazil; Brazil nut in Peru; Oak tree bark in Denmark; Venison in Scotland).[283] Certification standards for cork oak forests and pine resins have been developed in Spain, and several pilot cork certification initiatives are ongoing in Portugal, Spain, and Italy.

4.2. NTFPs in National Forestry Curricula

The use of NTFPs in forest landscape restoration programmes poses new challenges to the forestry sector traditionally orientated toward afforestation with a few fast-growing timber tree species in degraded areas. New expertise and know-how on managing, harvesting, growing, and planting a wide range of trees, shrubs, and herbal NTFP species is required to undertake a thorough assessment of the potential and opportunities for candidate NTFP restoration operations.

During the last two decades NGOs, private cooperatives, and research institutions have played an important role in raising awareness, developing NTFP production cooperatives, and assisting local communities and governments in developing pilot field experiences and restoration protocols for growth in tree nurseries and planting of a wide range of NTFPs. Currently, the forestry sector curricula and university study programmes are under revision for integrating ecological restoration and NTFPs' conservation and management in countries such as Spain and Morocco.

4.3. Legal Frameworks and Economic Incentives for NTFPs to Support Local Development

Government regulations about NTFPs' conservation, access rights, management, and commercialisation are not always well defined. Moreover, existing laws are occasionally contradictory and require resolution. In Latin America, for instance, most forestry concessions are granted for timber, while NTFPs are harvested without management plans through short-term permits and government-established quotas. In other cases, NTFP management falls under different ministries and legislations, making it a difficult issue to deal with for managers and certifiers. In the Mediterranean region, there is a cork oak forest conservation law in Portugal, while in North African countries local communities' rights of access for NTFPs in cork oak forest land are not always defined and the governments have the control of cork as a product.

International organisations and NGOs may play a greater role in advocating and assisting forest managers and governments to improve NTFP legislation and guidance, given that insufficient resources or incentives have been allocated to products that traditionally have generated small amounts of taxable income for states. Certification may serve to catalyse governments and multilateral organisations' nascent efforts to reinforce markets and legislation related to NTFPs.

References

Alexiades, M.N. 2002. Sangre de drago (*Croton lechleri*). In Shanley, P., Pierce, A., Laird, S.A., and Guillén, A. eds. Tapping the Green Market: Certification and management of non-timber forest products, Earthscan, London.

Brown, L., Robinson, D., and Karmann, M. 2002. The Forest Stewardship Council and non-timber forest product certification: a discussion paper. FSC, Mexico.

Campbell, B.M., and Luckert, M.K. 2002. Evaluando la Cosecha Oculta de los Bosques. Nordan-Comunidad Ed., Montevideo, 270 pages.

---

[283] Brown et al, 2002.

FAO. 1997. Non-Wood Forest Products Forestry Information Notes Handout, Rome.

Mallet P. 2001. *Certification Challenges and Opportunities*. Falls Brook Centre, Canada.

Moussouris, Y., and Regato, P. 1999. Forest harvest: Mediterranean woodlands and the importance of non-timber products to forest conservation, *Arborvitae* supplement. WWF/IUCN. Longer referenced version can be found at http://www.fao.org/waicent/faoinfo/forestry/nwfp/public.htm.

Moussouris, Y., and Regato, P. 2002. Mastic gum, cork oak, pine nut, pine resin and chestnut. In: Shanley, P., Pierce, A., Laird, S.A., and Guillén, A. eds. Tapping the Green Market: Certification and management of non-timber forest products, Earthscan, London.

Ndoye, O., and Ruiz-Perez, M. 1997. The markets of non-timber forest products in the humid forest zone of Cameroon. In Doolan, S. ed. African Rainforest and the Conservation of Biodiversity. Proceedings of the Limbe Conference, pp. 128–133, Earthwatch Europe, London.

Oliveira, R., and Palma, L. 2003. *Un Cordão Verde para o Sul de Portugal. Restauraçao de Paisagens Florestais*. ADPM ed., Portugal.

Pagiola, S., Bishop, J., and Landell-Mills, N. 2002. *Selling Forest Environmental Services*. Earthscan, London.

Shanley, P., Pierce, A., Laird, S.A., and Guillén, A., eds. 2002. *Tapping the Green Market:* certification and management of non-timber forest products, Earthscan, London.

Sunderland, T.C.H., and Dransfield, J. 2002. Rattan. In Shanley, P. Pierce, A., Laird, S.A., and Guillén, A., eds. 2002. Tapping the Green Market: certification and management of non-timber forest products, Earthscan, London.

Townson, I.M. 1995. *Income from Non-Timber Forest Products: Pattern of Enterprise Sciences*. Oxford University.

Vallejo, V.R., Serrasolses, I., Cortina, J., Seva, J.P., Valdecantos, A., and Vilagrosa, A. 2000. Restoration strategies and actions in Mediterranean degraded lands. In: Enne, G., Zanolla, Ch., and Peter, D., eds. *Desertification in Europe: Mitigation Strategies, Land-Use Planning*. European Commission, Luxembourg.

WWF. 2003. Conservation and Management of Biodiversity Hotspots in the Mediterranean. 10 Lessons Learned. WWF Mediterranean Programme, Rome.

Additional Reading

UNEP/WCMC. Nontimber forest products, Web site: http://valhalla.unep-wcmc.org/ntfp/biodiversity.cfm?displang=eng.

32
An Historical Account of Fuelwood Restoration Efforts

Don Gilmour

Key Points to Retain

Fuelwood is an essential component of people's livelihoods in developing countries, with average fuelwood requirement per family being estimated at 200 kilogrammes per person per year. Yet, fuelwood production and collection has been blamed for much forest loss and degradation.

Over the last few decades, different approaches have been taken to fuelwood production, from large-scale industrial plantations (1960s–1970s) to village woodlots (1970s–1980s) and a "people first" era (mid-1980s–1990s). The emphasis over these decades has shifted toward better understanding local people's needs and involving them in producing fuelwood.

The key constraints in addressing fuelwood shortages are social and political rather than technical, and relate to full engagement and empowerment of communities.

Future needs to improve fuelwood production include creating the right political and social conditions for people to make informed decisions about the sort of restoration objectives they have for their landscape.

1. Background and Explanation of the Issue

1.1. The Role of Fuelwood in Forest Loss and Degradation in Developing Countries

Forests in many developing countries are under heavy pressure to provide subsistence goods. The product that has received most attention is fuelwood, as it is often the major source of energy for cooking and heating. However, in many situations, particularly in parts of South Asia, forest products also provide the mineral nutrients that are essential for the maintenance of farming systems. In some cases the harvest of fodder (both grass and tree leaf material) can greatly exceed the biomass harvest of fuelwood. A common estimate of the average fuelwood requirement per family is about 200 kg per person per year, while the average off-take of fodder can be about 5000 kg per family per year.[284] Unrestricted biomass harvest has been blamed for much of the deforestation and forest degradation that has occurred in developing countries during recent decades. Whilst the role of fuelwood collection has sometimes been exaggerated, it has certainly contributed to forest degradation in some places and, particularly where collected commercially, has caused significant deforestation. As a consequence of fears about the impact of fuelwood

[284] Gilmour and Fisher, 1991.

and fodder collection, many development projects have focussed on forest restoration as a solution to both environmental and economic problems associated with forest loss and degradation.

In theory, forest restoration for fuelwood should be amongst the easiest forms of restoration, with its uncomplicated emphasis on rapid growth of a few species that burn effectively. Experience in places where forest restoration for fuelwood has worked show that there are few insurmountable technical difficulties. However, despite years of hard work and financial investment, efforts to restore forests for local human needs remain at best only partially successful in the main centres of activity in Africa and Asia. An understanding of why this occurred is essential if restoration efforts are to help provide energy and agricultural resources to many of the world's poorest communities.

It is possible to recognise three distinct eras that represent different approaches to the restoration of forest resources in these regions:

- Industrial plantation era: 1960s–1970s
- Woodlot era: 1970s–mid-1980s
- People first era: mid-1980s–1990s

The summary in Table 32.1 is drawn from the well-documented changes that have taken place in parts of South Asia and Africa. Fuelwood projects have been implemented successfully, but there has also been a depressingly long list of failures. Critical questions of equity and access remain even in some countries where there have been long-term and relatively successful programmes. Similar examples can be found in other parts of the world, although different countries have not followed the same time line. For example, most of Southeast Asia, Papua New Guinea, and the Pacific and large parts of central Africa, Latin America, and the ex–Soviet Union countries are only now coming toward the end of an era of major industrial focus for their forests. However, most (but not all) countries in these regions are converging rapidly toward embracing participatory approaches for many aspects of forest management.

In practice, many of the world's poorest people still rely primarily on wood products for their energy—about half of the global population. Forest landscape restoration projects are unlikely to be successful in areas where people need fuelwood unless they take this into account, and many communities will support restoration only if they can see clear benefits in terms of fuelwood resources. Natural forests managed primarily for fuelwood and fuelwood plantations can both be integrated successfully with wider efforts to restore forest area or quality, but require a detailed understanding of community needs, social structure, land tenure, and access and use rights.

TABLE 32.1. Three eras in fuelwood plantations.

Period	Characteristics
Industrial plantation era (1960s to 1970s)	Strong belief in importance of industrialisation of forestry for production of raw materials to meet needs of expanding populations and economies; belief that increased employment opportunities in rural areas would lead to decreased poverty
Woodlot era (1970s to 1980s)	Emphasis on afforestation and village woodlots based on scaling down of conventional forestry practices as a means to address fuelwood and desertification problems
People First era (1980s–1990s)	Increased understanding about the role of trees in livelihood strategies of rural people; less emphasis on firewood, more on management of existing forests, multiproduct species, integration of tree-growing with agriculture in agroforestry and farm forestry systems and on participation by target populations; an increased focus on nontimber forest products as sources of household income and welfare and a growing emphasis on devolution and increased participation, and on encouraging local management of forests as common property; stronger support for legislation to empower local users, and to protect the rights and lifestyles of forest dwellers

Adapted from Arnold, 1999, and Wiersum, 1999.

2. Examples

This section reviews fuelwood plantations through time.

2.1. Industrial Plantation Era: 1960s and 1970s

The key elements of this era are characterised by technical approaches to forest restoration and the creation of timber plantations for projected fuelwood and timber shortages; the assumption that industrialisation of all sectors including forestry would bring social and economic benefits to all sectors of society, with the benefits trickling down; and the application of technical and somewhat standardised approaches to management with little consideration of existing (local or indigenous) forest use systems and the local social and economic context. Despite heavy investment, most of these projects failed to deliver expected benefits. Furthermore, local people often suffered as a result of removal of natural forests, loss of rights and biodiversity, and because they missed out on any benefits that did occur. Examples of such plantations can be found in many parts of East Africa.

2.2. Woodlot Era: Late 1970s to Mid-1980s—From Industrial Forestry to Local Needs Forestry

As a result of the clear failures of the large-scale projects of the 1960s, a more localised and small-scale approach to forest development was introduced through major funding to woodlot programmes. These efforts were also boosted by fears of energy shortages, a perceived crisis in fuelwood, and fears that forest loss was causing major floods and droughts. The lessons of the previous era led to a major change in support to forestry as international donors sponsored a second generation of forestry activities based on more local participation and village woodlots established using local labour. Again, there was an assumption that local people would resolve long-term issues such as access and use rights. But again, villagers had little involvement in design of projects or how they were to be implemented, and as a result little attention was paid to which trees local people consider most useful, the long-term use of plantations, how benefits would be distributed, or the multiple roles that trees play in production systems. Furthermore, fast growing exotic species were generally used to meet perceived fuelwood needs. With increasing experience, it became apparent that woodlots across the world had also had only very limited success. There are several important reasons why they failed to meet their objectives. Projects often ignored the use and management of existing resources and multiple forestry products. Issues of tree and land tenure were not addressed and the presence of existing institutional arrangements for managing local forests such as forest user groups (particularly those involving women and poor people) were often not known or ignored. Local people would not invest labour to protect resources from which they had no certainty of benefiting, and the costs of participation in the programme and maintenance of the assets were generally too high. Control-orientated regulations often meant people had to travel great distances to get permits to cut, process, or sell wood products. The projects were also generally still outsider-driven, using a standardised technical approach imposed with poor consultation, dependent on external funding, and target driven, aiming at producing the maximum number of trees rather than at the quality of forest products. Woodlots were for instance established on a very large scale in parts of Pakistan.

2.3. People First Era: Late 1980s to 1990s

Following 15 years of uneven success, it became clear that much of the failure was due to a lack of involvement of local people in all phases of project development and implementation. This helped to stimulate a major shift in development philosophy and practice, with increasing pressure on governments to decentralise functions, growing support for participatory methodologies, and an emphasis on the importance of local determination of developmental priorities. However, problems remained, including those created by inequalities within and between communities, inadequate consideration of livelihood constraints, and the fact that participatory approaches are still used more in name than in practice. Governments have been reluctant to devolve power, and if community organisation is weak, devolution can lead to even greater inequities. A groundswell of interest created international support but sometimes pushed the rate of change beyond the capacity to implement. Some of the experiments in community driven forestry in parts of Nepal and northern India characterise this approach.

3. Outline of Tools

It is clear that the key constraints in addressing fuelwood and fodder shortages are social and political rather than technical; once a community is fully supportive of and empowered to implement local forest restoration, then the technical means are either already in place or can be easily learned. A wide suite of tools for community-based forest management already exists:

- Participatory approaches to resource and needs assessments
- Community mapping of land tenure and access
- Conflict resolution
- Small-scale forestry techniques

In the context of a broader forest landscape restoration programme, establishment of either plantations or seminatural forests for fuelwood will frequently be one part of a wider restoration effort. An important component of any approach, therefore, is the negotiating skills necessary to agree on where fuelwood will be prioritised within the landscape (see "Negotiations and Conflict Management").

4. Future Needs

Each of the three eras discussed in this chapter had problems associated with it. Some of the problems highlighted must be resolved in order to ensure that the processes being established can be sustained into the future and that the outcomes deliver the desired social and environmental benefits. Many of the challenges that are raised relate to broader issues of restoration within a landscape, for example, how to optimise land use within the landscape to include fuelwood plantations but also other land uses. Among the challenges that need to be addressed to ensure long-term sustainable outcomes are the following:

- Improved knowledge to manage forests for multiple products
- Mechanisms to manage trade-offs between multiple interests
- Full representation of all interest groups (particularly women and poorer people)
- Development of representative, accountable, and competent local organisations
- Development of representative, accountable, and competent government forest organisations
- Embedding forest restoration within an understanding of livelihood strategies
- Emphasis on quality of processes rather than rapid delivery of products irrespective of quality
- Top-to-bottom change in attitudes, behaviour, and commitment to participatory approaches within forestry and other land management organisations
- Devolution of power within forestry organisations to staff in the field
- Policy and legislation in support of new approaches to forest restoration

References

Arnold, J.E.M. 1999. Trends in community forestry in review. Community Forestry Unit, FAO, Rome.

Gilmour, D.A., and Fisher, R.J. 1991. Villagers, Forests and Foresters—The Philosophy, Process and Practice of Community Forestry in Nepal. Sahayogi Press, Kathmandu, Nepal.

Wiersum, K.F. 1999. Social forestry: changing perspectives in forestry science or practice? Thesis, Wageningen Agricultural University, The Netherlands (ISBN 90-5808-055-2).

Additional Reading

Hobley, M. 1996. Participatory forestry: the process of change in India and Nepal. Rural Development Forestry Study Guide No. 3. Overseas Development Institute, London.

Thomson, J., and Freudenberger Schoonmaker, K. 1997. Crafting institutional arrangements for community forestry. Community Forestry Field Manual No. 7, FAO, Rome.

Westoby, J. 1987. *The Purpose of Forests: the Follies of Development*. Basil Blackwell, Oxford.

33
Restoring Water Quality and Quantity

Nigel Dudley and Sue Stolton

Key Points to Retain

Water quality and quantity are decreasing, with direct impact on people's lives.

There appears to be a clear link between forests and the quality of water from a catchment, a more sporadic link between forests and the quantity of water, and a variable link between forests and the constancy of flow.

The potential role of restoration with respect to water supply needs to be considered on a case-by-case basis and on a long time-scale.

Far better tools and methodologies are needed for calculating net gains of different restoration and management actions from the perspective of water supply.

There is also a need to better understand the linkages between water supplies and forest cover to help use these links as arguments for restoration.

1. Background and Explanation of the Issue

Water is, in theory, a renewable resource. Yet, the profligacy with which it has been used, coupled with population growth and increasing per capita demands, means that provision of adequate, safe water supplies is a major concern.[285] World water withdrawals rose sixfold over the last century, and it is estimated that we already use well over half of accessible runoff. For several countries, reliance on non-renewable (or only slowly renewable) groundwater sources masks a problem that will become more acute as these are exhausted. In 1998, twenty-eight countries experienced water stress or scarcity (defined as being when available water is lower than 1000 cubic metres per person per year); by 2025, this is predicted to rise to 56 countries. Overall, our main water requirements are for crop irrigation, but the need for clean drinking water is also critically important. Today, around half of the world's population lives in urban areas, and of these an estimated one billion people live without clean water or adequate sanitation, principally in Asia, Africa, and Latin America. Annually, 2.2 million deaths, 4 percent of all fatalities, can be attributed to inadequate supplies of clean water and sanitation.[286] These problems will increase in the future as the rapid processes of population growth and urbanisation continue and as climate change makes rainfall more erratic and increases the regularity and severity of droughts.

1.1. The Role of Forests

Loss of forests has been blamed for everything from flooding to aridity. Although forests cer-

[285] De Villiers, 1999.
[286] United Nations Human Settlements Programme, 2003.

tainly play a critical role in regulating hydrology, this role is complex and variable. There appears to be a clear link between forests and the quality of water from a catchment, a more sporadic link between forests and the quantity of water, and a variable link between forests and the constancy of flow. What forests provide depends on individual conditions, species, age, soil types, climate, management regimes, and needs from the catchment.[287]

Forests in watersheds generally result in higher quality water than alternative land uses, because other uses—agriculture, industry, and settlement—are likely to increase pollutants entering headwaters, and forests also help to regulate soil erosion and sediment load. While there are some contaminants that forests are unable to control—the parasite *Giardia*, for example—in most cases forests will substantially reduce the need for treatment of drinking water. However, in contrast to popular understanding, many studies suggest that in both very wet and very dry forests, evaporation is likely to be greater from forests than other vegetation, leading to a decrease in water from forested catchments as compared with grassland or crops.[288] One important exception is cloud forest, where cloud water interception may exceed losses.[289] In addition, some very old forests apparently increase water, for instance mountain ash (*Eucalyptus regnans*) of 200 years or more in Australia.[290] The precise interactions between different tree species and ages, and different soil types and management regimes, are still often poorly understood, making predictions difficult. Opinion also remains divided about the role of forests in maintaining regular water flow. There is little evidence that forests regulate major floods, although flooding was the reason for introducing logging bans in, for example, Thailand and parts of China. One important exception is flooded forests, which do appear to help regulate water supply, this includes both lowland forests such as the Varzea forests of the Amazon and swamps in the uplands. Forested catchments have important local impacts in regulating water flow. Undisturbed forest is also the best watershed land cover for minimising erosion by water and hence also sedimentation. Any activity that removes this protection, such as litter collection, fire, grazing, or construction of logging roads, increases erosion. Suspended soil in water supplies can render irrigation water unfit for use, or greatly increase the costs to make it useful.[291]

The potential role of restoration needs to be considered on a case-by-case basis and probably also on a long time-scale. Establishing fast-growing plantations is unlikely to do much to help either the quantity or the quality of water, while carefully located and managed secondary forests can do much to regulate sediment load, other pollution, and erosion, and may in some situations also eventually affect flow. Restoration for water supplies should also look at options for reducing impacts from managed forests through, for instance, removing unnecessary roads or changing their location, camber, and drainage facilities.

2. Examples

The following examples show how restoration has been used to help water supply sources and also look at some situations where restoration is now needed to repair damage to forested catchments.[292]

2.1. Ecuador: Protection Remains a Primary Focus of Water Management, Although Many Protected Areas Also Need Restoration

About 80 percent of the capital city Quito's 1.5 million population receive drinking water from two protected areas: Antisana (120,000 hectares) and Cayambe-Coca Ecological Reserve (403,103 hectares). To control threats to these reserves, the government is working with

---

[287] Dudley and Stolton, 2003.
[288] Calder, 2000.
[289] Bruijnzeel, 1990.
[290] Langford, 1976.
[291] Dudley and Stolton, 2003.
[292] All examples from Dudley and Stolton, 2003.

a local nongovernmental organisation (NGO) to design management plans that highlight actions to protect the watersheds, including stricter enforcement of protection to the upper watersheds and measures to improve or protect hydrological functions, protect waterholes, prevent erosion, and stabilise banks and slopes, including restoration where necessary.

2.2. U.S.: Comprehensive Land Use Planning, Including Protection and Restoration, Helps to Protect Urban Water Supplies

The Catskill, Delaware, and Croton watersheds deliver 1.3 billion gallons of water per day to New York City and the metropolitan area, and the Catskill/Delaware watershed provides 90 percent of the city's drinking water. The Catskill State Park (IUCN Category V, 99,788 hectares) protects the watersheds. New York City has used a mixture of land acquisition and conservation easement payments to increase the level of protection and therefore avoid the need for building an expensive new treatment plant; this choice was backed by New Yorkers in a vote. Once land has been acquired, management will focus on maintaining water quality, although recreational uses like fishing, hiking, and hunting may be allowed in cases where it will not conflict with water quality and public safety. Here restoration focusses on restoring values for water across a whole catchment.

2.3. Sweden: Even in Commercially Managed Forests, Management and Restoration Can Be Tailored to Maintain High-Quality Drinking Water

Lake Mälaren and Lake Bornsjön supply Stockholm's water. The company Stockholm Vatten controls most of the 5543 hectares watershed of Lake Bornsjön, of which 2323 hectares, 42 percent, is productive forestland certified by the Forest Stewardship Council. Management is focussed on protecting water quality, and areas are left for conservation and restoration.

2.4. Panama: Reforestation in Catchments Is Starting to Be Seen as a Potential Way of Improving Water Quality

Panama City's and Colon's drinking water comes from the watershed of the Panama Canal. It has been estimated that if 1000 hectares/year of deforested land in the watershed were reforested, it would not be necessary to construct a proposed dam, and on this basis new laws were passed to promote forestation of the Panama catchments. However, World Bank consultants concluded that forests would not necessarily improve dry season stream flow and questioned whether the evidence justified using public funds to reforest pasture. Meanwhile the director of watersheds and the environment of Panama's canal ministry said that his department would support massive reforestation efforts to protect the canal's water supply.

2.5. Kenya: Degradation Can Undermine Forest Watershed Values, Thus Increasing the Need for Restoration

Nairobi has a population of three million residents and draws its water from several different sources, including the Ruiru, Sasumua, Chania II, and Ndakaini systems. Unfortunately, illegal logging is impacting on much of the region including the Aberdares National Park (IUCN Category II, 76,619 hectares), and Mt. Kenya National Park (IUCN Category II, 71,759 hectares), which are both important in supplying Nairobi with drinking water. According to the water resources minister, Martha Karua, the future for ensuring sustainable water supplies lies in harvesting rainwater, building reserves from dams, and replanting trees. This is a long-term vision, which "will not produce results in an instant, but we want to look back five years, ten years, fifteen years later and say our forest cover now is 40 percent—and this can be achieved."

The above examples show a growing understanding of the potential role of forests but also some continuing confusion, and it is clear that many governments—local and national—are faced with making decisions about the role of forests with respect to water supplies that draw more upon hearsay than strict science.

3. Outline of Tools

In general, watershed values are an additional argument for restoration rather than being associated with specific restoration techniques. Information for policy makers about the value of different forested watersheds remains scarce, and models for predicting responses in individual catchments are at best approximate. Restoration for water purposes within individual catchments will vary according to circumstances and will be able to draw on many of the tools outlined elsewhere. Two approaches may be particularly useful here:

- Protect, manage, restore: Using forest cover to maintain water supplies at a watershed scale often requires a mosaic approach, where protected areas, other protective forests, and various forms of management are combined depending on existing needs and land ownership patterns. Restoration then becomes a management option that can be used in any of the above. Agreeing on the mosaic and balancing different social, economic, and environmental needs on a landscape scale requires careful planning and negotiation. WWF and IUCN have developed a number of landscape approaches to help address this kind of broad-scale decision making,[293] and these or similar exercises could provide help in determining where restoration could be used most effectively (see more detail in "Why Do We Need to Consider Restoration in a Landscape Context").
- Payment for environmental services (PES): The central principles of the PES approach are that those who provide environmental services should be compensated for doing so, and that those who receive the services should pay for their provision. If particular management systems are needed in watersheds to maintain the quantity or quality of water supply downstream, the users—such as bottling plants or hydropower companies—should pay for these, which could in theory help to fund restoration in sensitive watersheds.[294] A team of researchers from the United States, Argentina, and the Netherlands has put an average price tag of $33 trillion a year on fundamental ecosystem services, almost twice the value of the global gross national product, and of this, water regulation and supply were estimated to be worth $2.3 trillion.[295] In Costa Rica users such as hydropower companies are sometimes paying farmers to maintain forested watersheds. Payment schemes work best when a relatively small amount of money can be used to support a particular management regime and result in major economic benefits to a small group of users—like a water company. In these cases it is relatively easy to identify reasonable payments and to negotiate amongst the buyers and sellers of the environmental service.

4. Future Needs

Many governments are making decisions about forests and water based on flimsy data and poor methodologies, leading to the type of disputes outlined in the case of Panama, above. Far better tools and methodologies are needed for calculating net gains of different restoration and management actions from the perspective of water supply, and WWF is currently planning to collaborate with the World Bank to help develop them. More basically, there is need for greater understanding of the links between forests and water, perhaps initially through better diffusion of existing research and case studies.

---

[293] Aldrich et al, 2004.
[294] Pagiola et al, 2002.
[295] Constanza et al, 1997.

References

Aldrich, M., et al. 2004. Integrating Forest Protection, Management and Restoration at a Landscape Scale. WWF, Gland, Switzerland.

Bruijnzeel, L. 1990. Hydrology of Moist Tropical Forests and Effects of Conversion: A State of Knowledge Review. UNESCO, Paris.

Calder, I.R. 2000. Forests and hydrological services: reconciling public and science perceptions. Land Use and Water Resources Research 2(2):1–12.

Constanza, R., et al. 1997. The value of the world's ecosystem services and natural capital. Nature 387:253–260.

De Villiers, M. 1999. Water Wars: Is the World's Water Running Out? Phoenix Press, London.

Dudley, N., and Stolton, S. 2003. Running Pure: The Importance of Forest Protected Areas to Drinking Water. WWF and the World Bank, Gland, Switzerland, and Washington, DC.

Langford, K.J. 1976. Change in yield of water following a bushfire in a forest of *Eucalyptus regnans*. Journal of Hydrology 89:87–114.

Pagiola, S., Bishop, J., and Landell-Mills, N., eds. 2002. Selling Forest Environmental Services: Market-Based Mechanisms for Conservation and Development. Earthscan, London.

United Nations Human Settlements Programme. 2003. Water and Sanitation in the World's Cities: Local Action for Global Goals. UN-Habitat and Earthscan, London.

34
Restoring Landscape for Traditional Cultural Values

Gladwin Joseph and Stephanie Mansourian

Key Points to Retain

Some values provided by forests can be essential to a culture. The restoration of these cultural values can be a major objective of restoration in a landscape.

Cultural values need to be considered along with economic and ecological values to make forest landscape restoration effective.

Often the restoration of traditional knowledge must go hand in hand with the restoration of certain species in order to sustain its continued protection and use.

Restoring for diverse cultural values encompasses a wide range of land holdings and tenure systems, and therefore needs to be culturally and geographically specific.

1. Background and Explanation of the Issue

People rely on forest products for basic subsistence but also for a range of other values.[296] Traditional cultural values that have coevolved with local ecosystem goods and services are integral to a community's health, food, livelihood, art, and spiritual needs. Degradation of ecosystems impacts the entire cultural lifestyles of these communities, generally leading to continued erosion of traditional knowledge systems.

Cultural traditions and values are as heterogeneous as ecosystems and their life forms. However, these values and traditions are under threat by external factors associated with global change such as globalisation, population growth, inequity in distribution of wealth and livelihood options, and climate change. These macro-drivers have cascading and complex impacts at the local levels on biodiversity and the traditional knowledge associated with it.

1.1. Cultural Values Are Lost with the Loss of Natural Forests

Cultural values provided by forests are both impacted by and impact on restoration. As forests are lost, so are the numerous values they provide. For instance, different wood essences that may be necessary for a community's religious rites may become more difficult or impossible to obtain. Thus the loss in forests could lead to the decline in local cultural values that have for centuries protected the land and its resources.

1.2. Cultural Values Can Help Promote Restoration

Specific cultural values can be used as a trigger for restoration. In degraded landscapes, a number of the identified forest functions and values to restore may be cultural. For instance, the forestry sector in Scotland has significantly

[296] Byron and Arnold, 1997.

evolved from a timber-based industry to a more community and culturally centred one, in response to demands from local people for recreational and aesthetically pleasing native woodlands representative of their own cultural identity (rather than nonnative plantations, with all that those implied).[297]

1.3. Cultural Keystone Species

In the same way that an ecosystem is dependent on ecological keystone species, an entire culture or society may be dependent on cultural keystone species (CKS).[298] These species are by definition central to the survival and essence of a culture for a number of reasons, including their link to the culture's myths, rituals, religion, etc. Identifying these CKS and using them to promote forest protection and restoration in a landscape can be a valuable contribution to the restoration of forest functions in a landscape.

Restoring ecosystems to strengthen traditional cultural lifestyles will follow the priorities and needs of the local communities. For example, medicinal plants can be incorporated into a kitchen herbal garden, a community-managed medicinal plant garden, or used to restore degraded lands. This would also imply the need to work with appropriate institutions. Food and nutritional needs could also be incorporated into these land-use systems depending on local preferences and needs.

2. Examples

2.1. Coca in the Amazon[299]

In various indigenous communities (Barasana, Desana, Uitoto, etc.) in the Amazon, coca is considered to facilitate cultural transmission of knowledge from elderly individuals to young adults. By chewing the powdered coca leaves, sages and apprentices attempt to please the Masters of Nature (semideities in their cosmology) with a valued gift. The importance of the coca plant for these communities lies in its essential role to allow communications with the supernatural beings governing nature, and thus it plays a central role in their very cultural identity.

Coca is also indispensable in major rituals such as the ritual of world healing and illness prevention (Yuruparí), the seasonal feasts offered by the community to the Masters of Nature to thank them for particular harvests, and the healing ceremonies led by the sage.

In this example, coca holds a unique value for local people provided by the Amazon forest, and it can be used as an objective to restore forest functions in the landscape. In other words, in an effort to meet different functions that forests provide in a landscape, the provision of coca can be one of these identified functions in order to satisfy a culturally driven demand.

2.2. Sacred Groves, Forests, and Gardens

Sacred groves, forests, and gardens are associated with places of worship in several traditions around the world. These patches of forests and diverse gardens are rich in biodiversity and are protected for their sacred value. All products available from these sacred groves are used for temple-related activities or structures. Cultural values have preserved and can also drive the restoration of these historic sacred groves. The *Devara kadus* in India[300] are an example of these sacred forests. Devara kadus are diversity-rich forest fragments ranging from 0.1 to 1000 hectares in size that are associated with places of worship across India. The sacred traditions and texts could provide the creative basis for promoting the conservation and restoration of these sacred land-use systems.

2.3. Socially and Economically Valuable Trees

Several species of trees have locally significant values that could be used to drive restoration in the landscape. Multiple economic and cultural values are historically linked to a specific ethnically defined region. For example, in certain regions in India, the common tropical dry-

---

[297] Garforth and Dudley, 2003.
[298] Cristancho and Vining, 2004.
[299] Drawn from Cristancho and Vining, 2004.

[300] Kushalappa and Bhagwat, 2001.

deciduous Neem tree (*Azadirachta indica*) symbolises a body of traditional values, knowledge, and uses. Almost all parts of the tree are used in medicine and agriculture. The leaves are used in traditional health systems, in religious rituals, and as green manure in agriculture. Oil is extracted from the seeds and has both medicinal and pesticidal properties. Neem cake which is a by-product of oil extraction is used as an organic fertiliser. The wood has a high calorific value as fuelwood. Neem wood is termite resistant and used to make door and window frames. Species with multiple values may be candidates to drive region-specific restoration of these species (and others) within the broader landscape.

2.4. Home Gardens

Home gardens have been described as "living gene banks" of indigenous varieties, rare cultivars, landraces, and species, as well as introduced species.[301] These multiple species have been conserved through generations. The selection of plants in these gardens is influenced by climate, soils, household preferences, and dietary habits. Home gardens in the tropics are a valuable land-use system to restore traditional fruits, nuts, medicinal plants, and other indigenous species of cultural value to local communities.

3. Outline of Tools

3.1. Toolkit for High Conservation Value Forests

WWF and ProForest[302] have developed a "toolkit" to identify high conservation value forests (HCVFs). This is an all-encompassing approach that recognises six different values the forests provide, one of which is cultural: "HCV6—Forest areas critical to local communities' traditional cultural identity (areas of cultural, ecological, economic, or religious significance identified in cooperation with such local communities)."

This methodology provides guidance on existing information at a global level, and direction for identifying forest values. For each of the six types of high conservation value (HCV), the toolkit identifies a series of elements that need to be considered. It then provides guidance for each element on how to decide whether there are HCVs within a country or region. When national HCVs have been defined, it is then possible to use this information so that specific forest areas can be evaluated for the presence or absence of the HCVs, in order to identify and delineate HCVFs.

3.2. A Participatory Process

If cultural values are to be used as an objective of forest restoration in a landscape, a participatory process will be necessary, and it may include the following steps:

- Document the traditional knowledge with local people to identify cultural drivers for the restoration of forest functions in a landscape.
- Together with local people, identify the current status of those cultural values.
- Through focus groups, discussions, and other locally applicable participatory tools, identify the links between those cultural values and other forest functions that may need to be protected and restored.
- In conjunction with stakeholders, set objectives for the protection and restoration of the identified cultural values.
- Develop locally adapted approaches, such as biodiversity-rich agroforestry, to restore cultural and other forest values in the landscape.
- Promote traditional knowledge pertinent to the local area through local schools and other local civic and user forums.

3.3. Clarifying Land Tenure and Access (Use) Rights

Processes that help clarify land tenure and access/use rights to valuable forest products are essential to protect and restore valuable forest areas. Appropriate protocols may be developed for restoring under different land-tenure regimes (also see "Land Ownership and Forest Restoration").

---

[301] Agelet et al, 2000.
[302] Jennings et al, 2003; also see www.proforest.com.

3.4. Ethnobotanical Surveys

Potential cultural keystone species may be uncovered through surveys. The results can then be used to promote adequate protection, management, and restoration of these resources.

4. Future Needs

Some identified needs for the future include the following:

- To document and exchange information about successful models of restoration for cultural values, and also where cultural values have driven restoration
- To increase understanding of potential cultural indicators and drivers of restoration, which requires more collaborative work among anthropologists, sociologists, and ecologists
- To integrate socioecological landscape-level approaches to culturally driven land-use systems such as home gardens and sacred groves to understand the process at a larger spatial scale
- To develop appropriate extension methods to enhance the diffusion of culturally driven restorative land-use systems
- To build capacity in adaptive and participatory research methods in restoration
- To develop/refine and use a holistic-systems approach to natural resource management; in most contexts, planning and management for conservation, sustainable-use, and restoration have to be developed together rather than as separate components.

References

Agelet, A., Bonet M.A., and Valles, J. 2000. Home gardens and their role as a main source of medicinal plants in mountain regions of Catalonia (Iberian Peninsula). Economic Botany 54(3): 295–309.

Byron, A., and Arnold, M. 1997. What Futures for the People of the Tropical Forests? CIFOR occasional paper 19.

Cristancho, S., and Vining, J. 2004. Culturally Defined Keystone Species. Human Ecology Review 11(2): 153–164.

Garforth, M., and Dudley, N. 2003. Forest Renaissance. Published in association with the Forestry Commission and WWF–UK, Edinburgh and Godalming.

Jennings, S., Nussbaum, R., Judd, N., et al. 2003. The High Conservation Value Toolkit. Proforest, Oxford (three-part document).

Kushalappa, C.G., and Bhagwat, S.A. 2001. Sacred groves: biodiversity, threats and conservation. In: Uma Shaanker, R., Ganeshaiah, K.N., and Bawa, K.S., eds. Forest Genetic Resources: Status, Threats and Conservation Strategies. New Delhi, India, Oxford and IBH.

Additional Reading

Baidyanath, S. 1998. Lifestyle and ecology. http://www.ignca.nic.in/cd_08.htm#BAIDH

Borthakur, S.K., Sarma, T.R., Nath, K.K., and Deka, P. 1998. The house gardens of Assam: a traditional Indian experience of management and conservation of biodiversity—I. Ethnobotany 10:32–37.

Fernandez, E.C.M., and Nair, P.K.R. 1986. An evaluation of the structure and functions of tropical homegardens. Agroforest Syst 21:279–310.

Malhotra, K.C. 1998. Anthropological dimensions of sacred groves in India: an overview. In: Ramakrishnan, P.S., Saxena, K.G., and Chandrasekara, U.M., eds. Conserving the Sacred for Biodiversity Management, pp. 423–438. New Delhi, India, Oxford and IBH.

Palni, L.M.S., Joshi, M., Agnihotri, R.K., and Sharma, S. 2004. Crop diversity in the home gardens of the Kumaun region of central Himalaya, India. PGR Newsletter, No. 138:23–28. FAO-IPGRI.

Ramakrishnan, P.S.R. 1998. Conserving the sacred for biodiversity: the conceptual framework. In: Ramakrishnan, P.S., Saxena, K.G., and Chandrasekara, U.M., eds. Conserving the Sacred for Biodiversity Management, pp. 3–15. New Delhi, India, Oxford and IBH.

Soemarwoto, O., Soemarwoto, I., Karyono Soekartadireja, E.M., and Ramlan, A. 1985. The Javanese home garden as an integrated agroecosystem. Food and Nutrition Bulletin 7:44–47.

Torquebiau, E. 1992. Are tropical agroforestry home gardens sustainable? Agriculture, Ecosystems and Environment 41:189–297.

Wiersum, K.F. 1982. Tree gardening and the examples of agroforestry techniques in the tropics. Agroforestry Systems 1:53–70.

Case Study: Finding Economically Sustainable Means of Preserving and Restoring the Atlantic Forest in Argentina

Stephanie Mansourian and Guillermo Placci

The Atlantic Forest of Brazil, Argentina, and Paraguay is one of the most threatened ecosystems on the planet, with only 7.4 percent of it remaining intact and large areas severely degraded and highly fragmented. Despite its current state, the Atlantic forest remains a rich repository of biodiversity. For example, inside the Atlantic forest, in the state of Bahia, 450 species of trees per hectare have been catalogued, a world record![303]

It is in northern Argentina, that one of the largest remnants of the Atlantic forest can still be found. In this area, Fundación Vida Silvestre Argentina (FVSA) is working with WWF to restore the landscape. One particular area, namely the municipality of Andresito, has been identified as a priority. It is a strip of land surrounded by four important strictly protected areas: the famous transboundary Iguazú National Parks of Brazil and Argentina, the Urugua-í Provincial Park, and the Foerster Provincial Park. The land in Andresito is divided into many small privately owned areas. The challenge is to work with the landowners and land managers to stop deforestation and forest degradation, to increase connectivity with the surrounding protected areas, and to establish buffer zones around them, while increasing local living standards.

The approach taken here by FVSA and WWF was first to map out clearly the different plots of land and identify the landowners and land uses. Second, a series of test sites were set up to identify the sorts of restoration techniques and mixes of species that would work best under local conditions. Then, a sustainable development and participatory planning learning process was mobilised gathering provincial and municipal officers, farmers, indigenous people, and members of NGOs and other private and public institutions. As a result of this, the participants committed themselves to working toward the accomplishment of a land-use plan, and a local commission was created with this goal.

Also, to ensure an income-generating activity alongside forest restoration for local populations, FVSA and WWF have been working on developing sustainable production of different crops. One such crop is the palm heart (*Euterpe edulis*), a native understorey palm tree that grows wild in the region and can generate significant income for local inhabitants while preserving the forest. Another alternative for small-scale farmers is planting yerba mate, a native plant that used to grow in patches throughout the forests.

So far, guidelines for the production of palm hearts have been developed and a cooperative of small-scale producers has been set up. Results are encouraging. If more small landowners can make a living through such sustainable restoration involving economically attractive measures, then the risk of them moving south and selling their land to big

[303] Di Bitetti et al, 2003.

logging companies can be removed once and for all.[304]

References

Di Bitetti, M.S., Placci, G., and Dietz, L.A. 2003. A Biodiversity Vision for the Upper Paraná Atlantic Forest Ecoregion: Designing a Biodiversity Conservation Landscape and Setting Priorities for Conservation Action. WWF, Washington, DC.

FVSA. 2004. Newsletter: News from the FLR Project in the Upper Paraná Atlantic Forest of Argentina. FVSA. Buenos Aires, Argentina.

[304] FVSA Nawsletter, 2004.

Section XI
A Selection of Tools that Return Trees to the Landscape

35
Overview of Technical Approaches to Restoring Tree Cover at the Site Level

Stephanie Mansourian, David Lamb, and Don Gilmour[305]

Key Points to Retain

There is no unique goal and trajectory for restoration.

Tools for restoration should be selected to achieve one or more targets depending on the specific context.

Various restoration tools could be used, some of which are presented in this chapter.

1. Background and Explanation of the Issue

A variety of approaches are available to carry out restoration at the site level, and this chapter provides an overview of them.

Interventions can be viewed along a continuum from passive to more active ones. The more passive the intervention, the more reliant one is on having sufficient germ plasm (seed sources or coppice material) available at or near the site. Passive interventions are the cheapest approaches, although the costs of preventing continued disturbances or degradation can sometimes be high. However, there are often circumstances requiring some form of direct intervention (i.e., active restoration). Examples of these situations are where topsoil has been eroded or soil has been heavily compacted by cattle, where invasive species have come to dominate the site or if some other disturbance (e.g., fire) has altered the natural balance and natural regeneration will either be extremely slow or will no longer occur.

More active forms of intervention are also needed when the passive approaches are likely to be slow or too risky. These interventions can take a variety of forms including enriching natural regeneration with species that may not be present (e.g., plants with large fruit that are often poorly dispersed) or planting a large number of different species, fertilising them, and carrying out weed control until the planted seedlings are established. The most appropriate approach depends on both the ecological and the socioeconomic circumstances prevailing.

Two prime considerations in determining what approaches to take to restore an area are

- the objectives set for the intervention, and
- the budget available.

Different objectives require different approaches. One could think of several quite different situations that would require very distinct approaches to site-level restoration. Several examples follow:

- Restoration of woodland to provide habitat for endangered fauna (see Example 1 in the next section)
- Restoration of an abandoned quarry for aesthetic purposes (see "Open-cast Mining Reclamation")

---

[305] This paper is based to a large extent on Lamb and Gilmour, 2003.

- Restoration of an endangered ecosystem (as is currently occurring with the dry forests in New Caledonia)
- Restoration of millions of hectares of degraded uplands primarily for economic development (as is currently occurring in Vietnam)

Similarly, the available budget will also be a key determinant when deciding what approach to take. For example, it might be economically necessary to use a variety of different approaches across a landscape, rather than using just the most effective biological approach, particularly if this is also very expensive. The most expensive approaches would normally be used to restore the most critical sites.

Before determining which action to take at the site level, a careful assessment needs to be made, based on ecological circumstances such as the fertility of soils, the extent of degradation, the proximity of remaining forest fragments, the types of species involved, the topography, rainfall, seasonality, etc. Social aspects need just as much attention as biophysical ones when determining what approach to take to restoration. For example, many local communities exercise usufruct rights over land adjacent to their settlements. Irrespective of the legal status of the land, unless the *de facto* situation is addressed effectively, it is unlikely that restoration efforts will be successful. We would generally recommend always opting for the least intervention possible. This is to (a) attempt to stay closest to natural processes but also (b) because the more active the intervention, the costlier it is likely to be.

2. Examples

2.1. Natural Regeneration Combined with Grazing in Corrimony (Scotland)[306]

In 1997 the Royal Society for the Protection of Birds (an NGO) acquired land in Corrimony, Scotland. The main objective was to increase habitat for capercaillie and black grouse. The long-term vision was to have at least two thirds of woodland cover restored with an emphasis on natural regeneration. However, because 99 percent of natural regeneration is broadleaf, it was decided to plant copses of Scots pine (*Pinus sylvestris*) in areas that were remote from seed sources. When the pines mature they will be able to regenerate naturally from their own seed. In addition, to achieve a habitat mosaic that also supports black grouse and other species of conservation interest, some grazing areas have also been retained. Preliminary observations suggest that this approach is effective.

2.2. Restoration of Temperate Forest Through Mixed Plantations in Canada[307]

Larson[308] presents one of the earliest modern examples of forest restoration in the deciduous hardwood forest region of eastern Canada, which started in 1886. The site was an old gravel pit in which 2300 saplings of 14 different species were planted in a mixture. These included local deciduous hardwoods and conifers as well as several exotics (*Acer platanoides*, *Fraxinus excelsior*, *Larix decidua*, *Picea abies*, *Pinus nigra*, and *Tilia cordata*). Some of these 14 species were planted in rows spaced 2.5 m apart. No subsequent site management was carried out apart from some early pruning. The nearest natural forest was 500 m away. By 1930 around 85 percent of the site had a sparse canopy, 31 percent of which was coniferous. By 1993 the canopy cover had increased to 95 percent, of which only 5 percent was conifer. The site, then 107 years old, contained 220 trees with a diameter at breast height exceeding 30 cm. Of the original 14 canopy-forming tree species, 10 were still present. Two new species had colonised. A diverse understorey of woody and herbaceous plants contained 36 species, most of which were reproducing. Some of the canopy trees were regenerating and were rep-

---

[306] Cowie and Amphlett, 2000.
[307] Lamb and Gilmour, 2003.
[308] Larson, 1996, in Lamb and Gilmour, 2003.

resented in the understorey but *Picea*, *Larix*, and *Pinus* were absent. Measurements suggest *Juglans nigra* (native) and *Acer platanoides* (exotic) will dominate the site in future. All new tree regeneration was found in areas with no conifers. The patterns of community structure that have evolved over time at the site are different from those in the native forests of southern Ontario but changes are leading to the development of a forest with a similar structure and appearance. One recent measure of the success of the planting is the fact that local authorities mistakenly listed the site as an important natural forest remnant within the local city boundary.

2.3. Restoring Tree Cover Through Agroforestry in Tanzania[309]

Studies in Tanzania have found that the *Shambaa* people use their traditional agroforestry and intercropping systems to improve both soil productivity and crop yields. The traditional agroforestry system consists of a multistorey tree garden, which involves the mixing of trees and farm crops in a spatial arrangement. The system includes a mixture of an understorey of coffee (and fruits), food crops such as maize/beans and a variety of pulses, a middle storey of *Grevillea robusta*, a multipurpose exotic species commonly used for timber, fuelwood, and building poles' production. The sites are not "restored" in the sense of reestablishing the original biodiversity. On the other hand, these sites have had key ecological functions such as nutrient cycling and net production restored. They are now floristically and structurally quite complex.

3. Outline of Tools: Approaches to Site-Level Restoration

Restoration at a site level often needs to integrate social approaches, such as agreements about land use to facilitate natural regeneration, with technical approaches to increase natural regeneration or enhance regeneration through planting where natural regeneration will no longer occur.

3.1. Reducing Degrading Influences

3.1.1. Removing the Cause of Degradation or Obstacles to Regeneration

In some situations restoration can be achieved through the use of natural regeneration simply by the removal of degrading influences such as cattle grazing or invasive exotic species. Technical interventions may also be needed, but often the emphasis needs to be on social processes, such as negotiating grazing rights with local cattle herders.

By protecting the area from any further disturbances (e.g., grazing, farming) natural colonisation may take over. However, this is only feasible in areas where

- general degradation is not extensive,
- soils are still of good quality, or
- seed sources or coppice materials are still available either from forests close by or in the soil (as evidenced by regrowth already present in the area).[310]

3.1.1.1. Advantages and Disadvantages

This approach is often cheaper since it requires little input, particularly if communities are able to eliminate grazing animals from the area. Costs can rise steeply if areas have to be fenced, but this is still generally cheaper than planting. It is also one of the few approaches that can be achieved over large areas. On the other hand, its disadvantages include that it may end up being unexpectedly expensive if fire, weeds, or pests need to be controlled. Likewise, the previous land users will have had to forgo their previous use of the site and may need compensation.

---

[309] Chamshama and Nduwayezu, 2002.

[310] Parrotta et al, 1997.

3.2. Initiating or Improving Tree Cover

3.2.1. Directing Ecological Successions

Directing ecological successions can be done in a number of ways, using different species and approaches. The aim and desire is to initiate a process whereby nature takes over. The following points need to be considered when attempting to stimulate natural succession:

- *The "founder effect"*: The initial species chosen will have a determining effect on the future succession in the landscape, which cannot always be anticipated.
- *Using nearby intact forests*: The nearer to an intact forest, the more chance of obtaining seeds via seed dispersers, and wind, and therefore the higher success rate. It should be noted, though, that different species from the intact forest will colonise at different rates.
- *Using wildlife to accelerate ecological processes*: It can be useful to use animals for processes such as pollination and seed dispersal, but in many instances this can be constrained by incomplete knowledge of the exact relationships. Alternatively, some key species may have disappeared from the region or be unable to move across the degraded landscape.
- *Using disturbances*: At some point in the restoration process the natural disturbance regime must be allowed to develop to prevent successions from being diverted or stagnating. For example, while restoration projects in fire-prone landscapes often require fire protection in the first few years to ensure seedlings become established, at some stage fires must be allowed or be reintroduced to ensure that normal successional processes can begin to develop.
- *Ecological "surprises"*: (1) Predators may harvest all the seed. (2) Successions may become dominated by a small number of aggressive species causing competitive exclusion and a decline in biodiversity. (3) Trees established to attract seed-dispersing wildlife may become focal points for weed colonisation. (4) The removal of exotic herbivores may allow grass fuel loads to increase and fire regimes to change. In all cases constant monitoring is needed to ensure that restoration continues as planned.

3.2.2. Stimulating Natural Successions

If natural regeneration does not occur or proceeds only slowly, it may be possible to accelerate the process. This might be done by removing weeds or reducing competition between existing species. Thinning to reduce tree density can open the canopy and can provide more opportunities for new species to colonise the site. Where soils are infertile, added fertilisers can enhance growth rates.

3.2.2.1. Advantages and Disadvantages

This requires relatively few inputs. However, it is less likely to work in areas where soils have been badly degraded and seed sources or coppice material are no longer available.

3.2.3. Direct Seeding

If sites are bare of trees it may be useful to overcome any dispersal problems by deliberately introducing certain species. Most reforestation is usually carried out by planting seedlings that have been raised in a nursery. The seedlings are commonly planted into a site that has been cleared of weeds and ploughed to ensure the seedlings develop quickly. The costs of raising seedlings, site preparation, and planting are high. Direct seeding bypasses these steps by sowing seeds directly on bare land. This can be done either manually or aerially.

3.2.3.1. Advantages and Disadvantages

Direct seeding is relatively cheap as it does not require nurseries to raise seedlings. Its disadvantages are that seeds are often subject to predators, and the young seedlings are very vulnerable to weed competition. The number of seedlings actually produced from seed can be very low. Therefore, a very large number will need to be sown, recognising that a large proportion will not survive. For this reason, this approach is not suitable for species for which seeds are not available in large quantities or where seed is expensive.

3.2.4. Scattered Tree Plantings

Trees may only gradually colonise some sites because they are poorly dispersed or because the competition (e.g., from grass) is too severe. Another way of accelerating successions is by planting single trees or clumps of trees across the landscape. The aim is for them to serve as perches for seed dispersers such as birds. Over time, they can become focal points for regeneration. Where species have wind-dispersed seed rather than animal-dispersed seed, such plantings can be arranged perpendicular to the prevailing wind and so assist seed dispersal across the landscape.

3.2.4.1. Advantages and Disadvantages

This approach is relatively inexpensive since it only requires a few plantings. However, it is dependent on wildlife being able to disperse seeds from intact forest remnants that remain nearby. The numbers of such wildlife that remain in degraded landscapes and their capacity to disperse seeds will vary with circumstance.

3.2.5. Enrichment Planting

In some situations the forest community developing from natural regeneration is missing certain key species. This may be because they have particular regeneration requirements or because they are poorly dispersed. The absence of these species may have economic consequences for the people dependent on these forests for their livelihoods. Alternatively the missing species may be important to the ecological functioning of the forest. In such cases it can be useful to try enriching the regenerating forest by planting seedlings of these species in appropriate microsites.

3.2.5.1. Advantages and Disadvantages

This approach enhances the capacity of the forest to provide commercial or social benefits by promoting the growth of certain key species. The disadvantage of the approach is the risk that any newly planted trees may be suppressed for some time by the overstorey. That is, the introduced species can be out-competed by taller trees, weeds, or vines. Some form of silvicultural treatment is often required for several years to remove this cover and ensure success.

3.2.6. Closely Spaced Plantings Using Limited Numbers of Species (the "Framework Species" Method)

This approach uses a small number of fast-growing species planted at close spacings (e.g., 1000 trees per hectare) to quickly form a closed canopy and so eradicate weeds. This new forest then forms a "framework" within which successional processes can operate. Over time seed-dispersing wildlife bring new species to the site and diversity is enhanced.

3.2.6.1. Advantages and Disadvantages

The advantage of this approach is that once the trees are established, they soon out-compete grass and weeds, making it easier for the species brought in by seed-dispersing animals to become established. The approach is especially suited to areas close to intact forest that can act as a source of seeds (and wildlife). The disadvantage is that successional development is dependent on the particular species that are dispersed into the site. Some species may be weeds so that monitoring is needed to maintain an appropriate successional trajectory. The initial cost can also be high.

3.2.7. Intensive Ecological Reconstruction Using Dense Plantings of Many Species (or Restoring a Biodiversity Island in a Degraded Landscape)

This involves intensive planting of a large number of tree and understorey species. The species used depend on the sites and soil types. Those that might be used include fast-growing species able to exclude weeds, poorly dispersed species, species forming mutually dependent relations with wildlife, and, possibly, rare or endangered species that might be present only in small numbers or in small geographic areas. Since the method bypasses the normal successional sequence the species used should come mostly from late successional stages, rather than early pioneer stages.

3.2.7.1. Advantages and Disadvantages

Because this is a good way to quickly establish a species-rich community, it is especially suitable for areas needing rapid restoration. On the other hand, it is comparatively expensive to raise and plant such large numbers of species and many may not survive if their site and habitat requirements are not fully understood.

3.2.8. Managing Secondary Forests

Careful management could allow the gradual improvement of economic resources as well as biodiversity and other ecological services at minimal cost. Another approach might be to foster the growth of certain tree or other plant species that are commercially attractive by removing or thinning competing trees. The choice of options depends on the origins of the forest and the range and abundance of the species it contains.

3.3. Reforestation for Productivity and Biodiversity

3.3.1. Monoculture Plantations Using Indigenous Species

Monoculture plantations are comparatively easy to establish and manage since all trees mature at the same time. Traditionally many such plantations have used exotic species. The timbers of these species are often of relatively low value. Some indigenous species can have much higher commercial values than fast-growing exotic species. Plantations of higher value timbers may be increasingly valuable in future once natural forests have been logged over.

3.3.1.1. Advantages and Disadvantages

Intensively managed plantations can yield a high commercial value. Plantations of indigenous species also provide some modest biodiversity benefits. The key disadvantage of using indigenous species is that little is usually known about their silvicultural requirements and most are comparatively slow growing.

3.3.2. Monoculture Plantations and Buffer Strips

Industrial plantations are often large and are established as continuous blocks. This leads to the simplification of landscapes. Breaking these extensive plantations up by using buffer strips of native vegetation or ecologically restored forests along streamsides and roads can add complexity and habitat diversity.

3.3.2.1. Advantages and Disadvantages

Buffer strips can help enhance conservation benefits by introducing more spatial complexity to a landscape and increasing connectivity allowing easier movement of plants and wildlife across landscapes. These strips or corridors can have a number of other benefits, including acting as fire breaks and streamside filters to enhance watershed protection.

3.3.3. Mosaics of Species Monocultures

Instead of using only one species in a plantation, an alternative could be to use more than one and create a mosaic of different types of plantations across the landscape. The landscape diversity could be further enhanced by surrounding each monoculture by buffer strips as described above.

3.3.3.1. Advantages and Disadvantages

The advantage of this alternative is that silvicultural management of each plantation remains simple; the disadvantage is that precise species-site relationships must be known if productivity in each of the different plantations is to be maximised.

3.3.4. Mixed Species Plantations

Site biodiversity may be enhanced if mixed species plantations are used instead of monocultures. These might be temporary mixtures where one species is used for a short period as some form of nurse or cover crop, or they may be permanent mixtures for the life of the plantation. Most mixed-species plantations usually have only a small number of species (under four), so biodiversity gains may be modest.

3.3.4.1. Advantages and Disadvantages

Mixtures can often generate benefits in addition to any biodiversity gain. These potential benefits include improved production, improved tree nutrition, and reduced insect or pest damage. There may also be financial gains from combining fast-growing species (harvested early in a rotation) with more valuable species that need longer rotations. Disadvantages include the fact that not all species' combinations are necessarily compatible and an inappropriate mix of species may lead to commercial failure. Also, having two or more species in a plantation necessarily leads to more complex forms of silviculture and management. This means that mixtures are likely to be more attractive to smallholders and farm forestry woodlots than large industrial-scale plantations.

3.3.5. Encouragement of Understorey Development

In many plantation forests, especially those near areas of intact forest, an understorey of native tree and shrub species will develop over time with many of the species being dispersed by animals. What began as a simple monoculture forest can acquire structural complexity and considerable biodiversity.

3.3.5.1. Advantages and Disadvantages

Such understories transform the range of services provided by the plantation. There can be benefits in terms of watershed protection and fire exclusion as well as biodiversity gains. However, they pose a number of dilemmas for managers who may find their original objectives being compromised or, at the very least, made more difficult to achieve. Difficult trade-offs may need to be made.

3.3.6. Agroforestry

Agroforestry is a form of agriculture that mixes trees and other crops in the same area of land (see "Agroforestry as a Tool for Forest Landscape Restoration"). Some forms involve mixtures of multipurpose trees and food crops; others combine scattered trees and pastures. In most cases a variety of species are used in the farm or "home garden" that differ in canopy and root architecture, phenology, and longevity.

3.3.6.1. Advantages and Disadvantages

Agroforestry has some particular advantages in landscapes where land for food production is limited and where human populations are large or increasing. Agroforestry creates spatial and structural complexity across landscapes and offers the prospect of agricultural sustainability and some biological diversity. On the negative side, biodiversity gains may be modest since many of the species used are relatively common agricultural crop species.

4. Management Considerations

4.1. How Many Species?

Restoration is often carried out to reestablish biological diversity and to also restore key ecological processes and functions. One unresolved question is the number of species needed to achieve this latter objective. Must all species be reestablished, or is there a point beyond which increases in species' richness may not provide any further benefits? The answer to the question is still unresolved, although it seems that species richness per se may not be as important as the structural or functional types of species that are used in reforestation. It is also clear that relationships present at small, local scales may not hold at larger landscape scales.

4.2. Trade-Offs

Inevitably some trade-offs may be required by managers needing to balance, say, promoting commercial timber production or fostering wildlife diversity. Production, at least in the short term, is usually favoured by developing plantations that use only small numbers of tree species. Most wildlife species, on the other hand, usually prefer species-rich and structurally complex forests. The final decision depends on such things as the preferences of the stakeholders involved, whether commercial timber production is the primary objective of reforestation, markets for the various goods that might be produced, and the degree of degradation across the landscape.

4.3. Intensity and Timing of Management Interventions

Managers concerned with maximising timber production will make decisions on a variety of interventions including whether or not to prune trees, when to carry out thinnings, and when to undertake a final clear-felling. All of these decisions have consequences for biodiversity and various ecological processes such as nutrient cycling. Biodiversity is usually favoured by enhanced spatial complexity. This means interventions that promote a mosaic of disturbances and recovery stages are preferable to large, spatially contiguous interventions.

5. Future Needs

While many approaches are available for restoring forest cover on degraded sites, it is often a challenge to gather adequate knowledge on the use of indigenous species. For this reason, a handful of exotic species (particularly pines, eucalypts, and acacias) are still favoured in many locations. These species often display superior growth characteristics compared with indigenous tree species. In addition, the seed of these species are often easily acquired and they come as a silvicultural "package" with established procedures and methodologies.

In most countries there is still insufficient knowledge on genetics, propagation techniques, competitive relationships between different species, and methods of raising most native species in nurseries.

A comprehensive framework that would help managers make choices based on the current situation but also based on funding, available human resources, size of the area, aim of the restoration, etc., is needed. This framework would also have to include socioeconomic elements, as these are often forgotten or left behind in technical issues dealing with restoration. Yet, without appropriate consultation, buy-in, and suitable social and economic reasons for engaging in restoration, success rates are unlikely to be high. Land tenure issues are particularly important to clarify before engaging in restoration.

Very importantly, there is a need for increased understanding and research on options to make restoration financially attractive. In many countries, long-term interests (restoration impact will only be felt in the long term) are not important as people face daily struggles. There is therefore a need to address this through short-term financial benefits from restoration (directly or indirectly). Institutional arrangements for restoration also need to be clarified. Restoration across a landscape requires a multidisciplinary and multisectoral approach, and relevant institutions and expertise need to be brought in with all stakeholders actively participating in the process.

References

Chamshama, S.A.O., and Nduwayezu, J.B. 2002. Rehabilitation of Degraded Sub-Humid Lands in Sub-Saharan Africa: A Synthesis. Sokoine University of Agriculture, Morogoro, Tanzania.

Cowie, N.R., and Amphlett, A. 2003. Corrimony: an example of the RSPB approach to woodland restoration in Scotland. In: Humphrey, J., Newton, A., Latham, J., et al. eds. 2003. The Restoration of Wooded Landscapes. UK Forestry Commission, Edinburgh, Scotland.

Lamb, D., and Gilmour, D. 2003. Rehabilitation and Restoration of Degraded Forests. IUCN, Gland, Switzerland, and Cambridge, UK, and WWF, Gland, Switzerland.

Parrotta, J.A., Turnbull, J., and Jones, N. 1997. Catalyzing native forest regeneration on degraded tropical lands. Forest Ecology and Management 99(1–2):1–8.

Additional Reading

Carnus, J.-M., Parrotta, J., Brockerhoff, E.G., et al. 2003. Planted forests and biodiversity. A IUFRO contribution to the UNFF Intersessional Expert Meeting on the Role of Planted Forests in Sustainable Forest Management, "Maximising planted forests' contribution to SFM," Wellington, New Zealand, March 24–30. In: Buck, A., Parrotta, J., and Wolfrum, G., eds. 2003. Science and Technology—Building the Future for the World's Forests and Planted Forests and Biodiversity. IUFRO Occasional Paper No. 15. International Union of Forest Research Organisations, Vienna.

Engel, V.L., and Parrotta, J.A. 2001. An evaluation of direct seeding for reforestation of degraded lands in central São Paulo State, Brazil. Forest Ecology and Management 152(1–3):169–181.

Lamb, D., Parrotta, J.A., Keenan, R., and Tucker, N.I.J. 1997. Rejoining habitat remnants: restoration of degraded tropical landscapes. In: Laurence, W.F., and Bierregaard, R.O., Jr., eds. Tropical Forest Remnants: Ecology, Management and Conservation of Fragmented Communities. University of Chicago Press, Chicago, pp. 366–385.

Parrotta, J.A. 1993. Secondary forest regeneration on degraded tropical lands: the role of plantations as "foster ecosystems." In: Lieth, H., and Lohmann, M., eds. Restoration of Tropical Forest Ecosystems, pp. 63–73. Kluwer Academic Publishers, Dordrecht, Netherlands.

Parrotta, J.A. 2002. Restoration and management of degraded tropical forest landscapes. In: Ambasht, R.S., and Ambasht, N.K., eds. Modern Trends in Applied Terrestrial Ecology, pp. 135–148. Kluwer Academic/Plenum Press, New York.

Parrotta, J.A., and Knowles, O.H. 2001. Restoring tropical forests on bauxite mined lands: lessons from the Brazilian Amazon. Ecological Engineering 17(2–3):219–239.

Sim, H.C., Appanah, S., and Durst, P.B., eds. 2003. Bringing back the forests: policies and practices for degraded lands and forests. Proceedings of an International Conference, October 7–10, 2002, FAO, Thailand.

36
Stimulating Natural Regeneration

Silvia Holz and Guillermo Placci

> **Key Points to Retain**
>
> Stimulating natural regeneration can be achieved in a number of ways, such as removing disturbances, enclosures, eliminating barriers, disperser management and spatial distribution of species within the restoration landscape.
>
> The art of restoring a forest landscape consists, to a large extent, of the strategic selection, combination, and adequate use of different methods for each stage and for each case.
>
> The principal needs for developing restoration projects based on stimulating natural regeneration are (1) to continually study the ecological processes, (2) to develop monitoring systems and statistical methods to compare different types of data at different scales, (3) to implement environmental education programmes, and (4) to develop strategies to decrease operative costs and to increase incentives for stimulating natural regeneration.

1. Background and Explanation of the Issue

Forests can regenerate in previously forested areas once the land ceases to be used for alternative purposes (e.g., grazing, agriculture, wood extraction). However, the recovery process can be extremely slow or inhibited in highly degraded ecosystems. The principal challenges for those working in forest restoration are to evaluate a forest's potential for recovery and, if necessary, to "accelerate" this process. Stimulating natural regeneration generally entails a lower financial cost than other restoration strategies, making it an attractive option for restoring large sections of land.

Natural regeneration can follow different trajectories and velocities according to how the different variables act in the system that is undergoing recovery. Variables such as light, humidity, temperature, availability of seeds and young trees, predation, and the structure of initial vegetation determine the successional trajectory of each site. This implies that, in general, succession in a region does not follow a linear and unique trajectory, but manifests itself in a whole range of stable and transitional states with different likely outcomes.[311] Thus, a great variety of restoration alternatives—modelled from the specific characteristics of the system and the specific objectives of the restoration project—can be proposed for a given system that are compatible with the likely outcomes of the natural succession that would otherwise occur.

The first step in the process of stimulating forest recovery involves identifying the principal factors that are acting as "barriers" or as "facilitators" to regeneration. Once these

---
[311] Vallejo et al, 2003.

factors have been identified, they can be manipulated to accelerate forest regeneration. Most studies have identified dispersion, competition with herbaceous plants, and poor soil conditions as being the most important barriers for tree settlement in abandoned farmlands[312] (also see section "Restoring After Disturbance" in this book). These studies highlight the importance of physical as well as biological barriers. On the other hand, trees, bushes, ferns, and fallen trees can also facilitate the natural recovery of an area.[313] The remaining vegetation attracts dispersers; microclimatic conditions that favour the regeneration of young trees develop underneath this vegetation, which can thus serve as "regeneration nuclei."[314] The relative influence of each factor on regeneration depends on each system and on the temporal and spatial scale in which the analysis is carried out. Restoration methods that use natural regeneration are based on barrier elimination, stimulation of facilitating factors, or the combined manipulation of both types of factors. In selecting the best methods for restoring the forests of a particular area, it is extremely important to study the forests intensively, in order to understand their behaviour at different scales.

Several factors limit the successful application of restoration methodologies based on stimulating natural regeneration:

- Lack of seed sources and dispersers: In many cases, there are no forest remnants that can behave as seed sources at restoration sites; therefore, natural regeneration possibilities remain restricted to the existing soil seed bank. In other cases, there are nearby forests but no seed dispersers due to the low number of animals (e.g., birds, mammals) in the area; thus, natural regeneration may be largely confined to species whose seeds are dispersed by wind.
- Uncertain directionality: Allowing natural regeneration to occur—without controlling the species' pool that is allowed to occupy a restored area—does not guarantee a high diversity of species in a forest. This may limit the success of restoration efforts in improving economic value for future wood exploitation or other specific activity.
- Difficulty in obtaining a high forest species' diversity: In addition to insufficient seeds, in areas with scarce or degraded forest remnants, there may be the added complication that some species will simply not be able to settle, thereby creating a forest with more limited diversity of species.
- Length of time required: A naturally regenerating forest goes through more successional states, and thus requires more time to reach a state similar to a mature forest than does a plantation composed of diverse species.

2. Examples

Natural regeneration can be used in very different ways when defining a landscape restoration strategy. Some examples of different methods are illustrated below:

2.1. Use of Diversity Nuclei

The littoral area of southern Brazil, formerly covered by Serra do Mar (Atlantic forest), is now severely deforested (Fig. 36.1). Currently, numerous actions are being carried out to preserve the remaining forests, and to restore the deforested areas.[315] Tree cover restoration in the Environmental Protection Area of Guaraqueçaba, is being aided by a strategy in which small stands of pioneer species (i.e., generally 1000 to 5000 young trees) are planted in the surrounding, more diverse stands (i.e., composed of pioneer species, initial secondary species, late secondary species, and climax species). The latter are either planted or are fragments of already existing forests in the area.[316] Plantations are carried out either in the whole area being restored or in half of this area, depending on the size of the area, its proximity

---

[312] Holl, 1999.
[313] Peterson and Haines, 2000.
[314] Guevara et al, 1986.

[315] See Sociedade de Pesquisa em Vida Selvagem e Educação Ambiental (SPVS). www.spvs.org.br.
[316] Ferretti, 2002.

FIGURE 36.1. Abandoned pastures in Antonina Reserve (Atlantic forest, Brazil), where the natural regeneration is limited by grass competition. (Photo © Silvia Holz.)

to forest patches, and the degradation of the system. Planted trees function as seed sources facilitating natural regeneration in the whole area. The treatment given to the soil (e.g., cleaning of grasses), the distance at which young trees are planted, and their size are selected according to site characteristics (e.g., type of soil, topography, and use history).

2.2. Framework Species Trees

Ecologically and socially appropriate methods for accelerating the forest recovery process within protected areas are being investigated in the seasonally dry tropical forest of the mountains of northern Thailand.[317] The "framework species" concept (i.e., the use of pioneer and climax species that strongly facilitate, more than other species, the natural regeneration of the area) has been adapted in this case. The main characteristics of framework species trees are: (1) high survival when planted at degraded sites; (2) rapid growth; (3) dense and spreading crown cover that shades out herbaceous weeds; (4) flowering and fruiting, or provision of other resources, attractive to wildlife at a young age; (5) resilience to burning (in systems with a dry season); and (6) reliable seed availability, rapid and synchronous seed germination and production of healthy seedlings in containers. Combinations of 20 to 30 species are used for plantations. These plantations significantly aid in the recovery of the basic structure of forests that grow naturally, resist disturbances, and attract seed-dispersing animals, thereby facilitating the natural regeneration of forests within the restoration area.

2.3. Remaining Vegetation as Facilitators of Regeneration

A great part of the Mediterranean Forests of Guadalajara in Spain has been transformed by wood extraction and grazing into scrublands with few tree species. In the Tonda de Tamajón woodland, native species are being introduced to increase biodiversity and accelerate the natural regeneration of the forest.[318] Tree and shrub species are selected using as criteria fruit type as well as the ecological niche that each one occupies. Efforts are made to increase the proportion of species that are used as food by wild boar populations (an important economic resource of the area). The remaining vegetation in the area is used as "nurse trees," whereby planting the young trees below the preexisting individuals protects them from sun exposure and against predation.

[317] See Forest Restoration Information Service. www.unep-wcmc.org/restoration/.

[318] See World Wide Fund for Nature, España. www.wwf.es.

FIGURE 36.2. Cattle pasture (left) and regenerating forest (right) 2 years after cattle was excluded in experimental plots in Andresito (Atlantic forest, Argentina). In this place, tree planting and grass cleaning was necessary during the first year of exclusion. (Photo © Silvia Holz.)

2.4. Elimination of Invasive Species by Planting Economically Important Native Species

The area of Andresito, in northeast Argentina, has been identified as a key area for the conservation of the Upper Paraná Atlantic Forest (Fig. 36.2); remaining forests there can guarantee the connectivity of the great forest masses of Brazil and Argentina.[319] In the framework of a project on Forest Landscape Restoration that involves a large number of people and institutions, different restoration methodologies are being investigated.[320] A particular problematic issue is the invasion of forests degraded by a native plant species behaving as an invasive, the *tala* (*Celtis* sp.), thus inhibiting natural regeneration. The strategy used in this case consists of the mechanical elimination of *tala*, followed by the plantation of *yerba mate* (*Ilex paraguariensis*)—a native tree species used in infusions, and a key product of the regional economy.[321] The fruit of yerba mate attracts birds, facilitating the natural regeneration of the area. Growth of canopy species is stimulated through selective cutting, in order to obtain a yerba mate production system under forest cover. Therefore, in addition to restoring a degraded area, an effort is also made to improve the financial opportunities of local farmers. This increases the likelihood that they will implement the restoration strategy, and that these restored areas will be preserved in the long term.

3. Outline of Tools

There is a wide variety of tools that can be used to stimulate natural regeneration. The art of restoring a forest landscape depends heavily on the selection, combination, and appropriate use of different tools for each stage and for each particular case.

- Management of early stages of natural regeneration in secondary forests: Natural regeneration is the most effective and economical way of restoring slightly degraded areas, with a good seed bank in the soil and forest remnants nearby. However, even these relatively intact systems should be monitored periodically to evaluate the need to carry out enrichment plantations.
- Closures: At sites with high numbers of herbivores, natural regeneration can be stimulated by limiting animal grazing, thereby allowing the growth of woody plants.
- Elimination of barriers using cattle and other animals: Cattle grazing can be an effective, easy, and inexpensive way to decrease the

---
[319] Di Bitetti et al, 2003.
[320] See Forest Landscape Restoration (FLR). http://www.panda.org/forests/restoration.
[321] Holz, 2003.

biomass of grasses that compete with young trees—in cases where the tree species are themselves not palatable to cattle.[322]

- Elimination of barriers through mechanical and/or chemical methods: The soil compaction that retards the settlement of young trees can be eliminated through, for example, ploughing. Grasses can be eliminated through herbicide application, manual weeding (e.g., using a cane knife), or mechanical weeding (e.g., with weeding machines).
- Installation of racks to facilitate regeneration: Where existing vegetation does not present a significant barrier to natural regeneration, artificial racks (e.g., crosses, sticks, or wires) on which birds can perch can be used to increase the seed rain in an area and, therefore, help accelerate site regeneration. In systems with grasses that retard regeneration, natural racks (e.g., trees, bushes) are often more effective, since they increase the seed rain as well as acting as shaders, decreasing grass coverage.[323]
- Planting a few species to stimulate regeneration: The selective planting of a few tree species can help stimulate natural regeneration by (1) offering additional perches for seed dispersers such as birds, and (2) shading out competing plants.
- Spatial distribution of species within the restoration landscape: The presence of species of different ecological groups—strategically located within the landscape—can help accelerate natural regeneration at this scale, as well as lowering significantly the costs that would be incurred by planting young trees throughout an entire restoration area. Planted stands with high species' diversity, as well as remnant forests in the landscape, can function as "diversity islands," providing seeds to the area throughout the restoration process.[324]
- Disperser management: Another possible tool for stimulating natural regeneration is to try to increase the number of dispersers (e.g., birds, mammals) in an area. This can be achieved by decreasing hunting activities and pesticide use, reintroducing species, and creating wildlife corridors.

4. Future Needs

4.1. Increase Current Knowledge

It is crucial to continue studying the following issues in order to be able to develop restoration actions based on natural forest regeneration:

- Species' ecology: Little is known about the phenology, reproductive biology, interactions with other species (e.g., pollination, seed dispersion, herbivory/predation) of many plant species.
- Dynamics of ecological succession: Restoration involves the manipulation of the natural succession process; therefore, it is necessary to know the factors involved in the natural regeneration of the system and the mechanisms through which they function.
- Behaviour of the system at different scales: For many systems, there is little information on patterns and processes operating at different scales.

4.2. Development of Monitoring Systems and Statistical Tools to Compare Different Types of Restoration

Monitoring systems, as well as statistical methods for comparing different types of data at different scales, are tools that need to be developed for adjusting current restoration methodologies. Detailed records of the history of site use and implemented restoration practices, as well as the use of standardised monitoring protocols, would facilitate such comparisons.[325] The use of nontraditional statistical methods (e.g., Bayesian methods) can allow for more efficient evaluation of restoration methods, because they are more robust when working with small samples, with no

---

[322] Posada et al, 2000.
[323] Holl et al, 2000.
[324] Kageyama and Gandara, 2000.

[325] Holl et al, 2003.

replicates, or with much noise in the system, and they also allow for the combination of different types of data.[326]

4.3. Implementation of Environmental Education Programmes

In general, recovery areas are perceived as nonproductive areas. If people can identify and appreciate the multiple functions of these areas, the potential for preserving the forest will increase, as will possibilities of implementing restoration projects in which natural regeneration will play a key role. This issue is particularly important in the development of educational programmes.

4.4. Financing of Restoration Processes

The development of strategies for decreasing operating costs and increasing incentives for stimulating natural regeneration is essential in applying the restoration methods developed at the experimental scale to the restoration of large areas. For example, it is important to consider the increase in the production capacity of the restored area, compensation for the opportunity cost for landowners, payment for environmental services, and the implementation of tax incentives.

References

Di Bitetti, M.S., Placci, G., and Dietz, L.A. 2003. A biodiversity vision for the Upper Paraná Atlantic Forest ecoregion: designing a biodiversity conservation landscape and setting priorities for conservation action. WWF, Washington, DC.

Ferretti, A.R. 2002. Modelos de plantio para a restauração. In: A Restauração da Mata Atlantica em Áreas de sua Primitiva Ocorrência Natural. Embrapa Florestas, Colombo, pp. 35–43.

Guevara, S., Purata, S., and Van der Maaler, E. 1986. The role of remnant forest trees in tropical secondary succession. Vegetatio 66:77–84.

[326] Marcot et al, 2001.

Holl, K. 1999. Factors limiting tropical rain forest regeneration in abandoned pasture: seed rain, seed germination, microclimate and soil. Biotropica 31: 229–242.

Holl, K.D., Crone, E.E., and Schultz, C.H.B. 2003. Landscape restoration: moving from generalities to methodologies. BioScience 53(5):491–502.

Holl, K.D., Loik, M.E., Lin, E.H., and Samuels, I.A. 2000. Tropical montane forest restoration in Costa Rica: overcoming barriers to dispersal and establishment. Restoration Ecology 8(4):339–349.

Holz, S. 2003. Atlantic Forest restoration in the buffer zone of Iguazú National Park (Argentina). Technical Report (not published).

Kageyama, P., and Gandara F. 2000. Recuperação de areas ciliares. Capítulo: 15. In: Rodriguez, R., and Filho, L., eds. Matas Ciliares: Conservação e Recuperação. Edusp, São Paulo, Brazil.

Marcot, B.G., Holthausen, R.S., Raphael, M.G., Rowland, M.M., and Wisdom, M.J. 2001. Using Bayesian belief networks to evaluate fish and wildlife population viability under land management alternatives from an environmental impact statement. Forest Ecology and Management 153: 29–42.

Peterson, C.J., and Haines, B.L. 2000. Early successional patterns and potential facilitation of woody plant colonization by rotting logs in premontane Costa Rica pastures. Restoration Ecology 8(4): 361–370.

Posada, J.M., Aide, T.M., and Cavelier, J. 2000. Cattle and weedy shrubs as restoration tools of tropical montane rainforest. Restoration Ecology 8(4): 370–379.

Vallejo, R., Cortina, J., Vilagrosa, A., Seva, J.P., and Alloza, J.A. 2003. Problemas y perspectivas de la utilización de leñosas autóctonas en la restauración forestal. In: Rey, J.M., Espigares, T., and Nicolau, J.M., eds. Restauración de Ecosistemas Mediterráneos. Universidad de Alcalá, Alcalá de Henares, pp. 11–42.

Additional Reading

Guariguata, M.R., and Ostertag R. 2001. Neotropical secondary forest succesion: changes in structural and functional characteristics. Forest Ecology and Management 148:185–206.

Guimarães Vieira, I.C., Uhl, C., and Nepstand, D. 1994. The role of shrub *Cordia multispicata* Cham. as a "succession facilitator" in an abandoned pasture, Paragominas, Amazonia. Vegetatio 115: 91–99.

Holl, K. 2002. Effect of shrubs on tree seedling establishment in an abandoned tropical pasture. Journal of Ecology 90:179–187.

Janzen, D.H. 1988. Guanacaste National Park: tropical ecological and biocultural restoration. In: Cairns, J.J., ed. Rehabilitating Damage Ecosystems, vol. 2., CRC Press, Boca Raton, FL, pp. 143–192.

Nepstad, D.C.C., Uhl, C., Pereira C.A., and Cardoso da Silva, J.M. 1996. A comparative study of tree of tree establishment in abandoned pasture and mature forest of eastern Amazonia. Oikos 76: 25–39.

Purata, S.E. 1986. Floristic and structural changes during old-field succession in Mexican tropics in relation to site history and species availability. Journal of Tropical Ecology 2:257–276.

Ramirez-Marcial, N., Gonzalez-Espinoza, M., and García-Moya, E. 1996. Establecimiento de *Pinus spp* en matorrales y pastizales de Los Altos de Chiapas, México. Agrociencia 30(2): 249–257.

Rey-Benayas, J.M., Espigares, T., and Castro-Diez, P. 2003. Simulated effect of herb competition on planted *Quercus faginea* seedlings in Mediterranean abandoned cropland. Applied Vegetation Science 6:213–222.

Slocum, M.G. 2000. Logs and fern patches as recruitment sites in a tropical pasture. Restoration Ecology 8(4):408–414.

Wunderle, J.M. 1998. The role of animal seed dispersal in accelerating native forest regeneration on degraded tropical lands. Forest Ecology and Management 99(1–2):223–235.

37
Managing and Directing Natural Succession

Steve Whisenant

> ## Key Points to Retain
>
> Carefully designed silvicultural strategies can accelerate growth, influence the direction of succession, increase the goods and services provided, or enhance diversity.
>
> Directing natural processes toward land use goals requires an understanding of the processes driving succession.
>
> Tools for managing and directing natural succession should be used as an imitation of natural processes rather than as a substitute for them.

1. Background and Explanation of the Issue

After regeneration begins on previously forested sites, carefully designed silvicultural strategies can accelerate growth, influence the direction of succession, increase the goods and services provided, or enhance diversity.[327] Selecting proper treatment options requires an understanding of the factors limiting successional change and increases in desired species. These treatments should be designed to assist natural processes rather than fight them. This is most likely to occur when forest restoration plans (1) consider and remove the underlying causes rather than the symptoms of degradation; (2) are based on an understanding of succession and threshold barriers that must be overcome through designed interventions; and (3) stimulate the desired successional behaviour with minimal interventions.

1.1. Consider Underlying Causes Halting Natural Succession

Many forest restoration programmes fail because they do not address the underlying causes of degradation. A number of social, political, and economic factors are often the underlying cause of forest loss or degradation. It is equally important to identify the biophysical barriers to recovery through natural successional processes. For example, livestock may contribute to degradation in some situations but be an important part of the recovery plan in other circumstances. Forests limited by excessive fire and invasive grasses may benefit from cattle that reduce fuel loads until the tree canopy begins to close. In contrast, forests limited by livestock that consume high percentages of developing seedlings benefit more from livestock exclusion than from control of unsustainable harvest of wood and nonwood forest products.

1.2. Understand Natural Succession and Potential Threshold Barriers

Having stimulated natural regeneration processes that establish forest species (see previous

---

[327] Lamb and Gilmour, 2003.

chapter), it is necessary to manage and direct succession processes toward the desired objectives. It is important to promote continued development of the vegetation to conserve soil, nutrient, and organic resources; restore fully functional hydrologic, nutrient cycling and energy flow processes; and create self-repairing landscapes that provide the goods and services necessary for biophysical and socioeconomic sustainability. Different stages of degradation require management actions that focus on different processes. Severely degraded sites require early repair of hydrologic, nutrient cycling, and energy capture and transfer processes. As the vegetation increases in biomass and stature, it reduces abiotic limitations of the site by improving soil and microenvironmental conditions. Directing natural processes toward land use goals requires an understanding of the processes driving succession. The rate and direction of succession is influenced by the availability of species, the availability of suitable sites, and by differential species' performance.

Previous land use has important and potentially long-lasting impacts on the rate and direction of natural succession.[328] Natural succession on abandoned farms and pastures is limited and directed by the available seed bank, sprouting ability of remaining stump and root systems, seed immigration, soil type and condition, and climatic conditions.[329] Natural recovery occurs most rapidly and completely following abandonment of pastures that were cleared by hand and received little weeding and light grazing. These areas benefit from diverse seed banks, nearness to seed sources, and sprouting from stumps and roots. Moderately grazed pastures are much less productive and diverse due to the loss of grazing intolerant species, diminished seed banks, and less organic matter in the upper soil horizons. Heavily grazed, mechanically cleared pastures are far more likely to remain dominated by grasses and forbs following abandonment, since they are completely dependent on seed immigration for successional development. Frequent burning prior to abandonment reduces the density of tree seed and sprouts. Large treeless areas are unattractive to most birds and bats that disperse small seeds. Monkeys and ground-dwelling mammals that disperse large-seed, late successional species are even more prone to avoid open areas. Thus, perching sites provided by isolated trees can accelerate succession.

1.3. Design Minimal Interventions to Achieve Goals

Will the site recover within an acceptable time frame in the absence of active restoration efforts? If so, will it provide the desired combination of goods and services? Answers to these key questions may be found by examining two types of reference sites. Selecting reference sites that have not been damaged provides an approximation of the potential goods and services. Reference sites that have been similarly damaged and allowed to recover naturally for different periods of time provide important information on the presence or absence of barriers to recovery. This provides critically important information about the passive intervention option. Active management interventions may be required where invasive species, damaged ecosystem processes, or other limitations halt natural recovery.

If the site is not seriously degraded and seed sources are adequate, the first few years of succession will be dominated by herbaceous vegetation and shrubs. This will typically be followed by early succession tree species and midsuccessional tree species will gradually become more dominant. In lowland humid forests, biomass peaks of early successional species occur at around 10 years. Mid-successional species may reach their peak biomass at 15 to 30 years, but remain dominant for many decades. These successional changes occur more slowly in less humid or very degraded environments.

Improving the management of ecosystem consumption (timber or wood harvest) is usually effective on relatively intact sites. Sites dominated by grasses may require vegetation control of the existing vegetation. This may be done with fire, herbicides, or mechanical or biological control methods. It may be necessary to

---

[328] Uhl et al, 1988.
[329] Kammesheidt, 2002.

add some species through seeding or transplanting. Denuded or depauperate sites that can neither stabilise nor achieve management objectives require enrichment plantings.

2. Examples

2.1. Restoring Dry Tropical Forests to Anthropogenic Grasslands in Guanacaste National Park, Costa Rica

Anthropogenic fire converted the dry tropical forest of Costa Rica to grasslands that continued to burn frequently. A programme begun in the 1980s effectively stopped fire and allowed the natural reforestation by trees. The initial forests, of species with wind-blown seed, rapidly covered the landscape. As these trees grew larger, seed-dispersing birds and mammals increasingly moved through the site and added new species to the developing forest.[330] This is an excellent example of removing barriers to natural succession and then allowing natural processes to operate over many decades to return an increasingly diverse forest to the landscape.

2.2. Plantation Trees as Nurse Plants to Increase Regeneration of Native Species

Tree plantations can sometimes facilitate the return of native vegetation. In Puerto Rico, tree plantations improved soil and microenvironmental conditions enough to facilitate the natural immigration of native species.[331] The plantation also accelerated the return of native species by attracting animals that brought additional seed. Tree plantations in the moist and wet tropics do not remain monocultures because native trees invade the understorey and penetrate the canopy of the exotic species. Unless site damage is extreme, native forests eventually dominate. Where damage is more severe, the resulting forests are likely to become a combination of native and exotic species.

FIGURE 37.1. Previously mined site in Hungary that has undergone natural regeneration for about 30 years. (Photo © Steve Whisenant.)

2.3. Spontaneous Regeneration of a Mine Site in Hungary

Mining is a drastic alteration of site conditions and processes. Planting trees on these sites is expensive and risky, thus they are often abandoned to natural processes. A mine site in Hungary received no active replanting, but 30 years following the cessation of mining, it shows numerous signs of spontaneous regeneration of herbaceous and woody vegetation (Fig. 37.1). The abundance of natural vegetation in the surrounding landscape provides seed sources. This site will take many more decades for recovery, but natural processes are operating in the absence of new disturbances.

3. Outline of Tools

Tools for managing and directing natural succession should be used as an imitation of natural processes rather than as a substitute for them. The tools described in the previous chapter focus on influencing natural regeneration. They remain appropriate throughout succession, but here is a list of tools for manipulating existing vegetation:

[330] Janzen, 1988.
[331] Aide et al, 2000.

Patience: Time can be used as a tool. Wait for signs and expression of successional trajectory. Understanding what drives and limits succession will make it easier to recognise the probable direction of successional change and the potential vegetation for that area.

Knowledge of potential successional pathways: Understanding how forest vegetation recovers following disturbances is a critical aspect of directing natural successional processes. Know what prevents improvement and remove that limiting factor.

Fencing: Where livestock delay, limit, or prevent successional development, fences that restrict livestock entry are one method for increasing seedling development. This may only be necessary until the seedlings grow out of reach of the livestock (or fences may also be more permanent for continued benefits).

Direct removal of invasive species: Invasive species may be killed or removed with herbicides, mechanical treatments, or hand removal to release native species. These tools may be expensive or very labour intensive, so their practicality is often limited to small or high priority sites.

Reducing invasive species with shade: Shade-intolerant invasive species are most effectively managed with tree species and management strategies that accelerate the occurrence of closed canopies. For example, establishing forests on fire-prone grasslands requires the prevention of fires until the forest canopy effectively excludes the grasses.

Thinning to reduce density or alter species' composition: Selective thinning may be used to provide products and income while increasing growth rates of the remaining trees. It may also be used to encourage regeneration and growth of certain desired species while reducing the abundance of more common species.

Enrichment plantings: Sites with no regeneration of shade-requiring late successional species may necessitate enrichment plantings under the canopy of earlier successional species. Enrichment plantings add species to sites where they are unlikely to enter through natural processes. They are most useful where the desired species, or suite of species, are neither present nor found in adjacent forests.

4. Future Needs

Priority areas for further development are:

Policies that encourage the development of natural, diverse forests: Government policies can accelerate destruction of natural forests or they can be crafted to encourage the development of natural and managed forests that combine production and conservation functions and reduce pressures on natural forests of high conservation value.

Improved understanding of successional processes and barriers to natural recovery: There are numerous gaps in our knowledge of succession and ways in which we might encourage and direct those processes. Many factors drive succession and similar impacts may have dramatically different results in different ecosystems. A more mechanistic understanding of the factors limiting or accelerating succession would greatly improve our predictive ability in new situations.

Novel strategies for payment of landscape forest restoration: New ways to fund forest restoration are essential. Programmes to plant trees are more easily funded than those designed to encourage and manage natural regeneration. This is unfortunate because natural succession often occurs more rapidly and at less risk than artificially planted forests.

References

Aide, T.M., Zimmerman, J.K., Pascarella, J.B., Rivera, L., and Marcano-Vega, H. 2000. Forest regeneration in a chronosequence of tropical abandoned pastures: implications for restoration ecology. Restoration Ecology 8(4):328–338.

Janzen, D.H. 1988. Tropical ecological and biocultural restoration. Science 239:243–244.

Kammesheidt, L. 2002. Perspectives on secondary forest management in tropical humid lowland America. Ambio 31:243–250.

Lamb, D., and Gilmour, D. 2003. Rehabilitation and Restoration of Degraded Forests. IUCN, Gland, Switzerland, and Cambridge, UK, and WWF, Gland, Switzerland.

Uhl, C., Buschbacher, R., and Serrao, E.A.S. 1988. Abandoned pastures in Eastern Amazonia. I. Patterns of plant succession. Journal of Ecology 76:663–681.

Feyera, S., Beck, E., and Lüttge, U. 2002. Exotic trees as nurse-trees for the regeneration of natural tropical forests. Trees 16:245–249.

Parrotta, J.A. 1995. Influence of overstory composition on understory colonization by native species in plantations on a degraded tropical site. Journal of Vegetation Science 6:627–636.

Whisenant, S. 1999. Repairing Damaged Wildlands: A Process-Oriented, Landscape-Scale Approach. Cambridge University Press, Cambridge, UK.

Additional reading

Ashton, M.S. 2003. Regeneration methods for dipterocarp forests of wet tropical Asia. Forestry Chronicle 79:263–267.

38
Selecting Tree Species for Plantation

Florencia Montagnini

Key Points to Retain

Plantations are a useful tool for restoration especially in areas where degradation is advanced, for instance in conditions of severe soil compaction, invasion by grasses, and advanced fragmentation.

In many cases information is lacking on local tree species that can be used for plantations: site adaptability, seed sources, germination and nursery requirements, and need for fertilisation.

Techniques for planting and tending of species are important to consider: need for fertilisers, mycorrhizae, irrigation, etc.

It is always preferable to use native species instead of exotic species, if a native species is available and grows well in the region.

1. Background and Explanation of the Issue

Tree plantations are sometimes the only alternative in restoring forest landscapes, at least in the short term, especially on very badly degraded soils. Low soil fertility, soil compaction after abandonment from cattle grazing, and invasion by grasses and other aggressive vegetation can be serious obstacles to natural forest regeneration. As the area of degraded lands expands, there is a greater need for tree species that can grow in such conditions and yield useful products (timber, fuelwood, and others) as well as environmental benefits (recovery of ecosystem biodiversity, soil conservation, watershed protection, carbon sequestration).

Tree species chosen for a plantation in the context of forest restoration can provide benefits from the tree products (timber, fuelwood, leaf mulches, etc.), and from their ecological effects, for example, nutrient recycling, or attracting birds and other wildlife to the landscape. The choice of a tree species depends on whether both productive and ecological advantages can be achieved in the same system, and in some cases one function, either productive or environmental, may be desired. Within a forest landscape, the preferred choice for restoration would be natural regeneration. Planting would only be a secondary option, to be used in cases where natural regeneration cannot proceed due to the obstacles mentioned above (poor soil conditions, long distances to seed sources, isolation, invasion by aggressive grasses). Within a landscape context, there should be a balance of socioeconomic goals (e.g., productivity) and biodiversity objectives for restoration.

The following factors influence species' choice for plantations:

1.1. Goals

1.1.1. Target Ecosystem Productivity and Biodiversity

Fast-growing, native pioneer species with high productivity are recommended for the initial stages of restoration of degraded lands. These species can help in facilitating the environment for later successional, longer-lived species whose end products are more valuable (better timber quality).

1.1.2. Saving Endangered Local Species

Preference should always be given to local species, especially those that are endangered. Fast-growing exotic species such as eucalypts, acacias, or pines should be used only when there are no available seeds of native species, or when environmental conditions are too harsh for any native species to survive. Exotic tree species predominate both in industrial and rural development plantations worldwide; however, native trees are more appropriate than exotics, because (1) they are often better adapted to local environmental conditions, (2) seeds may be more generally available, and (3) farmers are usually familiar with them and their uses. Besides, the use of indigenous trees helps preserve genetic diversity and serves as habitat for the local fauna.

Disadvantages of the use of native species are (1) uncertainty regarding growth rates and adaptability to soil conditions; (2) general lack of guidelines for management; (3) large variability in performance and lack of genetic improvement; (4) seeds of native tree species are often not commercially available and have to be collected; (5) high incidence of pests and diseases (e.g., the attack of the shoot borer *Hypsipyla grandella* to species of mahoganies and cedars); and (6) lack of established markets for many species. One of the strongest arguments for the use of native tree species in plantations is the high value of the wood and its increasing scarcity in commercial forests. Many native tree species of valuable timber grow well in open plantations, with rates of growth comparable or superior to those of exotic species in the same sites.[332,333]

1.1.3. End Use of Products

Most plantations whose purpose is to restore forest landscapes also have a productive purpose. Globally, half of forest plantations are for industrial use (timber and fibre), one quarter are for nonindustrial use (home or farm construction, local consumption of fuelwood and charcoal, poles), and one quarter are for nonspecified uses.[334] Among the nonspecified uses there are small-scale fuelwood plantations, plantations for wood to dry tobacco, etc. Therefore, species' choices reflect the end use of each plantation, while considering the purpose of forest restoration. For example, the native *Araucaria angustifolia* is used to replant deforested regions in Misiones, in North East Argentina, with the purpose of selling high-quality timber in a 40 to 45-year rotation. *Araucaria* thinnings are also a good fibre source. As they are native trees these plantations also hold local flora and fauna.

1.2. Issues Related to Use of Native Species

1.2.1. Genetic Selection

For several native species in developing countries there may not be enough genetic selection for the desired traits (fast growth, soil recovery, or other). Much research has been conducted by local institutions, universities, and ministries of agriculture and forestry. For example in Central America, Centro Agronómico Tropical de Investigación y Enseñanza (CATIE) has done genetic selection of local species such as *Cordia alliodora, Vochysia guatemalensis*, and other native species.[335]

---

[332] Piotto et al., 2003a,b.
[333] Montagnini et al., 2002.
[334] FAO, 2000.
[335] CATIE: www.catie.ac.cr.

1.2.2. Seed Availability

For many native species, studies on the phenology of trees may be needed (i.e., timing of flowering, fruiting, seed production, and seed collection). In addition, there must be enough seed storage capacity, which in some cases may require refrigeration, desiccation, and other procedures to accommodate seeds of tree species from mature forest. In the case of seeds from pioneer species, these are generally smaller, drier, and easier to store. At CATIE in Turrialba, Costa Rica, the seed bank has facilities suited to several native and exotic species that can be used in forest restoration, and this seed bank serves countries throughout Latin America. When the information is not known, specific tests have to be developed to understand the germination requirements and characteristics of each seed. Finally, growing requirements in the nursery must also be known, including need for fertiliser, inoculation with mycorrhizae, and time when they can be transplanted to the field conditions.

1.2.3. Preference by Local Farmers

Farmers most often prefer species whose silvicultural characteristics are well known, and species that have well-defined end uses and good markets. In many cases they also prefer native over exotic species. Seed or seedling availability in local nurseries is also an important factor defining farmers' preferences.

2. Examples

2.1. Plantations of Native Species for Restoration of Mine Spoils in Southeastern United States

In the Appalachian region of the southeastern United States, surface mining for coal has been extensive, coal being the main source of energy for power plants that generate electricity. Concern about the use of exotic species for mine soil reclamation has directed efforts toward native species, but the choices are narrowed considerably by the need for plant materials that can be established readily on adverse sites. For example, the Tennessee Valley Authority has planted sycamore (*Platanus occidentalis*) and sweet gum (*Liquidambar styraciflua*), which had performed the best in greenhouse studies in terms of growth, drought tolerance, and commercial value of products.[336] Surface coal mine lands are covered with grasses and other herbaceous species. The lands are reclaimed by returning mine spoil to the mined-out areas, grading when necessary, and planting with aggressive cover plants that will aid in preventing soil erosion as the trees mature. Drip irrigation is used in the initial establishment phases of the tree plantations. Replanting is done as needed one year after initial planting. These systems are successful in recovering mine spoil lands; however, substantial investments are needed to ensure tree establishment and growth. For example, sweet birch (*Betula lenta*) has also been used in mine reclamation because of its ability to grow on substrates that vary widely in tilth, concentrations of toxic metals, and fertility.[337] Inoculation with ectomycorrhizae (*Pisolithus tinctorius*) resulted in higher seedling biomass and better nutrient and water uptake. Inoculation with mycorrhizae is thus recommended to allow this species to flourish on surface mine spoils without heavy application of chemical fertilisers.

2.2. Mixed Plantations with Native Species for Restoration of Degraded Pastures at La Selva Biological Station, Costa Rica

Twelve native tree species were planted in mixed and pure plantations on degraded pasturelands at La Selva Biological Station in the Caribbean lowlands of Costa Rica, with the objectives of recovering soils and ecosystem biodiversity. There were three plantations, each with four species: Plantation 1: *Jacaranda copaia*, *Vochysia guatemalensis*, *Calophyllum brasiliense*, and *Stryphnodendrom microstachyum*; Plantation 2: *Terminalia amazonia*, *Dipteryx panamensis*, *Virola koschnyi*, and

---

[336] Brodie et al., 2004.
[337] Walker et al., 2004.

Paraserianthes sp., Plantation 3: *Hieronyma alchorneoides*, *Vochysia ferruginea*, *Balizia elegans*, and *Genipa americana*. In each plantation there was one nitrogen-fixing species, a relatively fast-growing species, and a slower-growing species. The criteria for species' selection were growth rate and economic value, potential impacts on soils and nutrient cycling, and seedling availability.[338] At 2 to 4 years of age, mixed plantations had greater growth and lower pest damage than pure stands for three of the 12 species tested, and there was no damage or no differences between pure and mixed conditions for the other species. The costs of plantation establishment were lower for the slower-growing species in mixed than in pure stands. When plantations were 9 to 10 years old, most species had better growth in mixed than in pure plantations. However, the slower-growing species grew better in pure than in mixed stands. Mixed plantations (combinations of three to four species) ranked among the most productive in terms of volume.[339] Mixed plantations had a more balanced nutrient stock in the soil: 4 years after planting, decreases in soil nutrients were apparent in pure plots of some of the fastest growing species, while beneficial effects such as increases in soil organic matter and cations were noted under other species. The mixed plots showed intermediate values for the nutrients examined, and sometimes improved soil conditions such as higher organic matter. The mixtures ranked high in terms of carbon sequestration in comparison with the pure plots of faster-growing species.[340,341] The mixtures of four species gave higher biomass per hectare than that obtained by the sum of a quarter hectare of each species in pure plots.[342]

2.3. Examples from Temperate Europe

During the last decade there have been increasing afforestation activities in several European countries. In Denmark, afforestation of former arable land with oak (*Quercus robur*) and Norway spruce (*Picea abies*) has been done extensively. An evaluation of soils under these plantations with ages ranging from 1 to 29 years, and a mixed plantation with both species (200 years of age) showed considerable accumulation of organic matter in the tree biomass and in the soil, especially in the older stands.[343]

In the southwestern Alps in France, the forest service has attempted forest restoration of badlands for erosion control since 1860, with the exotic *Pinus nigra* (Austrian black pine). The pines were expected to serve as nurse for the native broadleaved vegetation. A study done 120 years following reforestation showed that pines were too dense to allow for enough natural regeneration under their canopy: thinning and enrichment planting would be needed to accelerate regeneration of native species. The reestablishment of indigenous tree species was not inhibited by lack of nearby seed sources or by soil fertility. Thinning would facilitate the dissemination of seeds of the native species. Patches of native trees planted in enrichment could serve as additional seed sources of native species.[344]

3. Outline of Tools

3.1. Genetic Selection

Both tree breeding and silviculture have improved growth rates of several industrial species of eucalypts and pines. Good examples are *Eucalyptus grandis* and *E. urophylla* in Brazil. Much genetic improvement has been done by private companies, especially for the most frequently used species of pines and eucalypts. Research on other species, including indigenous trees, is underway at universities and other research institutions. For some native species, genetic improvement has advanced with trials of seed origin and progenies, the first step in the domestication of a species. For example, for *Cordia alliodora*, *Vochysia*

---

[338] Montagnini et al., 1995.
[339] Piotto et al., 2003b.
[340] Montagnini and Porras, 1998.
[341] Shepherd and Montagnini 2001.
[342] Montagnini, 2000.

[343] Vesterdal et al., 2002.
[344] Vallauri et al., 2002.

guatemalensis, and other native species in Central America, CATIE in Costa Rica has determined what are the best provenances (specific origin of the seed in a region or locality in a given country) that suit most planting conditions. In addition, progeny studies have helped to find what are the best sources of seed for *Acacia mangium, Eucalyptus grandis*, and other species.

3.2. Plant Ecology

Information on the following ecological characteristics of tree species will be useful in helping to select them for plantation purposes: light requirements, growth under different soil fertility conditions, resistance to drought, tolerance to low or high pH, tolerance to high concentrations of toxic metals, resistance against pest and disease, ability to sprout and to respond to pruning and coppicing, seed production, germination characteristics, need for inoculation with mycorrhizae, need for fertilisers, wood characteristics, and uses. In most cases basic ecological information on tree species can be found at universities, ministries of agriculture, or departments of forestry. Local information can also be obtained from nurseries, agricultural or forestry cooperatives, and from conversations with local producers. However, sometimes native species are poorly known, yet another reason for people's tendency to use exotics, which have been better studied.

3.3. Choosing Species, Designs, and Management to Stabilise Degraded Soils

Recent research in Costa Rica, Brazil, and Argentina investigated plantation tree species that could serve to ameliorate soil properties in degraded lands.[345] In Costa Rica, in just 3 years soil conditions improved in the tree plantations compared to abandoned pasture. In the top 15 cm, soil nitrogen and organic matter were higher under the trees than in pasture, with values close to those found in 20-year-old forests. The highest values for soil organic matter, total nitrogen, calcium, and phosphorus were found under *Vochysia ferruginea*, a species common in forests in the region. In Bahia, Brazil, values of at least five soil parameters under 15 out of the 20 species of the plantations were similar to or higher than those found under forest. Several species contributed to increased carbon and nitrogen, including *Inga affinis, Parapiptadenia pterosperma, Plathymenia foliolosa* (leguminous, N-fixing species), *Caesalpinia echinata, Copaifera lucens* (leguminous, non–N-fixing), *Eschweilera ovata, Pradosia lactescens* (of other families). Others increased soil pH and/or some cations, such as *Copaifera lucens, Eschweilera ovata, Lecythis pisonis*, and *Licania hypoleuca*. In Misiones, in North Eastern Argentina, the greatest differences in soil carbon and nitrogen levels under tree species and grass were found under *Bastardiopsis densiflora*, where they were twice those in areas beyond the canopy influence. The pH was higher under *Bastardiopsis densiflora* and *Cordia trichotoma*, while the sum of bases (calcium + magnesium + potassium) was highest under *Cordia trichotoma, Bastardiopsis densiflora*, and *Enterolobium contortisiliquum*. Most of the species identified in this research for their positive influence on soil properties are used in restoration projects, commercial plantations and agroforestry in each region.

3.4. Plantation Design—Pure or Mixed-Species Plantations

Mixed species' plantations have been established at several locations with varying results. However, results from a number of field experiments suggest that mixed designs can be more productive than monospecific systems.[346] In addition, mixed plantations yield more diverse forest products than pure stands, thereby helping to diminish farmers' risks in unstable markets. Farmers may prefer mixed plantations to diversify their investment and as a potential protection against pest and diseases, in spite of the technical difficulties of establishing and managing mixed plantations. Mixed stands may

---

[345] Montagnini, 2002.

[346] Wormald, 1992.

also favour wildlife and contribute to higher landscape diversity. As seen from the example presented above, mixed plantations can have many productive and environmental advantages over conventional monocultures. However, their main disadvantage lies in their more complicated design and management. Mixed plantations thus are often restricted to relatively small areas, or to situations when diversifying production is a great advantage, such as for small farmers of limited resources.

4. Future Needs

For forest landscape restoration, only native species should be used in plantations, except if, as in some of the cases mentioned earlier, there are good specific arguments for the use of exotics. Therefore, increased knowledge of characteristics and silviculture of native tree species is needed to assist in this objective. In particular, more information is needed on the performance of indigenous species in plantation conditions. In addition, silvicultural guidelines for plantations with indigenous species are needed to increase their adoption by local farmers. Market values are also an important factor influencing the adoption of native species by local farmers. A key question in species' choices with the dual purpose of restoration and production is how to balance economic objectives with biodiversity ones. Finally, there are some trade-off issues: Is it best to have smaller areas of exotic plantations or larger areas of native plantations? Again a balance between the two objectives—restoration and production—should give insights into the answer.

References

Brodie, G.A., Bock, B.R., Fisher, L.S., et al. 2004. Carbon Capture and Water Emissions Treatment System (CCWESTRS) at fossil-fueled electric generating plants. Third annual technical report 40930R03 (October 1, 2002–September 30, 2003) for U.S. Department of Energy/National Energy Technology Laboratory Award Number DE-FC26–00NT40930. Tennessee Valley Authority/Public Power Institute, Muscle Shoals, AL, in partnership with the Electric Power Research Institute, Palo Alto, CA.

FAO. 2000. Global Forest Resources Assessment 2000. Main report. http:/www/fao.org/forestry/fo/fra/main.

Montagnini, F. 2000. Accumulation in aboveground biomass and soil storage of mineral nutrients in pure and mixed plantations in a humid tropical lowland. Forest Ecology and Management 134:257–270.

Montagnini, F. 2002. Tropical plantations with native trees: their function in ecosystem restoration. In: Reddy, M.V., ed. Management of Tropical Plantation-Forests and Their Soil Litter System. Litter, Biota and Soil-Nutrient Dynamics. Science Publishers, Enfield (NH) USA, Plymouth, UK, pp. 73–94.

Montagnini, F, Campos, J.J., Cornelius, J., et al. 2002. Environmentally-friendly forestry systems in Central America. Bois et Forêts des Tropiques 272(2):33–44.

Montagnini, F., González, E., Rheingans, R., and Porras, C. 1995. Mixed and pure forest plantations in the humid neotropics: a comparison of early growth, pest damage and establishment costs. Commonwealth Forestry Review 74(4):306–314.

Montagnini, F., and Porras, C. 1998. Evaluating the role of plantations as carbon sinks: an example of an integrative approach from the humid tropics. Environmental Management 22(3):459–470.

Piotto, D., Montagnini, F., Ugalde, L., and Kanninen, M. 2003a. Performance of forest plantations in small and medium sized farms in the Atlantic lowlands of Costa Rica. Forest Ecology and Management 175:195–204.

Piotto, D., Montagnini, F., Ugalde, L., and Kanninen, M. 2003b. Growth and effects of thinning of mixed and pure plantations with native trees in humid tropical Costa Rica. Forest Ecology and Management 177:427–439.

Shepherd, D., and Montagnini, F. 2001. Carbon sequestration potential in mixed and pure tree plantations in the humid tropics. Journal of Tropical Forest Science 13(3):450–459.

Vallauri, D., Aronson, J., and Barbero, M. 2002. An analysis of forest restoration 120 years after reforestation of badlands in the south-western Alps. Restoration Ecology 10(1):16–26.

Vesterdal, L., Ritter, E., and Gundersen, P. 2002. Changes in soil organic carbon following afforestation of former arable land. Forest Ecology and Management 169:137–147.

Walker R.F., Mc Laughlin, S.B., and West, D.C. 2004. Establishment of sweet birch on surface mine spoil as influenced by mycorrhizal inoculation and fertility. Restoration Ecology 12(1):8–19.

Wormald, T.J. 1992. Mixed and pure forest plantations in the tropics and subtropics. FAO Forestry Paper 103, Rome.

Additional Reading

Carnus, J.-M., Parrotta, J., Brockerhoff, E.G., et al. 2003. Planted forests and biodiversity. An IUFRO contribution to the UNFF Intersessional Expert Meeting on the Role of Planted Forests in Sustainable Forest Management: "Maximising planted forests' contribution to SFM," Wellington, New Zealand, 24–30 March 2003. In: Buck, A., Parrotta, J., and Wolfrum, G., eds. Science and Technology—Building the Future for the World's Forests and Planted Forests and Biodiversity. IUFRO Occasional Paper No. 15. International Union of Forest Research Organisations, Vienna.

Evans, J. 1999. Planted forests of the wet and dry tropics: their variety, nature, and significance. New Forestry 17:25–36.

Montagnini, F., and Jordan, C.F. 2005. Plantations and Agroforestry Systems. In: Montagnini, F., and Jordan, C.F. 2005. Tropical Forest Ecology. The Basis for Conservation and Management. Springer-Verlag, Berlin–New York.

Parrotta, J.A. 2002. Restoration and management of degraded tropical forest landscapes. In: Ambasht, R.S., and Ambasht, N.K., eds. Modern Trends in Applied Terrestrial Ecology. Kluwer Academic/Plenum Press, New York, pp. 135–148.

Parrotta, J.A., and Turnbull, J.T., eds. 1997. Catalyzing native forest regeneration on degraded tropical lands. Forest Ecology and Management (Special Issue) 99(1–2):1–290.

Wadsworth, F.H. 1997. Forest production for tropical America. USDA Forest Service.

General Guidelines on Plantations

Cossalter, C., Pye-Smith, C. 2003. Fast-Wood Forestry. Myths and realities. Forest Perspectives. Center for International Forestry Research (CIFOR), Jakarta, Indonesia. www.cifor.cgiar.org/. Contains information on controversial issues regarding plantations such as social relevance, economic aspects, environmental effects.

Evans J. 1992. Plantation Forestry in the Tropics. Oxford University Press, Oxford, England. One of the most complete textbooks on plantation forestry for tropical countries.

FAO. 2000. Global Forest Resources Assessment 2000. Main report. http:/www/fao.org/forestry/fo/fra/main. The Food and Agriculture Organisation of the United Nations (FAO) publishes periodically statistics and information on plantations worldwide, area covered, uses, land-use changes, species, and other relevant information.

Forest Stewardship Council guidelines. www.fscus.org/. Contains materials related to certification of forest plantations; a full section on plantation forestry, principles, and criteria for sustainable management of plantation forestry.

International Tropical Timber Organisation (ITTO). 2002. Guidelines for the restoration, management and rehabilitation of degraded and secondary tropical forest. ITTO Policy development series no. 13. www.itto.or.jp. Gives detailed guidelines for how to assess a situation of forest degradation and to decide what is the best alternative for restoration.

Siyag, P.R. 1998. The Afforestation Manual. Technology and Management. TreeCraft Communications, Jaipur, India. Focusses on semi-arid regions. The book has a technical manual, explaining nursery techniques, site selection and preparation, fencing, soil and water conservation strategies, planting, care and maintenance of the plantations; a management manual, dealing with organisational aspects of afforestation, activity planning, monitoring, quality control and productivity, and record keeping; a section containing technical charts and tables to be used as models and reference; a section on management of charts and tables, and a tree planting guide.

WWF Web site on forest landscape restoration. www.panda.org/forests/restoration. Provides concepts, information on forest restoration projects in Africa, Asia/Pacific, Europe, and Latin America.

39
Developing Firebreaks

Eduard Plana, Rufí Cerdan, and Marc Castellnou

Key Points to Retain

Firebreaks are useful to stop low-intensity surface fires, as a line from which firefighters can operate, and as perimeters for prescribed fire projects.

Firebreaks vary in their effectiveness depending on adjacent hazards, the landscape to protect, and maintenance. When used alone, firebreaks cannot contain high-intensity head fires, but may serve as control points for extinction.

Firebreaks are expensive and more emphasis should be placed on understanding and managing fires.

1. Background and Explanation of the Issue

Restoration in fire-prone areas requires effective fire management policies if it is to be successful. In the major regions of the world fire services are usually able to stop medium- and low-intensity fires. Technological improvements along with increased fire prevention infrastructure (firebreaks, water points, etc.) have considerably increased the success in fighting wildfires. However, the ability to control fires is hampered in extreme climatic conditions, which are cyclical and are expected to become more extreme because of climate change. In these conditions, high load (biomass) leads to high-intensity, destructive fires, which cannot easily be controlled by infrastructure such as firebreaks.

Large wildfire events have occurred regularly in the last few years (e.g., Australia in 2002–2003, 1.3 million hectares (Mha); United States in 2000, 3 Mha; southwestern Europe in 2003, 0.5 Mha). This new phenomenon has encouraged major revisions to current fire risk management strategies, which have reached the following main conclusions:

- Fire is a natural element present in most forest ecosystems of the world, although in many regions, like the Mediterranean, for example, fire regime and frequencies have been greatly increased by human activity. Deeply disturbed secondary ecosystems are less resistant and resilient to fire than old-growth forest landscapes. Prevention infrastructures have to be based on an understanding of the current socio-ecosystem, ecological history, natural role of fire in the ecosystem, and on how forests (species, stand structures, etc.) are adapted to fire (improving tree community resilience and resistance).
- High fuel accumulations allow the development of high-intensity fires, which overwhelm prevention measures. Land-use changes (e.g., abandonment of agroforestry management) and, paradoxically, the success in controlling low-intensity fires, leads to fuel accumulation in the landscape.

- To improve the efficiency of prevention measures, local knowledge is needed of fire behaviour patterns in a particular region.
- Human settlements and infrastructure developed in fire risk areas need to tackle the phenomenon of fire from an environmental risk prevention perspective (in the same way as floods, avalanches, etc.) and integrate fire management into planning policies.

A number of alternatives exist in terms of fuel management including linear firebreaks (areas without vegetation or with low tree density); commercial forest management or selective thinning within forest areas to simulate resistance within the natural forest structure; prescribed burning to simulate the natural fire regime and fuel elimination; or fuel control through grazing. All of these measures are complementary and are focussed on reducing the fuel availability and fire severity, adapting the landscape structure to the natural fire regime or protecting the urban/forest interface.

Firebreaks must be established as barriers designed to stop surface fires (low-intensity fires), to be used as a line from which firefighters can operate, to set a backfire if necessary, and to facilitate the movement of people and equipment. They are also useful as perimeters for prescribed fire projects. Firebreaks prevent heat conduction, but not radiation, which may ignite fuel on the other side of the break (this commonly happens in high-intensity fires). Instead of total vegetation suppression, more and more firebreaks are designed as low tree density forest structures, which have less visual impact on the landscape and make it easier to control grass growth under the canopy. Fire behaviour models can be used to help us to place firebreaks in the optimal site, taking into account predominant winds, topography, and forest types among other factors. Using natural barriers like rivers or crests can be useful. When reducing tree density, it is also important to increase the canopy base height through pruning to create forests safe from crown fires. In commercial forestry, the choice of less combustible species such as *Acacia* spp. can be also considered in some cases.

No absolute standards for firebreak width or fuel manipulation are available. Firebreak widths have always been quite variable, both in terms of theory and on-the-ground practice. As the literature shows, the rule of thumb often adopted for firebreak widths is as follows[347]:

- Two to four times the height of adjacent trees
- Six to seven times the height of trees: wind regime passes from laminar to turbulent, letting flying embers and firebrands fall in the strip
- Average wind speed multiplied by time of flight of burning embers (about 15 seconds)
- Width greater than potential horizontal length of flames to be expected at the head of the fire. (For other recommendations see Table 39.1)

Unfortunately, for many reasons, firebreaks are sometimes not wide enough to be effective. Firebreaks may even sometimes act as chimneys, creating a route for the wind to blow and increase fire spread and intensity. The lack of shade on the ground creates good conditions for germination and growth of annual plants, which can themselves turn into dangerous fuels, characterised by a high rate of spread and high linear intensity. In windy sites firebreaks are not efficient because of the great flame length, which allows fires to jump across a complete network of firebreaks. A complete periodic cleaning is necessary for the proper maintenance of firebreaks. Prescribed fire can be effective but there is a potential risk of fire escaping along the edges. Mechanical treatments are an alternative but are quite expensive. A cheaper alternative can be promoting grazing into the area (sheep, goats, cows), but for this the forest owner's cooperation is needed. Grazing must also be managed carefully to avoid damage to trees and erosion of soil.

Another important issue is the urban/forest interface. Structural fire losses are increasing dramatically as more people build and live in proximity to flammable plant communities. A basic list for reducing the fuel load and therefore, risk in urban/forest interface is as follows[348]:

[347] Leone, 2002.
[348] Schmidt and Wakimoto, 1988.

TABLE 39.1. Recommended minimum distances needed in firebreaks with high-risk conditions.

Minimum distances needed in firebreaks with high risk conditions

Vegetation	Flat land	Land with 70% slope
Tree stand and low, dense brush	12 m	20 m
Tree stand and dense brush	25 m	35 m

Terrain	Width
Crests with slopes higher than 50%	60 m
Crests with high slope in one side (50%) and low slope in other (20%)	80 m
Crest with slow slopes (20%)	60–100 m
Flat land	100 m
Thin watercourse	150 m

Source: Vélez, 2000.

- Remove enough trees to reduce crown cover to less than 35 percent, leaving a minimum of 3 m of open space between crowns.
- Thin to a minimum of the height of two trees in each direction from home on level terrain (twice on slopes of 30 percent, four times on slopes higher than 55 percent).
- Prune with elimination of live and dead portions of crown up to 3 m from the ground to a minimum of twice the trees' height in each direction from home on level terrain, to reduce the incidence of surface fires getting into the tree crowns.
- Remove understorey trees or space them widely enough to reduce the chances of surface fires igniting them and in turn the main forest canopy.
- Clean up woody material including that accumulated in the above operations to reduce incidence and intensity of surface fires.

As a final conclusion, firebreaks and other spatially restricted fuel management zones vary in their effectiveness according to adjacent hazards, project construction (e.g., width), and maintenance. When used alone, firebreaks do not contain high-intensity head fires, but may serve as control points for indirect attack and flank fire containment. This is an important point given the high cost of constructing and maintaining firebreaks. Simulation studies in terms of such factors as fire spread, intensity, and the occurrence of spotting and crowning (fire of design) are basic for a cost-effective investment.

2. Examples

2.1. A Network of Firebreaks in Bages County, Catalonia (Spain) with Local Community Participation

The network of firebreaks project in Spain's Catalonia province had three main interrelated objectives. The first one, the assessment of risks, was intended to produce a spatial account of the potential forest fire risks occurring in the county by analysing each of the identifiable dimensions that contribute both to the increase in the likelihood of fire and in the negative impacts once the fire has started. This implied a detailed analysis, using a mixture of sources, of the distribution and causes and meteorological conditions of fire within the forested territory. The FARSITE programme generated risk analysis and Geographical Information System (GIS) maps were produced by the ARC/INFO© programme. The aim was to produce a territorial representation of risks and vulnerability in order to proceed with the assessment, as the second objective, of the human and technical resources available to minimise both the risk and the eventual harm due to forest fires in the county. The purpose, then, was to estimate the correspondence between fire risk and control capacity in the different locations. In turn, such resources were divided into fire prevention, detection, intervention, and infrastructure. The variables taken into account in this respect were (1) structure of fire protection barriers, (2) other measures to break fuel continuity (prescribed fire, grazing, green plantations), (3) forest management and selective thinning, (4) number and visibility of look out posts, and (5) forest mass accessibility. As the third objective, the overall aim of the project was to develop and implement a strategic plan to deal with such risks. This plan was the outcome of the integration of expert and relevant stakeholder knowledge carried out during the empirical research. In a series of 14 meetings, local managers, forest owners, and many other actors representing a large amount and diversity of the county's population were shown, and asked to respond to, the results of the expert GIS analysis of the situation. These maps and results were revised, modified, and enhanced as a result of the discussions. Eventually, specific measures were debated, actors' roles identified, and the actions to be pursued agreed upon with regard to fire prevention, fire prediction, and fire extinction. Thus, the crucial role of local populations was underlined during the whole process of the research and policy action. Participation was carried out at different stages, including for the assessments of fire risk, the estimation of control capacity resources, and, last but not least, at the implementation stage. The meetings were composed of individuals representing the following actors and agents: voluntary forest protection patrols, forest landowners, local public officials, fire brigades, the local environmentalist group, a local environmental consultancy company, a local expert on environmental issues, and the local media.[349]

2.2. Fuel Management Versus Fire Suppression? A Worldwide Overview

After years of investing in fire suppression, many developed countries have had to recognise that high-intensity fires are out of reach of suppression efforts, due to high fuel accumulation. Economic analysis shows easily that the cost per hectare of prescribed burning or thinning is cheaper than extinction,[350] but there is a lot of discussion about the optimum amount of treated forest surface, due both to the difficulties in analysing the fuel management productivity, and to the lack of completed data (cost of planning and monitoring). American and Australian fire control systems, which have had to deal with major fire problems in the last few seasons, have decided to increase the amount of fuel management, and the use of prescribed burning in particular (Victorian Bushfire Colonel Inquiry in 2004; Forest Healthy Initiative by USDA Forest Service in 2001), and even to let the natural fires do part of this job. Large and intense fires always take the majority of the costs of suppression. In California, some research simulating fire suppression scenarios using the fire growth model FARSITE have demonstrated how silvicultural treatment in strategic sites into forest areas (nonlineal firebreaks) can reduce the fire cost (damages and suppression costs) by 500 percent, with benefit-cost ratios of 2.94 and 1.47 in return intervals of 50 and 100 years, respectively.[351] Therefore, the priorities for investment in fuel management should be aimed at minimising these large-scale events, and fires of design are the best tool to do this.

3. Outline of Tools

- Landscape fuel management techniques and firebreaks maintenance measures: Management guidelines adapted to specific local conditions for silvicultural treatments (selective thinning and pruning), prescribed burning, or grazing are needed. Wherever possible, local agrarian activity should be used within fire prevention strategies as a means of promoting rural development and local stakeholders' involvement.
- Participatory methods with local stakeholders and policymakers: Agreement among all

---

[349] Tàbara et al., 2003.
[350] Agee et al., 2000.
[351] Finney et al., 1997.

the stakeholders involved is essential to ensuring the social sustainability of any fire prevention project.
* Territorial planning and legislative tools. In Italy, Spain and France for instance, grazing is legally recognised as a tool for fire prevention. It is highly desirable to include fire risk in urban and infrastructure planning.

4. Future Needs

The following three points are priorities for future work on firebreaks:

* Knowledge of the natural fire regime in each region and the forest structure is needed to avoid high-intensity destructive fires. Information tools such as fire behaviour models like FARSITE or geographic information systems should provide the information to design our infrastructures in the most cost-effective manner.
* Incentives are needed to ensure economic viability and cross-cutting legislation for the policy development of fuel management activities in a landscape, especially taking into account local stakeholders' participation.
* Awareness must be raised among society and policymakers showing the fire as a natural element of Mediterranean ecosystems, and the need to include fire risk management in landscape management and territorial planning. Improving the knowledge of fire as a natural risk shall improve the social viability of the measures adopted.

References

Agee, J., Baahro, B., Finney, M., et al. 2000. The use of shaded fuelbreaks in landscape fire management. Forest Ecology and Management 127:55–66.

Finney, M.A., Sapsis, D.B., and Bahro, B. 1997. Use of FARSITE for simulating fire suppression and analyzing fuel treatment economics. Symposium on Fire in California Ecosystems: Integrating Ecology, Prevention, and Management, 17–20 November 1997, San Diego, California. Association for Fire Ecology Misc. Pub. No. 1, pp. 180–199.

Leone, V. 2002. Forest management: pre and post fire practices. In: Pardini, G., and Pintó, J. eds. Fire, Landscape and Biodiversity: An Appraisal of the Effects and Effectiveness. Diversitas No. 29, Universitat de Girona, Spain.

Schmidt, W.C., and Wakimoto, R.H. 1988. Cultural practices that can reduce fire hazards to home in the Interior West. In: Fischerm, W.C., and Arno, S.F., eds. Protecting People and Homes from Wildfire in the Interior West: Proceedings of the Symposium and Workshop, 6–8 October 1987, Missoula, MT. Gen. Techn. Rep. INT-251, UT: USDA, Forest Service, Intermountain Research Station, pp. 131–141.

Tàbara, D., Saurí, D., and Cerdan, R. 2003. Forest fire risk management and public participation. In: Changing Socioenvironmental Conditions: A Case Study in a Mediterranean Region. Risk Analysis 23(2):249–260.

Vélez, R. 2000. La Defensa Contra Incendios Forestales. Fundamentos y Experiencias. McGraw-Hill, Madrid. ISBN: 84-481-2742-0.

40
Agroforestry as a Tool for Restoring Forest Landscapes

Thomas K. Erdmann

Key Points to Retain

Agroforestry systems that provide permanent tree cover should be promoted in forest landscape restoration initiatives where neither natural forest restoration nor full-sun crops are viable large-scale options.

An intimate knowledge of local livelihoods, forest use, and farming systems will be required for successful initiatives that aspire to restore forest landscapes and develop sustainable agriculture.

1. Background and Explanation of the Issue

1.1. Tree Cover, Soil Fertility, and Agriculture

Forest soils are often fertile, especially where forest ecosystems are relatively undisturbed and have been able to cycle and recycle essential plant nutrients and organic matter over long periods. Even where forest soils are poor, significant amounts of nutrients are often held in the above-ground biomass. In relatively young secondary forests or woody fallows, organic matter from tree litter (leaves, bark, branches, etc.) can quickly accumulate. Moreover, the deep root systems of trees are able to "pump" nutrients from the soil that are inaccessible to other plants.

1.2. The Dilemma of Shifting Agriculture

Farmers have targeted forest ecosystems for centuries. Usually this has taken the form of shifting agriculture, whereby a patch of forest is cleared, burnt, and then farmed for a few years until much of the soil fertility has been depleted and/or colonisation of the plot by weeds becomes too difficult to manage. In these traditional systems, the area is then abandoned and left fallow for a number of years.

If the duration of fallow periods is long enough, shifting cultivation is a sustainable system. However, in many tropical developing countries, high population growth rates have led to an increased demand for arable land that has, in turn, resulted in shorter and shorter fallow periods for these systems. The shorter fallow periods result in unsatisfactory soil fertility and declining yields. Over time, this has led to severely degraded lands no longer suitable for agriculture. Farmers are then forced to clear the primary forest again for the fertile soils needed for acceptable crop yields. The long-term result has been an accelerated rate of forest degradation and deforestation.

1.3. A Short Introduction to Agroforestry

Agroforestry is not a new practice and has, in fact, existed for as long as humans have practised agriculture. However, it is only during the past 30 years that it has received ample scientific attention and systematic study. The accepted definition of agroforestry is "a collective name for land-use systems and technologies where woody perennials are deliberately used on the same land-management units as agricultural crops and/or animals, in some form of spatial arrangement or temporal sequence."[352] The key word here is "deliberately," as the people employing these systems do it intentionally. Some forms of agroforestry are potential avenues for contributing to forest landscape restoration while also responding to agricultural needs and the shifting cultivation dilemma.

1.4. The Multipurpose Tree and Species' Choice: Domestication of Natural Forest Species and Biodiversity Considerations

One of the foundations of agroforestry is the multipurpose tree. This is a woody species (tree or shrub) that can furnish more than one product or service. Species that can provide these multiple benefits are preferable to those that furnish only one product or service and should be actively promoted in restoration efforts aimed at sustaining agriculture. For example, many nitrogen-fixing shrub species may enhance soil fertility while at the same time providing nutritious fodder for livestock and holding soil in place (combating erosion). Similarly, fruit trees can provide food while also contributing to soil conservation.

Which woody species to promote in restoration efforts is also important from a biodiversity standpoint. There may be natural forest species that provide sustenance for key threatened fauna in the landscape. Ideally, it would be preferable to encourage species that fill this niche while at the same time providing goods and services that are valuable for the local farming systems. Efforts to master propagation of these natural forest species so that they can be planted in densely populated landscapes may be a key ingredient to successful restoration. This is the first step in the domestication process whereby valuable, local native species are planted and incorporated into the farming system.

[352] Lundgren and Raintree, 1982, cited in Nair, 1993.

1.5. Competition Between Woody Plants and Herbaceous Crops

One critical issue to consider when planning forest restoration to sustain agriculture is competition for light, water, and soil nutrients between trees/shrubs and crops. Spatial trade-offs may need to be negotiated in order to achieve an acceptable balance of agricultural yields and forest goods and services. For example, it may be necessary to increase the spacing between hedgerows in order to achieve the desired agricultural yields in alley cropping (hedgerow intercropping) systems.

1.6. Trees Scattered Throughout the Landscape Versus Restoring a Closed Canopy Forest

In densely populated landscapes where arable land is in high demand, it may not be possible, from a socioeconomic standpoint, to restore significant areas wholly devoted to tree cover or forests. Alternatively, one may have to focus on planting trees and shrubs in "in-between" places on farms. These places could include farm or field borders, hedgerows along contour lines in sloping areas, or small clusters of trees and shrubs adjacent to homes. The goal would still remain restoration of the goods and services that woody plants provide.

1.7. Stakeholder/Client Needs and Forest/Tree Services

Lastly, but most importantly, it is critical to have a firm understanding of who the key stakeholders or clients are as well as what their land and natural resource use viewpoints and priorities are in any forest landscape restoration initiative. When it comes to restoration aimed at sustaining agriculture, the key stakeholder group will be local farmers who practise agriculture within the landscape in question. It will be of paramount importance to comprehend the local agricultural systems and their relationship to forest cover. Similarly, it is important to know how the forest is traditionally used, that is, which species provide products that are beneficial to the local population. This knowledge is required in order to design appropriate restoration interventions.

2. Examples

2.1. Extension of Home Gardens and Farming Under Natural Forest Fallow and Secondary Forests

Multistory home gardens that combine trees, shrubs, and shade-tolerant crops are found throughout the tropics.[353] Usually, they are diverse and can include fruit trees; nut trees; trees and shrubs that produce edible oils; high-value timber trees; woody and herbaceous plants that produce aromatic compounds; shade-tolerant tree crops such as rubber, cacao, and coffee; and shade-tolerant crops such as bananas, yams, cassavas, and spices. Due to their diversity, these systems are risk-averse and can provide economic and food products throughout the year. They are an important component of the livelihoods' strategies of uncountable poor, rural smallholders. Establishing, extending, and diversifying these home gardens offer enormous potential in many threatened and degraded forest landscapes throughout the world; the practice should thus be considered part of any forest landscape restoration (FLR) strategy. The following three examples of cash/tree crop systems could easily be combined with or connected to diverse home gardens. This can easily be practised in secondary forest and older fallow areas and, indeed is already an important practice of rain-forest colonists in Brazil, Peru, and Nicaragua.

2.2. Tree Crops and Forest Restoration

2.2.1. Rubber in Borneo[354]

Rubber is one of the principal tree crops for smallholders in southeast Asia. Despite claims to the contrary, rubber has actually led to increased tree cover in some areas of Borneo. This has happened as local farmers move from an extensive to a more intensive land use system. It is also coupled with the entry of the local population into a cash economy—rubber is a major cash crop—as well as an increased government presence and enforcement of legislation aimed at controlling forest encroachment and a switch from upland to irrigated rice production. Farmers in the cases studied actively created forests or rubber gardens in fallow or secondary forest areas or added rubber to traditional multistory home gardens. In both situations, the rubber trees are mixed with fruit trees and other trees that provide economic products as well as with spontaneous natural forest regeneration. These man-made forests are structurally complex and floristically diverse. The overall policy conclusion[355] was that it was preferable to promote tree or tree crop technologies when the maintenance of a forested landscape was desired.

2.2.2. Cacao in Côte d'Ivoire[356]

The introduction of cacao, coupled with influxes of migrants, has generally led to extensive deforestation in Côte d'Ivoire. More recently, however, land scarcity and better governmental enforcement against forest clearing has led to a change in this trend. Farmers are now adapting practices that lead to an overall increase in forest cover, planting grasslands and shrubby fallows with cacao in combination with fruit trees and high value timber trees (logging companies are now turning to valuable trees on older cacao plantations that were spared during plantation establishment). Old, unproductive, often shaded cacao plantations are being replanted with newer varieties of cacao and intercropped with yams and bananas. Deforested areas often pose new challenges to farmers who are forced to adapt and innovate; new practices can lead to restoration of forest or tree cover.

2.2.3. Shade-Grown Coffee in Central America

Coffee grown under the shade of natural forest trees or planted trees—often nitrogen-fixing

---

[353] Landauer and Brazil, 1990.
[354] de Jong, 2001.
[355] de Jong, 2001.
[356] Ruf, 2001.

leguminous species—is common throughout the coffee-producing areas of the world. As in the previous examples, these "coffee forests" can be both biologically diverse and diverse from an economic standpoint as they can be combined with fruit trees and high-value timber species. In Central America and Mexico, shade-grown coffee plantations are important habitats for migratory birds. They are often a significant livelihood component of poor farmers. Proposals are currently being developed to expand these systems and market them for their environmental services including watershed protection, biodiversity benefits, and carbon sequestration. This type of coffee production could be an important component of a forest landscape restoration strategy in many areas.

2.3. Improved Fallow

Improved fallow practices generally involve planting or directly seeding shrubby legumes in agricultural fields that have lost their soil fertility. Once the cropping cycle is ready to begin again, these shrubs are usually cut down and their biomass incorporated into the soil as "green manure." In some cases, the practice can commence in the last season or two of agricultural production if farmers retain regeneration of soil-enhancing woody plants in their fields during weeding, or even direct seeding of these species during hoeing or weeding operations. Another variation is that farmers spare a few widely spaced trees in their fields at the time of clearing; these trees contribute to maintaining soil fertility (and other products and services) during the cropping cycle and provide an immediate favourable micro-climate for the establishment of additional woody vegetation once the field enters the fallow cycle.

2.3.1. Using Nitrogen Fixing Species Sesbania Sesban and Tephrosia Vogelii in Zambia

Improved fallow systems have been tested and sometimes adopted throughout the tropics. One of the most successful examples is a system using the nitrogen-fixing species *Sesbania sesban* and *Tephrosia vogelii* in Zambia. Maize yields after 2 years of fallow with these species approach those of fully fertilised fields. These same species plus *Crotalaria grahamiana* also proved highly successful in western Kenya, doubling maize yields there.[357] Poor households tend to prefer this technology over the use of chemical fertilisers. However, the problem of farmers possessing insufficient land to place in fallow renders the potential widespread adoption of this practice problematic for many areas in the tropics where population growth rates are high.

Incorporating improved fallow systems in forest landscape restoration initiatives may be challenging, however. As indicated above, trees and shrubs are usually removed once the cropping cycle begins anew. The practice is thus only a temporary restoration of tree or shrub cover. One possible compromise would be to designate a contiguous shifting agricultural zone within a given landscape in which some of the land would always be covered by improved fallows. These improved fallow areas would shift from year to year within the designated zone.

2.4. Hedgerow Intercropping

Like improved fallow, hedgerow intercropping or alley cropping is a soil fertility maintenance or restoration practice. It involves establishing permanent hedgerows of shrubs and small trees—often species that fix nitrogen—in agricultural fields. The hedgerows are periodically pruned back and the biomass incorporated into the soil between them where crops are grown.

Despite promising results of experimental trials at many agricultural research stations, the practice has not been widely adopted by farmers. This stems from two major drawbacks. First, competition between crops and the hedgerow trees and shrubs is often severe, especially for water in semi-arid and subhumid areas. Second, the required periodic pruning represents a significant labour input that many small farmers cannot afford. Insecure land tenure, access to land and credit, and a focus by extension agents on soil conservation rather than economic returns are other often problematic issues that limit adoption.

[357] Place et al, 2003.

2.4.1. Potential Adverse Impacts

There are some risks associated with widespread adoption in a given landscape of the examples outlined above. The first one is that the agroforestry practices become too successful from an economic point of view and attract human migration to the landscape. Increased immigration could subsequently cause increased clearing of natural forest. Second, some of the exotic species used in these practices may become weeds and displace native woody species. This would likely have a negative impact on the landscape's natural biodiversity.

3. Outline of Tools

3.1. Rapid and Participatory Rural Appraisals

The rapid or participatory rural appraisal (R/PRA) method is now widely accepted and practised in rural development work. In general, it is a fairly quick and very useful means of gathering information on and engaging stakeholders. It is particularly appropriate for local communities. The method can be tailored to a wider variety of subjects. It usually consists of semistructured interviews that can be conducted with large, mixed groups or smaller, more homogeneous subgroups. In the context of restoration, R/PRAs can be used to understand local natural resource use, especially in relation to natural forests. They can also be critical tools for obtaining information on agricultural practices and the associated calendar of agricultural activities. It is common during R/PRAs to carry out a transect—walking across the landscape—noting pertinent information along the way and later assembling a visual summary of what was encountered. Similarly, participatory mapping exercises are commonly employed. Most importantly, R/PRAs can be used as a starting point for engaging stakeholders living in the landscape in question. After an analysis of problems associated with natural resource use, one can conduct a participatory brainstorming session on potential solutions to these problems. Restoration activities will often be proposed at this point.

3.2. Livelihoods' Analysis

Much of the following information is paraphrased from the Livelihoods Connect Web site[358] developed by the U.K. Department for International Development and the Institute of Development Studies. Livelihoods' analysis is a people-centred approach aimed at eliminating poverty. This approach is important for any forest landscape restoration initiative but particularly in those landscapes where agriculture is a major land use—after all, it is people who practise farming. Analysis is based on a sustainable livelihoods framework that includes an examination of assets (human, natural, financial, social, and physical capital), vulnerability, and how livelihood strategies can transform structures and processes. Besides being people-centred and always considering sustainability, the approach utilises other core concepts: dynamism, holism, macro–micro links, flexibility, and building upon strengths rather than needs. The approach also calls for a multidisciplinary team that covers environmental, economic, social, and governance aspects. Many tools can be used in livelihoods' analysis including R/PRAs. Other important tools cited in the literature include gender, macroeconomic and stakeholder analyses, as well as governance assessment.

3.3. Agroforestry Technologies for Forest Landscape Restoration

As seen in the examples in the preceding sections, there are a number of agroforestry practices or technologies that can be incorporated into restoration initiatives. These fall into three main categories:

- Technologies for restoring and maintaining soil fertility (e.g., improved fallows, hedgerow intercropping)
- Technologies for soil conservation (e.g., hedgerow intercropping on slopes, windbreaks)

[358] www.livelihoods.org.

- Cash crop technologies for income generation (e.g., home gardens)

The first two practices use trees and shrubs to provide essential agricultural services, while the third is more directly linked to maintaining and improving human well-being. Most of the technologies have been briefly described above.

4. Future Needs

Priorities for the future include:

- Negotiating land use trade-offs between agriculture and forests: One of the key, potential stumbling blocks in implementing forest restoration in landscapes where farming is a major land use is negotiating trade-offs between agriculture and forests. The success of these negotiations will be a critical determinant of stakeholder engagement and will ultimately dictate the success of the restoration initiative. In general, conservation practitioners have little or no experience in land use planning and stakeholder negotiations (also see more on this in "Negotiations and Conflict Management"). It is thus critical that guidance and training are provided in this area. It is also important to promote partnerships between conservation entities and those dealing with livelihood and development concerns.
- Propagation of indigenous tree species: Incorporating indigenous tree species into agroforestry systems can make any forest landscape restoration initiative more "biodiversity friendly" while at the same time providing goods and services desired by local farmers. Unfortunately, the biology of many of these species is little known or understood. Some basic, applied research may be needed to ascertain the most appropriate propagation techniques. The inclusion of these species in agroforestry systems is analogous to the domestication process; much has been written on this subject and this can presumably provide the foundation for guidance for restoration practitioners on this subject.

References

de Jong, W. 2001. The impact of rubber on the forest landscape in Borneo. In: Angelsen, A., and Kaimowitz, D. eds. Agricultural Technologies and Tropical Deforestation. CAB International, Wallingford and New York, pp. 367–381.

Landauer, K., and Brazil, M. eds. 1990. Tropical Home Gardens. United Nations University, Tokyo.

Nair, P.K.R. 1993. An Introduction to Agroforestry. Kluwer, Dordrecht, Boston and London.

Place, F., Franzel, S., Noordin, Q., and Jama, B. 2003. Improved fallows in Kenya: history, farmer practice and impacts. Paper presented at the InWEnt, IFPRI, NEPAD and CTA conference, "Successes in African Agriculture," Pretoria, South Africa.

Ruf, F. 2001. Tree crops as deforestation and reforestation agents: the case of cocoa in Côte d'Ivoire and Sulawesi. In: Angelsen, A., and Kaimowitz, D. eds. Agricultural Technologies and Tropical Deforestation. CAB International, Wallingford and New York, pp. 291–315.

Additional Reading

Angelsen, A., and Kaimowitz, D. eds. 2001. Agricultural Technologies and Tropical Deforestation. CAB International, Wallingford and New York.

Elevitch, C.R. ed. 2004. The Overstory Book: Cultivating Connections with Trees. Permanent Agriculture Resources, Holualoa, Hawaii.

Gladwin, C., Peterson, J., and Uttaro, R. 2002. Agroforestry innovations in Africa: can they improve soil fertility on women farmers' fields? African Studies Quarterly 6(1&2).

McNeely, J.A., and Scherr, S.J. 2003. Ecoagriculture: Strategies to Feed the World and Save Wild Biodiversity. Island Press, Washington, USA, Covelo, California.

Schroth, G., da Fonseca, G.A.B., Harvey, C.A., Gascon, C. Vasconcelos, H.L., and Izac, A.M.N. eds. 2004. Agroforestry and Biodiversity Conservation in Tropical Landscapes. Island Press, Washington, Covelo, London.

Young, A. 1989. Agroforestry for Soil Conservation. CAB International, Oxford, UK.

Part D
Addressing Specific Aspects of Forest Restoration

Section XII
Restoration of Different Forest Types

41
Restoring Dry Tropical Forests

James Aronson, Daniel Vallauri, Tanguy Jaffré, and Porter P. Lowry II

Key Points to Retain

Dry tropical forests have been overexploited by humans, and little remains now of this biologically rich and unique ecosystem.

There are a number of valid reasons to restore tropical dry forests, including their rates of endemism, their potential to yield medicines, aromatic herbs, and foods, recreational reasons, their genetic uniqueness, and their potential adaptability to climate change.

Case studies show that restoration of tropical dry forests in a landscape context, although a difficult undertaking, is highly possible and necessary.

1. Background and Explanation of the Issue

Vast expanses of the Earth's warm regions—perhaps 40 to 45 percent of all intertropical lands—were once covered with tropical dry forests (TDF).[359] These areas included the leeward coastal plains of tropical America and Madagascar, and many (or even most) islands of the Caribbean, the Pacific and the Indian Oceans, as well as many inland regions of Africa, Asia, and Australia. Today, TDFs are deeply, and perhaps irreversibly, transformed. Only 1 to 2 percent of the original (prehuman) area remains in a relatively intact and ecologically healthy condition. The remainder are so fragmented and subject to species' loss, habitat change, and genetic erosion that they must be considered in imminent danger of extinction.

1.1. Characteristics and Biological Wealth

Reflecting the very wide range of geological substrates on which they occur and the variable, unpredictable climate to which they are subject, TDFs harbour an astonishing variety of plants and animals that are remarkable in their structure, ecophysiology, chemistry, and ecology. They also show exceptionally high rates of endemism in all major groups of organisms. Sadly, however, the ecological importance and conservation value of TDFs only began to be recognised in the last 10 to 15 years, that is, much later than for tropical humid forests.

Tropical dry forests are characterised by continuous tree cover and a multitiered canopy. They also present a unique set of selective forces that have driven the evolution of a remarkable array of life forms. Unpredictable periods of sometimes severe water stress, followed by sudden and often spectacular increases in rainfall, lead to *pulses* in the availability of water, energy, and nutrients to plants and animals alike. This combination of interannual variation and unpredictability in resources, in areas where temperatures never

[359] Bullock et al, 1995.

drop below freezing, has catalysed the evolution of impressive arrays of deciduous, semideciduous, and evergreen trees, shrubs, and lianas, with very diverse chemistry, life forms, and reproductive systems. We speak of arrays in the plural because virtually every island, peninsula, or archipelago with TDF has its own unique set of species, many of which are locally endemic. Given the advanced fragmentation they have suffered, each surviving TDF community should be considered as a unique entity of the highest possible conservation value.

1.2. Attractiveness to People and Its Consequences

Due to their seasonality, gentle topographic relief, relatively rich soils, and proximity to tropical coasts where abundant food and water sources were available, TDFs attracted human settlers and hunters from very early times. Their rich and varied mineral deposits drew entrepreneurs and industrialists as well. As a result, the transformation and degradation of these forests often has gone on for long periods of time.

Prior to the onset of major human impact, TDFs were rich in tall canopy and emergent trees of great value for their dense, hard, and often beautiful and fragrant wood, such as Sandalwood (*Santalum album*). These were selectively harvested for local construction and, later, for international timber markets. Only relatively few people, rarely from the local community, benefited as a rule.[360]

Once the tree canopy giants were removed, the TDFs were usually subjected to progressive or wholesale cycles of transformation for cattle grazing or, more rarely, farmland or extractive production of fuel wood and charcoal (e.g., in southwest Madagascar, see below). This process—dating mostly from the late 1800s—often consisted of repeated burning and clearing until there remained little or none of the original assemblages of woody plants and soilborne seed banks. Faunal and microbial biota also changed as a consequence.

Nowadays, TDF fragments and adjacent areas are mostly used for extensive livestock grazing of limited economic value or biodiversity interest. In some areas, the surviving TDFs near cities are disappearing to make way for coastal hotel complexes and unplanned urban sprawl. In the few places where some TDF remains but is neither protected nor currently sought after for "development," TDF fragments are still subject to selective logging for their slow-growing but often exceedingly valuable timber [e.g., *Cordia*, mahogany, teak, sandalwood, and yellow wood (*Podocarpus* spp.)]. This short-sighted exploitation of the most valuable remaining trees constitutes a flagrant example of "artificial negative selection" which, in TDF and other endangered forests, surely should be controlled and re-legislated, or better yet halted altogether until natural regeneration or active restoration have had some time to permit forest recovery.

1.3. Reasons to Restore

It must be recognised, however, that what remains of TDF today are not especially attractive to most people, and only rarely do they capture the attention of tourists. Their low annual productivity makes TDF of minor interest to foresters or farmers. Therefore, lobbying for their conservation, and, more so still, their restoration, is problematic. However, biodiversity criteria alone more than justify the need for greater efforts, especially at the landscape and ecoregional scales. What's more, the economic perspectives for restored tropical dry forests are by no means negligible, even if the most valuable timber trees and game animals have in most cases long ago been removed.

Many plants in tropical dry forests are known to be of value for nontimber products, including medicines, biopharmaceuticals, food products, potential sources for crop improvement (e.g., an endemic wild rice species in New Caledonia), perfumes, cosmetics, etc. Also, TDFs have significant economic value if managed under multipurpose, multiuser forestry approaches, including the incorporation of innovative eco- and cultural tourism. Restoration should clearly play a major role in both scenarios, with community involvement built into these programmes.

[360] Roth, 2001.

Additionally, in urban or peri-urban zones, like those of Grande Terre, New Caledonia, restoration of native TDF is the obvious and most cost-effective approach to meeting growing demands for amenity plantings and green areas. The maintenance costs of climatically adapted ecosystems would surely be less than for conventional horticultural plantations of exotic species—and lawn grass!—and the aesthetic result could be well superior. Such garden forests, albeit confined to urban parks, roadside planting areas, and the like, could be a useful complement to educational efforts, and serve as gene banks for extra-urban or peri-urban restoration projects, where hectares of contiguous forest, or corridors among TDF fragments, are in need of seed and germ plasm.

Finally, with global warming and an overall trend toward drying in terrestrial systems, the plants, microorganisms and animals of tropical dry forests represent a wealth of genetic capital that should not be underestimated. These organisms can be anticipated to respond more readily to warming and desertification on a global scale than those adapted to humid tropical forests. Accordingly, they merit special attention from managers and engineers as well as public policy decision makers.

2. Examples

2.1. Area de Conservación Guanacaste, Costa Rica

An extensive and innovative landscape-scale restoration and management project has been underway in Guanacaste, northern Costa Rica, since 1985, under the direction of Dan Janzen.[361] This 110,000 hectare conservation area began as Santa Rosa National Park, and through the efforts of Janzen and successive, far-sighted Costa Rican governments, was gradually increased to a landscape scale that includes not only TDF but also wet forest and montane cloud forest, as well as 45,000 hectares of off-shore marine reserve, and integrates the people who live in the area. This effort may well be unique, and is certainly of considerable relevance and importance to worldwide efforts at TDF conservation. The key points are that ecological management, conservation, and restoration are approached conjointly and at a real landscape scale. Restoration is seen as biocultural and involves the development of highly innovative education activities and ecological economics.

2.2. New Caledonia (French Pacific Territory)

Following early initiatives of one of the authors (Jaffré), and his colleagues B. Suprin and J.-M. Veillon (as well as the Services Provinciaux de l'Environnement), attention began to grow about 15 years ago to the plight of the dwindling TDFs on the western coast of the largest island of New Caledonia—la Grande Terre. In 1998, WWF, the global conservation organisation, launched an effort to organise a consortium of nongovernmental organisations (NGOs), research institutions, and local government agencies to establish a multifaceted TDF programme in the context of the WWF forest landscape restoration programme. Underway since 2001, this programme has already carried out much of the preliminary reconnaissance and mapping of the many scattered TDF fragments, and has conducted valuable ecological, silvicultural, and horticultural studies for experimental restoration efforts slated to begin in 2005. At the time of this writing, a major effort is underway to secure the possibility of enabling the restoration of a significant pilot landscape in Gouaro Deva, one of the few remaining sites containing a relatively large area (450 hectares) of forest with the potential to conserve a representative piece of the formerly widespread dry tropical forests on Grande Terre. The prospects for an integrated protect, manage, and restore pilot project remain to be worked out with provincial and national policies, decision makers and, of course, local stakeholders.

Apart from the challenges of restoring a fragmented and degraded forest landscape, TDFs everywhere are facing very high and increasing

[361] Janzen, 2002.

pressures due to invasive species (ants, plants, deer, etc.), fire, and overgrazing. New Caledonia has perhaps the most endangered TDFs in the world,[362] which face all these threats and more. New Caledonia is one of the highest priority conservation hot spots in the world, with a very rich and highly endemic biota,[363] more than justifying the considerable effort being made to achieve lasting protection.

2.3. Western Madagascar

Together with many others NGOs, WWF has called attention to the alarming state and pressing need to initiate protect, manage, and restore efforts for what is left of TDFs in western Madagascar. Unlike New Caledonia and Costa Rica, relatively larges tracts still remain in Madagascar, from the Baobab-dominated forests north of Tuléar to the spiny forests in the extreme southwest. However, centuries-old Baobabs and all their extraordinary and endemic cohorts are increasingly being cut and cleared to make way for housing and hotels, while the other-worldly and unique Didieraeaceae/tree Euphorb-dominated spiny forest is being cut and transformed into charcoal by poverty-stricken people entirely dependent on local resources.

In this kind of socioeconomic context, the challenge of protecting and restoring TDF is intimately linked to the lives and livelihoods of the neighbouring human populations, who are the ones primarily impacting the environment. While the Malagasy government has strengthened its commitment to biodiversity conservation, its capacity to implement policy through "normal" administrative measures is very limited in isolated rural areas. Alternatives are required that make use of community-based conservation approaches in which natural resource management is tightly linked to local (traditional) economic and land tenure systems and to youth education aimed at instilling a basic understanding of the short- and long-term importance of natural ecosystems.

3. Outline of Tools

3.1. Monitoring Pressures

Controlling the pressures caused by livestock, invasive species, fire or land conversion is itself a restoration tool. For example, in northwestern Argentina, an innovative landowner and rancher named Carlos Saravia Toleda has developed techniques for controlled cattle grazing that actively favour reintroduction of selected native multipurpose trees, such as *Caesalpinia paraguariensis*, which has the special feature of flowering and fruiting over very long periods of the year, offering abundant, nutritious feed for livestock, while also providing habitat for birds, rodents, and other mammals, and a favourable canopy for the autogenic reestablishment of other trees and shrubs.

Passive control methods are usually preferable (see below), but in extreme cases direct action may be necessary, as in the volunteer-based initiative to protect TDF on the island of Hawaii. In other situations, costly tools such as fences or enclosures are required, for example in New Caledonia, where introduced deer otherwise prevent any regeneration of native dry forest species.

3.2. Promoting Natural Dynamics

Relatively inexpensive, passive restoration techniques are best suited to forests where, after controlling or limiting the sources of degradation, ecosystem resilience is high. This is the case in some overgrazed or severely burnt ecosystems, where the exclusion or complete restriction of livestock grazing or fire for several years is sometimes sufficient to promote self-recovery. Because plantations, especially in dry conditions, require considerable technical and financial investment, it is preferable to attempt passive restoration, evaluating its effectiveness and benefiting from innovative techniques developed. Doing so, however, requires knowledge of the functional ecology of tropical dry forests, and especially of the animals that disperse seeds of the main trees

---

[362] Gillespie and Jaffré, 2003.
[363] Lowry et al, 2004.

(birds, bats, etc.). Passive restoration has, for example, been used effectively in Costa Rica.

3.3. Active Restoration: Improved Planting Methods

In many instances restoration requires the introduction of woody species through planting, especially of the common and framework species of the original ecosystem, but also of rare or endangered species. The "Framework species" approach developed in Queensland, Australia, and applied with success in northern Thailand tropical dry forests[364] seems highly pertinent. Using this approach, 20 to 30 key tree species are selected that together seem to form the structural framework of the forest to be restored. Nursery work on germination and propagation is then required, followed by experimental plantations involving the selection and evaluation of individual species, mixtures of species, or presumed functional groups. This method is a large improvement on the classical approach of old forestry or revegetation efforts where, typically, only two or three fast-growing tree species are used. In long-term projects, the goal will often be to create islands or nuclei of framework trees with animal-dispersed propagules to catalyze the return of mammals, birds, and other mobile dispersers to the area.

Tree planting in seasonally dry areas with unpredictable rainfall obliges foresters, land owners, and restorationists to take into consideration the perennial risk of drought. This underscores the importance of selecting the right species, producing good-quality nursery stock, and carefully timing and effecting outplanting. In some situations direct seeding of dry or pregerminated propagules should be attempted. Inoculation with appropriate strains of rhizobia and/or mycorrhizae may also be advantageous or even necessary.

As mentioned, TDFs are characterised by very high levels of spatial heterogeneity, which has great impact on microscale differences in the availability of water, nutrients, and energy.

[364] Blakesley et al, 2002.

Planting in straight lines or prepared terraces is thus not necessarily the best way to proceed.

3.4. Soil Fertility and Amendments

Soils of badly degraded TDFs are frequently poor in organic matter and low in phosphorus availability. Thus, the adjustment and/or addition of organic or inorganic components is frequently essential to achieving plant establishment, even though the original soils may have been very rich.

4. Future Needs

The ecological economic valuation of dry tropical forests has rarely been evoked, let alone attempted. This represents a clear goal for the near future.

A better understanding of TDF biodiversity and ecosystem function is needed to reach meaningful restoration objectives. From early times, humans selectively removed the tallest, straightest, hardest trees for use in boat building, housing, and other activities that require dense, relatively long-lasting timber. A clear indication of the past removal of entire canopies may be found in the presence of remarkable numbers and diversity of lianas and vines representing a broad range of families, which clearly evolved to climb to the tops of trees taller than anything we see today. The remnant tropical dry forests we are now left with are truncated, so to speak, and restorationists must take this into account when setting structural, functional, and compositional objectives.

References

Blakesley, D., Elliot, S., Kuarak, C., Navakitbumrung, P., Zangkum, S., and Anusarnsunthorn, V. 2002. Propagating framework tree species to restore seasonally dry tropical forest: implications of seasonal seed dispersal and dormancy. Forest Ecology and Management 164:31–38.

Bullock, S.H., Mooney, H.A., and Medina, E. eds. 1995. Seasonally Dry Tropical Forests. Cambridge University Press, Cambridge, UK.

Gillespie T.G., and Jaffré, T. 2003. Tropical dry forest in New Caledonia. Biodiversity and Conservation 12:1687–1697.

Janzen, D.H. 2002. Tropical dry forest: Area de Conservación Guanacaste, northwestern Costa Rica. In: Perrow, M., and Davy, A. eds. Handbook of Ecological Restoration, Vol. 2 Restoration in Practise. Cambridge University Press, Cambridge, UK, pp. 559–583.

Lowry, P.P., II, Munzinger, J., Bouchet, P., Géraux, H., Bauer, A., Langrand, O., and Mittermeier, R.A. 2004. New Caledonia. In: Mittermeier, R.A., Robles Vil, P., Hoffman, M., Pilgrim, J., Brooks, T., Mittermeier, C.G., Lamoreux, J.L., and da Fonseca, G.A.B. eds. Hotspots Revisited: Earth's Biologically Richest and Most Threatened Ecoregions (in press).

Roth, L.C. 2001. Subsistence Farmers and Perverse Protection of Tropical Dry Forest. Journal of Forestry 99:20–27.

Additional Reading

Aronson, J., and Saravia Toledo, C. 1992. *Caesalpinia paraguariensis*: forage tree for all seasons. Economic Botany 46:121–132.

Dirzo, R. 2001. Forest ecosystems functioning, threats and value: Mexico as a case study. In: Chichilnisky, G., Daily, G.C., Ehrlich, P., Heal, G., and Miller, J. eds. Managing Human-Dominated Ecosystems. Monographs in Systematic Botany from the Missouri Botanical Garden, vol. 84. Missouri Botanical Garden Press, St. Louis, MO, pp. 47–64.

Elliot, S., Navakitbumrung, P., Kuarak, C., Zangkum, S., Anusarnsunthorn, V., and Blakesley, D. 2003. Selecting framework tree species to restore seasonally dry tropical forest in northern Thailand based on field performance. Forest Ecology and Management 184:177–191.

Gordon, J.E., Hawthorne, W.D., Reyes-Garcia, A., Sandoval, G., and Barrance, A.J. 2004. Assessing landscapes: a case study of tree and shrub diversity in the seasonally dry forest of Oaxaca, Mexico and southern Honduras. Biological Conservation 117:429–442.

Janzen, D.H. 1988. Tropical dry forests: the most endangered major tropical ecosystem. In: Wilson, E.O. ed. Biodiversity. National Academy Press, Washington, DC, pp. 130–137.

Lerdau, M., Whitbeck, J., and Hollbrook, N.M. 1991. Tropical deciduous forest: death of a biome. Trends in Ecology and Evolution 6:201–202.

Murphy, P.G., and Lugo, A.E. 1986. Ecology of tropical dry forest. Annual Review of Ecology and Systematics 17:67–88.

42
Restoring Tropical Moist Broad-Leaf Forests

David Lamb

Key Points to Retain

Three issues make tropical moist forests more difficult to restore: (1) the sheer diversity of plant and animal species that they usually hold, (2) very little is known about the ecology of most of these species, and (3) the human populations living in most degraded tropical landscapes are often poor and with few resources.

Some of the key questions to consider when restoring tropical moist forests are: (1) which species to use, (2) where to get the seeds, (3) how to raise the seedlings and establish them in plantations, (4) how to ensure animal and plant diversity, and (5) how to make restoration attractive to landowners.

All stakeholders must derive some benefits if restoration is to succeed.

It is likely to be difficult to restore all the original biodiversity and some more intermediate stage may be all that is possible. If particular key species are of interest, they may need to be restored separately.

1. Background and Explanation of the Issue

Degraded tropical landscapes now cover large areas. The nature and extent of these areas varies considerably, with some being so degraded that they have crossed an ecological threshold and been transformed into grasslands. Some of these grasslands are extensive and relatively homogeneous, and contain only a few remnant patches of undisturbed woody vegetation. Other tropical moist forest areas have been less disturbed but have lost their closed canopies and much of their previous structure and biological diversity. Many degraded landscapes now contain a mosaic of grassland and degraded forest together with patches of intact remnant forest. These degraded lands also differ in the extent to which they are occupied and used by human populations. Some are so degraded that only small human populations remain, while others are still heavily used by large numbers of farmers. These differences mean there are no simple prescriptions for restoring degraded tropical landscapes. The approach used in any location must take account of both the ecological and social circumstances present.

Of course, the same could be said of many degraded lands other than those occupied by tropical moist broad-leaved forests. And tropical forests are usually found in environments where plants grow quickly so that the potential for successional development and recovery is relatively rapid. But three particular issues make these ecosystems rather more difficult to restore than most. First is the sheer diversity of plant and animal species usually present in undisturbed tropical moist forests that must be considered if forests are to be restored. Second, very little is known about the ecology of most of these species. Third, the human populations living in most degraded tropical landscapes are often poor and have few resources. Indeed, their poverty may have been part of the reason the lands were degraded in the first place. If restoration is to be successful, it must help overcome this rural poverty. This often means complete biodiversity restoration is rarely achieved over large areas.

2. Examples

2.1. Restoration via Natural Succession

Large areas of tropical forest have developed on old farmland in Puerto Rico following the abandonment of farming on many areas across the island in the 1940s. This succession has occurred with little active intervention and represents a major increase in forest cover at little direct cost. The regenerated forest now has a density, basal area, above-ground biomass and species' richness similar to that of old-growth forests. However, the species' composition is different from that in old-growth forests, suggesting some intervention will be needed if the missing species are to be recovered.[365]

2.2. Intensive Restoration After Mining

One of the most intensive ecological restoration projects in the humid tropics is that which took place after bauxite mining in Brazil. In this case extensive research by the mining company had identified the plant and animal species present and revealed something of their ecology. Restoration was expensive and involved intensive site preparation (respreading topsoil, deep ripping) and replanting. Seedlings of 160 species were established at densities of around 2500 trees per hactare. Monitoring has also taken place to identify potential problems. Thirteen years after the project commenced, most of the original plant species are now present at the site and many wildlife species are beginning to recolonise the area.[366]

2.3. Restoration to Increase Landscape Linkages

Fragmentation is a common outcome of disturbance in many tropical areas. If these remnants can be linked by corridors, it should be possible to rejoin the isolated populations of plants and wildlife species. Such a corridor has been created in north Queensland. In this case the corridor is 1.5 km long and 100 m wide. The boundaries have also been "sealed" with an additional boundary of dense crowned tree species to minimise the so-called edge effect. The new forest was created using dense tree seedling plantings (less than 2-m spacing) and involved about 100 tree species. Intensive weeding meant that canopy closure was rapid. Additional plant species have colonised the site from intact forest at each end of the corridor.[367]

2.4. Single Species' Plantations Catalyse Restoration

Most traditional forest plantations use a single species grown in a monoculture. These are commonly planted at an original density of around 1100 trees per hectare, which means canopy closure is rapid and weeds are quickly excluded. Thereafter, thinning is carried out and the trees are harvested at the end of the rotation—commonly about 40 years. If these plantations are near intact forests they can acquire a significant understorey of native plant species. If no thinning is carried out and the plantations remain unlogged, a significant diversity of plant species may accumulate. This is often greater than would have occurred if the site had remained unplanted (because of the competitive abilities of weeds and grasses or because of recurrent fires that would have continued to burn the site). Several 60-year-old monoculture plantations (conifer and broadleaved hardwood) in northern Australia have

---

[365] Aide et al, 2000; Zimmerman et al, 2000.

[366] Knowles and Parrotta, 1995; Parrotta and Knowles, 1999.
[367] Goosem and Tucker, 1995; Tucker, 2000a,b.

acquired more than 350 species of trees, shrubs, epiphytes, vines, and herbs from nearby intact forest. Some of the trees have now grown up to join the canopy layers transforming the monoculture to a complex species-rich community. It should be noted, though, that in most monospecific plantations, active management for production prevents this from happening.[368]

2.5. Using High-Value Native Species

Malaysia has had a long silvicultural history. It is perhaps best known for the work carried out on devising silvicultural methods for natural forests, but significant areas of plantation have also been established. Much early work involved plantations of exotic species such as pine or *Acacia*. But more recently there have been a large number of species' trials to examine the silviculture of native species when these are grown in simple monoculture plantations as well as in more complex plantation designs.[369]

2.6. Reforestation in an Extensively Cleared Landscape

Large areas of Vietnam have been deforested. Extensive reforestation using mostly exotic species of genera such as *Eucalyptus* and *Acacia* has been carried out in recent years. Land is now being allocated to farmers and many are interested in reforestation. Very few of these farmers are interested in restoration because they cannot afford to be. This is despite Vietnam being a biodiversity-rich country. What is more likely to occur is that the landscape will evolve as a mosaic of agricultural land and small plantations. Many of these plantations will be composed of native species and some will contain simple mixtures of two or three species. The identity of these will vary from site to site. This means site diversity will remain modest, although landscape diversity will be enhanced. Opportunities for more species-rich plantations and more complex forms of silviculture may develop in the future as the standard of living increases, and are being tested in many rural areas within Vietnam.

3. Outline of Tools

3.1. Choosing a Method for Restoration

A variety of approaches have been used to restore tropical moist broad-leaved forests, and some of these are summarised in "Overview of Technical approaches to Restoring Tree Cover at the Site Level." Where funds are limited and regrowth forests are widespread it is probably more appropriate simply to protect these secondary forests from further disturbances and allow successional development to take place. Under most situations species-rich and structurally complex forests will then develop over time (see example 2.1 above). These forests will not necessarily regain all of the original plant or animal species. For example, poorly dispersed large-seeded plant species may be absent and wildlife with specialised habitat requirements may not be able to reach the regenerated forest. Determining which, if any, species have not reoccupied a particular site requires knowledge of the original forest biota and also necessitates that some form of monitoring is carried out to determine the extent of the recovery process. Once the identity of any missing species is known, action may be taken to attempt to remedy these losses.

Some more active form of intervention will be needed where regrowth forest is absent or where the opportunities for recolonisation are more limited (e.g., because fragments of the original forest are more distant). This may involve an initial planting with a short-lived fast growing tree species that shades out weeds and grasses. These trees can then be underplanted with specific target species. Alternatively, direct planting of all the target species can be done to initiate restoration. Active intervention like this requires significant funds, which are usually available only for purely restoration purposes

---

[368] Keenan et al, 1997.
[369] Akioka, 1999; Appanah and Weinland, 1993.

under certain conditions (see example 2.3). More commonly, reforestation will be carried out only where landowners expect to derive a benefit themselves, and in most cases this means some form of commercial harvesting will be required. Active intervention in these circumstances can range over a variety of methods and may involve enrichment planting of regrowth forests or some form of mixed-species' plantation establishment. Any biodiversity benefit from this reforestation will necessarily require the landowner to strike a compromise between optimising production and optimising the recovery of biodiversity present at that site. Under these circumstances "production" can involve timber trees as well as nontimber products (e.g., nuts, fruit, etc.) and the plantations may involve trees as well as understorey plantings of medicinal plants or cash crops. That is, there may be a range of possibilities available that offer different degrees of biodiversity gain as well as benefits for stakeholders.

3.2. Some Key Questions to Consider

Irrespective of which form of active intervention is used, several key problems commonly occur. These follow from the three issues referred to initially in the introduction.

3.2.1. Which Species to Use?

Moist tropical forests contain a variety of species and little is usually known about the ecology of most of these except for a comparative handful that might once have been harvested for timber. Since tree planting is mostly undertaken in the expectation of some commercial gain there is a tendency to use those species with the highest timber values. But these indigenous species often have particular site requirements and many are comparatively slow-growing. This means that plantations using these species have often failed—especially when the lands available for reforestation are poorer quality lands or where weeds are dominant. This has increasingly led plantation managers to use a relatively small number of faster growing and more tolerant exotic species such as pines, eucalypts, and *Acacia* that can grow well at these poorer sites. These offer production benefits but they contribute few ecological services. The reason for this choice is because managers are often unaware of the full range of options available to them or because they have been unable or unwilling to risk the various alternatives.

3.2.2. Where to Get Seed?

It is often difficult to get seed for many tropical forest species. Most species are usually present as scattered, isolated trees in relatively sparse populations, and most species have irregular fruiting patterns. Many also produce seed for only a short period and this seed can be difficult to store. This means it can be hard to collect seed from natural forests for large-scale plantings. But it may be even more difficult to collect seed from an adequate number of parent trees in heavily degraded landscapes.

3.2.3. How to Raise Seedlings and Establish These in Plantations?

Some species germinate readily and quickly reach a size suitable for planting. But other species germinate irregularly or need up to a year in a nursery before they can be planted in the field. Some species also depend on specialised mycorrhiza which may have been lost from the field when soil fertility has been depleted and sites have been degraded. This means that care needs to be taken to inoculate these species in nurseries prior to planting. In short, different species require different forms of nursery treatment in the nursery. This makes it difficult to raise seedlings of, say, 100 species to plant together in the field on a particular planting date. Species also differ in their capacity to become established in the field and tolerate acid soils, low nutrient levels, or full sunlight. Optimal conditions for one species may be suboptimal for another. Unfortunately, little is known about the attributes and tolerances of most moist tropical forest species.

3.2.4. How to Make Large-Scale Tree Planting Attractive to Land Managers?

Intensive restoration using large numbers of species to reestablish plant biodiversity rapidly over a large area is an expensive undertaking. Unless there is some kind of early financial return relatively few landowners are likely to be able to afford to use this approach. On the other hand, some individuals or communities may take the view that financial gains are less important than the provision of a range of forest services. In such cases reforestation that provides a production benefit whilst also generating some biodiversity or functional gain may be more attractive. The question in these circumstances may then be what kind of a production-biodiversity trade-off to make. Some of the site-based alternatives are outlined in the chapter on interventions (cited above), and the choice of which of these to use will depend on both ecological and socioeconomic circumstances. The most likely solution will be that the landscape will contain a mosaic of approaches, with some areas being devoted to intensive production while others such as riverine areas or steep slopes will be reforested largely for protection or biodiversity benefits.

3.2.5. How to Foster Animal as Well as Plant Diversity

It is commonly assumed that many wildlife species will recolonise reforested areas once successional development has generated sufficient habitat complexity. While this may be broadly true, many species require certain minimum areas to be reforested before they recolonise, and particular species sometimes have specialised habitat or resource requirements. Such species will require more detailed study before any restoration programme is successful. Of course, a more general prerequisite is that any wildlife remaining in undisturbed forest remnants in the region are able to reach the newly reforested areas. That is, reforestation should seek to provide linkages across the landscape to allow wildlife to move from residual forest areas into the newly restored forests.

4. Future Needs

There are several key issues that commonly limit the restoration of tropical moist forests:

4.1. Silviculture and Ecology of Key Structuring Species

There is little knowledge of the ecology of many of the key species needed to initiate successional development in tropical forests. This includes knowledge of fruiting and seeding phenology as well as information on where to obtain seed of these species, how to store this seed, how to raise seedlings, and how to establish these seedlings in the field.

4.2. Species-Site Relationships

There is often surprisingly little knowledge of the distribution patterns and site requirements of most tropical tree species. This problem is often even more acute because many sites at which a particular species was once found are now degraded in some way, for example, they have suffered a decline in soil fertility. This may mean a two-stage approach is needed in which the first plantings (e.g., a nitrogen-fixing species) modify the sites and make them more suitable for the target species. The preferred species might then be introduced as an underplanting or after the first forest has been harvested and removed (thereby paying for the cost of rehabilitation).

4.3. Methods of Enriching Degraded or Regrowth Forests

There are increasing areas of degraded or regrowth forests (regenerating after some disturbance such as agriculture or severe logging). These have lower levels of plant and animal biodiversity than the original forest. They often have a reduced ability to supply goods and services to communities living nearby. One way of overcoming both these problems is to accelerate their recovery by enriching these forests with certain target species (e.g., endangered or rare species; species providing commercially attractive nontimber forest products). But methods for doing this are often expensive or inefficient, and better, more effective means are needed.

4.4. Overcoming Impediments to Farm Forestry

Farm forestry is one means by which significant areas of land might be reforested and rural poverty might be tackled. Many farmers are interested in planting trees on land not needed for food production or other purposes. But these farmers may be prevented from doing so because of land tenure arrangements, financial constraints, limits on harvesting, or a lack of knowledge about the species best suited to the sites they have available. Such species must be ecologically appropriate and financially suitable. The impediments to farm forestry are often specific to particular sites and so will need specific solutions. A general principle, however, is that beneficiaries of reforestation (downstream land users, catchment authorities, conservation authorities, etc.) should assist landowners with the costs of reforestation.

4.5. Better Market Information for Farmers

Isolated traditional farming communities develop agricultural and silvicultural systems appropriate for their particular circumstances. But the arrival of roads and a cash economy usually means a major change is needed in the way they manage their crops and land. In many cases they become beholden to middlemen or timber buyers so that farming activities are carried out to suit these players rather than the farming community itself. As the areas of natural forests decline, better information is needed on the real value of certain tree crops and, potentially, the emerging market for ecological services.

References

Aide, T.M., Zimmerman, J.K., Pascarella, J.B., Rivera, L., and Marcano-Vega, H. 2000. Forest regeneration in a chronosequence of tropical abandoned pastures: implications for restoration ecology. Restoration Ecology 8:328–338.

Akioka, J. 1999. The Multi-Storied Forest Management Project in Malaysia. Forest Department, Peninsular Malaysia, Perak State Forestry Department, Japan International Cooperation Agency.

Appanah, S., and Weinland, G. 1993. Planting quality timber trees in Peninisular Malaysia: a Review. Malayan Forest record No. 38. Forest Research Institute of Malaysia, Kepong, Malaysia.

Goosem, S., and Tucker, N. 1995. Repairing the rainforest: theory and practice of rainforest re-establishment in north Queenslands Wet Tropics. Wet Tropics Management Authority, Cairns, Australia.

Keenan, R., Lamb, D., Woldring, O., Irvine, A., and Jensen, R. 1997. Restoration of plant diversity beneath tropical tree plantations in northern Australia. Forest Ecology and Management 99:117–132.

Knowles, O.H., and Parrotta, J. 1995. Amazonian forest restoration: an innovative system for native species selection based on phenological data and performance indices. Commonwealth Forestry Review 74:230–243.

Parrotta, J., and Knowles, H. 1999. Restoration of tropical moist forests on bauxite mined lands in the Brazillian Amazon. Restoration Ecology 7:103–116.

Tucker, N. 2000a. Wildlife colonization of restored tropical lands: what can it do, how can we hasten it and what can we expect? In Elliott, S., Kerby, J., Blakesley, D., Hardwick, K., Woods, K., and Anusarnsunthorn, V., eds. Forest Restoration for Wildlife Conservation. International Tropical Timbers Organisation and Forest Restoration Research Unit, University of Chiang Mai, Thailand, pp. 279–295.

Tucker, N. 2000b. Linkage restoration: interpreting fragmentation theory for the design of rainforest linkage in the humid wet tropics of north-east Queensland. Ecological Management and Restoration 1:35–41.

Zimmerman, J., Pascarella, J., and Aide, T. 2000. Barriers to forest regeneration in an abandoned pasture in Puerto Rica. Restoration Ecology 8:350–360.

Additional Reading

Banerjee, A. 1995. Rehabilitation of Degraded Forests in Asia. World Bank Technical Paper No. 270. World Bank, Washington, DC.

International Tropical Timbers Organisation 2002. ITTO Guidelines for the restoration, management and rehabilitation of degraded and secondary tropical forests. ITTO Policy Development Series No. 13. Yokohama, Japan.

Krishnapillay, B., ed. 2002. A Manual for Forest Plantation Establishment in Malaysia. Malayan Forest Records No. 45. Forest Research Institute, Malaysia.

Lamb, D. 1998. Large-scale ecological restoration of degraded tropical forest land: the potential role of timber plantations. Restoration Ecology 6:271–279.

43
Restoring Tropical Montane Forests

Manuel R. Guariguata

Key Points to Retain

Many characteristics of tropical montane forests make them a unique habitat for biodiversity, but they also have important economic and social values such as providing protection from landslides, and steady and clean water downstream.

Tools and approaches for restoring montane forests are not very different from those used in the lowlands; however, factors that may influence the outcome of a given restoration activity in montane areas are steep, erosion prone slopes, exposure to strong winds, and slow plant growth rates.

In the context of landscape scale restoration, there is a need to address the ecological and social linkages between tropical montane forests and their surrounding lowlands.

1. Background and Explanation of the Issue

1.1. Main Characteristics of Tropical Montane Forests

Drastic changes in elevation, precipitation, and direction of prevailing winds across small altitudinal ranges generate high levels of species' and habitat diversity in tropical montane forests. Also, because of their cool ambient temperatures, tropical montane forests serve as refugia of relict tree populations that are more typical of temperate latitudes. Moreover, tropical montane forests are home to unique vertebrate fauna—for example, mountain gorillas (*Gorilla beringei beringei*) in Africa, quetzals (*Pharomachrus mocinno*) in Central America, and spectacled bears (*Tremarctos ornatus*) in South America—and serve as elevational corridors for many bird species during times of seasonal food scarcity. Tropical montane forests are sometimes found as isolated patches within a matrix of either contrasting climate conditions (e.g., surrounded by desert vegetation such as in northwestern Venezuela) or vegetation types (e.g., surrounded by pine-oak forest in Mexico), which adds to their conservation value.

Other key characteristics of tropical montane forests are steep slopes with associated thin, infertile soils, chronic exposure to strong winds, low levels of solar radiation, and reduced rates of organic matter decomposition, all of which contribute to overall slow plant growth. From a restoration perspective, this means that recovering desired levels of forest structure and composition may take longer than in the surrounding lowlands.

1.2. Socioeconomic Rationale for Restoring Tropical Montane Forests

Restoration of tropical montane forests can fulfil both economic and conservation objectives. Landslides, for example, are a major source of damage to roads, dams, and human settlements in many montane areas. By restoring forest cover in deforested, landslide-prone sites, further mass erosion can be minimised through substrate stabilisation. In human-deforested areas, restoration of tropical montane forests may also be justified for the provision of environmental services as they play a critical role in the local hydrological cycle due to their role in cloud interception, especially in areas that do not receive much precipitation. Forest conservation elsewhere, however, may need to be actively linked to forest restoration in the uplands. For example, reduced forest cover in lowland areas could leave adjacent montane forests with not too many clouds to intercept.[370]

1.3. Restoring Montane Forests in the Face of Natural Disturbance

Although suppressing human disturbances such as fire and uncontrolled grazing is a key initial strategy of a given restoration initiative, taking into account the effects of natural disturbances on forest restoration may also be critical for success. For example, montane forests located in many tropical islands are usually prone to suffering severe hurricane damage as much as three times per century. In this case, options may include planting tree species with a known ability to resprout after stem breakage, with high stem wood density, or with specific architectural features; many palm species, for example, are known to survive hurricanes very well. Identification of naturally occurring, landslide-chronic areas may also help to prioritise or avoid investing in potentially costly restoration efforts that otherwise might be wasted.

2. Examples

2.1. Mount Kenya[371]

Mount Kenya is situated in the central highlands of Kenya. The national park is 715,000 hectares and it was gazetted in 1949. The surrounding forest reserves add another 1820 km^2 of protected area, making Mount Kenya the largest area of natural forest in the country.

The forest forms a major water catchment area from which two of the country's five river basins—the Tana and Ewaso Nyiro—rise, which together supply water to more than a quarter of Kenya's human population and more than half of its land area. Water users include the five main hydroelectric power sources, agricultural land, pastoralist range lands, and major urban centres.

Threats to the surrounding forests include illegal logging, charcoal production, cultivation of bhang, and encroachment. The glaciers on the mountain are also retreating because of global warming and climate change. A number of initiatives are now being undertaken together with communities to address the conservation and restoration needs of the montane forest. These are interesting examples of community initiatives of land management, restoration and protection of a unique environment in Kenya.

2.2. Sierra de las Minas, Guatemala

The Sierra de las Minas in Guatemala contains a biological treasure. At least 885 species of birds, mammals, amphibians, and reptiles, which amounts to 70 percent of all the species from these groups that are known to exist in Guatemala and neighbouring Belize can be found here. It is also an important tropical gene bank of conifers with 17 distinct endemic evergreen species. The area is thus considered an irreplaceable seed resource for reforestation and agroforestry throughout the tropics.

Besides its robust population of diverse flora and fauna, the Sierra de las Minas plays an important role in providing fresh, clean water to the many farms and villages in the Polochic and Motagua valleys below. More than 63 permanent rivers drain the reserve, making it the country's biggest single water resource. Local people depend on these small rivers for their agricultural crops (e.g., melon, tobacco, grapes, citric fruits, tomatoes). Bigger industries, such as soft drinks, fertiliser and paper-recycling plants, and hydroelectricity all rely on water

---

[370] Lawton et al, 2001.
[371] Carlsson and Lambrechts, 1999; Emerton, 1999.

generated at the Río Hondo station. A drop of 40 percent in water flow in the last 10 years has been attributed to forest loss.

Since October 1990 the reserve has been managed by a local nongovernmental organisation (NGO), Defensores de la Naturaleza. The reserve's managers are engaged in an environmental education programme designed to persuade local community leaders of the need to protect, manage, and restore the forests in Sierra de las Minas in such a way that they can continue to offer the services locally but also downstream. Payment schemes have been set up (see "Payment for Environmental Services and Restoration" for more information on such schemes) to ensure that those engaged in protecting and restoring the watershed, are paid by the beneficiaries downstream.[372]

3. Outline of Tools

3.1. Overcoming Barriers to Natural Succession

Assessing patterns of tropical montane forest succession following pasture abandonment, or after natural disturbances such as landslides, can provide important clues when designing restoration activities and when selecting what species to plant (or not) under a given level of site degradation. For example, in many tropical montane forests, those canopy tree species that dominate old-growth stands are the same colonisers of open, deforested areas.[373] Thus if a restoration goal is to re-create original species' composition, the selection of these particular species could be an appropriate choice.

Simple observations and experiments in sites that merit restoration can also help to discern what are main biotic and abiotic barriers that could be retarding natural forest recovery when designing a project. For instance (as in the lowlands), one of the main factors that retards forest recovery in tropical mountains is poor seed dispersal rates from adjacent forest.[374] Even when lack of seed supply is overcome, however, grasses and ferns that thrive in abandoned pastures tend to suppress growth and survival of tree seedlings; hence the removal of competing vegetation seems necessary during tree planting.[375] Controlled grazing can also facilitate both the establishment of planted trees and natural forest recovery through secondary succession.[376]

Another common barrier to the natural recovery of tropical montane forests is high rates of vertebrate seed predation in deforested areas. In other cases, reduced nutrient levels due to soil compaction or recurring fires can impede forest recovery even when seed survival is high. In short, strategies to restore tropical montane forests may need to be assessed on a case-by-case basis, and designed whenever possible for overcoming simultaneous barriers.[377]

3.2. Forest Plantations and the Role of Remnant Forest

Tree plantations in tropical montane areas can fulfil both conservation and production purposes as part of a restoration strategy. Yet, the choice of what species to plant must be made carefully, and it may be better to invest some time in selecting the appropriate species[378] rather than planting whatever is available in the local nursery. Tree species with high growth rates, prolific regeneration, or with any other desirable attributes can be easily identified after a few months of observations when published information is not readily available (Fig. 43.1).

Under conditions of severe soil degradation, for example, good candidate species are those that can quickly provide a closed forest canopy while improving soil fertility. However, in some cases, this alternative may be only part of an

---

[372] http://www.planeta.com/planeta/97/0897guatemala.html.
[373] Guariguata, 1990; Kappelle et al, 1996; Venegas and Camacho, 2001.
[374] Shiels and Walker, 2003.
[375] Pedraza and Williams-Linera, 2003.
[376] Posada et al, 2000.
[377] See an example in Holl et al, 2000.
[378] See an example in Knowles and Parrotta, 1995.

FIGURE 43.1. Establishment of a forest plantation for restoring tropical cloud forest in abandoned pasture in Xalapa, Veracruz, Mexico. The plantation consists of a mix of species typical of primary forest (*Quercus* and *Fagus*) and early successional species (*Heliocarpus* and *Trema*). (Photo © Guadalupe Williams-Linera.)

overall restoration strategy. For example, plantations of the fast growing, nitrogen-fixing tree *Alnus acuminata* in the Colombian Andes may not be the best long-term restoration tool as they seem to harbour fewer plant species in the understorey compared to similarly aged secondary forests following natural regeneration.[379] In severely degraded sites, however, planting nitrogen fixing trees such as *Alnus* can be an option in the short term as they help to recover soil productivity.

Planted windbreaks in montane agricultural landscapes are known to facilitate tree colonisation by increasing seed dispersal rates from nearby, remnant forest. The location and spatial arrangement of agricultural windbreaks as a restoration tool may be important in production landscapes where the enhancement of ecological connectivity and biodiversity recuperation is also a management objective. Planted windbreaks that are connected to forest may harbour more naturally dispersed seeds and contain higher diversity in their understoreys than those not connected.[380] This means that in some cases both the size and relative location of remnant forest fragments need to be considered when designing a given restoration strategy.

4. Future Needs

Currently, most tropical montane forests are highly fragmented. As a consequence, many of their component vertebrate species may be locally extinct either because of the small habitat area of the remaining fragments, or because those plant species that provide them with food resources are absent, or both.[381] In some cases, tropical montane forest restoration could focus on connecting existing fragments via forest plantations as a way to facilitate altitudinal bird migration, and therefore seed dispersal. More research is needed to support the selection of appropriate sets of plant characteristics, as well as the spatial arrangement of the planted trees in order to favour interpatch animal movement and habitat use—and not necessarily to restore forest cover per se.

References

Carlsson, U., and Lambrechts, C. 1999. Community initiatives and individual action on and around Mount Kenya National Park. Paper presented at the East Africa Environmental Network (EAEN)

---

[379] Murcia, 1997.
[380] Harvey, 2000.

[381] Cordeiro and Howe, 2001.

Annual Conference, 28–29 May 1999, Nairobi, Kenya.

Cordeiro, N.J., and Howe, H.F. 2001. Low recruitment of trees dispersed by animals in African forest fragments. Conservation Biology 15:1733–1741.

Emerton, L. 1999. Mount Kenya: the economics of community conservation. Evaluating Eden Series, discussion paper No.4. International Institute for Environment and Development, London.

Guariguata, M.R. 1990. Landslide disturbance and forest regeneration in the upper Luquillo mountains of Puerto Rico. Journal of Ecology 78: 814–832.

Harvey, C.A. 2000. Colonization of agricultural windbreaks by forest trees: effects of connectivity and remnant trees. Ecological Applications 10:1762–1773.

Holl, K.D., Loik, M.E., Lin, E.H.V., and Samuels, I.A. 2000. Tropical montane forest restoration in Costa Rica: overcoming barriers to dispersal and establishment. Restoration Ecology 8:339–349.

Kappelle, M., Geuze, T., Leal, M., and Cleef, A.M. 1996. Successional age and forest structure in a Costa Rican upper montane *Quercus* forest. Journal of Tropical Ecology 12:681–698.

Knowles, O.H., and Parrotta, J.A. 1995. Amazonian forest restoration: an innovative system for native species selection based on phenological data and field performance indices. Commonwealth Forestry Review 74:230–243.

Lawton, R.O., Nair, U.S., Pielke, R.A., and Welch, R.M. 2001. Climatic impact of tropical lowland deforestation on nearby montane cloud forests. Science 294:584–587.

Murcia, C. 1997. Evaluation of Andean alder as a catalyst for the recovery of tropical cloud forests in Colombia. Forest Ecology and Management 99:163–170.

Pedraza, R.A., and Williams-Linera, G. 2003. Evaluation of native tree species for the rehabilitation of deforested areas in a Mexican cloud forest. New Forests 26:83–99.

Posada, J.M., Aide, T.M., and Cavelier, J. 2000. Cattle and weedy shrubs as restoration tools of tropical montane rainforest. Restoration Ecology 8:370–379.

Shiels, A.B., and Walker, L.R. 2003. Bird perches increase forest seeds on Puerto Rican landslides. Restoration Ecology 11:457–465.

Venegas, G., and Camacho, M. 2001. Efecto de un tratamiento silvicultural sobre la dinámica de un bosque secundario montano en Villa Mills, Costa Rica. Serie Técnica No. 322. CATIE, Turrialba, Costa Rica.

Additional Reading

Bubb, P., May, I., Miles, L., and Sayer, J.. 2004. Cloud Forest Agenda. UNEP-WCMC, Cambridge, UK.

Guariguata, M.R., and Kattan, G.H., eds. 2002. Ecologia y conservacion de bosques neotropicales. Editorial Libro Universitario Regional, Costa Rica.

Kappelle, M., and Brown, A., eds. 2001. Bosques nublados del neotropico. InBio, Costa Rica.

Case Study: Conserving the Cloud Forests of Mount Rinjani, Lombok

Jeff Sayer and Triagung Rooswiadji

Rising majestically from lowland rice paddies to a height of 3726 m, Gunung Rinjani dominates the Indonesian Island of Lombok. The upper slopes of the mountain are clothed in cloud forest. The winds coming in off the sea cool as they are funnelled up the slopes of the mountain, moisture condenses onto the vegetation, and as a result the trees are permanently wet and are festooned in epiphytic orchids, lichens, and mosses. These forests are home to rare birds, black ebony leaf monkeys, barking deer, leopard cats, and palm civets.

The forests are now under intense pressure. Lombok is one of Indonesia's poorest and most densely populated islands. Pressure for land has always been intense but the problem has become much worse in recent years. First, following the Asian economic crisis in 1997 large numbers of Lombok people who had been migrant workers in Malaysia were sent home. Many of them returned to farming. Then the Bali bombing in 2001 had a huge impact on the tourist industry. As a result, the local Sassak people have fallen on hard times. A large part of their income came from work in hotels and restaurants, and from producing the beautiful handicrafts for which Lombok is renowned. Lack of cash employment is forcing them back onto the land. And with 2.9 million people crowded onto this $5625 km^2$ island, it is hard to make a living from traditional agriculture alone.

In theory, Gunung Rinjani's cloud forests—the only ones left on Lombok—are legally protected. But the Forest Department finds it difficult to enforce the laws when they cannot offer any alternative to the poverty stricken farmers. A large swathe of forest on the lower slopes has now been reduced to a patchwork of small fields, scattered trees, scrub, and grasses. Fires originating in these degraded areas are beginning to eat into the rich forests higher up the mountain.

This has implications for the entire island. Rinjani's forests act as water collectors for all of Lombok. Water flowing from the misty upper slopes irrigates the highly productive rice cultures of the plains and supplies domestic water to the towns and tourist resorts. Now the rice farmers in the lowlands are complaining that there is not enough water for their crops in the dry season, and they experience an increased number of floods when it rains.

In response to the crisis, Lombok's provincial government has linked up with the global conservation organisation WWF, and the U.K. Department for International Development to devise a strategy that can protect the forests and their vital watershed functions and still provide land and employment for the people.

As a contribution to this effort we have been developing a simple computer model to try and unravel the complexity of the Rinjani social-ecological system. The model uses the STELLA software and enables us to investigate the main drivers of land cover change and links between these changes and the livelihoods of the people. The model has been developed with local stakeholders and it has been useful in making their assumptions and interests more explicit.

We began by investigating the possibilities for making environmental payments to upland farmers in return for better farming and forestry practices. A bottled water company in the lowlands indicated that a modest amount of money could be available for this programme. The 42,000 water users in the provincial capital Mataram have agreed to a small levy to pay for watershed protection. However, the model suggested considerable difficulties in this approach. The number of farmers is very high—several hundred thousand—and payments that were high enough to have a real impact on their behaviour would cost more than the amounts that are likely to be available. Lack of legal clarity about land rights and the high diversity of farming systems that they use would combine to make the management of such payments very complicated.

The modelling exercise suggested that few solutions would be effective if they were not accompanied by more effective application of laws. But the difficult transition to democracy that Indonesia is now experiencing and the economic crisis are combining to make law enforcement very unpopular amongst the population.

So far one of the best options that has emerged has been to abandon government attempts to protect the watershed forests and, instead, to parcel out the land to poor people, who can use it on condition that they plant trees. This is a rather revolutionary idea. It is in fact saying that conventional approaches to watershed management are not workable in the present economic and social conditions found on Lombok. The compromise of encouraging the formation of a buffer zone of agroforestry plantations around the base of the mountain seems like a better option.

The initial trials have centred on the village of Sesaot. Farmers are given 0.1 hectare of land and are allowed to grow field crops for the first 4 years, until the trees grow. In the early years the farmers made money by growing crops such as chilli peppers between the tree seedlings. Now they are planting a wide variety of fruit and even timber trees. Mangoes, papayas, durians, jackfruit, custard apples, rambutans, and salak fruit are all being produced for sale to traders in the provincial capital Mataram. Jackfruit and macadamia are especially popular as they produce valuable fruit and nuts but also timber that is in high demand for the curio carvers in Bali.

The land remains under forest department "ownership" and the farmers have to pay a small rent for the right to cultivate it. On a pilot scale this programme has been an undoubted success, and previously degraded areas are now covered in profitable agroforests. However, the market for fruit and timber is limited, and unless the general economy picks up it will be difficult to extend the scheme to all the degraded areas of protection forest around the mountain.

The agroforestry trees protect the soils and the water supplies and the people earn a good living. These artificial forests do not have the same biodiversity values as the natural forests that used to exist in the protection forests, but they are better than the degraded scrub and farmland that covered the sites when the programme began. They offer the hope of providing stable and secure land use around the lower boundary of the forests.

The success of the agroforestry approach will be very sensitive to the incomes that farmers can obtain for their fruit and timber crops. We are going to continue to use our model of the Rinjani system to track how both the environment and people's livelihoods evolve over time. The model will provide a database and monitoring tool that will be used by the local stakeholder committee to help understand how the system is performing. It should help to determine how livelihoods change over time and how this is linked to changes in landcover.

The idea of payments for environmental services is still being pursued but as a complement to other approaches. The isolated hillside villages have few social services and the people's lives are still precarious. The people in the lowlands are richer and the rice farmers are making money out of the water that flows from the mountains, so there is some potential for a small water tax. This will not be given as cash to the upland farmers but will be used to build clinics and schools and improve the roads. The hillside people will get these services only if they respect the agreement and grow only tree crops. They will lose these social contributions if they grow tobacco, cassava, or other annual crops that are bad for soil erosion and do not conserve water.

The situation in Lombok, where valuable natural forests exist alongside poverty-stricken people desperate for more land, is typical of many developing countries in the tropics.

Rinjani National Park is one of Indonesia's most spectacular natural areas but there is no way that it can be protected if thousands of poverty stricken, land-hungry people live around the base of the mountain. Giving people rights to some areas of degraded natural forest may help save the national park.

44
Restoring Floodplain Forests

Simon Dufour and Hervé Piégay

Key Points to Retain

The extent, structure, and diversity of floodplain forests have been strongly modified by human pressures. Yet they are areas with a high biological diversity, and specificity, and riparian areas are important for fish, amphibians, and mammals and for fluvial system functioning.

Restoration of floodplain forests can be achieved at three scales: catchment, reach, and local scales.

Some important tools for restoring floodplain forests include assessment and inventories, monitoring, and integrated river basin management.

1. Background and Explanation of the Issue

1.1. Characteristics of Floodplain Forests

Floodplain forests are unique ecosystems that are located alongside rivers and streams. These systems derive their characteristics from periodic inundations. The extent, structure, and diversity of floodplain forests have been strongly modified by human pressures acting at the catchment, reach, and local scales. Even though many floodplains in Europe are characterised by natural forestation that began after the Second World War due to widespread changes in land-use practices, most European floodplain forests have disappeared.

Since the 1970s, the scientific community and land managers have recognised the ecological, economic, and social values of floodplain forest. These forests are very valuable because of their high potential in terms of wood production, protection of water quality, flood control, recreation, and improvement of the landscape. In addition, they are natural areas with a high biological diversity and ecological specificity due to the influence of water on habitat conditions. Riparian areas are important for fish, amphibians, and mammals (e.g., beavers). Additionally, the forests provide breeding habitat for birds, and are navigational aids and stopover sites for migrating species (e.g., the songbirds in the North Platte River). The need to preserve and restore them is now widely recognised.

Forest ecosystems that are under hydrological control evolved their original ecological processes in response to their proximity to and the dynamics of the river. Thus, the periodic water supply is a key process characterising floodplain forests. The land–water interfaces are important areas for biological exchanges, water supply and content, soil moisture, organic matter evolution, seed dispersal, and nutrient cycling.[382] Floodplain forests are part of

[382] Naiman and Décamps, 1990.

dynamic systems, and their conservation and restoration must take into account the hydrogeomorphic processes that structure the catchment and the landscape evolution.

In most cases, it is impossible to re-create pristine floodplain forest conditions, but mitigation measures can be developed to improve ecosystem quality. For this purpose, managers must identify practical strategies and tools.[383]

1.2. General Principles

The restoration of floodplain forest is often achieved at three scales:

1. Catchment scale: The improvement (e.g., more natural levels) of controlled factors (discharge, bedload supply) can be done at the catchment scale or in an upstream branch of the river network. Such hydrological and sedimentary river improvements have positive effects on floodplain habitats in terms of structure and diversity. The success of such "self-restoration" options, when they can be promoted, are difficult to evaluate because of multiple potential channel adjustments acting at various timescales.

2. Reach scale (10 to 100 km river length): The improvement of the hydrological connection between the active channel and the floodplain is an approach that can be accomplished at the reach scale by modifying the topography to lower the riparian surface in order to improve water flow across the floodplain, and also by raising the groundwater table.

3. Local scale (a few hectares of forest): The maintenance of the riparian structure slows down succession (preserves pioneer stages when the river has lost its capacity to do so) or favours specific assemblages of the modified ecosystems (removes exotic species, reforestation in cultivated areas, grazing control).

Restoration can be promoted at different scales depending on the target. The interventions at local scale usually generate fewer problems in terms of social acceptance, because plots are smaller in size and concern fewer users. The stakes are also less complex with fewer conflicts than those that must be managed when dealing with entire systems.[384]

1.2.1. Hydrological Connections

Reestablishment of hydrological fluctuation is a common topic in floodplain restoration, particularly reestablishment of the flood pulse that inundates forest patches according to their position within the riparian corridor. For this purpose, some actions must be promoted at a large scale, by specific management strategies controlling water diversion and storage for hydroelectric and pumping purposes. Increasing minimum flow downstream of dams is one of the most common options at this scale.

At the reach scale, various options can also be implemented to reestablish a more active hydrological connection, such as reinundating areas by dike removal or reconnecting side channels. Low-flow in groundwater levels should also be considered carefully, in particular downstream of dams and in reaches with active water pumping for agriculture and industries. Managers can then perform some measures to raise the groundwater level, such as favouring more flow in the floodplain's former channel network or artificial groundwater input from a reservoir or canal.

1.2.2. Bedload Transport

Restoration of sediment transport is another process-based option. Complete restoration of a dynamic system with all types of forest successional stages, when it has been affected by lateral and longitudinal disconnection (embankment, dams that interrupt sediment transfers), must include not only channel shifting, but also bedload transport preservation.

Bedload reintroduction and riparian zone re-dynamism can be accomplished at the reach scale by increasing levels of bank erosion and sediment remobilisation during floods, and by removing unnecessary dikes. Sediment reintroduction to maintain channel dynamics is being considered along the Ain River in France,

---
[383] FISRWG, 1998.

[384] Hughes, 2003.

where dam construction in the 1960s disrupted peak flows and the character of sediment transfers (through a Life Nature Programme).

Even within the framework of process-based restoration at the basin scale, the problem of dams and their possible removal sparks considerable debate within the scientific community. If the solution looks good from an ecological point of view (i.e., more natural hydrology, bedload transport, and biological connection), the reality is much more complex. It is advisable, in particular, to distinguish big dams from small dams that are located in the upper part of the channel network. Next, the socioeconomic context of each dam must be taken into account. Lastly, all the effects of dam removal are not known (for example, in the case of sediment contaminated by organic or inorganic components).

1.2.3. Forest Structure

Actions proposed at the catchment and reach scales can be achieved by interventions at finer scales by focussing on existing forested structures (which is cheaper and easier), through structural transformation of degraded woodlands or by creating new units.

For existing woodlands, forestry practices have to be adapted to their specificities. Generally, the ecological aims of restoration will be to improve biodiversity by respecting some basic rules that enhance or conserve near-natural functioning and structuring of the forest: high vertical complexity of different strata (uneven age structure), broad range of different successional states organised as a patchy mosaic, presence of woody debris, use of natural regeneration, etc. Such an approach is proposed in reaches where alluvial forest is still present but is no longer rejuvenated by channel processes (primarily bank erosion and flooding). The preservation of pioneer units is best accomplished artificially (cutting). Moreover, actions can also be performed to fight exotic species that themselves form monospecific communities on pioneer biotopes.

For highly disrupted forest structures like artificial plantations, modification of forestry practices is often not enough, except in the very long term. Instead, reconversion measures (defined as transformation of stand structure with a change of socioeconomic functions) have to be implemented. This often implies more intensive and expensive programmes (like plantations of indigenous species). In agricultural areas, plantation programmes can be promoted at a large scale for biodiversity purposes but also for flooding management (preserving areas of low vulnerability that can attenuate the peak flow), for water quality (buffer strip along agricultural-river contact), and for global warming (sequestration of carbon dioxide from the atmosphere).

2. Examples

Experiences in floodplain forest restoration are shaped by specific ecological problems, such as base flow decrease, peak flow cutting, sediment transport disruption, channel degradation and groundwater drop down, channel stabilisation, and diking and flooding protection, and by socioeconomic issues, industrial or agricultural water pumping, human pressure on forested corridor and landscape fragmentation. When looking at the European examples, a few cases use a process-based approach, such as on the Rhone, the Danube, the Elbe, and the Rhine (Table 44.1). In North America, the objectives for the Mississippi river and the Chesapeake Bay watershed (Potomac River, Susquehanna River) focussed more on water quality improvement (nutrient, pollutant, and sediment contents).

In other parts of the world, such as in Malaysia, the objective of floodplain restoration tends to be for the preservation of native fauna and flora. Finally, for many large rivers, in particular in recently industrialised countries, some restoration programmes are in place (River Ganga, River Yamuna in India, Amazonas/Solimoes River in the Amazonian watershed). In these cases the main priority, even if restoration is considered, often remains the conservation of natural areas and the decrease of physical and chemical water pollution.

TABLE 44.1. Examples of restoration measures proposed on different large rivers in Europe, America, and Asia.

	From catchment		To reach			To local options	
	Increase modified minimum flow	Re-inundate by dike removal or setting back	Reconnection of former channels	Raise groundwater	Lowering floodplain	Replanting woodland	Modification of forestry practices and laws
Danube River, Austria		x	x				
Danube River, Bulgaria							x
Elbe River, Germany		x					
Rhone River, France	x		x	x			
Rhine River, France		x	x		x		
Chesapeake Bay watershed, U.S.						x	
Lower Mississippi River, U.S.			x			x	
Middle Sacramento River, U.S.						x	
Kissimmee River Corridor, U.S.		x					
Chikuma River, Japan					x	x	x
Kinabatangan, Malaysia						x	x

Examples of different restoration measures proposed on large rivers in Europe, America and Asia are shown in Table 44.1.

2.1. Restoration of Physical Processes at the Reach Scale: The Rhone River (France) on the Site of la Platière

The Rhone River has been regulated since the middle of the 19th century to fight flooding, to improve navigation and irrigation, and to produce electricity. Along most of its French course the Rhone is characterised by a degraded landscape. In the reach of l'Île de la Platière (60 km south of Lyon), channel degradation and bank stabilisation caused by the installation of groins at the end of the 19th century, water pumping by chemical factories after 1950, and flow diversion to bypass canals after 1977 have all led to floodplain-channel disconnection and lowered the groundwater table (a loss of 2 m between the end of the 1960s and 1990). Consequently the forest has become drier, losing much of its alluvial characteristics. A restoration project has been in place since 1992 to re-inject water into the aquifer by reconnecting a side channel from which water can infiltrate and raise the groundwater table by half a metre. The hydrological connection is still infrequent for some forest patches, but functionality is greater today than it was 20 years ago. The next step to improve the hydrological connection is to increase the minimum flow that is not derived from the canal for electricity production.

2.2. Buffer Zone Restoration to Reduce Nutrient Pollution in the Chesapeake Bay Watershed

In 1983 federal, state, and local stakeholders established a programme to restore water quality and health conditions in the Chesapeake Bay watershed in Virginia and Maryland. The objective of the programme was to increase water quality and habitat resources within this formerly forested watershed (forest covered 95 percent of the watershed 300 years ago versus 6 percent today). One of the main measures was the restoration of streamside forests along the hydrographic network. After restoring almost

5000 km along the river bank, today the riparian forest buffers almost 60 percent of the channel network. This forest growth is complemented by a decrease in nitrogen and phosphorus utilisation, and has led to a significant decrease in nutrient pollution in the bay.

2.3. Actions on Riparian Cover Characteristics: Reforestation Along the Kinabatangan River (Malaysia)[385]

With the exception of the southeastern part of the United States, the issue of floodplain forest restoration in nontemperate areas is a more recent development than in industrialised regions. Thus, few projects exist. The restoration and conservation programme of the Kinabatangan River floodplain forest is one of the most advanced examples in the tropics. The forest, located in the Malaysian part of the island of Borneo, is highly impacted by the presence of palm plantations. This programme is carried out by the Sabah Wildlife department, Sabah's Department of Irrigation, and WWF Malaysia, and includes several actions, in particular reforestation along the riverbanks and reconnection of isolated forest fragments. At the regulatory scale, actions include modifying the legislation that enables the transformation of the natural forest patches into palm plantations, and campaigns that inform consumers of the origin of the palm oil and the forestry practices of the producer.

3. Outline of Tools

Two types of tools must be differentiated: (1) diagnosis tools to understand the status of the floodplain ecosystems in terms of diversity and connectivity, and (2) implementation tools and methods to use in restoration projects.

3.1. Assessment and Inventory

Before improving any landscape patch, one needs to understand how the landscape func-

[385] Teoh et al, 2001.

tions, how it has evolved to its present state, and the causes of human-induced modifications. Historical analysis is helpful in understanding forest cover evolution over the last century. Land-survey maps and aerial photos are useful documents to establish the structural state of alluvial forests over the past 50 to 100 years. Written forestry reports can be used for some large alluvial forest corridors, such as the Rhine or the Mississippi river that have both been managed for a few centuries.

Prior to acting at a local scale, it can be helpful to approach the problem at a larger regional scale to tailor actions to the right scale. An inventory at the national scale can be used as a preliminary step to identify possible project sites. Such inventories can be exhaustive for small areas, like in Switzerland or in Belgium, or more cursory for larger regions (for example, by satellite imagery). With either method, the inventory must include a database that contains some information on each site (percent of surface forested, stand structure, regrowth, plant diversity, river form, etc.).

3.2. Monitoring

Monitoring is important, as with all restoration programmes and should include both ecological and socioeconomic factors. Some socioeconomic factors that need to be taken into account for floodplain forests, but also for other large-scale restoration efforts, include ensuring legal protection status and property rights, and understanding and mitigating the impact on local stakeholders. Specifically, for floodplain forests, variables that need to be measured include hydrological, geomorphic, and biological characteristics (pre- and postrestoration survey).

3.3. Integrated River Basin Management

Integrated river basin management is one of the tools that can be used to attain objectives of water quality improvement, local development, flooding management, etc., and allow stakeholders to consider their options in

FIGURE 44.1. A reconnected channel in the Erstein natural reserve—Rhine river. (Photo © Simon Dufour.)

managing and implementing floodplain forest restoration. It involves looking at the entire basin when determining interventions (Fig. 44.1).

4. Future Needs

4.1. Improve Knowledge

During the last few decades, ecologists and geomorphologists have made important progress in understanding stream corridor response to river system evolution. A better quantification is now needed of the influence of site conditions on species' development and growth and on communities' composition, and diversity as well as better comprehension of the potential trajectories of the communities (i.e., rupture thresholds, lag of time response). To assess the value of floodplain forests, field-based studies are necessary to quantify realistically the influence of these forests on system fluxes (water and nutrient consumption, organic matter production) in a broad range of hydrogeomorphic conditions, for example, highly dynamic systems, incised or aggraded rivers, downstream dams, in cultural landscapes, etc. Physical and biological coupling models must be developed to evaluate better the efficiency of proposed management and restoration.

4.2. Apply the Idea of Acting Locally, but Thinking Globally

Most of the time, the restoration plan is developed at a local scale rather than at a larger scale. Managers should develop macromanagement strategies in order to make current environmental policies sharper. In Europe, the Water Framework Directive is a chance to promote such a large-scale approach. It is, for example, well known that the de-nitrification capacity of riparian units depends on connectivity conditions between the soil, root systems, and groundwater. However, these conditions do not exist all along the hydrographic network because of various channel geometry conditions. Before replanting forest along rivers to improve water quality, one must identify target reaches. It is possible to use Geographical Information System (GIS) analysis to identify sources of pollution and potential natural barriers to restoration.

4.3. Cost-Benefit Analysis

One of the most important issues is the assessment of the benefit provided by the alluvial forests, and also by the restoration measures in terms of resources, flood protection, water quality improvement, and heritage. For this purpose, there is a need to identify and to develop technical and methodological tools to quantify these benefits (the costs are easier to estimate). Economic studies should be conducted in different local demonstration programmes in order to validate the benefit of the measures for stakeholders.

References

FISRWG. 1998. Stream corridor restoration: principles, processes and practices. The Federal Interagency Stream Restoration Working Group, GPO item n° 0120-A.

Hughes, H.G. ed. 2003. The flooded forest: guidance for policy makers and river managers in Europe on the restoration of floodplain forests. FLOBAR2, Department of Geography, University of Cambridge, UK.

Naiman, R.J., and Décamps, H. eds. 1990. The Ecology and Management of Aquatic-Terrestrial Ecotones. MAB 4, UNESCO.

Teoh, C.H., Ng, A., Prudente, C., Pang, C., and Tek Choon Yee, J. 2001. Balancing the need for sustainable oil palm development and conservation: the lower Kinabatangan floodplains experience. Proceeding in ISP National Seminar, Strategic Directions for the Sustainablility of the Oil Palm Industry, Kota Kinabalu, Sabah, Malaysia, 11–12 June 2001.

Additional Reading

Alpert, P., Griggs, F.T., and Peterson, D.R. 1999. Riparian forest restoration along large rivers: initial results from the Sacramento river project. Restoration Ecology 7(4):360–368.

Goodwin, C.N., Hawkins, C.P., and Kershner, J.L. 1997. Riparian restoration in the Western United States: overview and perspective. Restoration Ecology 5(4):4–14.

Griggs, F.T., and Golet, G.H. 2002. Riparian valley oak (Quercus lobata) forest restoration on the Middle Sacramento River, California. USDA Forest Service; pp. 543–550.

Harris, R., and Olson, C. 1997. Two-stage system for prioritising riparian restoration at the stream reach and community scales. Restoration Ecology 5(4):34–42.

Hunter, J.C., Willett, K.B., McCoy, M.C., Quinn, J.F., and Keller, K.E. 1999. Prospects for preservation and restoration of riparian forests in the Sacramento Valley, California, USA. Environmental Management 24(1):65–75.

Landers, D.H. 1997. Riparian restoration: current status and the reach to the future. Restoration Ecology 5(4):113–121.

Moring, J.R., Garman, G.C., and Mullen, D.M. 1985. The value of riparian zones for protecting aquatic systems: general concerns and recent studies in Maine. In: Johnson, R.R., Ziebell, C.D., Pattern, D.R., Folliot, P.F., and Hamre, R.H., eds. Riparian Ecosystems and their Management: Reconciling Conflicting Uses. USDA Forest Service.

National Research Council. 2002. Riparian Areas, Functions and Strategies for Management. National Academy Press, Washington DC.

Piégay, H., Pautou, G., and Ruffinoni, C. 2003. Les Forêts Riveraines des Cours d'Eau: Écologie, Fonctions, Gestion. Institut pour le Développement Forestier, Paris.

Schoenholtz, S.H., James, J.P., Kaminski, R.M., Leopold, B.D., and Ezell, A.W. 2001. Afforestation of bottomland hardwoods in the Lower Mississippi Alluvial Valley: status and trends. Wetlands 21(4):602–613.

Tockner, K., and Schiemer, F. 1997. Ecological aspects of the restoration strategy for river-floodplain system on the Danube River in Austria. Global Ecology and Biogeography Letters 6:321–329.

45
Restoring Mediterranean Forests

Ramon Vallejo

Key Points to Retain

The Mediterranean region has been heavily modified by millennia of human intervention. This intervention has included different tree planting phases, with varying results.

Land abandonment and forest fires are common problems in the north of the Mediterranean, while demand for fuelwood and fodder are a key issue in the south.

Because of centuries of landscape modification, there are fewer reference ecosystems to guide restoration in the Mediterranean. There are instead three types of landscapes: highly degraded, cultural, and seminatural landscapes. The second type is also being modified under present land-use conditions.

The challenge lies in trying both to conserve key cultural landscapes and to restore the ecosystems that are the most degraded or under pressure.

1. Background and Explanation of the Issue

1.1. Forest Degradation in the Mediterranean: An Old Problem with a New Face

The Mediterranean basin has been enduring heavy and extensive human use for millennia. Throughout this long history, periods of resource overexploitation have led to significant forest loss and the reshaping of landscapes. Already in the fourth century B.C., Plato warned about both the degradation of Greek forests in the uplands and soil loss: "Hills that were once covered by forests and produced abundant pasture now produce only food for bees." In the past, fluctuations in human population were accompanied by fluctuations in land exploitation, with peaks of overgrazing, forest clearing for agriculture, forest overexploitation for firewood, charcoal production, and logging, intermingled with periods of land abandonment. Frequent wars often devastated the forests as well. The forest was especially overused in crisis situations. The consequent impact on forests was the degradation of vegetation, the reduction of forest surface, the degradation of soil quality, and the increase in soil erosion and flooding. The images of Mediterranean forests projected by enlightened travellers of the 17th to 19th centuries and the direct images from the early 20th century were discouraging. Most mountain areas were depicted as spoilt and the scarce preserved forests were hidden in remote, inaccessible areas, or belonged to wealthy families and/or the nobility, who used them as private hunting parks.

Socioeconomic and political circumstances drive land use and forest exploitation, and this is particularly marked in a region with a long history of human settlement such as the Mediterranean basin. In Southern Europe, economic development since the middle of the 20th century has resulted in a sharp change in tendencies, moving from thousands of years of steady degradation to a new phase of regeneration that is related to the loss of direct market profit from forests and woodlands and rural depopulation. Clearly, this general process has local exceptions in the less economically developed regions of southern Europe.

Meanwhile, in southern and eastern Mediterranean countries, resource exploitation mostly follows the same historical trends in relation to the increasing population growth and direct dependence of rural populations on natural resources. Poverty, now and in the past, is one of the main drivers of forest degradation forced by the primary need for food, fuelwood, and fibre.

Recent land use changes in southern Europe are resulting in the abandonment of less productive lands and substantial reductions in grazing pressure and forest exploitation. These changes are enabling spontaneous vegetation to recover, increasing connectivity in wildland areas and promoting fuel load accumulation in forests and shrublands. In addition, large afforestation programmes conducted during the 20th century significantly increased the forest surface, mostly with pine species and, to a minor extent, eucalyptus. A direct consequence of this dramatic modification in landscape structure and composition has been the spread of large wildfires in the Northern Mediterranean countries since the last quarter of the 20th century.[386] Wildfires have now become the major forest management problem in the region. We can expect the problem to become more and more acute in southern Mediterranean countries if the trend toward rural abandonment continues in the future.

1.2. Structural Problems

Ancient societies adjusted their lives to nature's pace. Industrialisation has caused the gap between both paces to increase dramatically. Present industrial and postindustrial societies change faster than forests. As a consequence, forest policies that respond to current demands from forests (or more generally from land-use interests) may become obsolete in only a few decades, leaving the next generations with a problem that may be difficult to reverse or that may even be irreversible. Examples of this time mismatch include (1) the clear-cutting of cork oak woodlands conducted in Portugal for wheat production during the 1930s, the later abandonment of many of these fields because of poor soil productivity, and the recent attempts to recover cork oak in these now degraded soils; and (2) the eucalyptus plantations established in dry areas of western Spain in the 1960s, which are now abandoned and no longer exploited, suffer wildfires, and, in some cases, are uprooted at a large economic cost to restore native forest.

Forest management and restoration is constrained by land tenure and traditional uses and rights, which are very diverse throughout the Mediterranean countries. There are countries where most of the forest land is private, such as Portugal with around 90 percent, and countries where practically all forest land is public, such as Turkey, Greece, and the Maghreb countries (*forêt domaniale*).

1.3. Reforestation Activities

Recognition of the need to preserve and enhance forests is very old. Already in 13th-century Spain, King Alfonso X promoted regulations to preserve forests against fires and uncontrolled clearing. Some relevant and documented pine afforestation dates back to the early Middle Ages in Spain. Throughout the Middle and Modern Ages, forests competed with grazing and agriculture, with rural people always trying to convert forests into pasture and cropland. Traditionally, grazing was considered by foresters as the prime enemy of forest conservation. The traditional pastured woodlands (*dehesa, montado, pascolo arbolato*) in the western Mediterranean can be considered multifunctional adaptations and compromise land uses given to these forests to solve the demands of rural population. Throughout the 18th and 19th centuries, there was an attempt to preserve and promote forests. Efforts began

---

[386] Pausas and Vallejo, 1999.

to crystallise in the afforestation of relevant surfaces by the end of the 19th century and became fully developed during the 20th century. In southern Europe, most of these afforestation efforts addressed watershed protection and dune fixation.

In relation to socioeconomic development and the decreasing dependence of the population on forest resources, a new perception of nature is growing in the European Mediterranean countries. This is generating new demands on the wildlands, more biased toward recreation, ecological, cultural, and landscape valuation. Of course, these new demands on forests and other wildland uses require the corresponding adaptation of forest restoration techniques to meet these demands.[387] With this in mind, recent afforestation measures for setting aside agricultural lands, promoted under the Common Agricultural Policy of the European Union, were conceived with the aim to recover native forest ecosystems.

2. Examples

The old reforestation projects conducted in the Mediterranean countries were not, strictly speaking, restoration projects as we understand this term nowadays. However, they share the main global aims of restoration, such as reducing soil erosion and runoff, or recovering natural forests, though sometimes exotic species were used as intermediate stages in the rehabilitation process.

2.1. Old vs. New Approaches[388]

2.1.1. Sierra Espuña (Murcia, Southeast Spain) in the Late 19th and early 20th Centuries

Frequent severe floods were chronically causing heavy casualties and large economic losses on the coastal floodplains in Eastern Spain. These were caused by torrential streams draining from the nearby mountain ranges. Most of these ranges were denuded of trees as a result of long-term overexploitation and the large logging activities pursued by the Navy for ship construction, especially during the 18th century. In the Segura basin (Murcia), after the devastating floods of October 1879 (761 casualties), the forest administration launched a reforestation project in 1886 called Defence Works Against the Floods in the Segura basin. The forest engineer R. Codorniu, one of the directors of this restoration project, wrote that in 1889 he did not see a single tree when crossing the hill slopes of the basin. This project started in 1892 and included the reforestation of almost 5000 hectares, accompanied by check dams, firebreaks, and temporary on-site forest nurseries. The climate of the site is dry to sub-humid. After studying the ecological conditions of the site, the species planted were mostly the native conifers *Pinus halepensis, P. nigra, P. pinaster,* and *P. pinea,* but with minor proportions of hardwoods (*Quercus faginea, Ulmus minor*) and other allochthonous or nonnative species in the site such as *P. canariensis, Acacia* sp., and *Abies pinsapo.* In 1902 some two million seedlings were produced for the project. In those times, most of the plantation work was manual and it took almost 30 years! (This would be difficult to repeat today.) Every year gaps were filled in order to achieve full survival of the stands. Nowadays, the site is covered with beautiful pine forests that have reached the second generation (Figs. 45.1 and 45.2), with a rich understorey and some scattered patches and individuals of hardwoods, mostly holm oak (*Quercus ilex*). Flood incidence in the basin has significantly decreased since the establishment of the forest. After several protection regulations, the site was declared a natural park in 1978 and a regional park in 1992. The site constitutes an island of green surrounded by agricultural lands, and desertified, hilly landscapes with a semiarid climate, and it is the main green recreational attraction in the whole region. The site has thus generated economic activities mostly related to ecotourism for the entire local population.

---

[387] Cortina and Vallejo, 1999.
[388] These projects are collected in the REACTION database: www.ceam.es/reaction.

FIGURE 45.1. Sierra Espuña example. Plantation works and general look of the site in 1895. (Photo © The Regional Ministry of Agriculture, Water and Environment, Murcia Region.)

FIGURE 45.2. Sierra Espuña example, present situation (2004). (Photo © Ramon Vallejo.)

2.1.2. Running a Pilot Project in Albatera (Alicante, Eastern Spain)

Some 50 km Northeast from the Espuña site, but at lower elevations and restricted to semi-arid climate (300–350 mm of precipitation per year), the Albatera site in the Crevillete Ranges consists of a pilot project of approximately 25 hectares to combat desertification under the initiative of the Spanish Ministry of Environment, and in the framework of the United Nations Convention to Combat Desertification (UNCCD) for the Northern Mediterranean countries. The area is covered with sparse vegetation and shows evidence of soil compaction and water erosion in the form of rills and gullies. Attempts to reforest the area with Aleppo pine were conducted through plantations in terraces in the 1970s and again in the 1990s, both times without success. Terraces show signs of advanced degradation. Under the initiative of the Spanish Ministry of Environment, the Regional Forest Administration of the Valencia Region conducted a pilot restoration project, with the aim of putting in practice the latest scientific and technical innovations developed through several research and development projects funded by the regional, national, and European Commission programmes. The project was carried out with scientific assessment from CEAM Foundation (Mediterranean Centre for Environmental Studies). The challenge for plantations in these degraded semiarid lands lies in improving plant survival rates (which are often lower than 50 percent) and growth. Irrigation is not applied in regular reforestation/afforestation projects in Spain. The main objective of the project was to enhance the recovery of woody vegetation and its diversity, and to stop land degradation, especially soil erosion. The project was based on previous field research in the same region and on a specific study on the physical and ecological characteristics, and degradation process occurring in the site. Restoration work was executed during the period 2002–2004. A relatively large number of native shrubs and trees were planted in the various habitats identified in the site: wild olives (*Olea europaea* var. *sylvestris*), mastic tree (*Pistacia lentiscus*), kermes oak (*Quercus coccifera*), juniper (*Juniperus oxycedrus*), oleander (*Nerium oleander*), Aleppo pine (*Pinus halepensis*), carob tree (*Ceratonia siliqua*), *Rhamnus lycioides, Tetraclinis articulata, Retama sphaerocarpa, Ephedra fragilis*, European palm (*Chamaerops humilis*), *Tamarix africana, Salsola genistoides*, and the alpha grass (*Stipa tenacissima*) for the most degraded soils. Seedlings were produced in the nursery using the latest criteria for quality control, promoting root development and good physiological performance. Soil preparation was designed to optimise water collection under the extremely dry conditions of the site. Therefore, micro-catchments for runoff collection were created, and complemented with mulching using forest debris. The soil was amended with good-quality compost from urban bio-solids, and the seedlings protected using tree shelters. Soil preparation techniques were efficient in collecting runoff, thereby significantly increasing water availability for the planted seedlings. As a consequence, seedling survival and growth was much higher than usual in these harsh, semiarid degraded lands. Two years after planting, some seedlings reached 70 cm in height. Although the project is in its very early stages of development, good seedling establishment in the critical transplanting shock provides promising perspectives for the recovery of mature and diverse native *macchia* in the medium-term. This recovery would entail more diverse ecosystems and improved protection against soil erosion and flooding risks.

2.2. National Mobilisation Project

In the 1970s, the Algerian government launched an ambitious reforestation programme to "stop the desert," called the Green Belt. The target area was a strip (1500 km, or around 3 million hectares) of steppes receiving between 200 and 300 mm of precipitation per year, and crossing the whole country from west to east parallel to the Sahara desert. These steppes were degraded because of overgrazing and inappropriate cropping promoting wind erosion and exacerbating the natural drought of the region. In its initial phase, the project was implemented by the Army using nearly exclusively Aleppo pine (*Pinus halepensis*). The local population, especially shepherds, reacted strongly against the plantations that obstructed their pastoral activities and in some cases destroyed natural pastures of alpha grass (*Stipa tenacissima*). Later on (from 1986 onward), and under the direction of the National Institute of Forest Research, the whole programme was revised and reshaped. The local population was involved in the afforestation work and rural development criteria were introduced, integrating afforestation with other activities. As a consequence, the species used were diversified, including both native and alien species: *Cupressus sempervirens, C. arizonica, Gleditsia triacanthos, Casuarina* sp., *Acacia* sp., *Pistacia atlantica, Eleagnus angustifolia,* and *Simmondsia chinensis.* In addition, seeding with herbs was conducted for dune fixation, and fodder shrubs (*Atriplex, Opuntia, Acacia, Prosopis*) and trees (*Tamarix gallica, Retama* sp., *Eleagnus angustifolia*) were planted for small family holdings. The initially ambitious target of 3 million hectares was revised down to around 300,000 hectares. The estimated survival rate for plantations was around 70 percent in the long term.[389] The programme received both positive and negative coverage. On the negative side, the initial lack of agreement with local populations, the extensive use of monospecific plantations of Aleppo pine, facilitating the expansion of pests (mostly pine processionary moth), and the little attention paid to biodiversity were cited. The positive aspects included the establishment of native Aleppo pine forests in the best sites and the national and international impact of the initiative.

2.3. The Pilot Experiences in Sidi Jaber: Approaching the Limits for Restoration

Sidi Jaber is located in southeast Morocco, with a precipitation between 200 and 300 mm per year, with large interannual variability. The region is considered to be at the threshold limit of having any productivity. As in the previous example, overgrazing and overcropping resulted in severe wind and water soil erosion. In the area there was competition between cereal cropping and the production of firewood and fodder. A project funded by the World Bank was set up with the objective to establish tree cover to produce fire wood, fodder, and shelter, and to reduce the drought effects on agricultural lands and pastures. For that purpose, adapted trees and fodder shrubs were selected, including both native and alien species. Seedlings were produced in local nurseries using on-site materials and applying reduced irrigation to pre-adapt the seedlings to water stress. Planting was carried out in winter, from November to February when the accumulated precipitation reached 50 mm. The surface of the project site was 22 hectares, and the project implementation was carried out during the period 1991–1993. Out of the 18 species tested, the best growth results were obtained with some exotics, especially *Acacia cyanophylla* (firewood species) that reached $2^1/_2$ m in height in 2 years in the field, and some eucalyptus. *Retama monosperma*, bridal veil broom, which is native in the region had a 100 percent survival rate after the first postplantation year. It is used for firewood in the region and cultivated as an ornamental plant in many warm areas of the world; *Atriplex nummularia* also yielded good survival and growth rates. This species accumulates salt from the soil and is used for fodder, although sheep and goats only consume it when no better palatable species are available. Therefore, its extensive use in Northern Africa has been questioned. Native species such as *Argania spinosa* (a species that is good for fodder), *Pistacia atlantica*, and *Acacia gummifera* (a North African endemic) also gave acceptable results. This pilot project proved that using appropriate species and plantation techniques may both promote ecosystem recovery and supply valuable resources for local people.

3. Outline of Tools

Hydrology and forest restoration projects have a long tradition in southern Europe.[390] Combining short-term stream correction engineering with reforestation for long-term watershed protection has resulted in the global improvement of degraded ecosystems and landscapes, and reduced floods and soil erosion. Nowadays, these projects have to be compatible with the social demands for biodiversity and landscape services. Recent research and development advances enable using a larger variety of native woody species for forest restoration.[391] One specific difficulty in the Mediterranean is the lack of original reference ecosystems

---

[389] Lahouati, personal communication.
[390] See, for example, Molina et al, 1989.
[391] Pausas et al, 2004.

to guide restoration. Instead, cultural landscapes that were created and were functional under past land-use systems are widespread but are being degraded under present land-use conditions. The challenge is trying to make the conservation of these cultural landscapes, and their diversity, compatible with stopping degradation.

New forest restoration techniques have been recently developed from several European Commission (EC) research projects. These include the procedures for cultivation of good-quality seedling, soil preparation techniques, including water harvesting with microcatchments, mulching and organic amendments, and the use of tree shelters to improve seedling survival and growth under harsh soil and climate conditions.[392] These techniques allow the use of local seeds and alternative materials, so they tend to be cheap and of widespread application.

Reforestation projects are traditionally weak in monitoring and evaluation. This deficiency limits the opportunities to learn from past successes and failures, and especially to take advantage of the unique source of information provided by old afforestation and reforestation programmes. For that purpose, evaluation tools and the inventory of old paradigmatic forest restoration projects in southern Europe are being undertaken within the European Commission's Research and Development Programme (see REACTION project: www.ceam.es/reaction).

4. Future Needs

Who pays the bill? Forest restoration is a very expensive activity. In the Mediterranean countries, it is usually carried out using public funds. The generalised decrease of direct profit from forest exploitation under semiarid and dry climates results in a negative cost-benefit balance in market terms. Therefore, the most relevant benefits from forest restoration derive from nonmarket goods and services provided by restored forest and shrublands, such as limiting soil erosion and floods, carbon sequestration, increase of diversity, aesthetic landscape values, and recreation. Public investments in forest restoration rely too much on political fluctuations, all the more so in developing countries. Economic internalisation of the goods and services provided by forests is clearly needed to progress in sustaining forest restoration activities.

References

Cortina, J., and Vallejo, V.R. 1999. Restoration of Mediterranean ecosystems. In: Farina, A. ed. Perspectives in Ecology. Backhuys, Leiden, pp. 479–490.

Molina, J.L., Navarro, M., Montero de Burgos, J.L., and Herranz, J.L. 1989. Afforestation Techniques in Mediterranean Countries (multilingual publication: Spanish. English and French). ICONA, Madrid.

Pausas, J.G., Bladé, C., Valdecantos, A., et al. 2004. Pines and oaks in the restoration of Mediterranean landscapes of Spain: new perspectives for an old practice—a review. Plant Ecology 171:209–220.

Pausas, J.G., and Vallejo, V.R. 1999. The role of fire in European Mediterranean ecosystems. In: Chuvieco, E. ed. Remote Sensing of Large Wildfires. Springer-Verlag, Berlin, pp. 3–16.

Vallejo, V.R., Bautista, S., and Cortina, J. 1999. Restoration for soil protection after disturbances. In: Trabaud, L. ed. Life and Environment in the Mediterranean. WIT Press, Southampton, pp. 301–343.

---

[392] Vallejo et al, 1999.

46
Restoring Temperate Forests

Adrian Newton and Alan Watson Featherstone

Key Points to Retain

While temperate forests tend to be lower in diversity of plant or animal species than tropical forests, the diversity of fungi, mosses, and lichens may often be very high, particularly in areas of high humidity.

Many temperate forests have been substantially modified by human activity, over periods of hundreds or even thousands of years, limiting our understanding of the original ecosystem and hindering the development of goals for restoration.

In many places where temperate forests are found, the value of the land is high, which limits opportunities for restoration.

The rate of recovery of temperate forests from anthropogenic disturbance tends to be very low.

Very little is known regarding the functioning of the soil fauna and microbial communities, which are likely to be of critical importance to ecosystem function and should be considered during development of restoration plans.

1. Background and Explanation of the Issue

1.1. Description of Temperate Forests

Temperate forests cover more than 20 million km^2 of the Earth's surface, including forest types such as boreal conifer forests, the mixed deciduous forests of the United States, Europe, western Asia, China and Japan, and the evergreen rain forests of Chile, New Zealand, and Tasmania.[393] In the Northern Hemisphere, dominant tree genera are typically members of the oak family (*Fagaceae*) or conifers such as pines (*Pinus*) and spruces (*Picea*). Southern Hemisphere forests are often dominated by southern beeches (*Nothofagus* spp.), mixed with conifers such as members of the Araucariaceae and Podocarpaceae. While temperate forests tend to be lower in diversity of plant or animal species than tropical forests, the diversity of fungi, mosses, and lichens may often be very high, particularly in areas of high humidity. Those of the Southern Hemisphere are characterised by many species that have restricted distribution. Temperate forests can be structurally complex, with up to seven distinct canopy layers. The largest trees can reach over 50 m in height with girths of 2 m or more. Spatial variation in forest structure and composition is influenced by the pattern of natural and anthropogenic disturbance, such as wind or

---
[393] Groombridge and Jenkins, 2002.

fire. When canopy trees die, the resulting gaps in the canopy are colonised by different elements of the forest flora. This process of "gap dynamics" is important in maintaining stand structure and diversity.

Temperate forests provide many services to people, including watershed protection and soil stabilisation, and also account for more than half of the carbon stored in forest ecosystems. In many areas they provide significant recreational use. Natural temperate forests are important reservoirs of genetic material of timber trees of economic importance, such as oaks, beeches, pines, and eucalypts. However, more than 500 temperate tree species are now threatened with extinction, often as a result of overexploitation.[394] Large areas of temperate forest have been cleared for agriculture. In Europe and parts of Asia, this process of deforestation has taken place over thousands of years, but continues to be a principal threat in many areas. Timber harvesting is also widespread. As a result many temperate forests are highly fragmented and old growth forests are now very restricted in extent. Other main threats to temperate forests include invasive species, urban development, browsing by vertebrates, mining, acid rain, and air pollution.

1.2. Restoration Issues

Forest landscape restoration depends on preventing forest loss and degradation caused by the above-mentioned threats, and enabling forest ecosystems to recover their functionality. Many of the issues relating to restoration of temperate forests are the same as those for other forest types. As elsewhere, the main focus of restoration will be to identify the main causes of forest loss and degradation, and to develop management responses to address them. Issues that are particular to temperate forests include:

Attributes of temperate forests: Keddy and Drummond[395] provided a detailed analysis of the properties or attributes of temperate deciduous forest ecosystems that could be used to define restoration objectives, or as the basis for monitoring restoration progress (Table 46.1). While providing a valuable first step, this analysis placed relatively little emphasis on landscape-scale attributes, and was restricted to temperate deciduous forest ecosystems in the eastern United States. The approach, therefore, could be usefully extended to other temperate forest types, such as conifer forests and Southern Hemisphere forests, and to the landscape scale.

Definition of restoration objectives: In some areas, such as central Europe and eastern Asia, deforestation has occurred over time scales of thousands of years. In such situations, the characteristics of pristine forest can be difficult or even impossible to define with precision, greatly complicating the development of appropriate restoration objectives.

Rate of forest recovery: Temperate trees, particularly those growing on infertile or marginal sites, display relatively low growth rates compared to tropical forests. Rates of forest recovery following the alleviation of disturbance generally tend to be low; it could take many centuries to fully restore the characteristics of old-growth forest ecosystems. Many conifers are particularly slow growing.

Restoration of key ecological processes: Ecological processes and natural disturbance regimes (e.g., occasional large-scale wildfires, wind throw, insect infestations, etc.) are important characteristics of temperate forests, particularly at a landscape level. The absence of such processes is a key difference between old-growth forests and the ecologically simplified plantations that have often replaced them. Restoration of these ecological processes presents challenges in many situations today, yet this may be critical to the recovery of fully functioning forest ecosystems.

Restoration potential of secondary forests: In some temperate areas where forests were previously cleared (e.g., the northeast United States and parts of Scandinavia), second-growth forests have become established naturally, and relatively minimal management is

---

[394] Oldfield et al, 1998.
[395] Keddy and Drummond, 1996.

TABLE 46.1. Ecological attributes for the evaluation, management, and restoration of temperate deciduous forest ecosystems.

Property	Potential values
Tree size	Old growth forests tend to be characterised by relatively high numbers of large trees. A mean basal area of 29 + 4 m^2 per hectare was recorded on 10 pristine sites.
Canopy composition	Mature forests tend to be dominated by only a few relatively shade-tolerant species. Successional forests tend to incorporate a larger number of tree species, including shade-intolerant species.
Coarse woody debris	Includes fallen logs, snags, and large branches. An important habitat component for many organisms including birds, mammals, invertebrates, and fungi. Highest volumes tend to be recorded in old growth stands (a mean of 27 mg per hectare recorded on 10 pristine sites).
Herbaceous layer	Many temperate deciduous forests are characterised by a diverse herbaceous flora, which may be sensitive to logging and especially grazing.
Epiphytic bryophytes and lichens	Diverse communities of cryptogams (mosses, and lichens) may typically be present on the trunks and branches of trees, particularly in undisturbed forests unaffected by aerial pollution, in humid environments.
Wildlife trees	Many birds, mammals, and invertebrates require trees with particular characteristics for habitat (e.g., as sites for nesting, perching, roosting, or foraging). Large-diameter snags (standing dead trees) and cavity trees (live trees with central decay) are of particular importance. Old growth forests tend to be characterised by ≥4 wildlife trees per 10 hectares.
Fungi	Temperate forests are often characterised by diverse communities of larger fungi, which play a critical role in decomposition and nutrient cycling. Many temperate trees form associations with ectomycorrhizal fungi, which assist in nutrient uptake and form an important food resource for many other organisms. The composition of fungal communities remains poorly documented, but diversity in old growth forests may exceed 100 species per hectare.
Birds	The composition of bird communities appears to be particularly sensitive to the area of forest patches, some species being dependent on large areas of intact forest.
Large carnivores	As large carnivores tend to be at the top of food chains, their presence indicates an intact food web. They may play an important role in keeping herbivore numbers in check, preventing overgrazing and browsing. Large carnivores have explicitly been exterminated in many temperate forests and therefore may need to be considered as an explicit objective of restoration action.
Forest area	In many areas, once-continuous tracts of forest have been highly fragmented as a result of human activity. Fragmentation reduces species' diversity and changes species' composition in remaining forests. Many mammals and birds are most affected because of their large territorial requirements. For a forest to contain the full complement of species, it must be large enough to accommodate, those species with largest area requirements (i.e. >100,000 hectares).

Adapted from Keddy and Drummond, 1996.

required to facilitate the further restoration of such sites toward an old-growth condition.

Socioeconomic context: Extensive temperate forest areas are situated within countries with a high level of economic development. While this can be of value in obtaining the necessary financial support for restoration action, it also creates difficulties. Land prices are often high, particularly in areas where the land has some agricultural value. Coupled with the high costs of human labour, this can make the cost of forest restoration prohibitive. Many areas are subject to intensive patterns of land use, which may themselves have long cultural traditions, such as in much of Europe. This greatly reduces the scope for large-scale forest restoration, which often can be achieved only through the development of partnerships with relevant landowners. In such circumstances economic incentives for forest restoration may be of critical importance.

Ecological complexity: Given that the ecological functioning of temperate forests is relatively well understood, and that temperate forests are relatively simple in terms of structure and composition, it could be argued that the restoration of temperate ecosystems should be technically simpler than in tropical regions. However, very little is known regarding the functioning of the soil fauna and microbial communities, which are likely to be of critical importance to ecosystem function.

Restoration methods: Forest restoration should ideally focus on encouraging natural regeneration and ecological recovery. However, many temperate forest areas are so degraded that artificial establishment of trees may be required to facilitate restoration efforts. Such planting has to be done with great care, and should seek to mimic natural regeneration as much as possible, if restoration objectives are to be achieved. Tree establishment approaches typically employed in commercial afforestation initiatives are generally inappropriate for use in forest restoration.

2. Examples

2.1. Caledonian Pine Forest, Glen Affric, Scotland

The native pinewoods of the Caledonian Forest in Scotland, characterised by Scots pine (*Pinus sylvestris*), comprise the westernmost extent of boreal forest in Europe, and originally covered 1.5 million hectares. By the late 20th century, their area had been reduced to 17,000 hectares, in isolated remnants consisting mostly of old trees, and there was a real danger of the forest disappearing completely. Situated west of Inverness in the northern Highlands, Glen Affric contains the third largest remnant of the native pinewoods, and this is also the largest extent of least-disturbed forest in Scotland. Most of the pinewood area there is owned by the U.K. government, and restoration work began in the early 1960s, when 800 hectares of forest were fenced off to exclude deer and sheep. This enabled a new generation of young trees to regenerate—the first to do so in 150 years (Figs. 46.1 and 46.2). Restoration work increased substantially from 1990 onward, and the main management techniques initially utilised included the following:

- Facilitating natural regeneration of the surviving native forest, through the exclusion of deer by fencing
- Extending the forest in areas where it had already disappeared by planting native trees, grown from seed of local provenance, in patterns that sought to replicate those of natural regeneration
- Felling of substantial areas of commercial plantations of exotic tree species, which were inhibiting the regeneration of the native forest

FIGURE 46.1. Athnamulloch. Planting Scots pine seedlings in a deforested part of Glen Affric in the Highlands of Scotland in 1991, as part of the restoration of the Caledonian Forest there. (Photo © Alan Watson Featherstone/Forest Light.)

FIGURE 46.2. Athnamulloch. By 2002, the planted pines were growing healthily and had been joined by naturally regenerating rowans. In the absence of overgrazing by deer, heather and blueberries have also flourished, covering up much of the exposed pine stump. (Photo © Alan Watson Featherstone/Forest Light.)

In recent years, the restoration work has entered a new phase, with greater emphasis on correcting imbalances in the diversity of tree species (due to the effects of past selective overgrazing and browsing by herbivores), linking up forest fragments throughout the watershed to provide an enhanced sense of a forested landscape, and paying greater attention to the restoration of other components of the ecosystem, such as scarce tree species, woodland insects such as wood ants, forest floor flowering plants, etc. A key factor for achieving further restoration of the forest community is reduction of the deer population, so that ongoing regeneration of trees and herbaceous plants becomes possible without the need for fences. Other significant work that will take place in the years ahead includes the conversion, or naturalisation, of the remaining plantations (many of which are of *Pinus sylvestris*) to a more natural forest structure. In recognition of Glen Affric's ecological importance and the progress made with restoration work, almost 15,000 hectares of land was declared a National Nature Reserve in 2002—the most stringent category of protected area in the U.K. A key feature of this restoration initiative has been the use of volunteer labour: work weeks in the forest have proved popular with a wide range of people keen to participate in practical forest restoration activities.

2.2. Temperate Rain Forests, Valdivian Ecoregion, Chile

The temperate forests of southern Chile account for more than half of the total area of temperate forests in the Southern Hemisphere, extending to a total of 13.4 million hectares. The forests are home to over 900 plant species, over 90 percent of which are endemic.[396] Clearance for agriculture, human-set fires, browsing, and logging have reduced the original forest cover of Chile by more than 50 percent. The temperate rain forests of the Valdivian ecoregion have been identified as a priority for conservation action by WWF. Although there is growing recognition of the importance of native forests within Chile, attempts at native forest restoration have only recently been initiated, primarily by collaborative partnerships between academic researchers and nongovernmental conservation organisations. A first attempt has been made to restore populations of alerce (*Fitzroya cupressoides*), a threatened conifer that produces a highly valued timber. This was achieved by first carrying out an intensive field exploration, which identified a number of remnant populations in an area where the species was thought to have become extinct. These provided a source of seed and cuttings that have been raised in local nurseries. Young plants have now been established on a number of sites near to remnant populations, primarily on agricultural land. As the species is very slow growing, and can live for thousands of years, it is clear that very long time scales are needed for restoration of alerce forest. However, the real value of this initiative may lie in the impact that it has had as a demonstration of how restoration can be achieved in practice, and in raising awareness about the potential for native forest restoration in the region. The participation of local private landowners in the initiative has been of particular importance in this context.

[396] Armesto et al, 1995.

Further restoration initiatives have been developed in Senda Darwin, a field station on the island of Chilöe, by the Fundación Senda Darwin. This area is typical of much of southern Chile, having suffered the combined effects of forest fire, logging, and browsing by livestock. Restoration is being achieved by removing livestock from remnant forest areas, and protecting them by fencing. Although recovery of the forest is slow, a noticeable increase in tree cover has been observed within the first 10 years of the initiative. Evidence suggests that loss of soil organic matter as a result of forest burning has resulted in soils becoming waterlogged, which has limited tree seedling establishment. Research has indicated that on such sites the presence of decaying logs or tree stumps is of particular importance in providing sites for seedling establishment. Recent activities have focussed on developing a nursery facility to raise native tree seedlings for artificial establishment, to assist the restoration process. Seedlings are being planted as linear corridors connecting forest fragments, to assist in the movement of plant and animal species between fragments. In this way, and by developing collaborative links with neighbouring landowners, the project is moving toward a landscape approach to forest restoration.

3. Outline of Tools

Restoration of temperate forests is greatly assisted by the extensive information resources that exist, based on many years of research and forest management, regarding the ecological requirements of different species and the processes of forest dynamics.

3.1. Geographical Information Systems

Geographical information systems (GISs) have proved to be of great value as a tool for planning and managing forest restoration projects. Their databases incorporating environmental information, such as soil, hydrology, and current land use, combined with maps of forest cover and associated biodiversity, can be used to prioritise areas for forest restoration and to develop restoration plans at the landscape scale.[397]

3.2. Spatial Modelling

Spatial modelling of forest dynamics is increasingly being used to explore management options and possible restoration pathways. Spatial modelling approaches coupled with GIS are also being used to analyse the habitat requirements and distribution of particular species.[398]

4. Future Needs

There is a general need for a shift from site-based restoration action to landscape-scale restoration. The development of forest habitat networks, linking forest fragments, is a useful concept in this context.

There is a need for increased research on the effectiveness of different restoration options in temperate forests, e.g., expansion of core area of forest fragments versus increasing connectivity between fragments. Research is also needed on identifying appropriate methods for monitoring progress toward restoration objectives.

A critical need is to identify how the restoration of forest landscapes can be achieved in areas of intensive, competing land uses, for example, through the development of partnerships of many stakeholders, supported by development of appropriate policy and funding mechanisms.

Increased emphasis is needed on restoring ecological processes in degraded temperate forests; many restoration initiatives currently focus solely on reestablishing tree cover, rather than on entire communities of plants and animals. In particular, practical methods are required for the reestablishment of microbial communities on degraded soils, as these may often be of critical importance for ecosystem function.

[397] Humphrey et al, 2003.
[398] Humphrey et al, 2003.

References

Armesto, J.J., Villagrán, C., and Arroyo M.K., eds. 1995. Ecología de los Bosques Nativos de Chile. Editorial Universitaria, Santiago de Chile, Chile.

Groombridge, B., and Jenkins, M.D. 2002. World Atlas of Biodiversity. California University Press, Berkeley, CA.

Humphrey, J., Newton, A., Latham, J., et al., eds. 2003. The Restoration of Wooded Landscapes. Forestry Commission, Edinburgh, UK.

Keddy, P.A., and Drummond, C.G. 1996. Ecological properties for the evaluation, management, and restoration of temperate deciduous forest ecosystems. Ecological Applications 6(3):748–762.

Oldfield, S., Lusty, C., and MacKinven, A., eds. 1998. The World List of Threatened Trees. World Conservation Press, WCMC, Cambridge, UK.

Additional Reading

Buckley, P., Ito, S., and McLachlan, S.M. 2003. Temperate woodlands. In: Handbook of Restoration Ecology, Cambridge University Press, Cambridge, UK.

Hunter, M.I. 1999. Maintaining Biodiversity in Forest Ecosystems. Cambridge University Press, Cambridge, UK.

Peterken, G.F. 1996. Natural Woodland. Ecology and Conservation in Northern Temperate Regions. Cambridge University Press, Cambridge, UK.

Case Study: The Ecological Restoration of Boreal Forests in Finland

Jussi Päivinen and Marja Hokkanen

Around two thirds of Finland's land area is covered by forest. For hundreds of years, slash-and-burn agriculture and tar burning have influenced the structure of forests. Also, the intensive forestry practised after the Second World War has caused significant changes in forest habitats. Few natural forests remain, and they are fragmented and now found mainly in protected areas.

In natural boreal forests, decaying wood of varying size and in various stages of decay is formed all the time. The decaying wood originates from various tree species, and is far more abundant than in commercial forests. As trees fall, they create small openings where new saplings grow. Deciduous trees, which demand more light, grow in the slightly larger openings, whereas spruces grow in the more shaded ones. Due to the constant changes, a natural forest is like a mosaic. Trees of differing size and species grow in random order; occasional small openings are found, as well as thickets.

As a result of effective fire prevention, extensive forest fires hardly occur anymore in Finland. In the past, there were frequent forest fires that left behind dead or dying charred wood. If a forest fire is limited to ground level, the entire tree stand may survive. If the fire reaches the tree tops, at least some of the trees die, and sometimes all of them. Forest fires usually increase the mosaic nature of forests. After the fire, dead and decaying wood is found unevenly distributed in the forest. Saplings grow in the openings formed, and the variation in the age and species' distribution of the trees, as well as the spatial variation of the forest, is often increased.

Forests are the primary habitat for 564 (38 percent) of Finland's threatened species. Furthermore, some 60 (33 percent) forest-dwelling species have already gone extinct in Finland. Many more species have gone extinct from parts of the country, especially from the southern part, which has been most influenced, and for the longest period of time, by humans. Particularly invertebrates, especially beetles, as well as fungi have become extinct.

Only a small fraction of the forests in protected areas are being restored. It has been estimated that the forest area on mineral soil that needs to be restored is approximately 29,000 hectares in protected areas in Finland. In addition, many extensions that are to be joined to existing nature protected areas are in need of ecological restoration. During the years 2003 to 2012, 16,500 hectares of forest are to be restored in protected areas in southern and western Finland. The need for ecological restoration of forests will diminish in the future, because natural processes that create habitats for endangered species begin to take place.

Increasing the Amount of Dead and Decaying Wood

The amount of dead and decaying wood is increased primarily in areas where the natural continuum of decaying wood is in danger of being broken, and in areas lacking decaying wood but with valuable species in the vicinity.

Dead and decaying wood can be produced by stripping the bark off trees while they are standing, or by cutting them down. Both stripping and felling are mainly done by chainsaw. Stripping irons or marking tools can also be used for stripping. Excavators can be used to fell trees together with their root clumps. The mineral soil thus exposed forms a good substratum for saplings.

Creation of Small Openings

Small openings are usually created in young, homogeneous conifer forests. The openings are created by felling all conifers within an area of a few hundred square metres. There are two main methods. Small openings, in which new deciduous trees may grow, can be created. Alternatively, conifers can be felled around the existing deciduous trees which are losing the competition for light and living space. The creation of small openings increases the amount of deciduous trees, and increases the mosaicity of the forest. The saplings growing in the openings also increase the diversity of the age distribution of the stand.

Burning

Burning is one forest restoration method. The sites picked for burning are usually of low or medium fertility, because highly fertile forests are usually too moist to be burned. When the forest is burned, some of the trees are charred, some die immediately, and some die over a period of years. As a result, wood in all stages of decay is continually produced in the area. The diversity of tree species usually increases after a fire. The new tree stands sometimes form in clusters, sometimes separately, with varying distances between the trees. The trees are of different ages, because part of the original stand survives the fire. Increased insolation caused by burning is a prerequisite for certain rare or endangered species.

The European Union (EU) supports boreal forest restoration in Finland. Several projects have received EU Life Nature funding for the ecological restoration of forests. The most extensive of the projects currently under way is the Restoration of Boreal Forests and Forest-Covered Mires project (www.metsa.fi/metsa-life), in which around 5000 hectares of former commercial forests belonging to Natura 2000 will be restored. The project will last until the end of 2007, and the state enterprise Metsähallitus and its partners are responsible for its execution.

Section XIV
Restoring After Disturbances

47
Forest Landscape Restoration After Fires

Peter Moore

Key Points to Retain

The fire situation needs to be analysed as well as possible with available data to support decisions about restoration.

Identifying and engaging with those who light fires, have fire responsibilities, or are impacted by fires is critical.

Protecting the restoration site from fire until species being used can withstand fire, if it is a natural disturbance, is essential.

1. Background and Explanation of the Issue

The need to restore a landscape for its conservation objectives after fire has impacted may appear to be clear and is often obvious. However, without an understanding of the causes of the fire and its role in the ecosystem, then what is "clear" and "obvious" may be totally misunderstood.

1.1. Short Historical Account of Fire

Throughout history there have been large fires that have damaged human assets and impinged on human perceptions. Some of these events have framed human response to fire. They continue to do so—Portugal, Spain, Los Angeles, and eastern Australia in 2003 and the Great Borneo fires of 1997–1998 are examples.

Fire is one of the oldest tools known to humans. It has been used as a management technique in land clearance and preparation for crops for centuries. For the thousands of farmers, ranchers, and plantation owners on the edge of the agriculture frontier pushing into forests, fire is the obvious mechanism. It is normally the least expensive and most effective way of clearing vegetation and of temporarily fertilising nutrient poor soils. In most cases the deliberate fire use we see in developing nations is an echo of what occurred historically in what are now developed countries such as the northeast United States in the 1700s where fire was used to clear forest and convert land to other uses, initially agriculture.

1.2. Short Introduction to Fire in the Landscape

Fire is a prominent disturbance factor in most vegetation zones throughout the world, the most ubiquitous after human urban and agricultural activities.[399] In many ecosystems fire is a natural, essential, and ecologically significant force, organising physical and biological attributes, shaping landscape diversity, and influencing the global carbon cycle. Fire has been part of the landscape since Mesozoic times. The combination of fires and grasses helped create the savannahs and open plains and provided

---

[399] Bond and van Wilgen, 1996.

opportunities for the proliferation of a wide range of grazing animals. For example, Australian vegetation has been subject to the influence of fire, by indigenous (aboriginal) burning and then by the burning practices of European settlers,[400] over a wide range of environments.[401] This pervasive fire presence has influenced a transformation in Australia to the current flora that are considered both fire tolerant and also in many cases are fire adapted requiring fire for regeneration and life-cycle stages.[402] This same story can be told for many ecosystems.

Forest fires occur because of either anthropological or natural causes. Lightning is the most common natural cause of fire. The majority of fires around the globe are caused by human activity. The extent and timing of fires differs between natural ignitions and fires by people, those by people generally being smaller. While it is difficult to compile precise figures, in the year 2000, a year that was not strongly associated with bad fires, the European Community's Global Burned Area Assessment Project identified 251,000,000 hectares of burn scars worldwide.[403]

In fire-sensitive ecosystems fire causes severe damage. One widely known example, tropical rainforest ecosystems, are characterised by high levels of humidity and moisture, they do not normally burn and are extremely prone to severe fire damage when they do. Damage from fire can be long lasting on a tropical forest ecosystem.[404]

Just as too much fire can cause problems, so can too little. Many fires in boreal forests are caused naturally by lightning. However, some countries, such as the United States, have had a policy of suppressing most fires that threaten to grow out of control. Under these circumstances fire suppression can lead to unnatural conditions in which forests, which have historically experienced small intermittent fires, no longer burn. Fire suppression can lead to a buildup of dead biomass, and altered tree species' composition, so when a fire does start, instead of being relatively small, it is much more intense and on a large scale. This conclusion seems to have been reinforced almost annually in the United States since 1986.

Understanding the reason fire is introduced to or suppressed from a landscape is critical. Should the reason not be addressed, restoring the landscape will be difficult and ultimately futile.

1.3. Brief Description of Fire Impacts

Fire has played, and will continue to play, a major role in shaping ecosystems throughout the world. Fires can produce local extinctions of species, alter species' composition and successional stages, and bring about substantial changes in ecosystem functioning (including soils and hydrology). In almost all forest ecosystems throughout the world, humans have altered the natural fire regimes by changing the frequency and intensity of fires. People have excluded or suppressed fires and changed the nature of the landscape so that a naturally occurring fire will not behave in the same way it would have done in the absence of human impact. The interrelationship between humans, fire, and forests is a complex one and has been the subject of many studies and reports.[405]

In some ecosystems, however, fire is an uncommon or even unnatural process that severely damages vegetation and can lead to long-term degradation. Such fire-sensitive ecosystems, particularly in the tropics, are becoming increasingly vulnerable to fire due to growing population, economic, and land-use pressures.[406]

In most developed nations the process of natural area loss and degradation has been slowed or reversed. Public responses to fire, generally viewing fire as negative and destructive, have led to a focus on fire suppression. This in its turn has had "profound effects on vegetation patterns."[407]

---

[400] Singh et al, 1981.
[401] Luke and McArthur, 1978.
[402] Gill, 1981.
[403] Joint Research Center of the European Commission, 2002.
[404] Cochrane, 2002.
[405] Jackson and Moore, 1998.
[406] Goldammer, 2000.
[407] Bond and van Wilgen, 1996.

1.4. The Fire Impact Cycle

The key variables of fire regimes are the following:

- Season in which the fire takes place
- The extent and "patchiness" of the fire
- The fire intensity—either too low or too high can create both negative and positive effects
- Fire frequency—too little time or too much time between fires can be negative

The cycle of fire impact hinges around these regime characteristics. The impact of a fire will be positive or negative depending on the degree to which the fire conforms to a regime that the landscape can accommodate. Wrong season, too small or too large, too high or too low an intensity, and too often or not often enough and the cycle may become out of balance leading to negative impacts. If the cycle remains too far out of balance with the landscape, then fire may lead to a long-term alteration to the ecosystem.

These characteristics of fire can create significant impacts if they hinder the ecosystem's capacity to absorb and harness their influence. So fire may not be intrinsically positive or negative but always has the potential to have a profound impact with potentially long-term effects. Fire is of specific concern where a particular landscape represents a significant or unique ecosystem of global importance. Under such circumstances it becomes even more important to evaluate and manage the role of fire to sustain those values.

Changes in the fire regime that fall outside the capacity of the landscape to contain them will possibly influence a cycle of impact that, depending on perspective, will be considered either negative or positive.

1.5. The Questions of Restoration After Fire

1.5.1. Why and When Restoring?

The natural and human created role of fires in landscapes sets up the context for decisions about restoring landscapes. The decisions need to be based very clearly on an understanding of the role of fire in a particular landscape. This in turn needs to be informed about the fire presence in the landscape—How many? How often? How large? How intense? What season? Also, the cause of fire in the landscape must be identified. Fires can be thought of as having the following characteristics:

- A source—the ignition means, such as lightning, matches, metal striking rocks
- A cause—the agent that lit the fire, such as farmer, tourist, or land-clearing contractor
- A motivation—the reason the fire was lit, such as negligence, livelihood, or accident

Armed with good knowledge of the fire characteristics, the reasons underlying the origin of the fire, and understanding the role of fire in a particular landscape, the following restoration questions can be answered:

- Is restoration likely to be successful or useful?
- Can/should the same species be used for restoration?
- Will restoration have to be "staged," with initial work creating the opportunities for later efforts?

1.5.2. Fire as a Natural Disturbance

The need for restoration will rest on the extent to which the fire regime is out of step with what the landscape can accommodate. Actions might include the following:

1.5.2.1. Controlling Fire to Bring It within the Regime that the Landscape can Absorb

- Reducing ignition sources
- Managing fuels
- Suppressing fires that do not meet the requirements for the landscape (a very difficult decision to make[408])

---

[408] It is far easier to suppress all fires than to make such a decision. Human assets may be impacted, perceptions of the role of fire in the landscape will differ, and hence the fires that should or should not be suppressed will vary. Conflict is likely, particularly when damage is caused.

- Replanting with local species to overcome losses, which will normally have to include protecting the replanting from fire that is inconsistent with the landscape fire regime
- Removing species that have been favoured by inappropriate fire or that have invaded, including the use of fire in some cases
- Undertaking physical works to protect, restore, or limit the degradation of the landscape features such as soil and drainage lines

1.5.2.2. Introducing Fire to Reestablish a Fire Regime Consistent with the Landscape

- Setting fires under prescribed conditions consistent with the fire regime
- Measuring and if necessary managing fuels
- Suppressing fires that do not meet the requirements for the landscape
- Removing species that have been favoured by inappropriate fire or that have invaded (including the use of fire in some cases)
- Undertaking physical works to protect, restore or limit the degradation of the landscape features such as soil, drainage lines.

1.5.3. Fire as a Degradation Factor

Where fire has no natural role in the landscape, then the steps are much clearer. Fire needs to be controlled to reduce its pressure on the landscape. Removing fire from a landscape entirely is generally impossible—accidents and very infrequently occurring combinations of factors will at some time create conditions that lead to fires.

1.5.4. Fire Used as a Tool

Where fire is being used as a tool in the landscape there is first a need to clearly establish the aspects of cause: ignition, source, and motivation. Depending on the insights developed there are likely to be a range of options for landscape restoration. If fire is not impacting negatively on the landscape, there may be no need to deal with fire and restoration to meet other objectives can continue. Fire may also be used as an active tool to accelerate restoration.

2. Examples

In general there are very few efforts to restore landscapes after fire anywhere in the world. Of the aspects of fire management, two—prevention and restoration—are notably absent and apparently ignored in most jurisdictions. Much of the work that is done on burnt areas has apparently been simplistic in origin (to stop erosion) and implementation (dropping grass seed from aircraft). Consequently in the literature and documentation there is little carefully considered fire-related restoration work described.

2.1. Attempting to Rehabilitate Rainforests in East Kalimantan, Indonesia

Following the severe fires that burnt through Grand Park Bukit Soeharto in East Kalimantan in the 1980s and early 1990s, the timber concession companies that had responsibility for areas elsewhere in the province were required to rehabilitate the park. This has taken the form of narrow plantings of an introduced *Acacia* species and roadside signs identifying the company responsible for each section of the rehabilitation. While it has reestablished tree cover, the vegetation is introduced and does not resemble the forest removed or lost to the fires in terms of species' mix, structure, or habitat.

As part of GTZ's Sustainable Forest Management Project, which was operating at the time of the fires, the following principles were developed for the rehabilitation of fire-affected forests:

- Maintenance of the forest area
- Sustainable management of forest resources: Economically sound management targets should be defined and agreed to by the concession's stakeholders, giving consideration to the local conditions and forest functions.

Appropriate silvicultural treatments should be performed to reach these management targets.
- Ecological sustainability: Management targets should be directed toward the type of forest that is native to the area. Silviculture activities should have minimal negative effects on the remaining stand and soil and should prioritise management of the residual stand, natural regeneration, and mixed planting using local species suitable to the site.
- Forest protection: The forest is the foremost asset so it must be protected from pests, disease, illegal logging, fire and other disturbances.
- Community participation to increase community welfare through benefits from forest resources and support efforts to protect the forest

2.2. Restoration in Giant Forest—Sequoia and Kings Canyon National Parks, California[409]

Development in giant forest in Sequoia and Kings Canyon National Parks altered the vegetation in several ways. Trees were cleared for buildings and parking lots, leaving distinct openings in the forest canopy. The forest overstorey was thinner because trees that threatened human safety and property were removed. Trampling and soil compaction reduced or eliminated the forest understorey, including grasses, wildflowers, shrubs, and tree seedlings. The soil seed bank, which influences the regenerative potential of the forest, was likely depleted. Small patches of wetland vegetation were lost where fill was placed over meadow edges or streams.

The disturbance caused by human development resembled that caused by natural, prescribed fire killing patches of mature trees, creating openings, or gaps, in the canopy. These fire-caused gaps were colonised by patches of abundant shrub and tree regeneration, particularly giant sequoia, with little regeneration beneath intact canopy.

Shrub and tree regeneration in fire-caused gaps was mapped and the patterns of regeneration were used as a model for restoring vegetation in Giant Forest Village. The short-term goal of vegetation restoration in Giant Forest Village is to reproduce the species' composition, density, and spatial pattern of regeneration that would result from a natural fire event. The long-term goal is to integrate the site into the natural fire regime typical of surrounding areas of giant forest, re-creating the range of natural variability and then allowing natural processes to thin the vegetation.

2.3. Restoration After Fires in Mediterranean Forest Landscapes[410]

Fires are part of the natural disturbances to which Mediterranean forests are adapted. Nevertheless, during the last decades the natural fire regimes have been altered and increasingly there are large-scale, very intense, and frequent human-induced fires. From experience in Portugal, where in 2003 WWF and the local nongovernmental organisation (NGO), Associação de Defesa do Património de Mértola (ADPM), developed plans to restore forest landscapes that were devastated by fires, a number of steps were taken:

- Geographical information system (GIS) assessment of soil degradation and hydrologic erosion risk of the different landscape components
- The GIS assessment of the fire incidence in the forest cover and mycorrhizal soil component in the mosaic of habitat types within the forest landscape
- Analysis of the socioeconomic impact, including forecasts in productivity loss and risk of abandonment of forest uses and rural exodus
- Planning the different technical options to be adopted within the landscape for preventing degradation and activating the natural recovery of burned areas, including burned vegetation management techniques;

[409] Source: http://www.nps.gov/seki/snrm/gf/ecology/vegetation.htm.

[410] This example was provided by Pedro Regato, WWF Mediterranean Programme.

- it is preferable not to remove burned vegetation from the forest area, as it provides protection to soil and to the natural regeneration.
- Active restoration in landscape areas with risk of soil erosion and little or no natural regeneration in the first years. As much as possible, it would be preferable to promote planting by combining root-sprouting species, such as evergreen oaks, small trees—strawberry tree, myrtle, mastic tree—with leguminous shrubs
- Management of sprouting trees, mainly oak species, through cutting operations to accelerate the establishment of healthy coppice woodlands
- Clearance of fire-prone monospecific shrublands, for example, rocky rose shrubs and plantation of scattered trees and shrubs, as well as pasture patches to increase plant diversity, accelerate succession, and reduce the risk of fires
- Nonintervention in areas with low fire impact where the natural regeneration has a good after-fire response
- Reducing the risk of fires recurring in the forest landscape
- Creation of natural firebreaks within the forest landscape, especially in areas where forest management options have simplified the landscape structure (see "Developing Firebreaks").
- Restoring riparian forest vegetation in ravines and river networks
- Redesigning tree plantations where timber/pulp commercial tree stands should be alternated with silvipastoral woodland stands—dominated by oak, ash, chestnuts, juniper, stone pine, etc.
- Restoring the economic and social potential of the burned forest landscape
- Activities should be participatory in order to understand and restore the economic and social values of burned forest landscapes
- Restoration should be designed and planned to reduce large-scale fire risk and may imply the need for funding schemes, such as governmental subsidies or environmental services payments, to support the establishment of natural and economically beneficial firebreaks, and to diversify the existing land-use options in private and public land

2.4. Potential Adverse Impacts

Adverse impacts of restoration after fires are most likely to result from the use of inappropriate (exotic) species, physical restoration efforts that change or impact soils or drainage features, or replanting that alters the preferred mix of local species. In the Bitterroot National Forest in Montana, wildfires burnt extensive areas in 2000. The amount of disturbance by both wildfires and fuel treatments before fires combined with the use of exotic seed in mixes applied for erosion control are suggested as factors in establishing invasive species in the landscape.[411] Conditions that potentially favour invasive species included increased light and nutrient levels, reduced plant competition, and exposed soil. In some sites, 2 years later, the fire weeds had increased in density and were present on plots that had previously been free of invasive species. Knapweed (there are several species) had increased in relation to the severity of fires—the more severe, the higher the density of this weed. There are cases of invasive species following wildfires that reduce the chance of native plant recovery identified in New Mexico in the United States.[412]

3. Outline of Tools

The major input required for framing restoration after fires is strong insight into the fires themselves. The facts, factors, and information that need to be gathered include those listed earlier. Collectively, fire-related data, identification of the fire regime, and clarity about cause (ignition, source of fire, motivation for fire) provide a solid foundation for dealing with the fires and then restoring the landscape if it proves possible and desirable. For developing nations, fire is often perceived as part of that

[411] Sutherland, 2003.
[412] Hunter et al, 2003.

development. Consequently analysis of livelihood requirements and sectoral use of fire in economic development is needed.

Analysing fires is essential and relatively straightforward if the data and information are available. The key information is simple and the focus is on the motivation for the fires—dealing with this is essential to identify the restoration strategy required and its components. Though there is no documented "formal" or "systematic" process for the analysis of fires, the process basically involves obtaining answers to a series of questions:

For fires:
- When did the fire start?
- Where did the fire start?
- When did the fire finish?
- How large is the area burnt?
- What ignited the fire?
- Why was the fire started?
- Where are the fires likely to be?
- What time of year/season are fires likely to occur?

For people:
- Who manages and influences land—communities, forest agencies, concessionaires, ministry of agriculture, ministry of transmigration, provincial and district leadership, others?
- Who is impacted—people, transport sector, tourism sector, health sector, agricultural sector, manufacturing industry?
- Who can assist with fires—fire services, communities, forest agencies, concessionaires, ministry of agriculture, ministry of transmigration, provincial and district leadership?

For those identified above:
- What role do they play?
- What is their motivation?
- Why should they be involved?
- Who is responsible and should fight the fire?
- Who is affected and will need/want to fight the fire?
- Who is responsible for fires that cause damage?
- Who is impacted by fires?
- Who should pay or undertake recovery?

For the landscape:
- What is the ideal landscape state, given the influences of fires and people?
- Is there an ecological role for fire in the landscape?
- Should/must fire have a role in the landscape?

By collating the answers to these questions as far as possible (informed guesses are sometimes the only information available), the fire "picture" can be framed.

Once the fire situation is understood, then decisions about restoration strategies and techniques can be made. If the fires are going to be repeated, then restoration itself may not be successful or require fire management to ensure restored areas are not burnt at all, not burnt before they can be, or are ready to be burnt.

4. Future Needs

There is increasing recognition of the often strong capacity communities have in fire management. Their reasons, skills, and understanding can be highly developed and should be harnessed. The community/local understanding of fire and its role as well as techniques for using fire should be the basis for improving fire management. Expanding the recognition of community-based fire management (CBFiM) and the core role people play through using fire in the landscape is essential in the context of nations where government structures and approaches are developing and resources and support may be limiting.

As discussed earlier it is critical to obtain, maintain, or initiate records of unwanted fires, fire use, and fire behaviour to enable analysis to support the refinement of techniques of deliberate fire use and targeting of information and inputs to reduce unwanted impacts of fires.

References

Bond, W.J., and van Wilgen, B.W. 1996. Fire and Plants. Chapman & Hall, London.

Cochrane, M.A. 2002. Spreading like wildfire—tropical forest fires in Latin America and the Caribbean. Prevention, assessment and early warning. UNEP, Regional Office for Latin America and the Caribbean, Mexico.

Gill, A.M. 1981. Adaptive responses of Australian vascular plant species to fires. In: Gill, A.M., Groves, R.H., and Noble, I.R., eds. Fire and the Australian Biota. Australian Academy of Science, Canberra.

Goldammer, J. 2000. Global Fire Issues. In: Saile, P., Stehling, H., and von der Heyde, B., eds. WALD-INFO 26. Special Issue—Forest Fire Management in Technical Co-operation. Gesellschaft für Technische Zusammenarbeit (GTZ). Eschborn, Germany.

Hunter, M.E., Omi, P.N., Martinson, E.J., Chong, G.W., Kalkhan, M.A., and Stohlgren, T.J. 2003. Effects of fuel treatments, post-fire rehabilitation treatments and wildfire on establishment of invasive species. Second International Wildland Fire Ecology and Fire Management congress and Fifth Symposium on Fire and Forest Meteorology, Orlando, Florida, 16–20 November.

Jackson, W.J., and Moore, P.F. 1998. The role of indigenous use of fire in forest management and conservation. International Seminar on Cultivating Forests: Alternative Forest Management Practices and Techniques for Community Forestry. Regional Community Forestry Training Center, Bangkok, Thailand.

Joint Research Center of the European Commission. 2002. Global Burnt Area 2000 (GBA2000) dataset: http://www.gvm.jrc.it/fire/gba2000/.

Luke, R.H., and McArthur, A.G. 1978. Bushfires in Australia. Australian Government Publishing Service, Canberra.

Singh, G., Kershaw, A.P., and Clark, R. 1981. Quaternary vegetation and fire history in Australia. In: Gill, A.M., Groves, R.H., and Noble, I.R., eds. Fire and the Australian Biota. Australian Academy of Science, Canberra.

Sutherland, S. 2003. Wildfire and weeds in the northern Rockies. Second International Wildland Fire Ecology and Fire Management congress and Fifth Symposium on Fire and Forest Meteorology. Orlando, Florida, 16–20 November.

Web Sites

US National Parks Service
http://www.nps.gov/fire/fire/fireprogram.html.
Global Fire Monitoring Centre
http://www.fire.uni-freiburg.de/programmes/natcon/natcon_5.htm.

Additional Reading

Bowman, M. 2003. Landscape analysis of aboriginal fire management in Central Arnhem Land, North Australia. Second International Wildland Fire Ecology and Fire Management Congress, Orlando, Florida, 16–20 November.

Ganz, D., Fisher, R.J., and Moore, P.F. 2003. Further defining community-based fire management: critical elements and rapid appraisal tools. Third International Wildland Fire Conference, October 6–8, Sydney, Australia.

Moore, P.F. 2001. Fires, community action and law enforcement in S.E. Asia. Paper prepared for the Forest Law Enforcement and Governance: World Bank East Asia Ministerial Conference, September 11–13, Denpasar, Indonesia.

Moore, P.F. 2001. Forest fires in ASEAN: data, definitions and disaster? ASEAN Regional Center for Biodiversity Conservation, Workshop on Forest Fires: Its Impact on Biodiversity, Brunei Darussalam, 20–23 March.

Moore, P.F., Ganz, D., Tan, L., Enters, T., and Durst, P.B., eds. 2002. Communities in flames: proceedings of an international conference on community involvement in fire management. FAO RAP Publication 2002/25.

Petty, A., Banfai, D., Prior, L.D., and Lehmann, C. (2003) Introducing the Kakadu Landscape Change Project: a multidisciplinary assessment of 50 years of landscape change in the tropical Savannah Region of Northern Australia. Third International Wildland Fire Conference, October 6–8, Sydney, Australia.

Reeb, D., Moore, P.F., and Ganz, D. 2003. Five Case Studies of Community Based Fire Management. FAO Headquarters, Rome.

48
Restoring Forests After Violent Storms

Daniel Vallauri

Key Points to Retain

After a violent storm, there is typically a move to restore, starting as soon as possible to implement salvage logging and replanting.

This leads to two paradoxes: an economic paradox, that financial profit is not always guaranteed but investment is facilitated; and an ecological paradox, that natural disturbances, including violent storms, are essential to the functioning and the preservation of biodiversity.

There is a good deal of information in the literature, and field experience includes large-scale use of natural dynamics.

Restoration questions after storms are a key topic in order to encourage forest management improvements, both on paper and in the field.

Careful lobbying, policy work, and communications are needed.

1. Background and Explanation of the Issue

Every year somewhere in the temperate zone violent storms damage forests and cause large economic losses for forest owners. In an average year in Northern Europe the area damaged is equivalent to the net increase in commercial forest area. The overall forest area affected is many times larger.

Forest damage due to violent storms is often described as a climatic and economic disaster. After a violent storm, there is typically a move by politicians, the general public, the media, and foresters to restore, starting as soon as possible to implement salvage logging and replanting. However, when considering responses to storms two main paradoxes should be considered:

- The economic paradox: Broken and uprooted trees have lost part of their timber value. Harvesting in forests damaged by violent storms is more difficult and dangerous, thus many trees do not cover the cost of logging operations. Artificial replanting (including soil treatments) is also expensive. As a whole, such a salvage logging/artificial replantation policy is extremely expensive for society, which generally supports these operations through European Union (EU) and national subsidies. These facts lead to a first paradox that even though financial profit is not always guaranteed in post storm operations, investment is facilitated and increased.
- The ecological paradox: Modern ecological theory asserts that natural disturbances, including violent storms, are essential to the functioning of old-growth forests and that they contribute positively to the preservation of biodiversity. Indeed, they drive greater species' diversity and sustain

never-ending forest cycles. This paradox is partly explained by the fact that, over recent decades, forest structure and composition have been increasingly modified for human uses. Management rules have sometimes weakened the resistance (e.g., large-scale, pure, even-aged spruce or poplar plantations) and resilience of forests (natural ability to regenerate without assistance), especially in central and western Europe.

In the aftermath of a violent storm, the main challenges for conservationists are the following:

- Avoid additional harsh human intervention, especially on soils or key habitats while logging. Numerous experiences prove that the direct impact of violent storms is often far less dangerous for biodiversity than poorly planned and implemented post-storm actions.
- Reintroduce forest productivity along with forest biodiversity and other social uses, if any of these functions have been damaged by the storms, and avoid restoration errors.
- Because it is one of the very few forest events that raise public interest about forest issues, the aftermath represents a key period for efficient lobbying and communication to improve field practices and above all forest policies (including subsidies).

Today forestry is facing the challenge of achieving sustainable and multifunctional management in complex ever-changing social and ecological environments. Storms are predicted to become more frequent in the temperate zone as a consequence of global climate changes. Thus, storms above all provide us with the opportunity to define management in closer harmony with nature's rules. Key questions to explore and answer to help with this process include: How can we better integrate natural disturbances in science-based forest management? How can we reduce forest vulnerability? How can we recover natural resilience? How can we help to restore?

2. Examples

There is a good deal of information in the literature on the effect of storms on forests and on restoration in various contexts. Here are a few recent examples from the temperate forests of Europe and North America.

2.1. Learning from Ecological Studies[413]

Nothing is permanent, except change.
Eraclite, 500 A.D.

Forest management ought to better integrate the consequences of ecological disturbances. This requires a deep understanding of natural disturbance regimes and forest resistance and resilience, which is also essential for forest landscape restoration.

Some of the key ideas about storm disturbance are listed below. We use as an example data from Fontainebleau National Forest, France:

- Time period, frequency, and intensity of events are variable. Climatic data on winds, ice storms, tornadoes, etc., and an analysis of past events that affected forests facilitate risk assessment. In Fontainebleau the periodicity of violent medium-size storms, for example, is evaluated as one event every 25 to 30 years (1938, 1967, and 1990 for the last century), more than half occurring between November and January.
- Resistance of forests to winds is a complex issue, expressed by a nonlinear and multiscale relationship among climatic, geographic, and ecological factors. Of the latter, the relationship between soil and forest stand structure is particularly decisive (type of root system, deciduous or evergreen, etc.). Oversimplified forest structures are dangerously sensitive to strong winds at landscape scales. In Fontainebleau, all stands are sensitive to wind speeds higher than 120 km/h, but pure evergreen trees with shallow root systems

---

[413] Pontailler et al, 1997; Rogers, 1996; Schaetzl et al, 1989.

(like spruce) on sandy or humid soils are more sensitive to damage.
- One of the consequences of the two previous points is that violent storms may result in very different levels of damage in terms of the proportion of uprooted trees or snags, and in terms of distribution (single-tree openings, medium-size gaps, or very large gaps). In Fontainebleau, violent winter storms in old-growth broad-leaved forest usually damage from 2.7 to 21.2 trees per hectare (a majority of beech trees with dbh (diameter at breast height) from 35 to 85 cm) and create a mosaic of small gaps (mean size 175 m^2) on 4 to 21 percent of forest area.
- Resilience depends on numerous factors, including biodiversity, ecosystem health, and structural complexity (forest stand and understorey). Depending on the size, characteristics, and context of gaps (seed availability, for example), natural regeneration occurs rapidly or not, with the expected target species or not. In Fontainebleau, single-tree gaps are rapidly closed by beech, whereas in larger openings oak could be dominant and birch colonises bare soil.
- Where forest is near natural in structure, storms support natural functioning which in turn supports biodiversity conservation, including species depending on open and humid habitats.

2.2. New York State: Banning Salvage Logging in Protected Areas[414]

In northern New York State, strong winds caused significant damage in July 1995 over approximately 400,000 hectares of private and public forests. Out of the approximately 175,000 hectares of public area designated as the Adirondack Park Forest Preserve, damage was particularly high (60 to 100 percent) over 9700 hectares and moderate (30 to 60 percent) over 25,300 hectares.

State policy following such events since the 1950s was technically based, focussed on forest health (threat of fire, deadwood, pests) and generally started with complete and rapid salvage logging, including in wilderness area (although it required a waiver from wilderness state legislation).

Considering the specific context of the 1995 storm, for which key elements were a well-prepared science-based expertise (including an information system that enabled rapid and reliable evaluation of scenarios), ecological pressure from society and weak economic demand for timber, a new official policy was adopted by the governor for the forest preserves. It corresponds to a near-complete reversal of preceding policy: no salvage logging, reinforcement of the "forever wild" statement for forest preserves; and operations limited to cleaning roads, trails, and campsite facilities. Salvage logging was specifically rejected as being uneconomic.

In Europe, another example of such a policy is the one from Bavaria National Park (Germany) following violent storms in 1983 and 1990.[415] Both examples are very relevant to the violent storm that damaged Tatra National Park (Slovakia) in November 2004. In Tatra National Park the restoration that began after the storm of 1915, which included salvage logging and artificial replanting of spruce, led 90 years later to the same catastrophic results, both ecologically and economically.

2.3. Restoration After the 1999 Storm in France: When Short-Term Subsidies Define the Strategy

The storms of December 1999 in France affected about 500,000 hectares, that is, 1/30 of the French forest area (140 million m^3 of downed wood). Apart from the importance of the damage, the sharp social debate following this storm forced forest stakeholders, including NGOs, to revise their strategy and to design restoration far more carefully than in the past.

WWF promoted a science-based strategy emphasising multifunctionality and sustainable

[414] Robinson and Zappieri, 1999.

[415] Fisher, 1992.

management. The strategy outlines seven main principles:

- Make a clear analysis of forest goals within the landscape.
- Define the priority of the actions (logging, planting, natural regeneration).
- Follow the time scale of nature (especially to allow natural regeneration).
- Reduce additional actions likely to lead to degradation while logging, such as using pesticides, etc.
- Use all the opportunities offered by nature (alternative natural successions).
- Closely mimic nature and facilitate its work.
- Avoid doing poorly and at high cost what nature could do better and at a lower cost (reduce artificial work, ploughing, spraying).

WWF and partner NGOs proposed detailed management rules, compiled into a published charter in 2000.[416] The Office National des Forêts, the manager of national and municipal forests, published also in 2001[417] a detailed guidebook for restoration.

However, despite important evolution in French forest management rules on paper, two main problems were driving the operations in the field:

1. Salvage logging was the norm and done in a hurry, sometimes with very little concern for soil sensitivity and biodiversity. It was even implemented in some protected areas or forest identified as being of high conservation value (e.g., forests inhabited by the last highly endangered capercaillie *Tetrao urogallus* in the Vosges mountains). Because of the storm's psychological shock and the will to sell damaged wood, forest managers and owners sometimes seek above all to work fast, which means very often work as usual, and they forget recent innovative rules and agreements.

2. The French forest subsidies' framework (including EU subsidies) after the storms of December 1999 was redefined nationally and adapted by each regional administration. Although some improvements were proposed at the national level, very little was in fact subsidised at the regional level. The result was that key operations like salvage logging and artificial plantations were relied on more than natural regeneration, for example. Salvage logging was subsidised for up to 1500 euros/hectares, without any precise rules for key environmental topics (like deadwood or habitat tree retention for example).

3. Outline of Tools

3.1. Learning About Storms, Forest Ecology, and Restoration

Storms, their impact on forests and biodiversity, and strategies for restoration are frequently written about in the scientific literature for various countries and forest types. Good syntheses of these reports also exist, but are not used enough as references to renew forest management and policies.

3.2. Forest Policies

There are three main reasons to support policy work that integrates natural disturbances into national forest laws and science-based management guidelines. First, forest managers are usually reactive to storms rather than proactive. We need to anticipate forest damage due to storms. Second, as stated earlier, national policies and subsidies tend currently to support rapid implementation of salvage logging in the field. Third, a rapid response to such disorganising, catastrophic, psychologically shocking events rarely produces good results unless there is already a deep understanding of forest ecology, firmly embedded in management rules and culture. It is important to be well prepared. Political lobbying helps to clarify questions about salvage logging, deadwood retention, logging in protected areas, management of pests, biodiversity, and sustainable management. Developing laws, subsidies, and technical tools in accordance with these issues is an important task.

---

[416] Vallauri, 2001; WWF et al, 2000.
[417] Mortier, 2001.

3.3. Restoration Guidelines

"Slow down the tractors," "Set wise restoration targets and trajectories in accordance with sustainable multifunctional forest management." "Take time to let nature do its work." "Help nature *only* when necessary." "Save nature as well as money." These could become the mottos of forest restorationists after violent storms. Or, to paraphrase, "Think and, only if needed, log and plant" should replace the common "Log, plough, plant, then think."

Good guidelines and experiences do exist in numerous regions, especially those hit by violent storms during the last 15 years, such as for example New York State, Switzerland, Germany, and France. However, a better promotion of existing guidelines and pilot experiences is important for the future. Key principles can be drawn from these examples. They include a deeper respect for forest ecology, forest functions, natural dynamics, and biodiversity, and thus wisely using what nature can provide for free, keeping subsidies for those silvicultural actions that may be needed in the medium term (such as thinning and additional planting).

3.4. Press and Communication

Forest issues suffer from low media interest, as they tend to be too technical and complex and not embedded in a strong political or social debate. They are not key financial issues for most developed countries, and are not appealing enough visually. They are based on too long term an agenda, with relatively rare, urgent and catastrophic events to catch people's attention; that is, they are not "sexy," except for forest fires, and violent storms! Recent debates in various countries have proven that the multifaceted questions raised by violent storms (drama, forest mismanagement, biodiversity, restoration) could be real topics for the media. It is also an important opportunity for foresters and conservationists to explain to society their ideas, choices, and field experiences. But as it becomes a hot issue, professionals should be prepared to deliver the right message at the right time, from the day after the storm to several months after the event. Rapid response packages, like the one initiated by the WWF European forest team, are very useful (also see "Marketing and Communications Opportunities").

4. Future Needs

4.1. Learning from Past Events, Adapting Guidelines, and Pilot Sites

In terms of scientific knowledge, the needs lie in synthesising and widely promoting key ideas, rather than developing new research, although some important questions, such as the comparative resistance to storms of mixed or uneven-aged forest stands vs. even-aged stands, and the economics of salvage logging, need some development. More could also be learnt from studying the old-growth forest ecology of protected forests.

Another important need is the adaptation of science-based management rules and tools (geographical information system, modelling), and ecological and economical expertise to different regional contexts. Thus, a wider exchange of experience after storms, together with a network of long-term pilot restoration sites, should be promoted.

4.2. Policy Needs

Restoration after storms is a key topic, especially in Europe, in order to encourage forest management improvements, both on paper and in the field, although the latter takes time. For Europe, part of the solution could be to improve guidelines for the use of EU subsidies in case of storm damages and for plantations. Careful lobbying at the time of changes in national forest law is needed.

References

Fisher, A. 1992. Long term vegetation development in Bavarian mountain forest ecosystems following natural destruction. Vegetatio 103:93–104.

Mortier, F. 2001. Reconstitution des forêts après tempêtes. ONF, Paris.

Pontailler, J.Y., Faille, A., and Lemée, G. 1997. Storms drive successional dynamics in natural forests: a case study in Fontainebleau forest (France). Forest Ecology and Management 98(1):1–15.

Robinson, G., and Zappieri, J. 1999. Conservation policy in time and space: lessons from divergent approaches to salvage logging on public lands. Conservation ecology [online] 3(1): 3, http://www.consecol.org/vol3/iss1/art3.

Rogers, P. 1996. Disturbance ecology and forest management: a review of the literature. USDA Forest Service Intermountain Research Station, report INT-GTR-336.

Schaetzl, R.J., Johnson, D.L., Burns, S.F., and Small, T.W. 1989. Tree uprooting: review of impacts on forest ecology. Vegetatio 79:165–176.

Vallauri, D. 2001. Si la forêt s'écroule. Quelle gestion forestière française après les tempêtes. Revue Forestière Française 54(1):43–54.

WWF, Greenpeace, RNF, FNE. 2000. Partnership charter for forest restoration after the December 99 storms in France. Paris.

Additional Reading

Armstrong, G.W. 1999. A stochastic characterisation of the natural disturbance regime of the boreal mixedwood forest with implications for sustainable forest management. Canadian Journal for Forestry Research 29:424–433.

Baker, W.L. 1992. The landscape ecology of large disturbances in the design and management of nature reserves. Landscape Ecology 7(3):181–194.

Bergeron, Y., and Harvey, B. 1997. Basing silviculture on natural ecosystem dynamics: an approach applied to the southern boreal mixedwood forest of Quebec. Forest Ecology and Management 92(1–3):235–242.

Dale, V.H., Lugo, A.E., MacMahon, J.A., and Pickett, S.T.A. 1998. Ecosystem management in the context of large, infrequent disturbances. Ecosystems 1:546–557.

Ennos, A.R. 1997. Wind as an ecological factor. Trends in Ecology and Evolution 12(3):108–111.

Faille, A., Lemée, G., and Pontailler, J.Y. 1984a. Dynamique des clairières d'une forêt inexploitée (réserves biologiques dc la forêt de Fontainebleau). I. Origine et état actuel des ouvertures. Acta Oecologica, Oecologia Generalis 5(1):35–51.

Faille, A., Lemée, G., and Pontailler, J.Y. 1984b. Dynamique des clairières d'une forêt inexploitée (réserves biologiques de la forêt de Fontainebleau). II. Fermeture des clairières actuelles. Acta Oecologica, Oecologia Generalis 5(2):181–199.

Foster, D.R., Knight, D.H., and Franklin, J.F. 1998. Landscape patterns and legacies resulting from large, infrequent forest disturbances. Ecosystems 1:497–510.

Larsen, J.B. 1995. Ecological stability of forests and sustainable silviculture. Forest Ecology and Management 73:85–96.

Peterson, C.J., and Pickett, S.T.A. 1991. Treefall and resprouting following catastrophic windthrow in an old-growth hemlock-hardwoods forest. Forest Ecology and Management 42(3–4):205–217.

Peterson, C.J., and Pickett, S.T.A. 1995. Forest reorganisation: a case study in an old-growth forest catastrophic blowdown. Ecology 76:763–774.

Pickett, S.T.A., Kolasa, J., Armesto, J.J., and Collins, S.L. 1989. The ecological concept of disturbance and its expression at various hierarchical levels. Oikos 54:129–136.

Romme, W.H., Everham, E.H., Frelich, L.E., Moritz, M.A., and Sparks, R.E. 1998. Are large, infrequent disturbances qualitatively different from small, frequent disturbances? Ecosystems 1:524–534.

Schaetzl, R.J., Johnson, D.L., Burns, S.F., and Small, T.W. 1989. Tree uprooting: review of terminology, process and environmental implications. Canadian Journal of Forest Research 19:1–11.

Sousa, W.P. 1984. The role of disturbance in natural communities. Annual Review of Ecological Systematics 15:353–391.

Ulanova, N.G. 2000. The effect of windthrow on forests at different spatial scales: a review. Forest Ecology and Management 135:155–167.

49
Managing the Risk of Invasive Alien Species in Restoration

Jeffrey A. McNeely

Key Points to Retain

Introduced species that become invasive can become a major concern as they can cause significant ecological and economical damage. Restoration may often equate to the removal of these species. On the other hand, in some cases, attempts to restore using inappropriate species has itself led to the problem of invasive alien species (IAS).

Restoration may often equate to the removal of these species.

Prevention and best practices for alien species are amongst the most important tools to contain the problem.

Because the problem is transboundary, it is necessary to create common protocols and to enhance the capacity to deal with invasive alien species.

1. Background and Explanation of the Issue

1.1. Overview of Invasive Alien Species

Globalisation has encouraged the free movement of goods but also of plants. On the one hand, plants are available from virtually anywhere in the world for various uses, but on the other hand, species that are moved by people from one part of the world to another can expand beyond the area where they were planted, and end up causing substantial damage to natural ecosystems. Further, global trade, transport and tourism also provides new opportunities for unintentional introduction of species, for example by introducing a nonnative species of beetle that can devastate plants being used to restore a forest.

Those alien species that become established in a new environment, and then proliferate and spread in ways that damage both ecosystem health and human interests, are considered invasive alien species (IAS). For example, a plant or animal transported beyond the ecosystem in which it occurs naturally may multiply out of control, endangering native species in the invaded ecosystem, undermining agriculture, threatening public health, or creating other unwanted—and often irreversible—disruptions.

Perhaps as many as 10 percent of the world's 400,000 vascular plants, have the potential to invade other ecosystems and harm native biota in a direct or indirect way.[418] Invasive species can transform the structure and species' composition of ecosystems by repressing or excluding native species, either directly by outcompeting them for resources or indirectly by changing the way nutrients are cycled through the system.

Invasive alien species have many negative impacts on human economic interests. Weeds

---

[418] Rejmanek and Richardson, 1996.

reduce crop yields, increase control costs, and decrease water supply by degrading water catchment areas and freshwater ecosystems. Pests and pathogens of crops, livestock, and trees destroy plants outright, or reduce yields and increase pest control costs.

1.2. Controlling Invasive Species

Removal of IAS often forms an important component of efforts to restore forest quality to existing forests.

Because of their adaptability and release from their natural prey or enemy, alien species are very difficult to control and can seriously hamper restoration efforts. Often a major factor of restoration is the removal of invasive species; for example, control of *Rhododendron ponticum* from the Himalayas is a major task in many U.K. nature reserves. In recent decades control has typically included herbicides and fire. However, both of these may in turn cause serious damage to the natural landscape unless properly supervised and managed.

In addition, some stakeholders may not wish for an invasive species to be removed, for example, if the species in question provides economic benefits. In such cases, it will be necessary to negotiate trade-offs and see how best to contain the species and ensure that its proliferation can be controlled.

2. Examples

2.1. Invasive Species Introduced Intentionally

In some cases, introduced species can be a significant problem, becoming established in the wild and spreading at the expense of native species and affecting entire ecosystems. Notorious forest examples of these IAS that have negative effects on native biodiversity include various species of Northern Hemisphere pines (*Pinus* spp.) and Australian acacias (*Acacia* spp.) in southern Africa, and *Melaleuca* from South America invading Florida's Everglades National Park. These and many other woody plants were introduced intentionally but had unintended consequences. Of the 2000 or so species used in agroforestry, perhaps as many as 10 percent are invasive.[419] While only about 1 percent are highly so, this includes some popular species such as *Casuarina glauca*, *Leucaena leucocephala*, and *Pinus radiata*.

2.2. Invasive Species Introduced Unintentionally

A worse risk may be the IAS that are introduced unintentionally, such as disease organisms that can devastate an entire tree species that is being used to restore a habitat. The Dutch Elm disease (*Ophiostoma ulmi* and *O. nova-ulmi*) and the American chestnut blight (*Cryphonectria parasitica*) in North America are notorious examples. Pests can have profound economic impact on native forests or plantations, such as gypsy moths (*Lymantria dispar*) or long-horned beetles (*Anoplophora glabripennis*). The economic impact of such pests amounts to several hundred million dollars per year.[420] Much of this economic toll is felt in forested ecosystems, even within well-protected national parks.

2.3. Controlling Invasive Grasses in Hawaii to Promote Restoration of a Unique Ecosystem

In Hawaii, the invasion of alien grasses has dramatically increased the frequency and intensity of fires in dry forests. This has contributed to the conversion of almost all native dry forests to grasslands dominated by alien species. A study was launched to investigate the role that landscape-level herbicide applications followed by native plant reforestation plays in reducing fire fuel load hazards and reversing the cumulative adverse ecosystem level effects of monotypic stands of invasive grasses.[420a] Successful small-scale restoration and alien grass control efforts at the Ka'upulehu Forest, located in North Kona on the Big Island of Hawaii, have provided baseline information necessary to expand restoration efforts to a landscape level. Fountain grass (*Pennisetum*

---

[419] Richardson, 1999.
[420] Perrings et al, 2002.
[420a] Cordell et al, 2002.

setaceum) cover has effectively been reduced from over 90 percent to less than 10 percent using weed-whacking and follow-up herbicide applications. Following this, natural regeneration can be observed in the following sequence: vines, followed by herbs, and then native canopy trees 2 to 3 years after grass removal. Furthermore, it has been documented that native tree canopy cover reduces fountain grass biomass by 50 percent, and native tree growth increases by 50 percent when fountain grass is removed from forested areas.

2.4. Controlling Invasive Species in New Caledonia's Dry Forests

Since Europeans arrived in New Caledonia 150 years ago, over 800 exotic plant species, 400 invertebrates and 36 vertebrates, have invaded the original ecosystem.[421] One notable example is an Indonesian deer (*Cervus timorensis russa*), which provides game for hunters on the island. Because this deer does not have any natural predator, it has multiplied rapidly and become a serious problem as it feeds on dry forest species. In doing so, this deer also hampers natural regeneration by eating the understorey and saplings. Fencing has been used to limit the damage caused by these ungulates. However, because of the high costs involved, this technique has only limited value. Research is also underway to identify more specifically which plants are preferred by the deer in order to better focus which species to use in restoration activities.

3. Outline of Tools

3.1. Prevention

Preventing damage requires predicting which species can cause harm and preventing their introduction, and dealing effectively with the cases in which a species is already causing problems. It is not always simple to distinguish an alien species from an invasive one; taxa that are useful in one part of a landscape may invade other parts of the landscape where their presence is undesirable. The first line of defence is to avoid introducing nonnative species in the first place, so forest restoration should use native species to the maximum extent possible. That said, it may well happen that a nonnative species has characteristics that are especially valued by the local people, for example producing valuable fruit, nuts, or gums. In such a case, special efforts (for example, see point 3.2, below) are required to ensure that the species does not become invasive.

3.2. Containing Purposefully Introduced Species

Great care is required to ensure that such species serve the economic purposes for which they were introduced, and do not escape to cause unanticipated negative impacts on native ecosystems and their biodiversity. One management option would be to plant only sterile forms, so reproduction and spread would be impossible. An even better option, especially when seeking to restore habitats, is to use only native species.

3.3. International Agreements

The 1951 International Plant Protection Convention was established to address some of the issues pertaining to invasive species, and new international programmes have been developed to respond to current serious problems.

3.4. Best Practices for Management of Invasive Species at the Site Level

Best practices for prevention and management of IAS have been designed.[422]

3.5. Global Strategy

A global strategy has been developed by the Global Invasive Species Programme (GISP). This has been widely circulated and provides guidance to countries. It includes aspects of research, capacity building, communications, international cooperation, and quick response.

---

[421] Gargominy et al, 1996.

[422] Wittenberg and Cock, 2001.

These elements are expanded in section 4, below.[423]

4. Future Needs

A comprehensive solution for dealing with invasive alien species as part of forest restoration is needed. Here is a suggested outline of this framework:

1. An effective national capacity to deal with IAS. Building national capacity could include:
- Designing and establishing a rapid-response mechanism to detect and respond immediately to the presence of potentially invasive species as soon as they appear, with sufficient funding and regulatory support
- Appropriate training and education programmes to enhance individual capacity, including customs' officials, field staff, managers, and policy makers
- Developing institutions at the national or regional level that bring together biodiversity specialists with agricultural quarantine specialists to collaborate on implementing national programmes on IAS
- Building basic border control and quarantine capacity, ensuring that agricultural quarantine officers, customs' officials, and food inspection officers are aware of the elements of the biosafety protocol.

2. Fundamental and applied research, at local, national, and global levels: Research is required on taxonomy, invasion pathways, management measures, and effective monitoring. Further understanding on how and why species become established can lead to improved prediction on which species have the potential to become invasive, improved understanding of lag times between first introduction and establishment of IAS, and better methods for excluding or removing alien species from traded goods, packaging material, ballast water, personal luggage, and other methods of transport.

3. Effective technical communications: An accessible knowledge base, a planned system for review of proposed introductions, and an informed public are needed both within countries and between countries. Already, numerous major sources of information on invasive species are accessible electronically, and more could also be developed and promoted, along with other forms of media.

4. Appropriate economic policies: New or adapted economic instruments can help ensure that the costs of addressing IAS are better reflected in market prices. Those responsible for the introduction of economically harmful IAS should be liable for the costs they impose. User rights to natural or environmental resources should include an obligation to prevent the spread of potential IAS, and importers of potential IAS should have liability insurance to cover the unanticipated costs of introductions.

5. Effective national, regional, and international legal and institutional frameworks: Coordination and cooperation between the relevant institutions are necessary to address possible gaps, weaknesses, and inconsistencies, and to promote greater mutual support among the many international instruments dealing with IAS. National, legal and institutional frameworks should be designed along the lines recommended by Shine et al.[424]

6. A system of environmental risk analysis: Such a system could be based on existing environmental impact assessment procedures that have been developed in many countries. Risk analysis measures should be used to identify and evaluate the relevant risks of a proposed activity regarding alien species, and determine the appropriate measures that should be adopted to manage the risks. This would also include developing criteria to measure and classify impacts of alien species on natural ecosystems, including detailed protocols for assessing the likelihood of invasion in specific habitats or ecosystems.

7. Public awareness and engagement: If IAS management is to be successful, the general public must be involved. A vigorous

---

[423] McNeely et al, 2001.

[424] Shine et al, 2000.

public awareness programme would involve the key stakeholders who are actively engaged in issues relevant to IAS, including botanic gardens, nurseries, agricultural suppliers, and others. The public can also be involved as volunteers in eradication programmes of certain IAS, such as woody invasive species of national parks.

8. National strategies and plans: The many elements of controlling IAS need to be well coordinated, and a national strategy should promote cooperation among the many sectors whose activities have the greatest potential to introduce IAS, including the military, forestry, agriculture, aquaculture, transport, tourism, health, and water supply. The government agencies with responsibility for human health, animal health, plant health, and other relevant fields need to ensure that they are all working toward the same broad objective of sustainable development in accordance to national and international legislation. Such national strategies and plans can also encourage collaboration between different scientific disciplines and approaches that can seek new options to deal with IAS problems.

9. Build IAS issues into global change initiatives: Global change issues relevant to IAS begin with climate change but also include changes in nitrogen cycles, economic development, land use, and other fundamental changes that might enhance the possibilities of IAS becoming established. Further, responses to global change issues, such as sequestering carbon, generating biomass energy, and recovering degraded lands, should be designed in ways that use native species and do not increase the risk of the spread of IAS.

10. Promote international cooperation: The problem of IAS is fundamentally international, so international cooperation is essential to develop the necessary range of approaches, strategies, models, tools, and potential partners to ensure that the problems of IAS are effectively addressed. Elements that would foster better international cooperation could include developing an international vocabulary, widely agreed and adopted; cross-sectoral collaboration among international organisations involved in agriculture, trade, tourism, health, and transport; and improved linkages among the international institutions dealing with phytosanitary, biosafety, and biodiversity issues related to IAS and supporting these by strong linkages to coordinated national programmes.

References

Cordell, S., Cabin, R.J., Weller, S.G., and Lorence, D.H. 2002. Simple and cost-effective methods control fountain grass in dry forests (Hawaii). Ecological Restoration 20:139–140.

Gargominy, O., Bouchet, P., Pascal, M., Jaffré, T., and Tourneur, J.C. 1996. Conséquences des introductions d'espèces animales et végétales sur la biodiversité en nouvelle-calédonie. Revue d'Ecologie (Terre et Vie) 51:375–402.

McNeely, J.A., Mooney, H.A., Neville, L., Schei, P., and Wagge J. eds. 2001. A global strategy on invasive alien species. IUCN, Gland, Switzerland.

Perrings, C., Williamson, M., and Dalmazzone, S. eds. 2002. The Economics of Biological Invasions. Edward Elgar Publishing, Cheltenham, UK.

Rejmanek, M., and Richardson, D.M. 1996. What attributes make some plant species more invasive? Ecology 77(6):1655–1661.

Richardson, D.M. 1999. Commercial forestry and agroforestry as sources of invasive alien trees and shrubs. In: Sandlund, O.T., Schei, P.J., and Viken, A. eds. Invasive Species and Biodiversity Management. Kluwer Academic Publishers, Dordrecht, pp. 237–257.

Shine, C., Williams, N., and Burhenne-Guilmin, F. 2000. Legal and institutional frameworks on alien invasive species: a contribution to the Global Invasive Species Programme Global Strategy Document. IUCN Environmental Law Programme, Bonn, Germany.

Wittenberg, R., and Cock, M. eds. 2001. Invasive Alien Species: A Tool Kit of Best Prevention and Management Practices. CAB International, Wallingford, UK.

50
First Steps in Erosion Control

Steve Whisenant

Key Points to Retain

Although natural erosion occurs on many landforms, accelerated erosion (caused by people's activities) is the focus of most restoration efforts.

Increasing the cover of vegetation or litter, preferably both, is the most effective strategy for reducing erosion.

1. Background and Explanation of the Issue

Forest landscape restoration requires the stabilisation of soil resources. The loss of soil to erosion leads to irreversible changes and degrades physical, chemical, and biological properties. Although natural erosion occurs on many landforms, accelerated erosion (caused by human activity) is the appropriate focus of most restoration efforts. Wind erosion is a serious problem that may occasionally be reduced by planting trees. Hill slope erosion, wind erosion, and mass movement (slump erosion) are common problems and are the primary focus of this chapter.

1.1. Understanding the Variety of Erosion Processes

Hill slope erosion is caused by the direct impact of raindrops on the soil surface, overland (inter-rill) flows, and small channel flows.[425] Overland flow begins as surface depressions are filled and when rain falls faster than water infiltrates into the soil. Although overland flow is often viewed as a sheet of water flowing over the surface, it typically includes numerous shallow, but easily definable channels, called rills. The relative amount of sediment detached and transported by inter-rill flow is small compared to splash and rill erosion. Rills are small enough to be removed by normal tillage operations, but may become too large (gully) to remove with tillage. Rill erosion is substantially more erosive than overland flow and is a function of hill slope length, depth of flow, shear stress, and critical discharge. Rill erosion starts when the eroding force of the flow exceeds the ability of the soil particles to resist detachment. Flow depth and velocity increase substantially where surface irregularities concentrate overland flows into rills. Once rills are established, the concentrated flow develops more detachment force, and the rill formation process is enhanced. Rill development moves upslope as headcuts. Some rills develop rapidly and become more deeply incised. These master rills become longer and deeper than their neighbours. Occasionally flows from adjacent rills break into master rills by eroding the boundary between them. As the rill flow becomes concentrated toward master rills, previously parallel rills develop a recognisable dendritic drainage pattern. As rills coa-

---

[425] Brooks et al, 1991.

lesce, flow concentrations and velocity increase until the more deeply incised rills become gullies.

Wind erosion is greatest on fine soil particles such as silt, clay, and organic materials. This wind-driven sorting increases the proportion of coarse materials in wind-eroded sites. Wind-blown particles are moved in three ways: (1) saltation, the bouncing of particles across the surface; (2) suspension in wind; and (3) surface creep, the movement of larger particles caused by the pushing action of saltating particles striking larger particles.[426] The amount of wind erosion is affected by soil erodibility, surface roughness, climate, unsheltered distance of soil exposed to wind, and vegetation cover. Thus wind erosion is reduced by rougher soil surfaces, lower wind speed at the soil surface, and more plant or litter coverage of the soil surface.

Mass movement is the downward movement of slope-forming materials without the primary assistance of a fluid. It occurs on steep slopes under the influence of gravity, often exacerbated by the weight of water in the soils. Mass movement occurs on steep slopes when deforestation, mining, fire, overgrazing, construction, or cultivation disrupts the landform–climate–vegetation equilibrium by removing the vegetation. Well-vegetated slopes generally move downward much slower than less vegetated slopes.[427] Plants, especially woody plants with strong, deep roots, greatly increase soil strength, providing a stabilising effect on the slope. In some cases, the plants also transpire significant quantities of water from the slope, thus reducing the weight that contributes to mass movements.

1.2. Protection Against Wind and Water Erosion

Increasing the cover of vegetation or litter, preferably both, is the most effective strategy for reducing erosion. Plants protect the soil with their canopy, add litter to the soil surface, and stabilise the soil with their roots. Litter on the soil surface reduces erosion. Soil erosion, from water or wind, is reduced with strategies that accomplish the following:

1. Maintain or establish a cover of vegetation, especially when erosion is most probable. Although perennial plants are most desirable, annual plants may provide critical, short-term seasonal protection.

2. Create a ground cover of litter, rocks, woody debris, erosion matting, or other materials until vegetation becomes established.

3. Increase soil surface roughness with above-ground structures or soil surface manipulations (such as pits or furrows) that are perpendicular to water or wind flows. This increases infiltration, reduces water velocity, and increases the wind speed necessary to initiate saltation.

4. Reduce fetch length of unobstructed slope surfaces. This reduces the ability of water or wind to detach and transport soil particles and minimises opportunities for overland flows to coalesce and form larger rills and gullies.

5. Incorporate biomass into the soil where possible. Like the previous strategies, it increases the rate and capacity of infiltration, thus reducing the amount of water available for erosion. Biomass incorporation also stimulates plant growth and soil biotic development that improve soil structure and nutrient cycling.

1.3. Additional Protection Against Mass Movement of Steep Slopes

Each of the previous strategies provides some protection against mass movement. Two additional strategies provide specific protection for slopes susceptible to mass movement.[428]

1. Steep slopes susceptible to mass movement are most effectively stabilised with trees and shrubs that have strong woody root systems. Significant taproot development below the slip surface greatly increases slope shear strength, which has a strong slope-stabilising influence.

---

[426] Toy et al, 2002.
[427] Morgan and Rickson, 1995.
[428] Morgan and Rickson, 1995; Whisenant, 1999.

FIGURE 50.1. Rock terraces constructed in Sichuan Province, China to reduce runoff and soil erosion during the establishment of trees. The availability of labour and the local presence of rocks made this scheme possible in this situation. (Photo © Steve Whisenant.)

2. High transpiration rates reduce susceptibility to mass failure by reducing the amount of water in the soil. Water increases the slope shear stress that causes mass movement of a slope. Transpiration increases as the leaf area of a particular species becomes higher. Thus, transpiration losses of new plantings are often increased with higher planting densities or larger trees. It is also important to select species that transpire during the highest water season when mass movement is most probable.

2. Examples

2.1. Slope Stabilisation in Sichuan Province, China

In the upper watershed of the Yangzi River, steep, deforested slopes of unconsolidated materials are very susceptible to mass movement. To reduce mass movement and soil erosion into the Three Gorges Reservoir, the Sichuan Forestry Institute, and several cooperating organisations initiated forest landscape restoration. The goal was to reforest cultivated fields and deforested slopes within this watershed. They created landscapes with fuel wood, medicinal plants, tree crops, and Chinese peppers around the villages. The landscape matrix, between villages, consisted of slope-stabilising trees that will provide wood resources in the future. Many of the long, steep slopes were terraced to increase both surface roughness and infiltration. Many of the terraces were reinforced with rock walls built by a readily available labour force in this region (Fig. 50.1). This created a stable environment for forest landscape restoration that should provide soil coverage, organic materials, and increased shear strength from the woody roots. With careful management, the forest vegetation will stabilise the slopes indefinitely.

2.2. Stabilising Mobile Dunes in Shaanxi Province, China

Highly mobile sand dunes were covering productive farms in northern China, near Yulin. These dunes, created by overgrazing of sandy lands to the north, were moving southward into productive agricultural lands. Local scientists developed a simple, practical strategy for dune stabilisation. Dormant willow (*Salix* spp.) branches cut to 1-m lengths were stuck vertically into the dune crests with only about 1 decimetre (dm) above the soil level. The willow branches set root and began a rapid growth that stabilised the dunes and captured additional wind-blown soil and organic particles (Fig.

FIGURE 50.2. Dormant willow (*Salix* spp.) stems (1–2 m long) were planted into active dunes near Yulin, Shaanxi Province, China, with only 5 to 10 cm remaining above the soil surface. They established rapidly and began to stabilise the dunes by capturing sand and other windblown materials. (Photo © Steve Whisenant.)

50.2). Combined with an effective ban on grazing by sheep and goats, this was a highly effective dune stabilisation programme that protected the farmland. Policies that improved grazing practices on the sand sources (in the northern desert) also diminished the volume of sand reaching the farms.

2.3. Reducing Off-Site Erosion with Watershed Restoration in Niger

Laterite plateaus in the Sahel of southwest Niger contain banded woody vegetation aligned on contours of gentle slopes. With degradation of these bands, caused by woody harvesting and browsing animals, less water is retained on the plateaus. This reduces vegetative growth and significantly increases runoff from the plateaus. This additional runoff during storm events leads to serious erosion and flooding in adjacent villages and farm fields. Reducing these off-site erosion problems required restoration of the vegetation and natural hydrologic regime of the plateaus.[429] This was accomplished by planting rapidly growing shrubs into microcatchments on the plateau. The catchments held sufficient water to allow establishment of shrubs. These shrubs produced ground cover, litter, shade, wind speed reduction, and root systems that fed soil organisms.

These changes dramatically increased infiltration, water retention, nutrient cycling, and energy flows into the soil. This effectively prevented erosion and flooding problems on the plateau as well as in the villages and farms surrounding the plateaus.

3. Outline of Tools

The most effective tools for reducing erosion are governmental policies and land management practices that maintain healthy vegetation and a cover of duff, litter, or woody debris.[430] Though conceptually simple, this protects the soil from raindrop impact, increases infiltration, reduces runoff, reduces saltation, and significantly reduces soil erosion. Once the area has been cleared, reestablishing a ground cover prior to the next erosion season is essential.

3.1. Grazing Management that Maintains Ground Cover

Poor grazing management probably contributes to more land degradation than any other practice, even in forested environments. Grazing practices that allow plants to periodically grow and reproduce will stabilise soil resources more effectively. Recently planted

---

[429] Manu et al, 1999.

[430] Whisenant, 1999.

FIGURE 50.3. Following a wildfire in Chipinque Ecological Park outside Monterrey Mexico, the remaining woody debris was used to create above ground obstructions to reduce erosion, hold water, and increase the natural recruitment of trees. (Photo © Steve Whisenant.)

forests may require protection from grazing animals for several years.

3.2. Wood Harvesting Schedules, Methods, and Spatial Patterns that Maintain Soil Coverage and Root Biomass

Fuel wood, timber, or any other type of wood harvesting must be scheduled and spatially arranged to maintain good soil coverage of plants and litter. Uneven aged and mixed species' forests are more easily harvested in small areas, which reduces the size of disturbed areas that can contribute to soil loss. Harvesting methods that reduce the presence of skid trails will reduce the concentration of water flows that increase erosion problems. Practices that leave more leaves, duff, and woody debris on the surface will reduce erosion hazards.

3.3. Local Materials for Soil Protection

Ultimately, perennial plants are the most effective and practical means of protecting the soil. However, it is often necessary to provide a "window of opportunity" during which plants can be established. Soil protection is essential and may be obtained with the use of locally available organic materials. Organic materials can be incorporated into the soil or placed on the surface to reduce erosion, increase infiltration, and moderate temperature extremes. Examples of organic materials include woody debris following wildfire (Fig. 50.3), animal waste, cotton gin trash, coconut fibre, olive pulp, and other readily available materials that can be used to protect the soil surface. Gravel or rocks may also be used as above-ground obstructions or to protect the soil surface.

3.4. Soil Surface Manipulations or Above-Ground Obstructions

Features that roughen the soil surface have the potential to reduce wind and water erosion while increasing soil water available for plant growth.[431] Pits, microcatchments, furrows, or cultivation may be used in appropriate circumstances to roughen the soil surface. Rocks, gravel, terraces, soil bunds, or plant materials are potential above-ground obstructions where available. These surface changes contribute to additional plant growth that establish positive feedback improvement systems that continue to increase infiltration, water storage, and nutrient cycling. This leads to still more functional improvements on the site.

[431] Whisenant, 1999.

3.5. Soil Conditioners (Polyacrylamides)

Polyacrylamides (PAMs) are synthetic polymers that bind soil particles and reduce surface crusting, thus increasing pore space and infiltration. They can produce dramatic, but short-lived, infiltration increases, with decreased erosion. They are still too expensive for widespread application during forest restoration, but may be practical in high-priority areas.

4. Future Needs

4.1. Policies that Discourage Degrading Forest Management Practices

Government policies may increase soil erosion from forests or they can be crafted to encourage the restoration and management of forest landscapes that provide important goods and ecological services without accelerating soil loss. Policies that prevent the complete removal of trees on the steepest slopes have the greatest impact on soil loss.

4.2. Improved Understanding of Watershed-Scale Processes

Forest restoration programmes are usually planned based on the attributes and objectives of specific fields, ownership units, or forest openings. This approach effectively assumes that the sites are functionally isolated from other parts of the landscape or watershed. This can lead to problems since each part of a landscape is continuously gaining and losing water, nutrients, soil, organic materials, and seed. Organic materials, landform, or microtopographic features control these movements of water, nutrients, and organic materials. A greater recognition and understanding of these resource fluxes can be used to great advantage in forest landscape restoration.

References

Brooks, K., Folliott, P.F., Gregersen, H.M., and Thames, J.L. 1991. Hydrology and the Management of Watersheds. Iowa State University Press, Ames, Iowa.

Manu, A., Thurow, T.L., Juo, A.S.R., and Zanguina, I. 1999. Agroecological impacts of five years of a practical programme for restoration of a degraded Sahelian watershed. In: Lal, R., ed. Integrated Watershed Management in the Global Ecosystem. CRC Press, New York, pp. 145–163.

Morgan, R.P.C., and Rickson, R.J. 1995. Slope Stabilization and Runoff Control: A Bioengineering Approach. E. and F.N. Spon, New York.

Toy, T.J., Foster, G.R., and Renard, K.G. 2002. Soil Erosion: Process, Prediction, Measurement, and Control. John Wiley and Sons, New York.

Whisenant, S.G. 1999. Repairing Damaged Wildlands: A Process-Oriented, Landscape-Scale Approach. Cambridge University Press, Cambridge, UK.

Additional Reading

Lal, R. 1990. Soil Erosion in the Tropics: Principles and Management. McGraw Hill, New York.

Satterlund, D.R., and Adams, P.W. 1992. Wildland Watershed Management, 2nd ed. John Wiley and Sons, New York.

Wu, X.B., Thurow, T.L., and Whisenant, S.G. 2000. Fragmentation and changes in hydrologic function of tiger bush landscapes, southwest Niger. Journal of Ecology 88:790–800.

51
Restoring Forests After Land Abandonment

José M. Rey Benayas

Key Points to Retain

Land that is abandoned for a number of ecological and socioeconomic reasons can regenerate either naturally or through management interventions.

Significant public and private funds are being invested in abandoned land reforestation, often without good planning.

Abandoned lands offer a huge potential for restoration.

Restoration of abandoned land must be viewed as an investment in ecosystem goods and services.

1. Background and Explanation of the Issue

Globally, degraded land due to agricultural activities is estimated at about 12,400,000 km^2.[432] In addition, large areas of cropland and pasture land have been abandoned during the last few years for different ecological and socioeconomic reasons. Ecological factors leading to land abandonment are in many cases ultimately the result of mismanagement at a landscape level (e.g., unadapted agriculture and overgrazing), and include productivity loss or the land exceeding cattle carrying capacity. Socioeconomic factors leading to land abandonment include a loss in farmland productivity, diversion of labour toward the industrial and service sectors, reduced subsidies for many crops and regions, and subsidised set-aside programmes.

These and other deforested areas can be (1) left to undergo secondary succession or passive restoration or (2) subjected to active restoration processes, mostly consisting of planting and managing native shrubs and trees. In the world, land abandonment and passive restoration have restored much more, and at a lower cost, than active restoration. However, active restoration is needed when the abandoned land suffers continuous degradation (e.g., soil erosion in dry regions), when the natural vegetation cover cannot recover in the area (e.g., abandoned cropland colonised by dense weeds in the tropics), and when accelerating secondary succession is desirable (e.g., reforestation of abandoned Mediterranean cropland). An additional benefit of active restoration is the creation of labour associated with ecosystem management in rural areas.

This issue is also important because public and private funds are being invested in abandoned land reforestation. From a holistic perspective, these actions must be viewed as the restoration of the world's natural capital, the services that ecosystems provide to humankind. Thus, research is needed to optimise the investment-benefit ratio.

[432] Bot et al, 2000.

FIGURE 51.1. A plot of abandoned agricultural land in a Mediterranean landscape that was actively revegetated with *Quercus ilex* 12 years ago. (Photo © Jose M. Rey Benayas.)

2. Examples

A large number of worldwide examples of ecosystem restoration are related to land abandonment and associated secondary succession. The scientific and technical literature reports a number of case studies that highlight both successes and failures. Typically, secondary succession has led to renewed functional ecosystems in scenarios where abandoned cropland and pastures had not been intensively used in the past, vegetation colonisation and growth was not limited by climate and/or soil constraints, the abandoned land was relatively small in size and there were remnants of natural vegetation nearby. Some examples are related to tropical slash-and-burn fields and paddocks that have turned to forest, Mediterranean mountain pastures and cropland that have turned to forest or shrubland, and abandoned rural areas in Africa that have turned to savannah or dwarf shrubland.

Failures are reported for abandoned lands where the environmental conditions are unfavourable to natural regeneration. Examples include all areas under desertification in the arid and semiarid regions of the world, large tropical paddocks with very compacted ground, and abandoned tropical cropland colonised by a dense carpet of weeds such as *Saccharaum spontaneum* that impedes the establishment of natural vegetation. For instance, seedling mortality in abandoned tropical pastures has been found to be above 50 percent, whereas it drops to less than 25 percent with appropriate management.[433] Active restoration is essential where ecosystem breakdown has occurred. The functioning of natural ecosystem processes such as seed dispersal are key factors to address when assessing restoration requirements. Many moist tropical forests depend on animal dispersal (as much as 90 percent of tree species). In the eastern rainforests of Madagascar, arboreal lemurs are essential for forest maintenance and regeneration. As lemur populations are decimated, most of the former rainforest regions in Madagascar are now severely degraded, representing an arrested succession dominated by alien species.[434]

2.1. Planting in Euro-Mediterranean Environments

In European Mediterranean environments public funds from the European Union have been available to encourage farmers to turn their cropland into forest plantations (Fig. 51.1). In these ecosystems, different abiotic and biotic factors hinder the establishment and growth of shrubs and trees, and some research

[433] Hooper et al, 2002.
[434] Holloway, 2000.

has been devoted to study how plantation projects benefit from appropriate management. The mortality of native *Quercus* species' seedlings during the first year is often above 60 percent if nothing is done to facilitate their establishment, and around 10 percent if management is applied.[435] Further, some studies have shown that appropriate management may provide a rapid plot cover by the introduced seedlings and reproductive saplings of slow-growing species by the seventh year. For instance, it has been reported for an experimental *Q. ilex* plantation in central Spain that, after 3 years of management—artificial shading and summer irrigation—and six additional years of interrupted management, the plot cover attained by the managed seedlings was 50 percent higher than that attained by the unmanaged seedlings; additionally, 15 percent of the managed seedlings produced acorns, whereas only 1.5 percent of the unmanaged seedlings were capable of producing seeds.[436]

2.2. Passive and Active Restoration in Mosaic Rainforest Landscapes of Latin America

Landscape mosaics are typical of many rainforest areas of Latin America, consisting primarily of a mix of cleared areas, secondary forest, and limited residual patches of primary forest. A portion of the cleared area is agriculturally marginal, and in many cases is being abandoned. Natural regeneration of forest cover from neighbouring seed sources on this land is typically rapid. For instance, in cloud forest landscapes in Oaxaca (Mexico), it has been reported that abandoned paddocks attain, after 35 years, an average of 63 percent of the tree basal area that is characteristic of the mature forests in the region.

However, species' diversity after natural regeneration is usually low, with stands typically dominated by a few fast growing pioneer species. Natural regeneration of a species' mix more typical of a primary forest will only occur over the long term. Planting seedlings of interior forest species after land abandonment could sharply accelerate the process of restoration of complex communities. Pioneer stands or monocultural plantations may be enriched with seedlings of late-successional animal-dispersed trees, or initial plantings could be done with mixes of late-successional and pioneer species. Active ecological research related to this topic is being undertaken in a few places such as the Highlands of Chiapas (Mexico). There, broad-leaved tree species have declined because they are intensively harvested by the local Mayan communities for firewood, and pines are consequently in expansion. Seedlings of the broad-leaved trees are being introduced at the fringe between the pine-dominated forests and clear cuts, with survival rates higher than 50 percent after 3 years due to the positive effect of pines on the introduced seedlings. However, pines may inhibit establishment of native vegetation in some environments.

3. Outline of Tools

The tools at hand for favouring restoration of abandoned land are a mix of ecological and socioeconomical actions (and sometimes "inaction") and techniques. Passive restoration is by far the main force that turns abandoned land into "original" or healthy ecosystems.[437] It has the advantage of being cheap. On the other hand, the disadvantages include that it can be very slow in low productive ecosystems, involves few people (no labour is needed), and may turn into a more degraded land or auto-succession loops. Secondary succession can be aided by simply eliminating grazing in certain areas after agreement with local users and land managers. Fencing can be used for this purpose, although this can add substantially to the cost in some situations.

3.1. Active Restoration Techniques

A number of techniques have been proposed in active restoration programmes in those parts of the world where shortage of water availability

---

[435] Rey Benayas, 1998.
[436] Rey Benayas and Camacho, 2004.
[437] Running, 2003.

is a major limiting factor for seedling establishment of native shrubs and trees. These techniques include artificial shading, irrigation in the dry season, elimination of herb competition, use of gels that absorb and very slowly release water, ground preparation to increase infiltration, and microtopography modification to canalise run-off toward the reforested plots. When nutrients are limiting, manure and compost from agricultural, industrial, or sewage plants' residues have been utilised. Another technique that has successfully been used is planting the seedling below the canopy of naturally established nurse shrubs, which provide an ameliorated microenvironment for the introduced seedlings. Many of these techniques are discussed in more detail in other chapters of this book. It should be noted that the choice of technique will need to be determined by the climatic, biophysical, and socioeconomic conditions.

3.2. Socioeconomic Tools

Socioeconomic tools can also be passive and active. In a free market economy, the ratio between benefits and costs of livestock or agricultural production has triggered the abandonment of large extensions of land throughout the world. In other cases, removal of perverse subsidies—such as elements of the Common Agricultural Policy in Europe that has encouraged farming on uneconomic and marginal lands—could help stimulate natural regeneration. Active financial tools that foster abandonment of livestock grazing and agricultural production also exist.

An innovative and promising tool is payment for the environmental services that forests provide to humans, which favours forest conservation first and encourages forest restoration second. This programme is already widely applied in Costa Rica (see "Payment for Environmental Services and Restoration").

Another tool is to subsidise set-aside programmes for agricultural lands and to convert those into forest plantations or restore the natural vegetation. This tool has been widely applied in the European Union (EU) Mediterranean countries. However, its success has been limited by the fact that the subsidies have encouraged some landowners to plough and reforest lands that had already been abandoned and were undergoing passive restoration.

Further socioeconomic tools—which are still very marginal—are related to the links between active restoration and environmental education and local sustainable development. For instance, the reforestation of vast extensions of abandoned land or the enrichment of secondary forests in developing countries requires the creation of a labour force and small industries such as specialised nurseries.

4. Future Needs

4.1. Evaluating Ecosystem Values

Before initiating any restoration programme after land abandonment, it is necessary to answer this question: Active or passive restoration? The answer necessarily goes through an evaluation of costs and benefits of the various options. We must never forget that the environmental benefits that humans receive from functional ecosystems or the loss of these benefits is part of the balance. We need better knowledge and awareness of what could enhance natural succession after abandonment, and the temporal terms, in various ecosystems. Natural regeneration should be properly monitored and mapped by field work and remote sensing and geographical information system (GIS) techniques. We must also take into account the potential social benefits of active restoration, particularly in developing countries. There is a need for scientific research to correctly assess such benefits.

4.2. Rethinking the Concept of Reforestation

It seems that we need a different concept of reforestation of abandoned cropland where plant production is limited as it occurs, for example, in dry Mediterranean regions. Nowadays, these reforestation efforts are based on extensive plantations of aligned trees, often of exotic species, that provide artificial monocultures that are rarely managed. Restoration ecology and forest landscape restoration present more integrated approaches to restoration. After land abandonment, the reforestation approach should be replaced by little, dense, diverse, strategically placed, and wisely managed reforested patches. These patches would actually be islands of functional ecosystems in a sea of intensively used or abandoned land, thus being compatible with other land uses (e.g., livestock grazing or crop production) and passive restoration in their surroundings. The islands would act as "sources and traps" of propagules of different species of plants and animals since many organisms would find refuge and food. These biodiversity reservoirs could function as nuclei for passive restoration of large extensions in the world. Such experiences need to be started rapidly and their lessons shared and replicated widely.

References

Bot, A.J., Nachtergaele, F.O., and Young, A. 2000. Land Resource Potential and Constraints at Regional and Country Levels. Land and Water Development Division, FAO, Rome, Italy (available on line at ftp://ftp.fao.org/agl/agll/docs/wsr.pdf).

Holloway, L. 2000. Catalysing Rainforest Restoration in Madagascar. In: Lorenco, W.R., and Goodman, S.M., eds. Diversity and Endemism in Madagascar. Orstom Editions, Paris.

Hooper, E., Condit, R., and Legendre, P. 2002. Responses of 20 native tree species to reforestation strategies for abandoned farmland in Panama. Ecological Applications 12:1626–1641.

Rey Benayas, J.M. 1998. Growth and mortality in *Quercus ilex* L. seedlings after irrigation and artificial shading in Mediterranean set-aside agricultural lands. Annals of Forest Sciences 55: 801–807.

Rey Benayas, J.M., and Camacho, A. 2004. Performance of *Quercus ilex* saplings planted in abandoned Mediterranean cropland after long-term interruption of their management. Forest Ecology and Management 194:223–233.

Running, S.W. 2003. Climate-driven increases in global terrestrial net primary production from 1982 to 1999. Science 300:1560–1563.

Additional Reading

Bakker, J.P., van Andel, J., and van der Maarel, E. 1998. Plant species diversity and restoration ecology: introduction. Applied Vegetation Science 1:3–8.

Perrow, M.R., and Davy, A.J. 2002. Handbook of Ecological Restoration. Vol. 2. Restoration in Practice. Cambridge University Press, Cambridge.

Temperton, V.M., Hobbs, R.J., Nuttle, T., and Halle, S. 2004. Assembly Rules and Restoration Ecology. Bridging the Gap Between Theory and Practice. Island Press, Washington.

52
Restoring Overlogged Tropical Forests

Cesar Sabogal and Robert Nasi

Key Points to Retain

Overlogged forests are degraded but nevertheless important. They may continue to be a source of timber and supply an important amount of forest products, particularly for local people whose livelihoods depend on their extraction.

One aspect of restoration would be to prevent adding more overlogged areas by implementing sound logging, silvicultural, and management practices.

There is an urgent need to appropriately disseminate the existing strategies, approaches, and techniques that are most appropriate for forest restoration of overlogged forests.

1. Background and Explanation of the Issue

1.1. Logged-Over Forests and Logging Impacts

Poor logging practices using heavy machinery are a prominent reason for the degradation of tropical forests. The term *overlogged forest*[438] is usually applied to this situation.

Logged over forests show a wide range of conditions according to the degree of direct or indirect disturbance, which depends on the logging system (Box 52.1), the intensity and frequency of timber extraction, and the quality of supervision and control. The amount of damage sustained by residual stands increases generally with the size of the machinery used and with increasing volumes of timber harvested.

Logging operations inevitably impact soils, stream flows, remaining vegetation, fauna, and biodiversity in general,[439] creating a more heterogeneous structure with patches of felling gaps, skid trails, etc. (Box 52.2). Soil impacts and damage to the residual forest all increase with increasing logging intensity. High extraction rates, by creating big canopy openings, favour fast-growing pioneer species or undesirable, weedy species (such as vines) and induce desiccating conditions. Moreover, large openings are subject to invasion by lianas that can be an obstacle to tree regeneration, and in heavily logged forests such openness also increases fire risks and propagation (particularly during long periods of drought). High extraction rates also result in a depleted residual stand that will not

---

[438] Overlogged forests in Asia are defined as natural primary or older secondary natural forests that have been badly damaged by overcutting and poor logging methods and have resulted in impoverished and ecologically unstable stands. If left untreated, these forests are unable to restore their original state within a reasonable period of time, or even to recover enough to provide the normal services of a forest (Banerjee, 1995).

[439] Bruijnzeel and Critchley, 1994; Fimbel et al., 2001; Frumhoff, 1995; Grieser Johns, 1997; Haworth, 1999; Putz et al., 2002; Stadtmüller, 1994; Thomson, 2001; Weidelt and Banaag, 1982; Woods, 1989.

> **Box 52.1. Logging Systems in Tropical Forests**
>
> Two main logging systems are usually distinguished:
>
> *Monocyclic logging* represents the removal of up to 100 percent of the commercially valuable stocking from a forest at relatively long intervals. The interval between harvesting operations is typically equal to the maturation period of the main species of trees felled, the so-called rotation period, which may be as long as 60 to 80 years or more. Because monocyclic logging removes not only mature but also semi-mature trees, a relatively large proportion of the forest may be affected. The volume of timber removed during monocyclic operations may be as high as $120\,m^3$/hectare in certain Southeast Asian forests, although more commonly the harvested volumes tend to converge around a value of about $60\,m^3$/hectare. The result of such intense logging is the creation of relatively large gaps in the canopy, stimulating light-demanding species in the regrowth.
>
> *Polycyclic logging* is the selective removal of only the largest individuals of desirable species. The objective is to wait for a sufficient number of trees to reach maturity, and then to remove these alone. Compared with monocyclic logging, fewer trees and a lower volume of timber is harvested, but the intervals between harvests are shorter. In some polycyclic systems, such as the CELOS (Centre for Agricultural Research in Surinam) system developed for Surinam, or the Tebang Pilih system advocated in Indonesia, this interval may be as short as 20 to 25 years. Volumes of wood removed are typically 20 to $30\,m^3$/hectare per coupe.
>
> Monocyclic logging inevitably causes more disturbances to the forest canopy and the soil surface than polycyclic systems. Typically, for every tree that is logged, a second is destroyed and a third is damaged beyond recovery. Under unimproved, standard management practices, polycyclic logging may cause damage to 15 to 35 percent of the remaining trees, whereas under monocyclic logging this figure may increase to 40 to 60 percent.
>
> (Adapted from Bruijnzeel and Critchley, 1994.)

> **Box 52.2. Biodiversity Impacts of Logging on Tropical Forests**
>
> The most severe impacts at the *landscape level* result from indirect consequences of logging such as increased access to remote areas, fragmentation, and altered fire regimes. Changes in the size, spatial distribution, and connectivity of habitat patches alter species' distribution patterns, forest turnover rates, and hydrologic processes. Most *ecosystem-level* impacts are a direct consequence of logging activities. The structural impacts of logging change the relative proportions of life forms and biogeochemical stocks, as well as nutrient and hydrologic cycling, productivity, and energy flows. At the *community level*, logging can substantially change the characteristics, composition, and trophic structure of forest stands. The most obvious impact is the change in proportions of successional stages in forest stands. Key ecological processes such as pollination, herbivory, seed dispersal, and predation are all affected by logging especially when it is more intensive. The most obvious *species-level* impact of logging is on the abundance and age/size distribution of harvested and damaged trees. The *genetic* component of biodiversity is likely to be the most sensitive of all components to logging because of reductions in effective population size and interruptions in gene flow.
>
> (Source: Putz et al. 2002)

be able to recover an acceptable timber yield within a reasonable and economically profitable harvesting cycle period.[440] The extraction pressure on a set of high-value species may cause a dysgenic trend (removal of large trees with each cut leaving genetically inferior trees for future crops and seed sources).[441]

Other dramatic, indirect impacts are associated with logging wherever social pressures (e.g., by colonists) and institutional weaknesses (e.g., law enforcement) prevail. Under these conditions logged-over forests are frequently subject to further disturbance, leading to increased degradation or even conversion to other land uses. Land invasions, illegal logging, poaching, and fire are amongst the most serious threats faced by forest owners/managers. This tragedy is at the crux of most of the debate on sustainable forestry in tropical regions and will certainly last for a while.

1.2. Why Restore Logged-Over Forests?

Overlogged forests are degraded but nevertheless important. They may continue to be a source of timber and supply an important amount of forest products, particularly for local people whose livelihoods depend on their extraction. Such forests may still provide special biodiversity conservation services or be important for other environmental services (e.g., water, carbon). With alarming rates of landscape fragmentation, these remnant forest resources—more and more frequently found as patches of logged-over/degraded primary forests—are becoming critical components of restoration strategies. Logged-over forests may also represent a valuable means of stabilising small-scale colonists in agricultural frontier areas.

Objectives for restoration of overlogged forests must be set by societal demand and encompass both social and ecological goals. They will depend on the degree of degradation, the desired future condition as defined by the landowner or land user, and the (biophysical and socioeconomic) context at the landscape level. The restoration work can either emphasise the protection functions for biodiversity recovery and other environmental services (e.g., water, carbon uptake) or privilege the potential for production functions of the ecosystem (safety net functions, commercial production, or multiple-use) or both.

1.3. Improving Logging Practices

One aspect of restoration would be to prevent adding more overlogged areas by implementing sound logging, silvicultural, and management practices. Good planning and careful implementation of timber harvesting operations substantially contribute to reduce the negative impact of bad logging. *Reduced-impact logging* (RIL), a term now widely used, encompasses the implementation of a series of pre- and postlogging guidelines designed to protect advanced regeneration (i.e., seedlings, saplings, poles, and small trees) from injury, to minimise soil damage, to prevent unnecessary damage to nontarget species (e.g., wildlife and nontimber forest products), and to protect critical ecosystem processes (e.g., hydrology and carbon sequestration). The Model Code of Forest Harvesting Practices published by the United Nations Food and Agriculture Organisation[442] has been widely used as a reference to elaborate similar sets of harvesting guidelines.

The RIL techniques constitute a substantial step toward sustainable management. A further improvement in RIL is the integration of silvicultural principles, guidelines, and practices.[443] These techniques should in particular aim to keep extraction rates below an acceptable threshold compatible with timber yield capability, limit the impact of harvesting on tree species' diversity and composition, and maintain timber species' populations by reducing the impact of logging on their ecology.[444]

---

[440] Applegate et al., 2004.
[441] ITTO, 2002.
[442] Dykstra and Heinrich, 1996.
[443] Wadsworth, 1997.
[444] Sist et al., 2003.

2. Examples

2.1. Restoration of Degraded Forests by Enrichment Planting

Overlogging and forest fires, or combinations of the two, have created millions of hectares of medium to heavily disturbed forests in many parts of Southeast Asia.

One of the main technical approaches for restoration in these forests has been the establishment of enrichment plantings, either in lines or in gaps. Line planting has been used if the surrounding trees are small (≤ 10 m). The gap planting method is especially suitable when the surrounding trees are taller (>10 m). In practice, line and gap planting methods using artificial or natural regeneration complement each other.

In Indonesia, where degraded forests resulting from unsustainable management and wild fires account for over 20 million hectares, the International MOFEC (Ministry of Forestry and Estate Crops)—Tropenbos Kalimantan Project developed in 1987 a research programme based on indigenous Dipterocarpaceae species aiming at rehabilitating these heavily disturbed forests.[444a] Line planting experiments were conducted in the Wanariset Research Forest (East Kalimantan), mainly consisting of enrichment plantings with dipterocarps. Several techniques were employed for the production of planting stock, including seedling production in nurseries from seeds, wildlings collected in the forest, and seedlings derived from cuttings raised in the nursery. Vegetative propagation of dipterocarp species, especially stem cuttings' production, gave promising results and is being used for large-scale plantations, for instance, Meranti (*Shorea* spp.) plantations with stem cuttings in Long Nah, East Kalimantan.

Other practical experiences with enrichment planting were conducted in Banjarmasin, South Kalimantan, under the Reforestation and Tropical Forest Management Project financed by the Finnish International Development Agency.[445]

2.2. Rehabilitation of Log Landings and Skid Trails[446]

Unplanned logging using heavy machinery causes excessive damage to the soil, watercourses, and vegetation, particularly through the opening up of harvesting infrastructure (roads, log landings, and skid trails). The rehabilitation of most impacted areas has been attempted in different ways. In heavily logged dipterocarp forests, skid trails and log landings represent a significant proportion (up to 40 percent) of the total area. This level of disturbance also affects the recovery of the residual stand, prolonging the next cutting cycle from 20–30 to 40–50 years.

In Sabah, Malaysia, two rehabilitation techniques were tried for planting dipterocarps on log landings and skid trails: direct open planting of seedlings, and planting a nurse crop with subsequent underplanting of dipterocarps. For open planting, in general species with drought and heat tolerance and resistance to pests and diseases should be used. The major drawback of this option is that the dipterocarp seedlings grow too slowly to provide protection from erosion or to rehabilitate the damaged soil. Therefore, the technique is most suitable for skid trails where flanking vegetation provides some remnant canopy and where natural regeneration of pioneer tree species along the skid edges provides organic matter and helps ameliorate the soil.

An alternative to open planting, especially for large open areas, is to plant fast-growing native pioneer trees on the site first and then underplant with dipterocarp seedlings. Pioneer trees are better adapted to the open conditions of degraded sites, and they grow much faster than dipterocarps. Once the dipterocarp seedlings have established, the nurse trees should be thinned to allow increasing amounts of light to reach the seedlings.

Using this system has several advantages: (1) the nurse crop trees are fast-growing, allowing them to compete well with vines and climbing bamboo, and reduce soil erosion; (2) rapid

---

[444a] Effendi et al. 2001.
[445] Adjers et al., 1995; Korpelainen et al., 1995; Tuomela et al., 1996.
[446] Source: Nussbaum and Hoe, 1996.

growth and production of organic matter will improve the soil's physical and chemical properties, particularly if nitrogen-fixing species are used; (3) as the dipterocarp seedlings are planted under a partially established canopy, a wider range of species can be used, and mortality due to heat and water stress will be reduced.

3. Outline of Tools

Restoration interventions to attain the defined objectives may range from simply protecting the site from further disturbances (e.g., illegal logging, fire) and allowing natural regeneration and successional processes to restore ecosystem functionality, to intensive silvicultural practices to improve species' composition and commercial productivity, and even soil and water conservation measures to prevent and control erosion.

Most tools and technologies needed for restoration of logged-over (and also secondary) forests can be found in the extensive literature on silviculture and forest management.[447]

Four broad steps may be considered for restoration: secure protection of the area; plan for restoration; implement restoration interventions; and monitor and evaluate them. The sections below mainly focus on some of the tools and technologies for planning and implementation.

3.1. Secure Protection of the Area

A precondition of investing in restoration work is to secure the protection of the area against further undesired disturbance (illegal logging, poaching, fire, grazing, etc.). This entails an assessment of the local conditions (e.g., exploitation practices and consequences, past and existing agreements) and the analysis of its outcomes, as well as the capacity to effectively control or reduce stress and risk factors. There is an ample suite of participatory techniques (approaches, tools, and methods) that can be used for this purpose.[448] The ITTO restoration guidelines[449] also provide some principles and recommended actions (see principles 8, 11, 12, 15, 16, 20, and 22).

As a result of the field assessment, some preventive or corrective measures will need to be put in place. Most critical in many situations are fire prevention and control measures. Bad logging creates favourable conditions for fire outbreaks (e.g., accumulation of biomass, invasion by weed species, and desiccation of organic soil matter, all of which can increase fire risks). Other threats frequently result from external forces such as illegal extraction activities, invasion by settlers, and the expansion of agricultural activities. Fire prevention and control are therefore critical for any sustainable use of the area to be restored. These involve a range of active and passive measures, including consultation and training of local people, buffer zones of green firebreaks (especially comprising species valued by the local people), and systems for early detection and suppression. (More information on restoration and fires can be found in "Forest Landscape Restoration After Fires.")

3.2. Plan For Restoration

Protection measures and restoration interventions should be adequately planned. Drawing up a medium-term management/restoration plan may be necessary.

A management plan requires information such as an inventory of the standing stock and its condition, including composition, size, and stem quality. An assessment of the regeneration (seedlings, saplings, and advanced growth of marketable or preferred timber and nontimber species) should be considered. Information on nontimber forest products (NTFPs) can be collected as part of this inventory. Important for planning (zonation and mapping purposes) is also the systematic assessment of the physical conditions affecting the restoration work

---

[447] Useful references for tropical forests include Dupuy, 1998; FAO, 1998, 2000; Higman et al., 1999; Hutchinson, 1988; Lamprecht, 1989; Peters, 1996; Thomson, 2001; Wyatt-Smith, 1963.

[448] For instance: Carter, 1996; Jackson and Ingles, 1998; and Shell et al., 2002.

[449] ITTO, 2002.

(watercourses, topography, soils, vegetation types, etc.).

The advanced regeneration of current and potential commercial or useful tree species should be the first target for interventions. To guide decisions on silvicultural intervention a simple assessment method called *diagnostic sampling* can be used. Diagnostic sampling is a rapid and inexpensive method intended to estimate the potential productivity of a forest stand and decide whether treatment is necessary or not, and if necessary, whether it can be delayed or not, and what type of treatment should be given. Steps and field procedures for using this method can be found in Hutchinson[450] and FAO.[451]

For monitoring purposes, permanent plots or continuous forest inventory plots should be established in order to provide the necessary baseline data of forest growth and response to the interventions.

Based on the medium-term plan, an annual plan (at the compartment level) is usually done. This is an operational tool for guiding the implementation of the planned activities. It may entail measures for erosion control and/or to protect/enhance biodiversity (of particular vegetation types or species), demarcation of riverine corridors to be retained for hydrological reasons or of wildlife corridors, etc.

3.3. Implement Silvicultural Interventions

Silvicultural interventions are generally necessary to overcome the relative depletion of commercial tree species, to compensate for the slow growth rate, and to ensure a future commercial timber value of the forest.[452] Options that can be applied, depending on the condition of the forest stand and the objectives (what major products are expected), include improvement treatments, treatments to stimulate natural regeneration, enrichment planting, and direct planting.

Working with preexisting natural regeneration is the cheapest and safest way to recover the original forest, provided there is plenty of the desirable (e.g., current and potential commercial) species. This is usually the case with forests that have only been lightly degraded through uncontrolled timber exploitation. In more degraded conditions, however, the lack of adequate regeneration or an uneven distribution over the area entails difficult silvicultural work, making it necessary to resort to more costly interventions.

Some examples of interventions are given below. The interested reader will find more detailed information in the various dedicated chapters of this volume (see Section XI, A Selection of Tools that Return Trees to the Landscape).

3.3.1. Improvement Treatments

Improvement treatments (or tending operations) basically aim to provide more space for trees of desirable species. This is done first through the application of an operation called *overstorey removal*, by which overmature, defective noncommercial individuals (called *relics*) are removed (usually by poison-girdling) from the upper levels of the forest canopy. A second phase consists of *liberation thinning*, a treatment that releases young growth from the competition from commercially less desirable species. The prescriptions for liberation may easily be altered to accommodate changes in market demand or alternative management requirements (e.g., maintain keystone food resources for animals).

Timber stand improvement (TSI) is a well-known silvicultural treatment used by preference in dipterocarp forests. Usually conducted 5 to 10 years after logging, it basically involves the cutting or killing of unwanted trees and climbers to improve growing conditions for the remaining trees and species' composition of the stand. A detailed description of procedures is found in Weidelt and Banaag.[453]

---

[450] Hutchinson, 1991.
[451] FAO, 1998.
[452] ITTO, 2002.
[453] Weidelt and Banaag, 1982.

3.3.2. Treatments to Stimulate Natural Regeneration

The lack of advanced regeneration (or its unsatisfactory spatial distribution), particularly of the desirable species, is a main constraint usually found in more heavily disturbed forests. If the objective is to restore populations of these species, treatments to stimulate their natural regeneration thus become a priority as part of the post-logging interventions.

3.3.3. Enrichment Planting

Enrichment planting (also known as underplanting) is defined as the introduction of valuable species on degraded forests without the elimination of valuable individuals already present. Enrichment of logged-over forests may be appropriate in areas where natural regeneration of desired species is insufficient or soil characteristics are not conducive to other uses, or even when the interest is to introduce high-value species that do not regenerate easily, keystone food species or even fruit trees or other species with commercial or local value.[454]

3.3.4. Direct Planting

Direct tree planting in logged-over forests is sometimes used for rehabilitating localised areas that were more heavily impacted by harvesting infrastructure (roads, log landings). These patches of trees or shrubs are planted primarily for erosion control (e.g., slope stabilisation). Planting in log landings and other open areas for growing commercial trees is another option.

4. Future Needs

We probably know enough about the general impacts of timber harvesting on tropical forests, and also about the main courses of action for restoring these ecosystems. We certainly need to know more, but above all we need to apply what is already known and learn as we go along. This entails the need to substantially increase efforts to appropriately disseminate the strategies, approaches, and techniques most appropriate for forest restoration. Awareness-raising, training, and technical assistance are preconditions to the actual application of restoration in practice.

There are many challenges posed to improve restoration of overlogged forests. Some of the most pressing are as follows:

- Analyses of financial and environmental costs and benefits of restoration options and their effects on forest productivity, species' recovery, biodiversity, and carbon sequestration
- Development of enrichment planting guidelines that are species- and site-specific
- Development of cost-effective fire control measures with minimal biodiversity impacts
- Development of an adequate and supportive legal framework for overlogged forest restoration.

References

Adjers, G., Hadengganan, S., Kuusipalo, J., Nuryanto, K., and Vesa, L. 1995. Enrichment planting of dipterocarps in logged-over secondary forests: effect of width, direction and maintenance method of planting line on selected *Shorea* species. Forest Ecology and Management 73:259–270.

Applegate, G., Putz, F.E., and Snook, L.K. 2004. Who pays for and who benefits from improved timber harvesting practices in the tropics? Lessons learned and information gaps. CIFOR, Bogor, Indonesia.

Banerjee, A.K. 1995. Rehabilitation of degraded forests in Asia. World Bank Technical Paper Number 270. World Bank, Washington, DC.

Bruijnzeel, L.A., and Critchley, W.R.S. 1994. Environmental impacts of logging moist tropical forests. International Hydrological Programme/IHP Humid Tropics Programme Series No. 7. UNESCO–IHP–MAB. Paris, France.

Carter, J. 1996. Recent Approaches to Participatory Forest Resource Assessment. Rural Development Forestry Study Guide 2. ODI, London.

Dupuy, B. 1998. Bases pour une Sylviculture en Forêt Dense Tropicale Humide Africaine. Document 4.

---

[454] Lamprecht, 1989; Montagnini, 1997; Weaver, 1993.

Projet FORAFRI, CIRAD, CIFOR, Montpellier, France.

Dykstra, D., and Heinrich, R. 1996. FAO—Model code of forest harvesting practice. Food and Agriculture Organisation of the United Nations. FAO, Rome.

Effendi, R., Priadjati, A., Omom, M., Rayan, M., Tolkamp, W., and Nasry, E. 2001. Rehabilitation of Wanariset secondary forest (East Kalimantan) through dipterocarp species line plantings. *In:* Hillegers, P.J.M., and Iongh, H.H., eds. The Balance Between Biodiversity Conservation and Sustainable Use of Tropical Rain Forests. Tropenbos International, Wageningen, The Netherlands, pp. 31–44.

FAO. 1998. Guidelines for the Management of Tropical Forests–1. The Production of Wood. FAO Forestry Paper 135. Rome.

FAO. 2000. Management of Natural Forests of Dry Tropical Zones. FAO Conservation Guide 32. Rome. Prepared by R. Bellefontaine, A. Gaston and Y. Petrucci.

Fimbel, R., Grajal, A., and Robinson, J., eds. 2001. Conserving Wildlife in Managed Tropical Forests. Columbia University Press, New York.

Frumhoff, P.C. 1995. Conserving wildlife in tropical forests managed for timber. BioScience 45(7): 456–464.

Grieser Johns, A. 1997. Timber Production and Biodiversity Conservation in Tropical Rain Forests. Cambridge Studies in Applied Ecology and Resource Management. Cambridge University Press, Cambridge, UK.

Haworth, J. 1999. Life After Logging: The Impacts of Commercial Timber Extraction in Tropical Rainforests. Friends of the Earth Trust, London.

Higman, S., Bass, S., Judd, N., Mayers, J., and Nussbaum, R. 1999. The Sustainable Forestry Handbook. A Practical Guide for Tropical Forest Managers on Implementing New Standards. IIED–SGS. Earthscan Publications Ltd., London.

Hutchinson, I.D. 1988. Points of departure for silviculture in humid tropical forests. Commonwealth Forestry Review 67(3): 223–229.

Hutchinson, I.D. 1991. Diagnostic sampling to orient silviculture and management in natural tropical forest. Commonwealth Forestry Review 70(3): 113–132.

International Tropical Timber Organisation (ITTO). 2002. ITTO guidelines for the restoration, management and rehabilitation of degraded and secondary tropical forests. ITTO Policy Development Series No. 13. ITTO, Yokohama, Japan.

Jackson, W.J., and Ingles, A.W. 1998. Participatory Techniques For Community Forestry. A Field Manual. IUCN, Gland, Switzerland and Cambridge, UK and World Wide Fund for Nature, Gland, Switzerland.

Korpelainen, H., Adjers, G., Kuusipalo, J., Nuryanto, K., and Otsamo, A. 1995. Profitability of rehabilitation of overlogged dipterocarp forest: a case study from South Kalimantan, Indonesia. Forest Ecology and Management 79:207–215.

Lamprecht, H. 1989. Silviculture in the Tropics. Tropical Forest Ecosystems and Their Tree Species—Possibilities and Methods for Their Long-Term Utilization. GTZ, Eschborn, Germany.

Montagnini, F. 1997. Enrichment planting in overexploited subtropical forests of the Paranaense region of Misiones, Argentina. Forest Ecology and Management 99:237–246.

Nussbaum, R., and Hoe, A.L. 1996. Rehabilitation of degraded sites in logged-over forest using dipterocarps. *In:* Schulte, A., and Schöne, D., eds. Dipterocarp Forest Ecosystems. Towards Sustainable Management. World Scientific Publ., Singapore, pp. 446–463.

Peters, C.M. 1996. The Ecology and Management of Non-Timber Forest Resources. World Bank Technical Paper Number 322. Washington, DC.

Putz, F.E., Redford, K.H., Robinson, J.G., Fimbel, R., and Blate, G.M. 2002. Biodiversity conservation in the context of tropical forest management. The World Bank, Environment Department Papers, Paper No. 75. Washington, DC.

Sheil, D., et al. 2002. Exploring biological diversity, environment and local's people's perspectives in forest landscapes. Methods for a multidisciplinary landscape assessment. CIFOR, Bogor, Indonesia.

Sist, P., Fimbel, R., Nasi, R., Sheil, D., and Chevallier, M.-H. 2003. Sustainable management of mixed dipterocarp forests needs more ecological rules than a minimum diameter for harvesting. Environmental Conservation 30(4):364–374.

Stadtmüller, T. 1994. Impacto hidrológico del manejo forestal de bosques naturales tropicales: medidas como mitigarlo. Una revisión bibliográfica. CATIE, Proyecto silvicultura de Bosques Naturales, Turrialba, Costa Rica.

Thomson, L. 2001. Management of natural forests for conservation of forest genetic resources. *In:* FAO/DFSC/IPGRI, Forest Genetic Resources Conservation and Management. Vol. 2: In Managed Natural Forests and Protected Areas (In Situ). International Plant Genetic Resources Institutte, Rome pp. 13–44.

Tuomela, K., Kuusipalo, J., Vesa, L., Nuryanto, K., Sagala, A.P.S., and Adjers, G. 1996. Growth of dipterocarp seedlings in artificial gaps: an experiment in a logged-over rainforest in South Kalimantan, Indonesia. Forest Ecology and Management (81):95–100.

Wadsworth, F.H. 1997. Forest Production for Tropical America. USDA Forest Service. Agriculture Handbook 710. USDA, Washington, DC.

Weaver, P.L. 1993. Secondary forest management. In: Parrotta, J.A., and Kanashiro, M., eds. Management and Rehabilitation of Degraded Lands and Secondary Forests in Amazonia. Proceedings of an International Symposium. Santarem, Para, Brazil, 18–22 April 1993, International Institute of Tropical Forestry, USDA Forest Service, and UNESCO Man and the Biosphere Programme. Rio Piedras, Puerto Rico and Paris, pp. 117–128.

Weidelt, H.-J., and Banaag, V.S. 1982. Aspects of Management and Silviculture of Philippine Dipterocarp Forests. Philippines–German Rain Forest Development Project. GTZ, Eschborn, Germany.

Woods, P. 1989. Effects of logging, drought, and fire on structure and composition of tropical forests in Sabah, Malaysia. Biotropica 21:290–298.

Wyatt-Smith, J. 1963. Manual of Malayan silviculture for inland forests. Malayan Forest Record No. 23, part III.

Additional Reading

Hutchinson, I.D. 1996. Techniques for silviculture and management in natural tropical forests, logged and secondary. In: Parrotta, J.A., and Kanashiro, M., eds. Management and Rehabilitation of Degraded Lands and Secondary Forests in Amazonia. Proceedings of an International Symposium. Santarem, Para, Brazil, 18–22 April 1993, International Institute of Tropical Forestry, USDA Forest Service, and UNESCO Man and the Biosphere Program. Rio Piedras, Puerto Rico and Paris, pp. 142–152.

Louman, B., Quirós, D., Nilsson, M., eds. 2001. Silvicultura de Bosques Latifoliados Húmedos en América Central. CATIE, Turrialba, Costa Rica. Serie técnica: Manual técnico/CATIE no. 46.

Weaver, P.L. 1987. Enrichment plantings in tropical America. In: Figueroa, J.C., Wadsworth, F.H., and Branham, S., eds. Management of the Forests of Tropical America: Prospects and Technologies. Proceedings of a Conference. San Juan, Puerto Rico, September 22–27, 1986, pp. 259–278.

53
Open-Cast Mining Reclamation

José Manuel Nicolau Ibarra and Mariano Moreno de las Heras

In addition to science, imagination is needed to see the potential of the land and to relate it to the need of the local region.

Bradshaw, 1988

Key Points to Retain

Application of an inadequate conceptual framework is often behind the failure of mining reclamation projects, including insufficient understanding of reference ecosystems, short-term planning, and insufficient consideration of contingencies.

Cooperation between mining companies and environmental institutions is necessary to integrate reclaimed areas into conservation programmes at a regional scale.

Good erosion models for reclaimed areas as tools for land-form design have been developed.

One major area in need of improvement is the application of laws that require rehabilitation of mined sites.

1. Background and Explanation of the Issue

Human activities involving major soil removal, such as open-cast mining, urban development, civil works, and so on, are the first source of sediment reaching the oceans via rivers. At a local scale, mining impacts on biodiversity, water quality, and land use are frequently very high. Mining is one of the anthropic activities causing some of the most dramatic disturbances on nature. In fact, there is a positive feed-back interaction between nonenergetic and energetic mineral extraction, which also contributes to greenhouse gas emissions. Technology for mining reclamation has been widely developed in the last two decades for most regions of the world. However, in practice, most of the "reclaimed lands" have achieved poor results.[455]

Application of an inadequate conceptual framework is often behind the failure of mining reclamation projects. There are two types of driving forces in mining reclamation: determinism and contingency.[456] Usually only deterministic processes are considered. In addition, reclaimed areas must be recognised as open ecosystems interacting with their surrounding environment. A conceptual model including its practical consequences on mining reclamation planning is shown in Figure 53.1. This model assumes that change more than equilibrium is the essence of nature, following the new paradigm in ecology.[457]

Reclamation success depends on several contingent or circumstantial events, which are often unpredictable: (1) initial conditions

---
[455] Haigh, 2000.
[456] Pickett et al, 2001.
[457] Kolasa and Pickett, 1991.

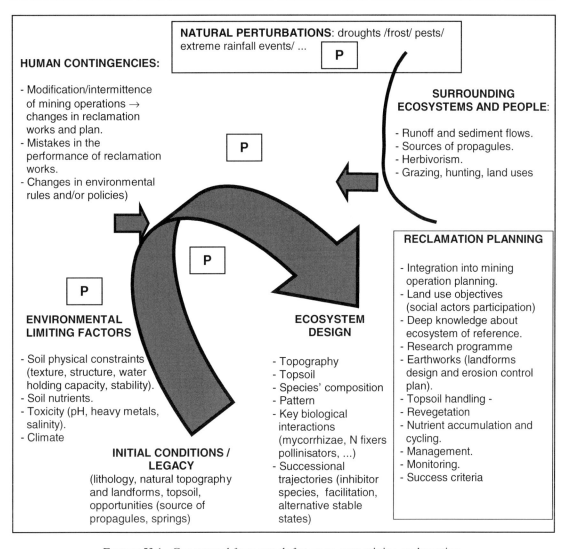

FIGURE 53.1. Conceptual framework for open-cast mining reclamation.

(natural climate and topography, type and abundance of topsoil); (2) natural perturbations (droughts, extreme rainfall events, frost periods, pests); (3) influence of the surrounding ecosystems and people (runoff and sediment flows, sources of propagules, herbivorism, grazing, hunting, land uses); and (4) human contingencies (modification/intermittence of mining operations; mistakes in the performance of reclamation works; changes in legal rules, etc.).

Deterministic processes involved in mining reclamation have been well studied and a wide set of reclamation techniques and tools have been developed. Most typical of them in mining reclamation are abiotic limiting factors and nutrient cycling. Bradshaw[458] identified the main physical and chemical problems that can be found in mine soils and their short and long-term treatments, which are shown in Table 53.1.

Following the proposed conceptual framework, the Reclamation Planning box in Figure 53.1 shows the main issues that should be considered, from the practical perspective, in order

[458] Bradshaw, 1988.

TABLE 53.1. Specific problems of mine soils and their treatments (Bradshaw, 1988).

Category		Problems	Immediate treatment	Long-term treatment
Physical	Structure	Too compact	Rip or scarify	Vegetation
		Too open	Compact or recover with fine material	Vegetation
	Stability	Unstable	Stabiliser/mulch	Vegetation
	Moisture	Too wet	Drain	Drain
		Too dry	Organic mulch	Vegetation
Nutrition	Macronutrients	Nitrogen	Fertiliser	Legume
		Others	Fertiliser + lime	Fertiliser + lime
	Micronutrients		Fertiliser	—
Toxicity	pH	Too high	Organic matter or pyritic waste	Weathering
		Too low	Lime	Lime
	Heavy metals	Too high	Organic mulch or tolerant cultivar	Inert covering or tolerant cultivar
	Salinity	Too high	Weathering or irrigate	Tolerant species or cultivar

to improve the performance of open-cast mining reclamation.[459] In addition:

1. Both mining and reclamation activities must be carried out simultaneously in an integrated way in order to optimise the opportunities offered by mining operations. This makes reclamation works cheaper, quicker, and more successful.

2. Reclamation projects must be designed and developed by companies and social actors together. It is critical to get an agreement about the final objectives for the reclaimed areas as well as their use and maintenance.

3. Although general protocols for reclamation are available, it is always necessary to carry out specific research in order to adapt or develop them to the local conditions and to obtain in depth knowledge about the reference ecosystem. Cooperation between companies and conservation organisms and nongovernmental organisations (NGOs) is valuable for this phase.

4. A plan of monitoring and survey is essential for checking, improving, or redirecting the applied practices.

2. Examples

2.1. Fire Management in Jarrah Forest Restoration on Bauxite Mines in Western Australia[460]

Alcoa World Alumina Australia commenced mining bauxite in the Jarrah forest of western Australia in 1963. Since then, 10,600 hectares have been rehabilitated. The climate is typically Mediterranean with winter rainfall and summer drought. Early restoration efforts were based on imported species of pine and eucalypt from Eastern Australia. This exotic vegetation is very resilient to natural forms of disturbance, so plant richness remains low and ecological succession runs slowly. The current rehabilitation objective is to reestablish a functional Jarrah forest ecosystem that will fulfil the forest land uses (conservation, timber production, water catchment protection, and recreation). Rehabilitation began with the reshaping of the 2- to 5-m-high pit walls. Topsoil was re-spread. As topsoil returned, a few tree stumps, logs, and rocks were returned to the mined areas to provide habitat for fauna. The ground was ripped to a depth of 1.5 m. A seed mix of 70 to 100 local species was broadcast on the freshly cultivated ground. Other plant species were

---

[459] Adapted from Australian Environmental Protection Agency, 1995.

[460] Smith et al, 2004.

planted. A mixed fertiliser (nitrogen, phosphorus and potassium (NPK) and micronutrients) was applied at 500 kg per hectare by helicopter.

In 1997 Alcoa and the Department of Conservation and Land Management (CALM) developed completion criteria and standards. Specifically, the completion criteria require restored areas to be resilient to fire and capable of integration into CALM's Jarrah forest fire management programme. Alcoa supported research to determine how the vegetation and associated faunal communities respond to fire, in order to define when and under what conditions fire should be reintroduced into rehabilitated areas.

2.2. Restoring Tropical Forests on Lands Mined for Bauxite—Examples from Brazilian Amazon[461]

Since 1979, the Brazilian mining company Mineraçao Rio do Norte (MRN) has developed a reforestation programme aimed at restoring the evergreen equatorial moist forest destroyed at a rate of 100 hectares per year during bauxite ore extraction at Trombetas in western Pará State. The Trombetas bauxite mine is located in the Saracá-Taquera National Forest on an upland mesa at an elevation of 180 m. Restorationists working in most tropical settings are usually hampered by lack of basic information on the wide variety of native tree species that characterise the pre-disturbance forests, as well as insufficient understanding of the ecology of disturbance and natural recovery to design effective restoration programmes. A notable exception is MRN, which has used a systematic nursery and field research strategy to develop a reforestation programme based on mixed plantings of more than 70 native old-growth forest tree species.

Two main research programmes were carried out in the last 11 years, and a number of reforestation methods as well as site preparation and topsoil replacement protocols were tested.

Native forest species' propagation and performance assessment programmes involved evaluations of fruiting phenology, seed viability, seed germination treatments, propagation methods (direct seeding, use of stumped saplings, wildings, and nursery-grown seedlings), and early survival and growth during the first 2 years after outplanting. A total of 160 species were evaluated. The standard reclamation and site preparation sequence was followed, which includes levelling of the clay overburden, replacement of approximately 15 cm of topsoil and woody debris (removed from the site prior to mining and stockpiled for up to 6 months prior to application), deep-ripping of lines to a depth of 90 cm (1 m between lines), and planting along alternate rip lines at 2- by 2-m spacing (2500 trees per hectare) using seeds, stumped saplings, or potted seedlings, depending on species and treatment. The total cost came to approximately $2500 per hectare.

The following conclusions can be drawn: Careful site preparation practices, particularly judicious topsoil handling and reapplication prior to tree planting, are essential for the establishment of forest cover, elimination of competing grasses, and acceleration of natural forest succession. Floristic enrichment of the reforested areas is largely dependent on seed-dispersing wildlife, so restoration managers need to be cognizant of the critical role of wildlife, actively encourage wildlife conservation in the surrounding landscape, and design restoration treatments that will provide suitable habitats for a variety of target wildlife species.

2.3. Open-Cast Coal Mining Reclamation in Utrillas-Teruel (Spain) in a Semiarid (Mediterranean-Continental) Environment[462]

Minas y Ferrocarril de Utrillas, SA (MFUSA) company commenced open-cast mining in the Utrillas coalfield in the early 1980s. The area is

---

[461] Parrotta and Knowles, 2001.

[462] Nicolau, 2003.

located in central-eastern Spain at 1100 m of altitude. A major limiting factor is water deficiency in soil, and therefore reduced water availability for plants. Mean annual rainfall is 466 mm, 28 percent falling in June and May and 20 percent in September. The water deficit is 292 mm from June to October. Restoration of the mines was orientated toward agricultural uses in agreement with social actors.

Improving soil moisture content was the key success factor in the Utrillas region. The MFUSA company developed a restoration protocol in which the three elements of the ecosystem, namely, landform, soil, and plants, were designed in an integrated fashion to optimise the supply of water and nutrients and to control the abiotic exploitation of erosion.

Land forms based on the platform-bank model with slopes of about 30 degrees had to be abandoned because rainfall infiltration is low in steep slopes, and runoff leads to high rill and sheet erosion. In turn, rill erosion increases water deficiency at the slope scale by reducing opportunities for runoff reinfiltration into the soil downslope.[463] The best-identified topography was that based on the hydrological basin as unit for reclamation. This is composed of slopes with natural vegetation, flat areas for agricultural use, and a drainage network including watercourses, pools and sediment ponds. Topsoil was carefully selected for its physical properties (water-holding capacity).

Characteristics of constructed slopes were as follows: gradient between 18 degrees and 21 degrees; insulation from runoff from platforms, tracks, and upper berms; topsoil spreading (50 cm thick); tillage transverse to the slope; supply of organic fertiliser; sowing with herbaceous species at the end of winter; surface tillage to bury seeds. Three years later, in winter, woody species were planted.

This protocol has been successful to get grass back, which controlled soil erosion and started soil formation. However, ecological succession is proceeding slowly. In fact, introduced grass community have inhibited natural colonisation.

2.4. Problems in the Reclamation of Coal-Mine Disturbed Lands in South Wales Coalfield[464]

Reclamation in South Wales started in Pwll Du mine in the 1940s. Three surface mines were reclaimed during the 20th century.

More recent land reclamation practice often involves applied topsoil (100 to 150 cm) and the establishment of seeded grass covers to allow sheep to graze. Reclaimed areas are managed by Commoners Associations.

However, large tracts of land, officially listed as "reclaimed" from former mineral operations, are in very poor condition. On-site problems include gullying, poor vegetation cover, erosion, and poor soil structure. Off-site they cause problems due to accelerated runoff and, more occasionally, chemical and sediment pollution. Some of these problems are due to poor engineering and poor land husbandry, but they are magnified by natural processes. Some mine spoils/soils include a high proportion of friable shales. These break down rapidly, when exposed to disturbance/weathering, releasing clays, which clog up soil pores and impede the infiltration of water. This causes a progressive deterioration of the land with symptoms that may include water logging, replacement of grass by moss/lichen/bare ground, dieback of soil microbiota, increases in soil bulk density, and decreases in soil aggregate stability.

Remedies that are being applied by the Oxford Brookes University group include developing a large/active soil microbiota capable of transforming clays in water-stable soil aggregates. This is done by introducing deep-rooting tree species because they are vigorous and reliable soil formers and because, with a little help, they can support large and active populations of microorganisms.

3. Outline of Tools

A wide set of tools can be found in the references below. The following tools are more specific to mining reclamation:

---

[463] Nicolau, 2002.

[464] Haigh, 1992.

- The first measure for protecting the most valuable ecological areas from mining impacts should be the use of geographical information systems (GIS) plus environmental planning methodologies at the regional scale.
- In relation to topography design, Evans[465] affirms that "to successfully incorporate the design of relief forms, the stability of the final forms must be predicted, which implies the use of hydrological and erosion models." In recent years, some erosion models for reclaimed areas have been developed, which are now being used in relief design. We suggest using the RUSLE (Revised Universal Soil Loss Equation) 1.06 (for mined lands, construction sites, and reclaimed lands), which is a model that estimates the annual surface erosion by water[466] and can be used for slope design. This model is available free on the Web site http://sedlab.olemiss/rusle.
- As off-site impacts on aquatic ecosystems are among the heaviest disturbances produced by open-cast mining, an erosion and runoff control plan is essential. Several software packages are available on the market. We recommend evaluating the effectiveness of erosion and sediment control plans.[467] This can be acquired through the International Erosion Control Association at http://www.ieca.org.
- Topsoil handling is a key but easy issue when it is planned. A critical point is storage. It should be stored for a short period of time and in small stockpiles. A second point is the spreading of topsoil on the reconstructed topography. To avoid soil compaction, such an operation must be carried out with topsoil that is neither too dry nor too wet.
- Soil amendment is a quite general matter in land reclamation. Table 53.1 shows a number of remediation procedures proposed by Bradshaw.[468]
- A very useful "tool" from the practical point of view is to count on an environmental expert working in the field as mining and reclamation projects are going on. This person—in addition to being responsible for the fulfilment of the reclamation project—should foresee the contingencies and should profit from the opportunities offered by the physical environment, mining operations, local administration, and social actors.

4. Future Needs

Performance of surface mining reclamation shows high heterogeneity depending on the countries, the environments, and the companies; consequently, the needs are very different. In developed countries the main task is to reclaim again thousands of "reclaimed" hectares, which do not fulfil minimum requirements.

From the technical point of view the weakest points are land-form design and ecosystem dynamics knowledge. Erosion and hydrological models should be incorporated into reclamation planning. Also the reference ecosystem has to be used for reclaimed ecosystem design and to identify a number of successional trajectories, stable states, and thresholds of irreversibility.

In developing countries, efforts in research must be intensified as has been seen in the example of the Brazilian bauxite mine. Reclamation laws must be enhanced or enacted in some cases, but most importantly, laws must be observed and enforced. However, often in practice, this may seem utopian. In many cases mineral deposit discovery and exploitation means deep environmental impacts, social and political conflicts, corruption, and even armed violence. The imbalance is so high that often neither society nor the politicians are sufficiently prepared to have a positive relationship with the transnational mining corporations. Given such conditions, an international mining code of good practice would be useful.

We think that NGOs can be very helpful in: (1) promoting experimental research, (2) training local restorationists, (3) favouring local communities' participation, and (4) advising governments of developing countries.

---

[465] Evans, 2000.
[466] Toy and Foster, 1998.
[467] Fifield, 1997.
[468] Bradshaw, 1988.

References

Australian Environment Protection Agency. 1995. Rehabilitation and revegetation. Best Practice Environmental Management in Mining. Commonwealth of Australia, Barton.

Bradshaw, A.D. 1988. Alternative Endpoints for Reclamation. In: Cairns, J.R., ed. Rehabilitating Damaged Ecosystems, vol. 1. CRC Press, Boca Raton, Florida, pp. 70–85.

Evans, K. 2000. Methods for assessing mine site rehabilitation design for erosion impact. Australian Journal of Soil Research 38:231–247.

Fifield, S.J. 1997. Field Manual for Effective Sediment and Erosion Control Methods. Hydrodynamics, Inc., Parker, CO.

Haigh, M. 1992. Problems in the reclamation of coal-mine disturbed lands in Wales. International Journal of Surface Mining and Reclamation 6:31–37.

Haigh, M. 2000. The aims of Land reclamation. In: Haigh, M., ed. Reclaimed Land. Erosion Control, Soils and Ecology. A.A. Balkema, Rotterdam, The Netherlands. pp. 1–20.

Kolasa, J., and Pickett, S.T.A. 1991. Ecological heterogeneity. Ecological Studies, No. 86. Springer-Verlag, New York.

Nicolau, J.M. 2002. Runoff generation and routing on artificial slopes in a Mediterranean-continental environment: the Teruel coalfield, Spain. Hydrological Processes 16:631–647.

Nicolau, J.M. 2003. Trends in relief design and construction in opencast mining reclamation. Land Degradation and Development 14:215–226.

Parrotta, J.A., and Knowles, O.H. 2001. Restoring tropical forests on lands mined for bauxite: examples from Brazilian Amazon. Ecological Engineering 17:219–239.

Pickett, S.T.A., Cadenasso, M.L., and Bartha, S. 2001. Implications from the Buell-Small Succession Study for Vegetation Restoration. Applied Vegetation Science 4:41–52.

Smith, M.A., Grant, C.D., Loneragan, W.A., and Koch, J.M. 2004. Fire management implications of fuel loads and vegetation structure in jarrah forest restoration on bauxite mines in Western Australia. Forest Ecology and Management 187:247–266.

Toy, T., and Foster, G. 1998. Guidelines for the Use of the Revised Universal Soil Loss Equation (RUSLE) version 1.06 on Mined Lands, Construction Sites and Reclaimed Lands. Office of Surface Mining, Denver, CO.

Additional Reading

Barnhisel, R.I., Darmondy, R.G., and Daniels, W.L., eds. 2000. Reclamation of drastically disturbed lands, No. 41 Agronomy series. ASA, CSSA, SSSA Publishers. Madison, WI.

Bradshaw, A.D. 2000. The use of natural processes in reclamation—advantages and difficulties. Landscape and Urban Planning 51:89–100.

Haigh, M., ed. 2000. Reclaimed Land. Erosion Control, Soils and Ecology. A.A. Balkema., Rotterdam, the Netherlands.

Harris, J.A., Birch, P., and Palmer, J. 1996. Land Restoration and Reclamation. Principles and Practice. Longman Higher Education, Horlow, Essex, UK.

Hobbs, R.J. 1999. Restoration of disturbed ecosystems. In: Walker, L.R., ed. Ecosystems of Disturbed Ground. Series Ecosystems of the World, 16. Elsevier, New York, pp. 673–689.

Parker, V.T., and Pickett, S.T.A. 1997. Restoration as an ecosystem process: implications of the modern ecological paradigm. In: Urbanska, K., Webb, N., and Edwards, P., eds. Restoration Ecology and Sustainable Development. Cambridge University Press, Cambridge, UK, pp. 17–32.

Urbanska, K., Webb, N., and Edwards, P., eds. 1997. Restoration Ecology and Sustainable Development. Cambridge University Press, Cambridge, UK.

Section XIV
Plantations in the Landscape

54
The Role of Commercial Plantations in Forest Landscape Restoration

Jeffrey Sayer and Chris Elliott

> **Key Points to Retain**
>
> Plantations can represent an opportunity for the restoration of landscape functions, but they can also represent a threat to natural systems.
>
> This chapter illustrates how commercial plantations can be part of the solution to the challenge of restoration and not always part of the problem.
>
> A basic principle to be agreed to is that plantation forestry should provide multiple production and environmental functions.
>
> Considerable work has been done on more environmentally friendly approaches to tree establishment.

1. Background and Explanation of the Issue

A rapidly increasing proportion of the world's wood is coming from plantations. Many of these are large-scale industrial plantations and they are often established on degraded lands. Such plantations can represent an opportunity for the restoration of landscape functions but they can also represent a threat to natural systems. Tree planting has been seen as the solution to many environmental problems as witnessed by national tree planting campaigns, programmes to re-green deserts, etc. Elsewhere environmental groups campaign against all plantation forestry on the grounds that it replaces native vegetation and often intrudes on land used by local people. Plantations are often viewed as sterile monocultures with little biodiversity or other environmental value yet many studies have shown that even intensively managed industrial plantations often support surprisingly high biodiversity values.[469] In addition, industrial plantations can form parts of landscape mosaics in ways that help to provide a mix of production and environmental functions. The European Union has pioneered the use of environmental payment systems to achieve these "multifunctional landscapes."

Forest plantations are defined by the United Nations Food and Agriculture Organisation (FAO) as "forest stands established by planting and/or seeding in the process of afforestation or reforestation." The FAO does not restrict its definition to timber or pulp plantations. Because of their increasing significance as a supply of fibre for wood industries, rubber (*Hevea* spp.) plantations are now included in global assessments of forest plantations. Recent figures from FAO show that new forest plantation areas are being established at a rate of 4.5 million hectares per year, with Asia and South America accounting for more new plantations than any other region. About 70 percent of new plantations, or 3.1 million hectares per year, are successfully established; in the remainder, an

---
[469] IUFRO, 2003.

astonishing 30 percent, trees are planted but they are often not cared for and die.

Of the estimated 187 million hectares of plantations worldwide, Asia has by far the largest area, accounting for 62 percent of the world total. In terms of composition, *Pinus* (20 percent) and *Eucalyptus* (10 percent) remain the dominant genera worldwide, although the diversity of species planted is increasing. Industrial plantations (producing wood or fibre for supply to wood processing industries) account for 48 percent of the global forest plantation estate and nonindustrial plantations (e.g., for provision of fuelwood or soil and water protection) for 26 percent. The purpose of the remaining 26 percent is unclear.

The extent of plantations in industrialised countries is harder to measure than in developing countries. Most forests in Western Europe contain some planted trees, so the distinction between plantations and natural forests is less clear cut than in the new plantations in the tropics. Industrialised countries tend not to distinguish between plantations and natural forests in their inventories.

The FAO has identified the 10 countries with the largest plantation development programmes (as reported by percentage of the global plantation area): China, 24 percent; India, 18 percent; the Russian Federation, 9 percent; the United States, 9 percent; Japan, 6 percent; Indonesia, 5 percent; Brazil, 3 percent; Thailand, 3 percent; Ukraine, 2 percent; and the Islamic Republic of Iran, 1 percent. These countries account for 80 percent of the global forest plantation area. All of them are countries with large extents of degraded landscapes.

Global interest in forest landscape restoration was partly triggered by environmental concerns about plantation forestry. Public criticism of large-scale Sitka spruce (*Picea sitchensis*) plantations in Scotland led the U.K. Forestry Commission to reverse its policies on upland tree planting. The emphasis is now given to planting native woodlands for amenity and wildlife values. Not only the species planted but also the spatial layout of the plantations is designed to imitate natural woodlands.[470]

Large commercial plantations subsidised by the World Bank were a cause célèbre for the environmental movement in India in the 1980s. Rural people complained that the exotic species planted did not provide fodder for their animals or supplies of the nontimber products that they needed for their daily subsistence. Tree-hugging campaigns were launched to prevent the clearing of natural forests by the plantation agencies.[471]

Pulp plantations in Indonesia have been strongly opposed by environmentalists because they often replace natural forest and deny access to the land to local people. Similar controversies have surrounded commercial plantations in Chile and government sponsored plantation schemes in Vietnam.[472]

However, forest landscape restoration almost always involves reestablishing trees, and the purpose of this chapter is to illustrate how commercial plantations can be part of the solution to the challenge of restoration and not always part of the problem.

2. Examples

2.1. Environmentally Beneficial Commercial Plantations: Plantations in Brazil

The plantations established by the American billionaire Harvey Ludwig at Jari in Brazil[473] are an excellent example of how sensible management has turned what started as a major environmental threat into a model of good landscape management. The scheme started with the planting of large areas of a single exotic species. Many trees died and the plantations failed to achieve their commercial objectives, but their establishment did cause the loss of large areas of natural forests. The Jari plantations have changed hands twice and are now owned by a Brazilian family company. A greater diversity of trees is now planted in 300,000 hectares of plantations and large areas

---

[470] See Smout, 2000.
[471] Carrere and Lohman, 1996; Cossalter and Pye-Smith, 2003.
[472] Lang, 2002.
[473] See www.metsopaper.com.

of natural forest have been set aside for protection within the plantation area. Additionally, 700,000 hectares of natural forest in the immediate areas have been brought under sustainable management for timber. The Jari operations are now certified by an internationally accredited certification scheme. The area now represents an environmentally sound balanced landscape containing protected, managed, and plantation forests.

2.2. Environmentally Beneficial Commercial Plantations: Pulp Plantations in Sumatra[474]

Pulp plantations in Sumatra have been under a lot of criticism for their negative environmental and social impacts. They often replaced natural forest of high biodiversity value, and many local people were displaced by their establishment. Indonesian law required that plantation companies set aside up to a third of their land as natural forest set asides, but this rule was largely ignored or the set asides were neglected and illegally logged, often by subcontractors who sold the logs to the pulp mills. Under pressure from environmental NGOs, one of the companies, APRIL, has now supported the establishment of a national park to conserve the remaining forests located within its plantation estate. The infrastructure of the plantation company provides access for park managers, and profits from the plantation operation help to pay for park protection costs.

2.3. Environmentally Beneficial Cosmmercial Plantations: Conifer Plantations in the United Kingdom

In the United Kingdom exotic conifer plantations have long been opposed by the public, which often preferred the open treeless landscapes of upland Scotland and Wales even though these were the result of overgrazing by sheep in the 19th century. A good account of the controversy surrounding the issue of upland conifer plantations is given in Smout.[475] As commercial conifer plantations began to be phased out, a new problem arose. It was discovered that the conifer plantations when they were newly planted provided the habitat for a large proportion of the U.K. population of the rare falcon, the merlin (*Falco columbarius*). Early successional woodlands that occur after commercial plantations have been logged were providing the only habitat for a rare species. In this case, keeping some of the land under commercial plantations was contributing to landscape functionality.

3. Outline of Tools

In many landscapes commercial plantations will have a potential role in restoration. Much will depend on where in the landscape they are located and how they are managed.

Plantations do not always have to be of a single species. It is not always necessary to keep the land under the trees bare; weeds and spontaneously colonising local trees can be encouraged. Mixed local species can be planted along water courses or around the periphery of the plantation to soften the visual impact of the plantation and provide habitat for wildlife. Plantations can be used to provide corridors between patches of natural woodlands. Plantations can provide many products and thereby reduce the pressure on natural forests. Plantations can sometimes be used as nurse crops to help improve the soil and create conditions so that native species can become established.

Plantations are often established using industrial techniques that tend to result in uniform stands that are relatively low in biodiversity and other environmental and social values. But considerable work has been done on more environmentally friendly approaches to tree establishment.[476] In any use of commercial plantations to contribute to landscape restoration objectives, it is essential to ensure that the plantations are managed to the highest possible standards. The International Tropical Timber Organisation (ITTO) Guidelines for the

---

[474] APRIL, 2004.
[475] Smout, 2000.

[476] Good accounts of this work are given in Nilsen, 1991, and Whisenant, 1999.

Establishment and Sustainable Management of Planted Forest[477] remains a good source of information on the important issues. But those guidelines were issued 11 years ago, and they give only passing attention to landscape and biodiversity issues. These are the areas of current concern, and the rest of the chapters in this volume address issues that are pertinent to this issue. The more recent ITTO Guidelines for the Restoration, Management, and Rehabilitation of Degraded and Secondary Tropical Forests[478] go further in addressing these larger scale issues. They probably constitute the best technical document currently available on the role of plantations in restoring landscape functions.

The key to harnessing the potential beneficial roles of plantations will be to develop a vision of what the ideal configuration of the landscape would look like. This vision needs to be based on an understanding of the uses that all stakeholders will make of the landscape. Public participation in the process of developing this vision is important. Commercial plantation companies must be brought into this process as early as possible and be convinced that the commercial viability of their enterprises will be enhanced through developing their plantations in an environmentally sustainable way. Arguments for this might include the avoidance of local opposition or even sabotage of the plantations, the possibility of achieving green certification and thus better market access, and the general advantages that come with being seen as good corporate citizens.

The basic principle needs to be agreed on—that plantation forestry can and should provide multiple production and environmental functions. This multifunctionality can be achieved through diversification within the plantation or by the development of landscape mosaics that are designed in such a way that production and environmental functions are spatially distributed so that the "whole is greater than the sum of the parts." Achieving optimal landscape mosaics is often difficult because it requires coordinated land allocation by different land managers and owners. Formal spatial planning can often achieve this, but informal negotiations amongst local land owners can also be effective. Some large plantation operators control enough land to establish mosaics within a single land-holding.

A number of publications deal with the issue of how plantation management can support biodiversity conservation objectives. Several of these are listed in the references to this chapter. Many of them focus on the biodiversity that can be encouraged within the plantations themselves. There is now more interest in the landscape ecology of plantation forestry. Significant recent experience comes from Western Europe and the Mediterranean, and the books on landscape ecology listed in the references begin to describe these experiences.

4. Future Needs

Much still has to be learned about how emerging understanding of landscape ecology can be used as a tool for forest landscape restoration. This is one of the challenges of conservation for the coming decades.

A new challenge is emerging that will play a major role in the future of plantations and landscapes. This is the prospect of significant funding for afforestation in attempts to sequester carbon. These forest plantations will be acceptable to the conservation community only if they provide multiple environmental benefits. This means that forests established to sequester carbon will have to provide landscape and biodiversity benefits as well. They will have to contribute to forest landscape restoration.

References

APRIL. 2004. Sustainability Report 2004. Asia Pacific Resources International Holdings ltd. Jakarta, Indonesia.

Carrere, R., and Lohmann, L. 1996. Pulping the South: Industrial Tree Plantations and the World Paper Economy. Zed Books, London.

---

[477] ITTO, 1993.
[478] ITTO, 2002.

Cossalter, C., and Pye-Smith, C. 2003. Fast Wood Forestry: Myths and Realities. CIFOR, Bogor, Indonesia.

ITTO. 1993. ITTO Guidelines for the Establishment and Sustainable Management of Planted Tropical Forests. ITTO, Yokohama, Japan.

ITTO. 2002. ITTO Guidelines for the Restoration, Management and Rehabilitation of Degraded and Secondary Tropical Forests. ITTO, Yokohama, Japan.

IUFRO. 2003. Occasional paper No. 15. Part 1: Science and technology—building the future of the world's forests. Part ll: Planted forests and biodiversity. ISSN 1024–1414X, IUFRO, Vienna.

Lang, C. 2002. The pulp invasion; The international pulp and paper industry in the Mekong Region. World Rainforest Movement, Moreton-on-the-Marsh, UK.

Liu, J., and Taylor, W.W. 2002. Integrating Landscape Ecology into Natural Resource Management. Cambridge University Press, Cambridge, UK.

Nilsen, R., ed. 1991. Helping Nature Heal: An Introduction to Environmental Restoration. A Whole Earth Catalogue, Ten Speed Press, Berkeley, CA. (Deals with restoration in a U.S. context.)

Smout, T.C. 2000. Nature Contested: Environmental History in Scotland and Northern England since 1600. Edinburgh University Press, Edinburgh, UK.

Whisenant, S.G. 1999. Repairing Damaged Wildlands—A Process-Oriented, Landscape-Scale Approach. Cambridge University Press, Cambridge, UK.

Buckley, G.P., ed. 1989. Biological Habitat Reconstruction. Belhaven Press, London.

Cairns, J. Jr., ed. 1988. Rehabilitating Damaged Ecosystems, Vols 1 and 2. CRC Press, Boca Raton, FL.

FAO. 2001. Global Forest Resources Assessment 2000—Main Report. Forestry Paper 140. ISBN 92-5-104642-5. FAO, Rome.

Gobster, P.H., and Bruce Hull, R., eds. 1999. Restoring Nature: Perspectives from the Social Sciences and Humanities. Island Press, Washington, DC.

Holl, K.D., Loik, M.E., et al. 2000. Tropical montane forest restoration in Costa Rica: overcoming barriers to dispersal and establishment. Restoration Ecology 8(4):339–349.

Jordan, W.R. III, Gilpin, M.E., and Abers, J.D., eds. 1987. Restoration Ecology: A Synthetic Approach to Ecological Research. Cambridge University Press, Cambridge, UK.

Lamb, D. 1998. Large scale ecological restoration of degraded tropical forest lands: the potential role of timber plantations. Restoration Ecology 6(3): 271–279.

Luken, J.O. 1990. Directing Ecological Succession. Chapman and Hall, London.

Reiners, W.A., and Driese, K.L. 2003. Propagation of Ecological Influence Through Environmental Space. Cambridge University Press, Cambridge, UK.

Walker, L.R., and del Moral, R. 2003. Primary Succession and Ecosystem Rehabilitation. Cambridge University Press, Cambridge, UK.

Additional Reading

Aide, T.M., Zimmerman, J.K., et al. 2000. Forest regeneration in a chronosequence of tropical abandoned pastures: implications for restoration ecology. Restoration Ecology 8(4):328–338.

Web Sites

www.metsopaper.com.
www.developments.org.uk/data/issue21/amazon.htm.

55
Attempting to Restore Biodiversity in Even-Aged Plantations

Florencia Montagnini

Key Points to Retain

While even aged plantations offer much less biological wealth than natural forests, they may prove more valuable than severely degraded lands and may even be a step along the way to restoring a forested landscape.

Plantations can help recovery of biodiversity by (1) attracting seed dispersers, (2) reducing grasses and favouring the growth of seedlings, and (3) ameliorating the microclimate.

Plantations can be designed to improve biodiversity by (1) planting at low densities, (2) using mixed-species designs, (3) using native species, (4) planting close to a natural seed source (forest), and (5) thinning to allow more native vegetation to come through.

Further work is necessary on how to achieve better plantation connectivity with forests across landscapes, and on improving legislation related to plantations.

1. Background and Explanation of the Issue

Even-aged plantations (i.e., plantations that were established by planting tree seedlings all at the same time, or within a few months of each other) are the most frequent plantation type in both tropical and temperate regions. In general these plantations are monospecific (i.e., planted with a single species in large blocks). Frequently, they are composed of exotic species (for example, pine plantations in the Southern Hemisphere; plantations of eucalypts in any temperate or tropical region except Australia; teak in Indonesia or Latin America). The majority of plantations are established for industrial purposes (timber or fibre). However, in addition to providing wood products, plantations could have a function in combating desertification, providing fuelwood, protecting soil and water resources, rehabilitating degraded lands, providing rural employment, and absorbing carbon to offset carbon emissions.[479] Tree plantations can also be a source of cash, savings, and insurance for local farmers.

With regard to biodiversity conservation or restoration, plantations are often viewed in a negative light.[480] It has been claimed that monocultures of exotic plantations are no more diverse than monocultures of soybeans or other agricultural crops. Some authors do not even want to use the term *forest plantations*, claiming that monospecific plantations are not truly forests.

However, while plantations in general support fewer native wildlife species than a natural forest, they may sometimes hold more diversity that other land uses in the same region (e.g., agricultural land, pastures, degraded land).

---

[479] Keenan et al, 1999; Lamb, 1998; Montagnini, 2001.
[480] Carnus, 2003.

1.1. Even-Aged Plantations and Biodiversity

Plantations may serve biodiversity under certain conditions:

1. In severely degraded areas: Plantations can support a greater diversity of native plant species in their understoreys than agriculture or pasture systems. Plantation composition, design, and management will vary according to the objectives of the plantation, and so will the factors that influence biodiversity within and around them.

2. In areas where natural regeneration is very slow or very difficult: In some areas, natural forest regeneration may be significantly delayed by physical or biological barriers (e.g., distance from seed source, heavily compacted terrain, etc.). The establishment of plantations may overcome some of these barriers by attracting seed dispersal agents into the landscape and by ameliorating local microclimatic conditions within the area, thereby accelerating the recovery of biodiversity. Plantations may help local biodiversity by facilitating regeneration of native tree species and providing habitat for forest animals.[481]

If large-scale, monospecific plantations are in full production, concern for biodiversity by company owners is often restricted to the conservation areas that they maintain by law or as a result of pressure from society. Nonetheless, there are exceptions, such as when plantations are managed to address particular conservation pressures. The prime interest in a plantation will not be biodiversity; however, conservation or restoration of biodiversity may become a secondary objective. In general, there are cases in which restoration of biodiversity and naturalness in existing plantations is justifiable and should be actively sought, for instance:

- Where plantations are established on degraded land that could be restored into native forest
- Where plantations have been abandoned
- When even quite unnatural plantations can still provide habitat for a specific and important species through quite simple management changes—for example, managing plantations for specific nesting birds or mammals that can utilise them, such as the establishment of *Dipteryx panamensis* plantations, a species whose seeds are feed for the endangered green macaw in North Eastern Costa Rica
- When plantations are situated close to biologically important areas, and where changes in the management of the plantation can help maintain or support these areas
- When part of the plantation land, either by law, economics, or feasibility, is not under plantation and could be managed in such a way as to counterbalance the effect of the plantation on biodiversity

To make judgements about when and where these approaches might be applicable, it is important to understand the context in which the plantation exists and the factors that alter biodiversity. In all these cases, a key question is whether the desired changes should come about by allowing or encouraging natural regeneration or whether some more active type of intervention is needed. In some cases, restoration may result in a more natural forest overall; in others, the plantation may remain as a highly unnatural crop but with specific elements that support a small number of desired species (which can also be important to maintain a functional landscape).

1.2. Factors that Alter Biodiversity in Even-Aged Plantations

The following factors can alter biodiversity in even-aged plantations:

- The use of nonnative species: Although they do not always become invasives, nonnative species are often less adapted to environmental conditions, could disturb the ecological balance between functional groups of species, both vertebrates and invertebrates, and could result in ecosystem viability problems in the long run. They may also thrive out of control because of the absence of their traditional predators.
- Tree species' diversity, pure or mixed plantations: Diversity is clearly less in monospecific

---
[481] Cusack and Montagnini, 2004; Parrotta, 1992; Parrotta and Turnbull, 1997.

than in mixed plantations. In contrast, in mixed plantations there is a greater variety of habitats both in the vertical and horizontal dimensions of space that can also attract a larger number of animals (birds, bats, and other mammals), which can act as seed dispersers of species from nearby natural forests.
- Loss of forest habitats and microhabitats: If a plantation replaces natural forest, there is a loss of species. That is the case with many reforestation projects in the tropics, where plantations of a single species are established in areas that once supported rain forest.
- Loss of other natural habitats: Sometimes plantations are established in regions that have never supported forest in historical times (afforestation), for example, pulpwood and timber plantations in the delta of La Plata river in Argentina, and in Uruguay, where the natural ecosystems are prairies. In these, plantations result in loss of specific biodiversity and landscape naturalness.
- Status of plantation exploitation: When a plantation is no longer productive, due, for example, to market changes that have affected the prices of tree products, plantation owners may not manage the plantations for production, but may let natural regeneration proceed under the plantation canopies. For example, several plantations were established in Puerto Rico by the U.S. Forest Service and the Department of Natural Resources in the 1960s. Management of these plantations was limited and abundant understorey biomass and species' diversity is found under the canopy of Caribbean pines, mahoganies, and other exotic species.
- Chemical influences on soils by tree species: Eucalypts have been claimed to have negative effects on understory vegetation[482]; however, effects may vary according to the species and sites. For example, in highland ecosystems in Ethiopia, richness and biomass of herbaceous species in plantations of eucalypts and pines were as high as in natural forest (most of the species found under plantations were widespread species, mainly weeds invading from montane or wooded grassland).[483]

[482] Cossalter and Pye-Smith, 2003.
[483] Michelsen et al, 1996.

FIGURE 55.1. Understory regeneration under the native tree species *Vochysia guatemalensis* in a 12-year-old plantation at La Selva Biological Station, Costa Rica. (Photo © Florencia Montagnini.)

2. Examples

2.1. Increasing Biodiversity in Tropical Plantations by Mixing Indigenous Tree Species (Costa Rica)

At La Selva Biological Station, mixed plantations that integrated native tree species had a relatively high abundance and high numbers of regenerating species in their understory, as opposed to pure plantations.[484] Higher plant species' richness accumulated under *Vochysia guatemalensis*, *Virola koschnyi*, *Terminalia amazonia*, *Hyeronima alchorneoides*, and *Vochysia ferruginea*—all species commonly planted by farmers in the region (Fig. 55.1). Natural regeneration was higher in understoreys with low or intermediate light availability. Most of

[484] Cusack and Montagnini, 2004.

FIGURE 55.2. Regeneration of woody species was very low in areas not used by plantations, in comparison with regeneration under plantations of native tree species at La Selva Biological Station, Costa Rica. Seeds were collected from under each plantation species and in areas not covered by trees for comparison of seed dispersal by birds and bats. (Photo © Florencia Montagnini.)

the seeds entering open pastures were wind-dispersed, while most seeds entering the plantations were bird- or bat-dispersed. This suggests that the plantations facilitate tree regeneration by attracting seed-dispersing birds and bats into the area (Fig. 55.2). The different species of the plantations created different conditions of shade and litter accumulation, which in turn affected forest regeneration.[485] Competition from grasses is a major factor influencing woody invasion under these plantations. High accumulation of litter on the plantation floor may help diminish grass growth and thus encourage woody invasion under the species' canopies. Farmers who manage their plantations with the purpose of restoring local biodiversity may have as an option, after harvesting the timber, the tending of the natural regeneration of useful species. In this manner, they obtain the profits from selling the timber from the plantation, and later they will have valuable timber species in the regenerated forest.

2.2. Thinning to Restore Biodiversity in Pure Plantations of Teak (Costa Rica)

In the Parrita valley, seven teak stands of three to 12 years and one 49-year-old stand that had been planted on old pasture or agricultural land were surveyed.[486] Soils were acid Ultisols and Inceptisols. Initial spacing was 2½ by 3 m or 3 by 3 m. Plots were set along transects where basal area of trees, open canopy percentage, leaf litter, percent plant cover, number of individuals, and biomass of understorey were measured. A total of 66 plant families and 132 genera were recorded. Teak density was the strongest predictor of understorey development; therefore, it was concluded that thinning is the most important management strategy to increase understorey biodiversity in these plantations.

2.3. Restoring Indigenous Biodiversity While Dealing with Invasive Species in Plantations

In several cases, a previously forested area is invaded by aggressive grass, for example, *Imperata cylindrica* in Indonesia, *Imperata brasiliensis* in Brazil, *Saccharum spontaneum* in Panama, and *Pennisetum purpureum* in Africa, or by ferns. The competitive advantage of grasses, combined with degraded soils and lack of nutrients, prevents germination and initiation of tree seedlings. These grassland areas are often maintained by fires that inhibit colonisa-

---

[485] Carnevale and Montagnini, 2002.

[486] Luoma, 2002.

tion by tree species.[487] In many cases it is not feasible to plant tree seedlings without first removing the invasive vegetation. Following treatment to eliminate or reduce the invasive vegetation, fast-growing tree species, often exotic, are planted to initiate tree cover, suppress the grass, and ameliorate the environment.[488] This facilitates the establishment of other tree seedlings that may be brought later to restore the original forest, or to start a mixed or a monospecific plantation, depending on the objectives.

2.4. Fighting Invasive Species in a Plantation in the Eastern United States

In the eastern U.S., one of the most challenging invasive plants for forest restorationists is the nonindigenous shrub, Amur honeysuckle (*Lonicera maackii*), which has an ability to resprout after cutting and possibly has allelopathic effects on native vegetation, turning invaded sites into shrublands.[489] In southwestern Ohio, glyphosate herbicides were used to eliminate honeysuckle and facilitate the establishment of planted seedlings of native tree species (*Fraxinus pennsylvanica, Quercus muehlenbergii, Prunus serotina, Juglans nigra, Cercis canadensis, Cornus florida*). The end result was successful restoration with an increase in native woody plant diversity.

3. Outline of Tools

3.1. Role in Attracting Seed Dispersal Agents into the Landscape

Positioning of plantations in the landscape influences the movements of seed-dispersing birds. For example, plantations attract more dispersers if they are set between forest patches to facilitate bird movement. Tree recruitment may be higher in plantations that are connected to forests by long and narrow patches. Some species are better as "perches" due to their architectural characteristics. For example, at La Selva Biological Station in Costa Rica, more abundant regeneration was found under the canopy of *Vochysia guatemalensis* than under other native species of the same plantation.[490] The result was attributed in part to the architecture of this species, whose branching pattern is particularly suited to birds and bats. In addition, the architecture of this tree species allowed for a more varied light environment that could accommodate a larger number of species.

3.2. Planting to Improve Local Microclimatic Conditions

As mentioned in the examples above, plantations create better light conditions for seedlings that are shade-tolerant. Plantation shade suppresses grass and fern growth, thus favouring the growth of woody seedlings. Temperature fluctuations are also ameliorated under the canopy. Litter production can also help suppress the growth of grass.[491]

3.3. Factors Influencing Natural Regeneration Under Plantation Canopies

3.3.1. Plantation Type

A low-density plantation may favour growth of grasses instead of a varied understorey. An initial tight tree planting density (2×2 m, 3×3 m) will ensure early shading of grass, thus favouring competition by shade-tolerant woody seedlings. Thinning will be needed later to free up the growing tree seedlings.

3.3.2. Plantation Design

Mixed plantations have a higher variety of environments for seed dispersers and create greater variety of ecological niches allowing for more diverse regeneration.

---

[487] Chapman and Chapman, 1996.
[488] Ashton et al, 1997; Fimbel and Fimbel, 1996; Kuusipalo et al, 1995; Otsamo et al, 1999; PRORENA, 2003.
[489] Hartman and McCarthy, 2004.
[490] Guariguata et al, 1995.
[491] Lamb, 1998; Parrotta and Turnbull, 1997.

Planting at different times so as to have a mosaic of plantations of different ages is often done to suit different market demands. This offers a more varied environment that can help recruitment of other species and can create different niches and habitats that may favour some wildlife.

Planting at wider distances and thinning can allow greater light penetration in the understorey. At the same time, early shading by a rapidly developing plantation canopy may help suppress aggressive grass vegetation, therefore favouring broad-leaved species in colonising the understorey and thus increasing biodiversity.

3.3.3. Distance to Natural Forest or Other Sources of Seeds

Regeneration may be seriously prevented by lack of seed and other propagules, if plantations are set like islands in a sea of pasture or other degraded vegetation.

3.3.4. Species' Choices

Native pioneer species should be the first choice because fast-growing pioneer species shade out grasses sooner. Native species are in better balance with the rest of the ecosystem. However, in extreme cases, when the land has been too damaged for native species to grow on, exotics are an option as shown in the examples.

3.3.5. Plantation Management

Thinning is probably the most important management intervention to favour regenerating trees in plantations. For example, an analysis of forest restoration after 120 years of reforestation with the exotic *Pinus nigra* in the Alps in France, showed that in order for the pines to serve as a true nurse for the native broad-leaved vegetation, thinning and enrichment planting were needed. Thinning facilitates the dissemination of seeds of the native species. Gap openings or even small clear cuts in the pine plantations were recommended in areas affected by infestation with mistletoe. Planting patches of native trees can serve as seed sources for future regeneration of native species.[492]

There are a variety of management strategies that can be used to increase diversity in plantation ecosystems, even those including exotic species. These strategies include thinning, as mentioned above; decreasing the intensity of management operations (fertilisation, weeding); diversifying the number of tree species planted; planting so as to have a mosaic of plantations of different ages; and leaving forest remnants in the landscape.[493] Management strategies that fall within the guidelines needed for forest certification (according to schemes such as the Forest Stewardship Council scheme, FSC) help to ensure that plantation forests as well as native forests are managed in a way that promotes wildlife habitat.

4. Future Needs

More experiences are needed on plantations and connectivity across landscapes. For example, connectivity can be obtained through the use of lines or even isolated trees in the landscape, serving to buffer the actual plantation area, changing the "shape" of the plantation, etc.

There needs to be more work on the relationship between the plantation itself and its surroundings. Taking a landscape approach helps deal with both the area inside a plantation and the area around it.

More information is needed on the long-term dynamics of tree regeneration in plantations; most studies focus on young plantations.

Specific management guidelines are needed to favour biodiversity, especially thinning and enrichment. For example, Ashton et al[494] designed a comprehensive set of guidelines suited to the forests of Sri Lanka. The guidelines indicate silvicultural treatments needed for a number of understorey and canopy species, including size of the canopy openings

---

[492] Vallauri et al, 2002.
[493] Carnus et al, 2003.
[494] Ashton et al, 2001.

needed for each species, and mode of planting (isolated seedlings or in groups or patches), as well as the economic value of each species. See next chapter "Best Practice for Industrial Plantations" for other management interventions to promote biodiversity.

Attention should be given to alternatives that can help farmers to increase biodiversity while maintaining a profitable system, by enquiring into farmers' goals and preferences for tree species.

Finally, many countries need to improve legislation related to subsidies and establishment and monitoring of plantations, and their influence on biodiversity.

References

Ashton, P.M.S., Gamage, S., Gunatilekke, I.A.U.N., and Gunatilekke, C.V.S. 1997. Restoration of a Sri Lanka rainforest: using Caribbean pine *Pinus caribaea* as a nurse for establishing late-successional tree species. Journal of Applied Ecology 34:915–925.

Ashton, P.M.S., Gunatilleke, C.V.S., Singhakumara, B.M.P., and Gunatillcke, I.A.U.N. 2001. Restoration pathways for rain forest in southwest Sri Lanka: a review of concepts and models. Forest Ecology and Management 525:1–23.

Carnevale, N.J., and Montagnini, F. 2002. Facilitating regeneration of secondary forests with the use of mixed and pure plantations of indigenous tree species. Forest Ecology and Management 163:217–227.

Carnus, J.-M., Parrotta, J., Brockerhoff, E.G., et al. 2003. Planted forests and biodiversity. In: Buck, A., Parrotta, J., and Eolfrum, G., eds. Science and Technology—Building the Future of the World's Forests. Planted Forests and Biodiversity. IUFRO Occasional Paper No. 15. IUFRO, Vienna, Austria, pp. 33–49.

Chapman, C.A., and Chapman, L.J. 1996. Exotic tree plantation and the regeneration of natural forest in Kibale National Park, Uganda. Biological Conservation 76(3):253–257.

Cossalter, C., and Pye-Smith, C. 2003. Fast-wood forestry. Myths and realities. Forest perspectives. Center for International Forestry Research (CIFOR), Jakarta, Indonesia.

Cusack, D., and Montagnini, F. 2004. The role of native species plantations in recovery of understory diversity in degraded pasturelands of Costa Rica. Forest Ecology and Management 188:1–15.

Fimbel, R.A., and Fimbel, C.C. 1996. The role of exotic conifer plantations in rehabilitating degraded tropical forest lands: a case study from the Kibale forest in Uganda. Forest Ecology and Management 81:215–226.

Guariguata, M.R., Rheingans, R., and Montagnini, F. 1995. Early woody invasion under tree plantations in Costa Rica: implications for forest restoration. Restoration Ecology 3(4):252–260.

Hartman, K.M., and McCarthy, B.C. 2004. Restoration of a forest understory after the removal of an invasive shrub, Amur honeysuckle (*Lonicera maackii*). Restoration Ecology 12(2):154–165.

Keenan, R.J., Lamb, D., Parrotta, J., and Kikkawa, J. 1999. Ecosystem management in tropical timber plantations: satisfying economic, conservation, and social objectives. Journal of Sustainable Forestry 9:117–134.

Kuusipalo, J., Goran, A., Jafarsidik, Y., Otsamo, A., Tuomela, K., and Vuokko, R. 1995. Restoration of natural vegetation in degraded *Imperata cylindrica* grassland: understory development in forest plantations. Journal of Vegetation Science 6:205–210.

Lamb, D. 1998. Large scale ecological restoration of degraded tropical forest lands: the potential role of timber plantations. Restoration Ecology 6:271–279.

Luoma, J. 2002. Understory vegetation characteristics along teak (*Tectona grandis*) plantation/natural forest ecotones in Costa Rica. In: Tropical Resources: The Bulletin of the Tropical Resources Institute. Yale University, School of Forestry and Environmental Studies, New Haven, CT, pp. 11–16.

Michelsen, A., Lisanework, N., Friis, I., and Holst, N. 1996. Comparison of understory vegetation and soil fertility in plantations and adjacent natural forests in the Ethiopian highlands. Journal of Applied Ecology 33:627–642.

Montagnini, F. 2001. Strategies for the recovery of degraded ecosystems: experiences from Latin America. Interciencia 26(10):498–503.

Otsamo, A., Hadi, T.S., Kurniati, L., and Vuokko, R. 1999. Early performance of 12 *Acacia crassicarpa* provenances on an *Imperata cylindrica* dominated grassland in South Kalimantan, Indonesia. Journal of Tropical Forest Science 11(1):36–46.

Parrotta J.A. 1992. The role of plantation forests in rehabilitating degraded tropical ecosystems. Agriculture, Ecosystems and Environment 41:115–133.

Parrotta, J.A. and Turnbull, J. 1997. Catalizing native forest regeneration on degraded tropical lands. Forest Ecology and Management 99:1–290.

PRORENA. 2003. The Native Species Reforestation Project (PRORENA) Strategic Plan 2003–2008. Center for Tropical Forest Science (CTFS) (Smithsonian Tropical Research Institute) (STRI), and Tropical Resources Institute at the Yale School of Forestry and Environmental Studies, New Haven, CT. (Unpublished document.)

Vallauri, D., Aronson, J., and Barbero, M. 2002. An analysis of forest restoration 120 years after reforestation of badlands in the south-western Alps. Restoration Ecology 10(1):16–26.

Additional Reading

Montagnini, F., and Jordan, C.F. 2005. Plantations and agroforestry systems. pp. 163–215. In: Montagnini, F., and Jordan, C.F. 2005. Tropical Forest Ecology. The Basis for Conservation and Management. Springer-Verlag, Berlin-New York.

Piotto, D., Montagnini, F., Kanninen, M., Ugalde, L., and Viquez, E. 2004. Forest plantations in Costa Rica and Nicaragua: performance of species and preferences of farmers. Journal of Sustainable Forestry 18(4):57–77.

56
Best Practice for Industrial Plantations

Nigel Dudley

> **Key Points to Retain**
>
> Forest plantations have been a major threat to forests and forest biodiversity because of poor management practices and little or no planning for their location within landscapes.
>
> Well-managed and appropriately located plantations, however, can sometimes play an important role in healthy, diverse, and multifunctional forest landscapes.
>
> There is an urgent need for capacity building with respect to good social and environmental management for plantations.

1. Background and Explanation of the Issue

The area of forest plantation in the world has increased by 17 percent in the last decade, half from the conversion of natural forests to plantations and half from afforestation or reforestation on previously nonforested or deforested lands. Timber plantations often impose significant environmental and social costs, particularly when they are established through the conversion of natural forests, as has often been the case, for example, in Indonesia and Chile. Indiscriminate forest clearing, uncontrolled burning, and disregard for the rights and interests of local communities have often been associated with plantation establishment. Unless there are significant changes in policies and practices, in many regions the expansion of plantations will continue to threaten forests of high conservation value, freshwater ecosystems, forest-dependent peoples, and habitats of endangered species. However, well-managed and appropriately located plantations can play an important role in healthy, diverse and multifunctional forest landscapes, for instance, by providing a sustainable source of timber and freeing up other areas to be set aside as reserves. The plantation industry can also, if properly managed, generate valuable foreign exchange earnings and employment opportunities for producer countries. The principles of forest landscape restoration recognise that plantations can play a role in a sustainable forest landscape, if they are well managed and have the support of local communities and are well-sited within the landscape (e.g., not in areas of high or potentially high biodiversity). Key elements of sustainability within the plantation forest industry are the following:

- Maintenance of high conservation value forests: plantations should not replace high conservation value forests. This will normally require well-informed negotiations among a wide range of stakeholders to integrate plantations with the mosaic of other land uses.
- Multifunctional forest landscapes: plantations should enhance environmental values by providing corridors between, and buffer zones around, natural forest areas and should

enhance social values by providing benefits to local communities.
- Sound environmental management practices: the industry should adopt management practices that minimise environmental impacts such as air and water pollution, forest fires, soil erosion, pest invasion, and biodiversity loss.
- Respect for rights of local communities and indigenous peoples: the industry should recognise legal and customary rights of local and indigenous communities to own, use, and manage their lands, territories, and resources.
- Positive social impacts: the industry should maintain or enhance the social and economic well-being of plantation workers and communities.[495]
- Proficient regulatory frameworks: regulatory frameworks should encourage best practices. At a minimum, the industry should respect all national laws. Responsible behaviour will often require performance standards exceeding local and national laws, especially where regulatory frameworks are underdeveloped or governance is weak.
- Transparency: the industry should adopt and make public, policies, practices, and implementation plans pertaining to their social and environmental performance. They should encourage independent, publicly available performance monitoring, involving local stakeholders in both development of standards and performance monitoring.

2. Outline of Tools

Assuring that plantations play a positive rather than a negative role depends on two factors: locating plantations in places where they do not destroy valuable natural habitat or undermine people's livelihood options, and managing them in ways that minimise detrimental impacts.

2.1. Locating Plantations

Many plantations are badly planned. Baseline surveys and consultation with local communities can help to reduce problems. A number of tools exist:

- Initial cost-benefit analysis: draws on desk studies, remote sensing, and initial site surveys to determine whether further investment is justifiable, and covers government policies and regulations; tenure; social issues relating to local communities; geography (soil, climate, topography); existing land use; nearby protected areas; existing and planned infrastructure (roads, rivers, etc.); options for plantation species; and economics.
- Feasibility study: provides the information needed to make the decision about whether or not to go ahead with the project, covering topography; vegetation/land cover; ecology and biodiversity; soils; hydrology of major watercourses and ground water sources; land use and land rights; socioeconomics; interest in investment projects; field trials of possible plantation species if necessary; and economics.
- Principles for plantation establishment: several existing principles provide the basis for site location and should include minimising impact on important natural habitats and minimising detrimental impacts on local human communities.

2.2. Managing Plantations

Once a suitable site has been identified, care needs to be taken to minimise the environmental and social costs of the plantation, with particular emphasis on groundwater contamination, soil erosion, and fire disturbance. Several codes of practice and detailed guidelines exist[496] and it is possible to apply for a credible third-party certification scheme. An outline guide to best practice is given in Table 56.1, designed to be used as a site-level rapid assessment tool.[497]

---

[495] Davis-Case, 1990.

[496] Dykstra and Heinrich, 1996; FAO, 1977, 1978.
[497] In addition to the references given immediately above it also draws on Burrough and King, 1989; Hamilton, 1988; Hurst et al, 1991; Sedlack, 1988a,b.

TABLE 56.1. Guide to helping plantation managers.

PLANNING
Has a feasibility study been carried out?
Has an environmental impact assessment been carried out?
Does a management plan exist?
Does the management plan include biodiversity and environmental issues?
Does the management plan include social concerns?
SOCIAL VALUES
Protecting peoples' rights
Have stakeholders been consulted?
Have efforts been made to include all relevant stakeholders?
Have vulnerable human communities been included in the consultation?
Has information about the plantation been distributed in the vicinity?
Have efforts been made to find out opinions about the plantation?
Are local people involved in management decisions?
Rate level of involvement (check one)
    Active consultation
    Seeking consensus
    Negotiating
    Sharing authority
    Transferring authority
Benefits to the local community
Is there a local community liaison officer employed by the plantation?
How many jobs does the plantation provide?
    Permanent
    Temporary
What proportion of jobs goes to local people?
    Permanent
    Temporary
Are wage levels equivalent to national standards?
Does the plantation provide the following benefits to the local community:
Preferential access to its products?
Improved roads and other infrastructure?
Opportunities for community involvement in management?
Recreational opportunities?
Hydrological services (improved freshwater and fisheries downstream)?
BIOLOGICAL VALUES
Provision for biodiversity
Is there a biodiversity conservation officer for the plantation?
Is there a biodiversity plan for the plantation?
Are workers instructed regarding biodiversity conservation?
Is the plantation established in place of
    Primary forest?
    Secondary forest?
    Scrub?
    Farmland?
    Deforested land?
    Unforested land?
Does the plantation contain adequate provision for the protection of the following habitats:
Remaining natural or seminatural forest fragments?
Protection forests, e.g., to protect degraded sites, slopes, and landscape values?
Riparian woodland and other natural vegetation?
Wetland areas, peat, and marshes?
Individual trees in the landscape (e.g., for raptor nests)?
Other microhabitats (corridors, nest sites, lairs, etc.)?
Has there been restoration of natural forests within the plantation?
Is biodiversity conservation adequate within the plantation?
Rare or threatened species?

TABLE 56.1. *Continued*

Protection of protected areas
Is the plantation within a protected area?
Does the plantation directly border onto a protected area?
Has the plantation increased access to a protected area (e.g., for bush meat hunting or illegal logging)?
Protection of cultural sites and aesthetic values
Is there a staff member specifically responsible for protection of cultural and aesthetic values?
Has an integrated management plan been developed that incorporates cultural values?
Is provision made for protection of the following artefacts:
Archaeological sites (e.g., earthwork fortifications)?
Historical sites (e.g., buildings, pathways, etc.)?
Spiritual sites (e.g., sacred groves, graves, etc.)?
Burial sites?
Readily identifiable cultural sites such as buildings?
Cultivated areas (e.g., fruit gardens)?
Areas of local distinctiveness and importance?
Areas where vegetation management has important historical associations (e.g., ancient coppice)?
ENVIRONMENTAL VALUES
Does the plantation have a detailed policy for minimising environmental damage during site preparation, planting, fertiliser use, thinning, and harvesting?
Is there a staff member specifically responsible for environmental management?
Site preparation
Does site preparation include some or all of the following:
Steps to avoid sensitive soils?
Soil erosion control measures?
Contour ploughing on steep slopes?
Elimination of heavy machinery in wetland areas?
Provision of cut-off drains on steep slopes?
Construction of settling pools in drainage systems?
Steps to avoid using heavy machinery when soil moisture is high?
Seepage buffer zones along the contour and alongside natural watercourse?
Planting
Are the following areas avoided in planting:
Cliff edges?
Steep slopes?
Caves and sinkholes?
Buffer areas around watercourses and wetland areas?
Sites of historical and cultural value?
Fertiliser use
Are the following steps taken to minimise damage from fertiliser run-off:
Matching applications to the needs of sites and species?
Use of slow-release fertilisers or slow-release application methods?
Use of application methods that avoid broadcasting fertilisers over the whole area?
Application at the period of maximum growth?
Avoiding application in periods of low growth and/or heavy rainfall?
Avoiding application next to watercourses or near groundwater sources?
Monitoring losses including monitoring of algal blooms near the plantation?
Including alternative methods such as use of tree residues, composts, mulches, and manures?
Harvesting and extraction
Does the plantation take any of the following steps to avoid damage during harvesting:
Avoiding times when soil conditions will encourage erosion?
Planning of compartments and coupes?
Planning extraction routes?
Avoiding felling areas of biodiversity importance?
Avoiding felling areas of cultural importance?
Use of a range of extraction techniques depending on soil and climatic conditions?
Liaison with local people to identify the least disruptive times for harvesting?
Ensuring sufficient supply of safety equipment?

TABLE 56.1. *Continued*

Road building and use
Does the plantation have a staff member especially responsible for road building and maintenance?
Does the plantation take any of the following steps to minimise impacts of road building and use:
Have a plan to minimise length, width and gradient of roads?
Avoid building roads in high erosion risk areas?
Compact roads after construction and ensure revegetation?
Install bridges, ditches, and culverts as needed?
Install cut-off drains, silt traps, and pools?
Use and enforce speed limits?
Limit the size and weight of vehicles using the roads?
Close secondary roads when they are not needed?
Close roads during the wet season or other unsuitable climatic conditions?
Minimise pollution and noise for local communities?

PEST AND WEED CONTROL
Reducing risks from invasive species
Does the plantation take any of the following steps to reduce invasive species:
Avoid likely invasive species?
Practice hygiene in seed and other imported material to avoid introducing pests and diseases?
Planning roads to minimise the spread of invasive species?
Training staff to recognise invasive species?
Have a pest control programme?

Controlling weeds
Does the plantation take any of the following steps to reduce weeds and impacts of weed control:
Instructing all staff in the identification of the main weed species?
Hand weeding?
Flame weeding?
Use of small-scale mechanical weeding equipment?
Spot treatment with herbicides?
Use general herbicides?

Controlling pests and diseases
Does the plantation take any of the following steps to reduce pests and diseases:
Select trees that are resistant to pests and diseases?
Use planting strategies to minimise pest attack (e.g., mosaic of different species and/or ages, including natural forest)?
Train workers to spot pest and disease attack and key pests?
Use cultural and biological controls?
Use pesticides?

Use of pesticides
Does the plantation take any of the following steps to reduce detrimental impacts from pesticides:
Choosing the least toxic and least persistent pesticides?
Ensure that workers are properly trained in safe use of pesticides?
Ensure that safety equipment is available and is used?
Take steps to avoid spray drift or contamination of watercourses?
Store pesticides in secure places?
Minimise the number of occasions on which pesticides are used?

FIRE CONTROL AND MANAGEMENT
Does the plantation have a staff member especially responsible for fire management?
Does the plantation take the following steps to avoid fire:
Liaison with local people to ensure that there is minimal resentment toward the plantation?
Have pubic educational material about fire hazards (e.g., posters or leaflets)?
Planning to minimise fire risks through use of fire breaks, choice of tree species and use of?
Build and staff watch towers?
Appoint local fire prevention officers?
Train and equip staff to combat fires?

STAFF TRAINING
Does the plantation offer any of the following training opportunities:
Relevant written information (translated into the local language if necessary)?
Laminated cards for use in the field (e.g., pest identification charts, pictures of areas to avoid planting)?

TABLE 56.1. *Continued*

Videos of health and environmental safety procedures?
Training courses for permanent and temporary staff?
Relevant training for contractors?
Does the plantation provide information on the following topics:
Social relations regarding the plantation?
Biodiversity management?
Care of the environment during operations?
Pest, disease, and weed control?
MONITORING AND EVALUATION
Does a monitoring and evaluation programme exist for the plantation?
Is the plantation independently certified (e.g., by a certifier affiliated with the Forest Stewardship Council)?

3. Future Needs

There is an urgent need for capacity building with respect to good social and environmental management for plantations, which needs to go beyond the minority of companies that embrace best practice through certification and include pressure on all companies, including through the marketplace, to meet minimum best practice standards. From a technical perspective, better guidelines for site selection are required, as are tools to help plan the retention of natural vegetation within plantations.

References

Burrough, E.R., Jr., and King, J.G. 1989. Reduction in soil erosion on forest roads. USDA Forest Service, General Technical Report INT-264, Ogden, UT.

Davis-Case, D. 1990. The community's toolbox: the idea, methods and tools for participatory assessment, monitoring and evaluation in community forestry. Community Forestry Field Manual No. 2. UN Food and Agriculture Organisation, Rome.

Dykstra, D.P., and Heinrich, R. 1996. FAO Model Guide of Forestry Practice. Food and Agricultural Organisation of the United Nations, Rome.

Food and Agriculture Organisation of the United Nations. 1977. Planning Forest Roads and Harvesting Systems, FAO Paper No. 2, Forestry Department. FAO, Rome.

Food and Agriculture Organisation of the United Nations. 1978. Establishment Techniques for Forest Plantations. FAO Forestry Paper No. 8. FAO, Rome.

Hamilton, L.S. 1988. Minimising the adverse impacts of harvesting in humid tropical forests. In: Lugo, A., Clark, J.R., and Child, R.D., eds. Ecological Development in the Humid Tropics. Winrock International Institute for Agricultural Development, Morrilton, AR.

Hurst, P., Hay, A., and Dudley, N. 1991. The Pesticides Manual. Journeyman Press: London and Concord, MA.

Sedlak, O. 1988a. Principles of Forest Road Nets. Food and Agriculture Organisation of the United Nations, Rome.

Sedlak, O. 1988b. Maintenance of Forest Roads, Food and Agriculture Organisation of the United Nations, Rome.

Part E
Lessons Learned and the Way Forward

57
What Has WWF Learned About Restoration at an Ecoregional Scale?

Nigel Dudley

Key Points to Retain

Forest landscape restoration is a process that should ideally be integrated with protection and sustainable management of forests at a landscape scale.

A suite of different responses is required for successful restoration, depending on circumstances, ranging from policy changes through negotiation, stakeholder processes, research, capacity building, and practical interventions.

Monitoring and the associated evaluation are both critical but present real challenges in addressing forest restoration on a landscape scale.

1. Background and Explanation of the Issue

Although there has been a long history of individual forest restoration projects, until recently few attempts have been made to integrate restoration into either broad-scale conservation or wider sustainable development initiatives. In 2000, WWF the global conservation organisation set a target to run a number of forest landscape restoration initiatives around the world—"at least ten forest landscape restoration initiatives underway"—to test out ideas and approaches to restoring multiple forest functions over a landscape. The target was achieved, providing initial experience of successes and failures, and at the same time those involved were actively learning from the actions of others involved in restoration: conservation and development organisations, governments, and research bodies. The experiences of WWF's partner organisation, IUCN, the World Conservation Union, is particularly relevant here. This chapter summarises some of the main experience to date.

1.1. A Growing Recognition of Need

Until recently, the need for restoration has been more clearly recognised by the development community than by conservation professionals. Many conservation biologists believed that protecting remaining natural or seminatural habitat was a far higher priority than restoring degraded habitat, and that, in any case, restoration could seldom achieve anything of great significance from a conservation perspective. This means that restoration projects have tended to focus on human needs—fuelwood, fodder, windbreaks, etc.—rather than potential conservation benefits. There was resistance to a restoration target even within WWF. Over the 5-year period of the programme, and at least in part as its result, many of these objections have declined or disappeared. Research showed the extent to which many high biodiversity ecosystems are already in need of restoration, either because natural habitat has declined below critical levels or because forest loss is causing wider problems such as siltation of freshwater or mangroves[498]. One implication has been increased support for restoration activities within conservation programmes, including by the Convention on Biological Diversity.

1.2. Restoration Needs to Be Integrated with Protection and Management

Restoration is generally a time-limited process, albeit often a lengthy one, that will eventually result in an ecosystem that either can function by itself, perhaps in a protected area, or requires some level of continual management. One important element in planning restoration is to decide how a restored forest will be managed in the long term, which itself helps to decide what type of restoration activities are required. The transition between "restoration" and "management" can sometimes be quite subtle; for instance, removal of alien invasive species may involve a single operation or a long-term management task. Restoration may sometimes be an intervention in a landscape that is already protected or managed for some other purpose. For example, efforts to increase the deadwood component in some Finnish protected areas involve artificially creating deadwood to help maintain a few endangered saproxylic species (see "Restoration of Deadwood as a Critical Microhabitat in Forest Landscapes"); it is assumed that in the future natural processes will maintain this microhabitat.

1.3. Restoration Should Be Regarded as a Process

Restoration, being a time-limited intervention, is different from other forms of "permanent" management, including protection. Specific restoration projects, therefore, need to identify an end point. This raises philosophical and practical questions about what such an end point could be; many conservation organisations implicitly assume that restoration should seek to re-create a "natural forest" such as might be found in the absence of humans. But many of the world's forests have only developed since *Homo sapiens* evolved and have never existed in a "pristine" prehuman state. More specifically, the social goals of many restoration activities mean that some useful forests may be profoundly unnatural if they are primarily aimed at, for instance, supplying food or energy. This is sometimes also the case from the perspective of biodiversity conservation, for instance, when forests are suppressed by fire to provide savannah habitat or conversely where forests are already so small and fragmented that fire is artificially suppressed to protect remnant species. Setting end points for restoration remains a challenge in many cases and one that involves asking larger questions about the long-term aims of both conservation and development within a landscape.

1.4. A Suite of Responses is Required

Experience from WWF's project portfolio and from other restoration initiatives suggests that the traditional focus of restoration projects on establishing tree nurseries and tree planting is usually irrelevant in terms of creating major changes to forest cover or forest quality, although there are exceptions to this general rule. Large-scale tree planting is also too costly an option for most situations. The programme has experimented with five different responses:

1. Policy changes that can increase the proportion of natural regeneration or near-natural forest management on a major scale—for example, work with the Vietnamese and Chinese governments aiming at making strategic changes to policy initiatives like the Chinese "Grain for Green Programme" and Vietnam's "5-million Hectare Programme", which both currently focus almost exclusively on plantations, to increase the proportion of natural regeneration within these programmes (see "Perverse Policy Incentives" and case study "Monitoring Forest Landscape Restoration in Vietnam").

2. Stakeholder involvement and negotiation at a landscape or ecoregional scale to create

---

[498] Dudley and Mansourian, 2003.

conditions conducive to natural regeneration—for example, work with local organisations in New Caledonia and Madagascar (see case study "Madagascar: Developing a Forest Landscape Restoration Initiative in a Landscape in the Moist Forest") that aims to agree on priorities and actions that will benefit both human society and wildlife

3. Management interventions to change the nature of forest management and thus increase forest quality—for example, initiatives being undertaken by WWF's European Forest Team in terms of responses following major storms or policies toward management of dead timber in secondary forests (see "Restoring Forests After Violent Storms" and the chapter cited above on deadwood.)

4. Use of specialist knowledge in the development and dissemination of technical expertise to facilitate restoration—for example, the guidance being developed in Portugal with the aim of helping improve use of European Union grants (see case study "The European Unions Afforestation Policies and their Real Impact on Forest Restoration") or the use of economic analysis to make the case for natural regeneration of endangered island forest ecosystems in the Danube (see "Practical Interventions that will support Restoration in Broad-scale Conservation").

5. Small-scale strategic tree planting, linked to identification of need through, geographical information system (GIS) mapping and field surveys—for example, to reconnect elephant habitat through oil palm plantations along the banks of the Kinabatangan River in Sabah, Malaysia, to allow natural movement of elephant herds, and to reduce other impacts of forest fragmentation (see "Restoring Quality in Existing Native Forest Landscapes")

1.5. Policy Changes are Often the Most Urgent Challenge

A succession of national and international commitments, practical projects, and workshops have demonstrated general support amongst governments, businesses, and communities to look seriously at the question of restoration. However, most large-scale restoration projects are currently still focussed on a very narrow band of options, including a predominant emphasis on large-scale exotic monocultures. While these may well have a role in the landscape, they are only one fairly small part of what makes up a forest estate. Work with governments in countries as diverse as Vietnam, China, Madagascar, Morocco, the United Kingdom, and Portugal has shown that there is also a willingness to look at new approaches. Progressing from words to actions, including changing well-funded schemes that have already developed some momentum, is a considerable challenge, but is probably the way of making the largest impact. However, policy work is seldom as popular as practical projects with donor agencies or other bodies that might support restoration, as the latter provide instant results for reporting, whereas the impacts of changes in policy, whilst often more profound, are harder to report. Building support for long-term policy work on restoration is an urgent priority.

1.6. Success or Failure is Hard to Measure

Work on Integrated Conservation and Development Projects (ICDPs) suggests that a good monitoring and evaluation system is often the key to success, giving project staff the information needed for the adaptive management that is always needed in a complex project[499]. Development of a monitoring programme, therefore, was the first discrete piece of work undertaken by the WWF restoration programme and this has been tested and applied but is still a long way from capturing all relevant data (see "Monitoring Forest Landscape Restoration in Vietnam"). Many of the changes aimed for by restoration programme are inevitably subtle, may be slow to emerge, and are not easy to capture in simple statistics. Monitoring of impacts or outcomes is inevitably a long-term process. Yet these are precisely the kind of data that many governments and funding agencies require, and much work needs to be done on better monitoring systems.

[499] McShane and Wells, 2004.

1.7. Most Existing Restoration Projects Have Made Little Attempt to Reconcile Ecological and Human Needs

Indeed, as mentioned above, most restoration projects have focussed on human needs, and in fact often on an outsider's perception of what those needs might be, so that, for instance, numerous fuelwood projects have failed because their instigators did not understand the energy needs of local communities, which may have been better served at least in the short term by burning dried dung or other materials than by giving valuable land to tree crops.[500] On the other side, many conservation-based restoration projects have ignored what other stakeholders might require from the landscape altogether, with the result that the pressures causing forest degradation remain and undermine restoration efforts. The need to reconcile social and conservation needs, particularly in landscapes where people are most directly reliant on forest resources, is reinforced by analysis of existing work.

1.8. Many Fundamental Questions Remain Unanswered

When WWF's forest restoration programme began, we assumed that we would draw on a large body of experience. In fact we found more questions than answers. They include quite basic issues relating to, for instance, where natural regeneration might work, the efficacy of biological corridors, how to carry out stakeholder assessments over wide landscapes, and the sustainability of nontimber forest harvests. Many important restoration precepts are based more on assumption than on research, which in part reflects funding difficulties. Restoration needs the injection of research cash that was created for sustainable forest management. Organisations like the Society for Ecological Restoration International can help to spread information, but there is also an urgent need for better coordination between researchers and those involved in practical restoration.

1.9. The Need for a Movement

Social change seldom comes from a single individual or organisation, however much they might like to think so, but instead when impetus for change builds to the extent that it can carry along doubters and overcome opposition. So far, restoration, at least from the perspective of its role as a major part of conservation strategies, has remained the enthusiasm of a minority rather than a widely supported priority. The general lack of restoration programmes within large conservation organisations is an indication of this. The early experience now needs to gain momentum, more support, and, in particular, far more widespread government commitment.

1.10. Lots of Enthusiasm but Little Cash

It has proven surprisingly difficult to raise funding for restoration, which remains outside the experience or the targets of most large donor agencies and even governments. The kind of mass movement for restoration that is now required will also need realistic amounts of money. Building support amongst donor agencies, multilateral lending banks, and government departments, therefore, is also an essential factor in future success.

References

Dudley, N., and Mansourian, S. 2003. Forest Landscape Restoration and WWF's Conservation Priorities. WWF International, Gland, Switzerland.

Leach, G., and Mearns, R. 1988. Beyond the Fuelwood Crisis. Earthscan, London.

McShane, T.O., and Wells, M.P. 2004. Getting Biodiversity Projects to Work: Towards more effective conservation and development. Columbia University Press, New York.

---

[500] Leach and Mearns, 1988.

58
Local Participation, Livelihood Needs, and Institutional Arrangements: Three Keys to Sustainable Rehabilitation of Degraded Tropical Forest Lands

Unna Chokkalingam, Cesar Sabogal, Everaldo Almeida,
Antonio P. Carandang, Tini Gumartini, Wil de Jong, Silvio Brienza, Jr.,
Abel Meza Lopez, Murniati, Ani Adiwinata Nawir,
Lukas Rumboko Wibowo, Takeshi Toma, Eva Wollenberg, and Zhou Zaizhi

Key Points to Retain

Three key lessons have emerged from a Centre for International Forestry Research (CIFOR)-led study on reforestation/rehabilitation/restoration in six countries:

1. It is necessary to strengthen local organisation and participation in restoration projects.
2. It is necessary to consider local socioeconomic needs in choices of approaches and options.
3. In the long run, it is necessary to ensure that clear and appropriate institutional support and arrangements are in place.

1. Background and Explanation of the Issue

In many tropical countries, government agencies, international agencies, the private sector, and civil society have expended much effort and resources in forest rehabilitation activities to meet rising demands both for forest products and environmental services.[501] The projects have differed in scale, objectives, background conditions, and implementation strategies, and results have been variable. It is critical to draw strategic lessons from these experiences and use them to plan and guide future efforts to increase their chances of success and long-term sustainability. The key lessons and examples in this chapter are based on the preliminary results of the study Review of Forest Rehabilitation Initiatives—Lessons from the Past, undertaken by CIFOR in collaboration with national partners in six countries: Peru, Brazil, Vietnam, the Philippines, Indonesia, and China. The study involved a comparison of a full range of forest rehabilitation projects in each country, an assessment of the technical, ecological, and socioeconomic outcomes of selected case studies, and workshops to obtain the inputs of concerned stakeholders (http://www.cifor.cgiar.org/rehab/).

The review focussed on initiatives that aimed to establish trees on formerly forested land to enhance productivity, livelihoods, or environ-

---

[501] Sim et al, 2003; Sayer et al, 2004.

mental services through deliberate technical, socioeconomic, or institutional interventions. Integrated projects with forest rehabilitation components were also included. The assessment looked at any rehabilitation methods that involved trees, including agroforestry, plantations, and assisted natural regeneration.

Countries have chosen a variety of approaches and incentives to rehabilitate degraded land driven by many different considerations. The four Asian countries in the study have a long history of forest rehabilitation, and the governments played a major role in providing funds and implementing projects, particularly in early efforts. International donor–funded forest rehabilitation increased in importance in recent decades. The trend is now toward more private sector, community-based, and local government rehabilitation efforts for production, livelihoods, or environmental benefits. In the Philippines and China, this translates into a diversity of tenurial and institutional arrangements with the involvement of multiple actors and a range of objectives. Project outcomes on the ground are unclear, but China and Vietnam report success in terms of increased forest cover. In Vietnam, China, the Philippines, and recently Indonesia, political motivations and policy changes have led to intermittent large-scale efforts. Planting trees, in particular fast-growing exotic species, has been the predominant method in Asia, although natural regeneration through protection is also important in China and Vietnam.

In contrast to the larger role played by government in Asia, small-scale farmer rehabilitation efforts appear more important in Brazil and Peru, with colonist agriculture and livestock production being the major land degradation factors. The government mainly provides incentives and schemes for farmers' participation. In Brazil, farmers' associations play an important role in project discussion and support. Rehabilitation efforts are also more recent, since the 1990s, and fewer in number, although growing. Projects are small in size and involve agroforestry cash crops, fast-growing native tree species, and integration with other livelihood activities like bee keeping or fish production.

1.1. Three Key Lessons Learned from Past Rehabilitation Projects

Three lessons have been learned on sustaining rehabilitation efforts of degraded tropical forest lands across the six countries reviewed:

1. Strengthen local organisation and participation in projects. More attention should be given to involve, work with, and strengthen local participation from project conceptualisation to implementation and management. Active participation of the key actors taking into account local knowledge and practices is essential for sustaining the effort. Agricultural and forestry policies should aim to develop and strengthen local organisations and promote appropriate strategies for technology transfer. (Fig. 58.1) A well-organised group has higher possibilities of succeeding, particularly during the phases of product harvesting, processing, and commercialisation. Numerous positive and negative cases exemplifying this lesson exist across the Peruvian and Brazilian Amazon, the Philippines, and Indonesia.

2. Consider local socioeconomic needs in choices of approaches and options. Livelihood-enhancing activities must be part of the plan, and projects developed should address the needs of people in the area in order to ensure their participation and interest in sustaining the project. In some instances, rehabilitation projects have actually deprived people of their original livelihoods (such as agriculture on the lands to be rehabilitated), while not providing viable alternatives. Many cases were observed across the Philippines and Vietnam where the project beneficiaries subsequently burned the project area so that they could be reemployed in the process of replanting or rehabilitation. It is imperative to carry out a socioeconomic analysis of promising production systems and small-scale trials before promoting them. It helps if local farmers and communities benefit directly from the rehabilitated forests. Technologies to be promoted should match the situation and capacity of the producers. Tree-based production systems that incorporate tree species with shorter harvesting cycles and good market prospects tend to be more adoptable. Processing and commercialisa-

FIGURE 58.1. Social forestry programme by the Ministry of Forestry with local farmer participation on private lands in East Kalimantan. The planted species, teak, was selected by the farmers. (Photo © Takeshi Toma.)

tion of products should be considered from the start if rehabilitation aims at economic objectives. Integrated production systems (e.g., agroforestry, livestock, and fish) can help increase food security and overcome market instability. Positive and negative cases exemplifying this lesson exist in all six study countries.

3. Ensure clear and appropriate institutional support and arrangements. Strong and appropriate institutional support is critical for promoting investment and local participation in rehabilitation projects, and ensuring their sustainability. This includes clear and undisputed land-tenure status, a facilitating legal framework and policies, and good coordination among agencies at different levels. Also important are formalised institutional arrangements with clear division of tasks, rights, costs, and benefits among multiple stakeholders as a result of thorough and mutually acceptable negotiations. Clear and mutually accepted institutional arrangements help to avoid conflicts, support coordinated project management and fulfilment of assigned tasks, and ensure agreed-upon benefit flows to different stakeholders and their stake in the long-term success of the project. Enforcement of agreements is an important part of such institutional arrangements. Positive and negative cases exemplifying this lesson exist in Vietnam, China, and Indonesia.

These three factors that contribute to successful forest rehabilitation are highly interrelated and occurred across different project types with different implementing actors, project scales, objectives, funding sources, and socioeconomic conditions. Project types ranged from government-driven reforestation to community-based forest management, joint management, state or private company plantations, company–community partnerships, cooperative or group activities, integrated livelihood projects, and private tree farming or agroforestry. Each of the three lessons is illustrated below with cases from different countries. Some cases are illustrative of more than one of the specified lessons, but have been placed under the major lesson to which they relate.

2. Examples

2.1. Strengthen Local Organisation and Participation in Rehabilitation Projects

2.1.1. KMYLB (Farmers Association for Forest Land Inc.) Agroforestry Development Corporation, Brgy, Nugas, Alcoy, Cebu, Philippines

KMYLB is a community-based forest management (CBFM) project of the government of the Philippines' Department of Environment and Natural Resources, located in a public forest area in southern Cebu. The project area of 1651 hectares was occupied by settlers early on and subject to a government-led social forestry programme in the 1980s with many farmers granted the Certificate of Stewardship Contract. This was followed by the issuance of a reforestation contract in 1996 for people to develop the remaining open areas. As part of the reforestation contract, there were community organising activities that gave birth to KMYLB as a people's organisation. The people's organisation was then given the CBFM agreement in 1999 by the government, consolidating the many stewardship contract areas, the plantations, and the remaining natural forests in the area. Community organising was one of the major activities that enabled active community participation in forest development and protection. High levels of cooperation and interest in CBFM activities have been observed among community members. Each member is assured of continuous benefits from the forest through individual forest gardens and community plantations. Many organisational problems did occur, but these were transitory and helped the organisation mature and strengthen its internal policies. The strength of the people's organisation and its successful development and protection of the CBFM area also makes it a magnet for supportive infrastructure and livelihood programmes from international nongovernmental organisations (NGOs) and others.

2.1.2. Agroforestry Development in the Rio Cumbaza Basin, Peru

The San Martín region, with a land area of 1.9 million hectares, is the most deforested area in the Peruvian Amazon. Deforestation and land degradation are mainly due to short-rotation slash-and-burn agriculture and the production of illegal crops. The project Management, Conservation, and Productive Development in the Rio Cumbaza Basin (1997–2001) executed by the NGO CEDISA (Centro de Desarrollo e Investigación de la Selva Alta), promoted agroforestry systems for rehabilitating and maintaining soil productivity (Fig. 58.2). These systems were well received by farmers because they were based on species of economic importance such as coffee, and incorporated promising short-rotation forest tree species (such as *Schizolobium amazonicum*, *Calycophyllum spruceanum*, and *Colubrina glandulosa*) and other species (mainly fruits) traditionally used for subsistence and the local market. Families actively participated in the design and establishment of the rehabilitation areas. The project also promoted the formation of organised farmers' groups to strengthen their negotiation capacity in local and regional markets and with development agencies. One of these is a

FIGURE 58.2. Agroforestry trial for rehabilitating degraded lands and improving farmers' livelihoods in Peru. (Photo © Takeshi Toma.)

committee of ecological farmers who adopted low-impact production strategies (including agroforestry and management of naturally regrowing forests) in buffer zones of protected areas. The project promoted community involvement in conserving and managing their natural resources, in generating added value for their products, and in developing markets for nontraditional timber species.

2.2. Consider Local Socioeconomic Needs in Choice of Approaches and Options

2.2.1. The Bai Bang Pulp and Paper Mill, Vietnam[502]

The Bai Bang Pulp and Paper Mill Project in Vietnam costing $360 million was implemented between 1974 and 1992. The project was designed by the Vietnamese government and Swedish Development Assistance with little consideration of how sufficient wood supply could be obtained from the surrounding region, where there was high pressure on the land from small farmers who subsisted on low-technology agriculture and grazing. As a result, the mill operated at less than full capacity for a long time. The local population challenged the monopoly on the wood and forest land claimed by the forestry sector. Only a minor part of the wood and bamboo cut by forest enterprises could be used in the mill, as some 50 percent was diverted, for instance, to Hanoi as fuelwood. Population pressure on the forest lands increased with the construction of new roads and loss of jobs in the forest enterprises. However, in recent years private farmers have been selling wood to the mill, thereby altering the supply situation dramatically, and the mill is now producing at capacity. Some state forest enterprises are still in operation and producing wood for Bai Bang, but much of the current supply of mostly bamboo is grown and sold by farmers. One important failure of the whole process was inadequate project planning that led to the adoption of inappropriate strategies.

The mill, however, provided a stable market where people could sell wood products, and they responded by starting to grow trees.

2.2.2. Rehabilitation of Degraded Pasture Lands Project— Alternative Association of Producers, Brazilian Amazon

The Alternative Association of Producers (APA) in the Municipality of Ouro Preto D'Oeste, Rondônia, Brazilian Amazon, was funded in 1992 by small-scale farmers in the region with the objective of providing land-use alternatives to slash-and-burn agriculture and cattle ranching. With the support of government-sponsored programmes (Type A– Ministry of Environment, Brazilian Fund for Biodiversity) and NGOs (Movement Laici Latin American, Group of Research and Extension in Agroforestry Systems of Acre-Pesacre), APA focussed work on rehabilitating degraded pastures and secondary regrowth through integrated production systems involving the planting of various fruit and forest tree species along with aquaculture and bee keeping. With around 300 participating families, the association has improved the infrastructure for processing and commercialisation of the diverse products coming out from the rehabilitated areas, which include fruit pulp and syrups, canned palm hearts, honey, guarana powder, medicinal oils, and furniture from wood residue. Labour conditions and quality of life of the families have improved significantly, contributing to the sustainability of this project.

2.2.3. Project in Vila de Novo Paraíso, Municipality of São Geraldo do Araguaia, Pará State, Brazilian Amazon

AGROCANP (Associaçao dos Pequenos Productores do Grotão dos Caboclos de Novo Paraíso), an association of small-scale farmers and residents of the community of Novo Paraíso, started a project to rehabilitate degraded areas in several farmers' lands in 1996. The project was supported by an NGO

---

[502] Ohlsson et al, 2004.

and funding from a government programme (Type A–Ministry of Environment). The activities proposed by the project included the introduction of production systems based on the agroforestry practice known as "agriculture in stages," which consists of establishing herb, shrub, and woody species together with small, medium-sized, and large tree species in the same area. This project experienced the same problems already found in various other projects implemented in the Amazon in the 1970s and 1980s. Farmers did not participate directly in the initial project proposal and even less in the selection of species to be included in the agroforestry modules. There was no market prospecting or planning for the products to be grown. Labour investment was too high, and there was little security of production and income. Given this situation, families abandoned the agroforestry modules and returned to their only income source, livestock rearing for milk production, despite much criticism.

2.3. Ensure Clear and Appropriate Institutional Support and Arrangements

2.3.1. Farm Forestry in Gunung Kidul, Yogyakarta Province, Indonesia

Gunung Kidul used to be a dry area with limited water supply that made it a poor region. The local community started rehabilitating the degraded land in the 1970s. The local government then supported community efforts through formal recognition of the community initiative, the provision of facilitating local regulations, and funding support. The community and the local forestry agency successfully rehabilitated the area using participatory approaches. The dry landscape of 11,072 hectares has been afforested with mainly teak and some *Acacia* sp., and now provides both wood and ecological benefits. Land productivity, forest cover, and water availability in the area have increased, sedimentation rates have decreased, and the microclimate has improved. All of the above have in turn resulted in increased supply of timber, fodder, and fuelwood. Community income and access to education, health, and other services have also improved.

What differentiates this case from numerous others is that the effort was not a top-down approach with the government forcing an initiative on the community. Rather, the government acted appropriately in response to local needs and provided strong institutional and financial support for the local initiative. Local institutions were recognised and empowered, technical support was provided, and the community was allowed to sell timber and to continue its activities. The community itself was highly motivated to transform the area and its livelihoods, and were also supported by strong leadership from within. Rights and responsibilities were clearly divided among the government, the forestry agency, and community groups in the implementation of this effort.

2.3.2. Diversified Institutional Arrangements in Guangdong, China

The province of Guangdong in southern China has had considerable experience in recent years with formalising institutional arrangements, and clarifying rights and roles of different stakeholders to ensure the success and sustainability of its extensive rehabilitation efforts. With these efforts, Guangdong has increased its forest cover from 27 to 57 percent of the land area from 1985 to 2003. The province's experiences with diverse institutional arrangements are serving as models for the rehabilitation of degraded forest lands nationwide.[503] Tenure stabilisation, institutional reform in the rural areas, and opening up of wood markets helped to stimulate the involvement of different stakeholders in rehabilitation. Diversified institutional arrangements among stakeholders appeared, such as cooperative and joint afforestation by different levels of government, state forest farms with village committees, and village committees with private individuals; stock sharing; and private investment on leased land. From 1999 to 2000, Guangdong issued a series of favourable policies further encourag-

---

[503] SFA (State Forestry Administration), 1999.

ing and facilitating the development of private commercial afforestation. There have been 540,000 private entities (including private individuals, and private, civil, and foreign enterprises) that have invested in afforestation in Guangdong using a wide range of institutional arrangements since 1993, and they have contributed to rehabilitation of 1.04 million hectares of degraded lands with fast-growing and high-yielding plantation forests by 2003.[504]

The development of different types of management options involving multiple institutions in Guangdong was accompanied by a clear division of responsibilities, rights, and benefits of the different stakeholders through formal contracts. For example, in the 30-year joint afforestation projects of the Chikan and Xiangang towns of Kaiping city, the state forest farms offer funds and technology, the village committees provide the degraded forest land, and the town forestry stations guarantee supervision. Rights, responsibilities, and cost- and benefit-sharing arrangements are first decided by negotiation among the three stakeholders and then spelt out in a contract. Net profits from the fast-growing high-yielding timber and resin plantations within the 30-year contract period would be shared by these stakeholders in agreed proportions—50 percent due to the investing party, 40 percent due to the land-owning party, and 10 percent to the management party. The investing party has decision-making rights from project planning to implementation, and responsibilities for afforestation and plantation protection. The land-owning and management parties have consulting rights from project planning to implementation, and responsibility for protecting the plantations from man-made or natural disasters. The land is to be delivered back to the village committees within half a year after the project's expiration.

2.3.3. Three KfW-Funded Afforestation Projects, Northern Vietnam

Three afforestation projects funded by the German Development Bank (KfW) operated in Bac Giang, Quang Ninh, and Lang Son provinces in northern Vietnam. Since their start (in 1995, 1999, and 2001, respectively), the projects have established some 23,000 hectares of new forest through plantation and natural regeneration and have established 17,000 deposit accounts with a total savings of 2.5 million Euros.[505] The projects have had positive results because they effectively implemented early on the national forest land allocation programme such that participant farmers had clear rights over their land. The project worked in 80 communes (each with several villages) and established forest farm groups and completed village land use planning in 75 of them. In addition, funds invested into the project were carefully directed to generate benefits for participating farmers, while strict responsibilities were agreed upon. This combination of three essential factors—clear tenure, benefits for participating farmers, and agreements on roles and responsibilities—explains the success of this project.

3. Outline of Tools

3.1. Strengthen Local Organisation and Participation in Projects

The literature is replete with tools to strengthen local participation and collaboration in resource management. Key volumes include Borrini-Feyerabend[506], the Food and Agriculture Organisation's (FAO) series for community forest management, and training materials from the Regional Community Forestry Training Center for Asia and the Pacific, in Bangkok. These include participatory tools and processes for social communication, information gathering and assessment, local organisational development, planning, implementation, considering local knowledge, conflict management, and monitoring and evaluation. CIFOR has developed interactive tools (Co-learn[507]) for collaborative learning and creating shared visions and pathways to reach these visions. General

---

[504] Deng Huizhen, 2003.
[505] KfW Project in Brief, 2003.
[506] Borrini-Feyerabend, 1997.
[507] CIFOR, ACM Team, 2003.

criteria and indicators or guidelines are available for community participation and organisation, conflict management, and use of local knowledge in community managed landscapes[508], plantation landscapes[509] and restoration of degraded landscapes.[510] Tools have also been designed to engage local forest dwellers in collaborative development of criteria and indicators for sustainable forest management using their local knowledge.[511] Many of these tools are directly applicable or can be easily adapted to strengthen participation in rehabilitation projects.

3.2. Consider Local Socioeconomic Needs in Choices of Approaches

DFID's (the UK Department for International Development) sustainable livelihoods toolbox provides numerous tools for using sustainable livelihoods approaches at different stages of the project cycle, from planning to implementation, monitoring, and evaluation. The FAO[512] has a manual on selecting tree species based on community needs. Ames[513] describes methods for comparing the economic value of producing commercial forest products with other local income earning opportunities. The ITTO restoration guidelines[514] provide numerous suggestions on livelihood-enhancing activities, including evaluating prospects for forest products and environmental service payments, evaluating different rehabilitation options and trade-offs with other land uses, adding value to rehabilitation products, and developing partnerships for processing and marketing.

Various tools have been outlined and assessed for processing and commercialisation of forest products including business planning, the enterprise development approach, and market analysis and development.[515] The latter combines ecological sustainability and social and financial objectives in small-scale, low capital, low-skills enterprises. Networking especially between technicians working on forest products and potential producers and markets is also mentioned as a possible approach.

Numerous sets of indicators have been developed within CIFOR and elsewhere for assessing and evaluating socioeconomic impacts of different projects, processes, or policy changes. The current rehabilitation review study has a set of such indicators specifically tailored for assessing the impacts of rehabilitation initiatives.

3.3. Ensure Clear and Appropriate Institutional Support and Arrangements

The FAO[516] provides a rapid appraisal tool for tree and land tenure. Participatory mapping can be used to develop and affirm agreements among stakeholders about tenure boundaries.[517] Other tools available to design and assess institutional arrangements and support include group and key informant interviews, Venn diagrams, matrices, flow diagrams, cost-benefit analysis of different institutional options, stakeholder analysis[518], and the "4 Rs" approach, which attempts to define stakeholders by their respective rights, responsibilities, returns from a given resource, and relationships.[519] The 4 Rs approach draws attention to tenure issues as crucial in shaping people's differentiated concerns with and capacities to manage land and trees. Relationships among stakeholders comprise various facets: service, legal/contractual, market, information exchange, and power. CIFOR has developed general criteria and indicators for institutional agreements, land tenure, and legal frameworks to ensure sustainability of community-managed and large-scale plantation landscapes.

---

[508] Ritchie et al, 2000.
[509] Poulsen et al, 2001.
[510] ITTO, 2002.
[511] Haggith et al, 1999.
[512] FAO, 1995.
[513] Ames, 1998.
[514] ITTO, 2002.
[515] Lecup et al, 1998.
[516] FAO, 1994.
[517] Wollenberg et al, 2002.
[518] Grimble and Chan, 1995.
[519] Vira et al, 1998.

4. Future Needs

Based on the results of this research project, the following needs have emerged:

- Adapting available participatory approaches and tools for rehabilitation projects with different management objectives, socioeconomic and ecological conditions, and stakeholder groups.
- Simple technical guidelines for target groups on how to design, implement, and monitor rehabilitation efforts, incorporating participatory approaches and tools for different rehabilitation objectives and site conditions.
- Participatory planning process to generate simple validated management plans for degraded forest landscapes. Such management plans include mapping; identifying tenure arrangements; choosing appropriate rehabilitation and livelihood options; developing a management strategy; establishing a monitoring framework; clearly assigning rights, responsibilities, costs, and benefits; and formal arrangements for coordination of activities and enforcement of agreements.
- Evaluating prospects for forest products and environmental service payments to communities. This includes the feasibility of producing high-value timber for industries; timber, fuelwood, and other forest products for local needs and markets; and payments for biodiversity, watershed, and carbon functions at the local to international levels.
- Framework for assessing potential contribution and impact of different rehabilitation approaches to communities, in comparison with other local income-earning opportunities and alternative land uses.
- Market research and viable marketing strategies adapted to the specific conditions offered by different types of degraded forest lands. By promoting local-level and value-added production and processing, and developing partnerships to enhance processing and marketing efforts prospects for improving local incomes can be improved.
- Boosting policy, donor, and implementer support for genuine local participation and consideration of local needs in rehabilitation projects. It is important to integrate rehabilitation activities with regional development strategies and community development activities based on local conditions and needs.
- Institutional and political instruments including incentives to support different rehabilitation objectives.

References

Ames, M. 1998. Assessing the profitability of forest-based enterprises. In: Wollenberg, E., and Ingles, A., eds. Incomes from the Forest: Methods for the Development and Conservation of Forest Products for Local Communities. CIFOR, Bogor, Indonesia, and IUCN.

Borrini-Feyerabend, G., ed. 1997. Beyond Fences: Seeking Social Sustainability in Conservation. IUCN, Gland, Switzerland.

CIFOR. ACM Team. 2003. Co-Learn: collaborative learning. CIFOR, Bogor, Indonesia. CD ROM and manual.

Deng Huizhen. 2003. To confidently development "non-public-system" forestry. Guangdong Forestry 2003(1):8–9.

FAO. 1994. Tree and land tenure: rapid appraisal tools. Community Forestry Manual No. 4. FAO, Rome.

FAO. 1995. Selecting tree species on the basis of community needs. Community Forestry Field Manual No. 5. FAO, Rome.

Grimble, R., and Chan, M.K. 1995. Stakeholder analysis for natural resource management in developing countries. Natural Resources Forum 19: 113–124.

Haggith, M., Prabhu, R., Purnomo, H., et al. 1999. CIMAT: a knowledge-based system for developing criteria and indicators for sustainable forest management. In: Cortes, U., and Sanchez-Marre, M., eds. Environmental Decision Support Systems and Artificial Intelligence: Papers from the AAAI Workshop. Technical Report. No. WS-99-07 CIFOR, Indonesia. pp. 82–89.

ITTO. 2002. ITTO guidelines for the restoration of degraded forests, the management of secondary forests degraded and rehabilitation of degraded forest lands in tropical regions. ITTO, CIFOR, FAO, IUCN, WWF International, Yokohama, Japan. ITTO Policy Development Series. No. 13.

KfW Project in Brief. 2003 KfW Afforestation Project Circular, Hanoi.

Lecup, I., Nicholson, K., Purwandono, H., and Karki, S. 1998. Methods for assessing the feasibility of sustainable non-timber forest product-based enterprises. In: Wollenberg, E., and Ingles, A., eds. Incomes from the Forest: Methods for the Development and Conservation of Forest Products for Local Communities. CIFOR, Bogor, Indonesia, and IUCN.

Ohlsson, B., Sandewall, M., Sandewall, R.K., and Phon, N.H. 2004. Government plans and farmers intentions—a study on forest land use planning in Vietnam. Ambio; in press.

Poulsen, J., Applegate, G., and Raymond, D. 2001. Linking C&I to a code of practice for industrial tropical tree plantations. CIFOR, Bogor, Indonesia.

Ritchie, B., McDougall, C., Haggith, M., and Burford de Oliveira, N. 2000. Criteria and indicators of sustainability in community managed forest landscapes: an introductory guide. CIFOR, Bogor, Indonesia.

Sayer, J., Chokkalingam, U., and Poulsen, J. 2004. The restoration of forest biodiversity and ecological values. Forest Ecology and Management 201:3–11.

SFA (State Forestry Administration). 1999. Forestry development of China. Chinese Forestry Publishing House, Beijing.

Sim, H.C., Appanah, S., and Durst, P.B., eds. 2003. Bringing back the forests, Policies and Practices for degraded lands and forests. Proceedings of an international conference, 7–10 October 2002, Kuala Lumpur, Malaysia, FAO Regional Office for Asia and Pacific, Bangkok, Thailand.

Vira, B., Dubois, O., Daniels, S.E., and Walker, G.B. 1998. Institutional pluralism in forestry: considerations of analytical and operational tools. Unasylva 49(194):35–42.

Wollenberg, E., Anau, N., Iwan, R., van heist, M, Limberg, G., and Sudana M. 2002. Building agreements among stakeholders. ITTO Tropical Forest Update 12(2):1–7.

59
A Way Forward: Working Together Toward a Vision for Restored Forest Landscapes

Stephanie Mansourian, Mark Aldrich, and Nigel Dudley

1. Context

The primary aim of this book has been to gather knowledge and experience from a number of practitioners around the world in order to assist conservationists and others in their efforts to restore forests. Restoration has been presented here in the context of a landscape approach, which we believe is a more practical scale for making decisions about returning healthy forest cover and functions to areas where they have been lost or degraded. We have been fortunate in persuading many leading experts to help us in putting the book together, and some of the key lessons or current state of knowledge are summarised briefly below.

It has also become apparent during our research that a large number of unknowns remain. Another emerging purpose of the book is therefore to highlight areas for further development and to call on the conservation community, and others, to address these needs. One important gap that has appeared in different chapters is the need for a comprehensive framework. Using the information gathered through this extensive book, we have attempted to sketch out a framework for the restoration of forests in landscapes. It is hoped that this framework will serve as a guide for practitioners, although it is not meant to be a rigid template. It will need to be used, tested, and refined. Many gaps and research needs have also emerged through this book and the most salient of these are highlighted and summarised under the framework below. More specific ecological research needs can also be found in Appendix 1.

2. Lessons Learnt

As a starting point, we consider some of the key lessons emerging from this book:

1. A lot of experience exists on site-based aspects of restoration; we need to harness it, learn from it, share it, and disseminate it. However, there is much less experience on larger scale restoration interventions (see, for instance, Chapters 19, 20, 27, 48, and 52)

2. Social, political, and economic elements are fundamental to successful forest restoration, yet they are often not part of restoration initiatives (see, for instance, Chapters 4, 6, 17 and 57).

3. The underlying causes of forest loss and degradation are often not addressed in restoration, and contribute to the failure of restoration attempts (see, for instance, Chapters 10 and 11).

4. Policy change can be a powerful lever for large-scale restoration that can yield much more significant results than a large number of small-scale initiatives (see Chapters 17, 48, and 50, for example).

5. There is still a tendency for a lack of communication between disciplines: economists analyse the costs of deforestation, while

foresters look at the potential for restoration, and development organisations promote sustainable agriculture (see, for instance, Chapters 10, 13, 18, and 21).

6. Restoration is a moving target with no ultimate end state; rather, the most preferable end state is for the landscape to be nudged into the tracks of a natural trajectory. While reference landscapes and forests are essential to help set a target for restoration, they are not the only element to consider, as long-term human interaction with forests and the evolution of cultural landscapes, and anticipation of future changes, such as climatic patterns, all need to be factored in when setting goals for restoration of forest landscapes (see Chapters 14 and 15).

7. Environmental, socioeconomic, and political circumstances evolve during the (lengthy) duration of a restoration initiative, thus adding complexity to the planning of a restoration initiative. Climate change is another factor adding complexity and uncertainty to the process (see, for instance, Chapters 4, 5 and 9).

8. To achieve a restored landscape that can satisfy different stakeholders' needs, negotiation and trade-offs will be essential (see Chapters 8 and 18).

9. Incentives for maintaining and/or restoring forests are limited by insecure ownership to forest land or unclear access to forest products (see Chapter 12).

10. Restoration is implemented to reverse not only forest loss but also forest degradation. In response, the improvement of forest quality requires addressing forest composition, pattern, functioning, the process of renewal, resilience, and continuity (see Chapter 26).

11. Persistent challenges for forest landscape restoration relate to planning at large scales, the integration of social and ecological dimensions, and monitoring within large areas (see, for instance, Chapters 9, 13, 20, and 21).

12. Restoration need not always be done in the most direct or obvious manner; for instance, promoting alternative income generation practices may help relieve pressure on land and thus support natural regeneration (see Chapter 19).

13. Even with pure biodiversity conservation aims, forest protection is no longer sufficient, and it would appear that for restoration to make a difference, it usually needs to be planned and implemented at the landscape scale in the context of forest protection and management and other interrelated elements in the landscape (see, for instance, Chapter 7).

14. Financing restoration is a challenge. A number of possible sources exist: the public sector (through subsidies and incentives), the private sector (through payments for environmental services and ethical investments), and multilateral and aid agencies (through grants). Through the Kyoto protocol there is potential to finance restoration, although there remains some uncertainty and concerns over these "carbon sink" projects as critics argue that funds and efforts should go toward reducing fossil fuel emissions at their sources rather than absorbing carbon (see Chapters 22, 23, and 24).

15. Agriculture and forests often compete for land. Restoring landscapes using agroforestry systems can help manage trade-offs between the two (see Chapter 40).

16. For restoration purposes, it is important to understand the role of fire presence in the landscape. In some cases fire is an important element, while in others it is wholly unnatural (see Chapters 39 and 47).

17. Restoration after storms has often not been well managed. As storms are predicted to become more frequent because of climate change, a challenge is to use the media attention they create to lobby for better policies and improved enforcement (see Chapter 48).

18. Well-managed industrial plantations may have a role to play in the restoration of forest landscapes as one element in a landscape mosaic that provides a mix of production and environmental functions (see Chapters 54 and 56).

19. Three key lessons that have emerged from a comprehensive study led by CIFOR of past afforestation/reforestation efforts in six countries show that there is need to strengthen local organisation and participation; there is a need to consider local socioeconomic needs in choices of approaches and options; and there is a need to ensure clear and appropriate institu-

tional support and arrangements (see Chapter 58).

3. An Emerging Framework for Forest Landscape Restoration

As a result of compiling this book and the key lessons identified, it appears that there is an urgent need for a comprehensive framework that will help managers make choices (providing options) based on state of degradation, impact of forest loss/degradation, funding, available human resources, political and institutional considerations, size of the area, aim of the restoration, etc.

This section outlines such a framework for restoring forests in landscapes and includes under each element the identified gaps in current knowledge, tools, and approaches.

Once refined and tested, this framework could form a companion set of tools to existing conservation frameworks, such as WWF's ecoregional methodology, the Nature Conservancy's 5-S approach, or the systematic conservation planning pioneered in New South Wales. Many of the elements drawn from this book provide the basis for such a framework, although we are aware that much remains to be developed over the next few years.

This framework would entail the following:

1. *A systems approach*, reflecting the complexity of the overall system (landscape) and the relationship between its parts—both ecological and social. A landscape needing restoration is a dysfunctional system where the components are unable to fulfil all their potential roles. Therefore, taking a systems approach allows a better understanding of the whole and helps to ensure an integrated approach to the restoration of functions of the different parts. For instance, many restoration initiatives currently focus solely on reestablishing tree cover, rather than on entire communities of plants and animals, or fail to address issues such as environmental services or original landscape patterns.

2. *An adaptive management approach*: Given the long-term nature of restoration, and the level of uncertainty involved as well as changing conditions, it is important to ensure that there is leeway in the system for adaptive management. It is also important to promote an experimental approach or a "learning by doing" approach. This will be effective only with appropriate monitoring and tracking tools in place.

3. *An integrated approach*: It is important to consider restoration not in isolation from other conservation and development projects, but rather as an integral part of joint efforts to achieve a sustainable ecosystem or landscape. This implies better integration of restoration within current planning approaches, including, for instance, those related to protected area selection or forest management, but also development-oriented projects, species conservation, freshwater projects, etc. It is also important to approach forest protection, management, and restoration as elements of a holistic approach to forests.

3.1. The Elements of the Emerging Framework for Forest Landscape Restoration

Thirteen elements are proposed for this framework, each of which is explained in further detail below.

1. Assessment of impacts of forest loss and of restoration
2. Addressing underlying causes of forest loss and degradation
3. Supportive political environment
4. Negotiation and prioritisation
5. Setting multiple objectives for restoration in the landscape
6. Empowerment and engagement
7. Multiple scales of implementation
8. Implementation through multidisciplinary teams
9. Modelling and decision-support tools
10. Sustainable financing
11. Measuring changes in landscape values (monitoring and evaluation)

12. Capacity building/dissemination and exchange
13. A focussed programme of research

3.1.1. Assessment of Impacts of Forest Loss and of Restoration

Unless the impacts of forest loss and degradation are truly understood, it will be difficult to engage the necessary stakeholders fully and to understand the likely evolution of a long-term restoration programme. Often the beneficiaries of restoration are not those living near the forest but rather are downstream users of services; the distribution of costs and benefits of restoration, therefore, need to be carefully considered. Not all costs and benefits can be quantified in monetary terms, however, and issues of equity, including with future generations, also need to be taken into account.

Outstanding needs include:

- More effective ways of measuring forest values in order to promote their restoration (through payment systems for instance)
- Ways of evaluating and describing the differential importance of forest products and services to different people and therefore the differential impacts of changes in forest quality and extent (see, for instance, Chapters 4 and 12).

3.1.2. Addressing Underlying Causes of Forest Loss and Degradation

Failures in past restoration projects can be traced back to inadequate consideration of the original causes of the forest loss and degradation. Careful allocation of resources is needed to ensure that relevant data are collected to advance understanding of the causes of forest loss and degradation to help frame the planning of future restoration interventions.

Outstanding needs include:

- More effective integration of relevant threats' analyses in restoration programmes
- The gap between threats' assessment, and implementation of project activities, needs to be more effectively breached (see Chapter 10).

3.1.3. Supportive Political Environment

All too often those implementing restoration have not taken into account the political and legal environment in which they operate. Yet, policies have the power to either contribute to the failure of restoration interventions or on the other hand to become a major tool in support of large-scale restoration efforts.

Outstanding needs include:

- To convince governments and decision makers of the necessity, importance, and urgency of ecologically and socially sound forest restoration (see, for instance, Chapters 7 and 14)
- To encourage improvements in forest management (that reduce the need for restoration), both in theory and in practice (see, for instance, Chapters 48, 50, and 56)
- Development of an adequate and supportive legal framework that emphasises forest restoration (see, for instance, Chapters 52, 53, 56 and 58)
- Major policy changes to improve restoration, including removal of perverse subsidies and introduction of positive incentives for responsible restoration (see Chapters 11, 17, and 45, for example)
- The presence of representative, accountable, and competent local organisations and institutions that can support integrated restoration programmes (see, for instance, Chapter 58)
- Policies that encourage the development of natural, diverse forests
- Strengthening compliance with and increasing the respect for different key laws related to restoration (see, for example, Chapters 48 and 53)
- Understanding better the complex issues of land rights and how they interact with various factors, such as incentives and policy environments.

3.1.4. Negotiation and Prioritisation

The move from site to landscape entails a similar move from one stakeholder to many. And each stakeholder is likely to have different needs and expectations from the landscape. For this reason it becomes essential to negotiate restoration interventions and their outcomes as they will impact on many people. Questions to address include:

- How do those initiating a restoration project agree with other stakeholders on priority areas for restoration?
- More specifically, how do they determine core areas, minimum viable areas, the type of forest to be restored, etc., within the constraints of those living in the landscape?
- How can stakeholders reach agreement on trade-offs between social, economic and ecological priorities?

Outstanding needs include:

- Identifying how the restoration of forested landscapes can be achieved in areas of intensive, competing land uses (see, for instance, Chapters 40 and 45)
- Processes to negotiate and manage trade-offs between multiple interests (including specifically agriculture and forest restoration) (see Chapters 8 and 40)
- More practical experience in negotiating trade-offs when looking at restoring forest functions in a landscape (see Chapters 8 and 18).

3.1.5. Setting Multiple Objectives for Restoration in the Landscape

The tendency has been to limit restoration projects to one or two objectives, yet the reality is that in complex landscapes with different stakeholders, successful restoration will need to have a number of objectives. In practically all circumstances it will be particularly important to achieve both ecological and socioeconomic goals for restoration.

Outstanding needs include:

- Much better understanding of the likely process of forest restoration itself, along with more accurate methods of measuring progress (see, for instance, Chapters 9 and 14)
- Improved knowledge about how to manage forests for multiple products and objectives

- Guidance on the evaluation of ecological and social aspects within the concept of high conservation value forests and on the role of restoration techniques in addressing them.

3.1.6. Empowerment and Engagement

A necessary element of the framework will be to ensure that the right people have a say in decisions that will affect their future and the land they live on. Although there is a wealth of experience in participatory approaches to conservation and development, most of these are implemented on a relatively small scale (village or community) and much still needs to be learned about effective participation across a whole landscape.

Outstanding needs include:

- Tools to engage stakeholders in restoration efforts effectively across a wider landscape (see, for instance, Chapter 18)
- A better understanding of the role of forests in both poverty prevention and poverty reduction (see Chapter 4).

3.1.7. Multiple Scales of Implementation

As it appears that many factors beyond simply the technicalities of, for example, seed propagation affect restoration, planning a restoration effort needs to be done at large scales and at different levels, with many different people. Nonetheless, ultimately that large-scale plan will need to translate into a series of site-based efforts that contribute to the overall landscape effort.

Outstanding needs include:

- More experience about making the transition from planning to execution within large-scale restoration efforts (see, for instance, Chapter 57)

3.1.8. Implementation Through Multidisciplinary Teams

To address social, economic, political, and institutional aspects of restoring a landscape, restoration efforts will need to involve more disciplines than they have to date. The establishment and systematic use of multidisciplinary teams will be critical to successful restoration in landscapes.

Outstanding needs include:

- Refined approaches for undertaking integrated and multidisciplinary analyses and project implementation
- Improved cooperation at local and international levels between different agencies and nongovernmental organisations (NGOs) (see, for instance, Chapters 13 and 58).

3.1.9. Modelling and Decision-Support Tools

Improved modelling techniques can assist in the formulation of a concerted and shared plan for restoring a landscape. Whilst sophisticated modelling approaches have been developed for other aspects of conservation, such as protected area selection, they remain poorly developed for restoration decision making.

Outstanding needs include:

- Participatory GIS-based decision-support tools to guide choices (of restoration intervention, of species' mixes, of locations, etc.) related to restoration within landscapes (see Chapter 16).

3.1.10. Sustainable Financing

To promote restoration, we need arguments that can, where possible, also be described in economic terms. This can be achieved through better valuation of the range of goods and services that forests provide.

Outstanding needs include:

- The development of strategies for decreasing operating costs and increasing incentives for stimulating natural regeneration in applying the restoration methods developed at the experimental scale to the restoration of large areas. For example, it is important to consider the increase in the production capacity of the restored area, compensation for opportunity costs to landowners, payment for environmental services, and the implementation of tax incentives (see Chapters 36 and 40)
- New and innovative ways to fund forest restoration including more alternative options to make restoration financially attractive (see, for instance, Chapters 23, 24 and 31)
- A better understanding of what mechanisms need to be in place for different payment for environmental services' (PES) systems to work; and also better understanding about the impacts of PES schemes on poor people and how the poor can really benefit from PES (see Chapter 23)
- Information on regrouping or "bundling" different ecosystem services
- Analyses of financial and environmental costs and benefits of restoration options and their effects on forest productivity, species' recovery, biodiversity, and carbon sequestration (see Chapter 52).

3.1.11. Measuring Changes in Landscape Values (Monitoring and Evaluation)

A number of monitoring needs have been repeatedly identified throughout this book. Despite expertise in survey methods, there is still much to be learnt about accurate ways of monitoring of both biodiversity and, more critically, ecological integrity, but also the socioeconomic dimension of forest restoration in landscapes that will allow proper assessment of restoration outcomes over time. Monitoring is also necessary to help guide the choice of the best restoration method under different conditions. Lessons learnt from many past restoration efforts are still being gathered and these are important to guide future interventions and reorientate current ones.

Outstanding needs include:

- Improvement in methodologies for monitoring and evaluating human well-being in the context of restoration (see Chapters 20 and 21)

- A unified procedure for monitoring restoration programmes
- Adequate funds to support long-term monitoring, evaluation, and adaptive management
- Translating the results of both ecological and socioeconomic indicators effectively to inform a landscape-level restoration effort
- Best practices on how to design, implement, and learn from monitoring work that involves multiple stakeholders.

3.1.12. Capacity Building/ Dissemination and Exchange

There already exist a number of tools, approaches, instruments, and experiences related to restoration and what is and is not working. These need to be better used, shared, and widely disseminated as a matter of urgency. Existing organisations such as the Society for Ecological Restoration International (SERI) are obvious repositories for such knowledge, although innovative vehicles such as the clearing house mechanism set up by the Convention on Biological Diversity and the PALNET system of the World Commission on Protected Areas could broaden the coverage. Community and traditional knowledge should not be ignored; specifically, this issue has been raised in this book when it comes to fire management (see Chapter 47) or traditional medicines (see Chapter 34) or nontimber forest products (see Chapter 31). Recognising and learning from community knowledge appears even more important in the context of nations where government structures and approaches are developing and resources and support may be limiting.

Outstanding needs include:

- Substantially increased efforts to disseminate the strategies, approaches, and techniques most appropriate for forest restoration (see Chapters 48 and 52, for example)
- Awareness-raising, training, and technical assistance, as these are preconditions to the application of restoration in practice
- Capacity building for conflict management and negotiation within conservation and forestry organisations in terms of building the ability to work across broad scales and disciplines. Most of the tools and expertise are known but have been applied in only a very limited way within the field of natural resource management (see Chapter 18)
- Adaptation to different regional contexts of science-based management rules and tools (GIS, modelling) and ecological and economical expertise
- In addition, specific training programmes will be necessary to disseminate current knowledge, tailored for different audiences, for example:

  Farmers: Farmers may need encouragement and training to adopt better farming techniques that contribute to the restoration of wider benefits across the landscape (as explained in Chapter 40).

  Local forestry officers: Local forestry officers may need to see beyond the strict forestry objectives of replanting hectares of forests, for instance, without addressing quality issues and without necessarily engaging local communities.

  Plantation companies: Another identified training need is for plantation companies to understand and implement minimum social and environmental management standards for plantations (see, for instance, Chapters 55 and 56).

  Conservationists: Biologists and conservationists involved directly in restoration projects may require training in adaptive and participatory research methods in the context of restoration.

3.1.13. A Focussed Programme of Research

This book has outlined a large amount of existing knowledge on forest restoration, but it has also raised a large number of research needs. It is hoped that through this publication, a sharper and more defined research programme in the field of forest restoration can be initiated. The appendix highlights the most important and urgent research priorities.

4. Working Together Toward a Vision

In the face of growing threats to the world's forests, and more generally to the natural resources that life depends on, we urgently need to be restoring a greater area of forest ecosystems and their functions with increased efficiency. However, as we know from the experience we do have, the process takes time, can be costly, and there are still many unknowns.

Therefore, it is even more urgent and important to share existing knowledge related to restoration more effectively, and to integrate restoration more thoroughly into relevant conservation and development work. The contents of this book, and other available resources like it, provide us with a good start. However, as this chapter has shown us, just disseminating current knowledge is not enough, as there is still much that we need to understand.

For its part, in 2001 the Forests for Life Programme of WWF added a third focal theme of forest restoration within a landscape context—forest landscape restoration—to the longer-standing commitments to protected areas and improved management of production forests, particularly certification.

This was done in direct response to requests from some parts of the WWF network and their partners (particularly in South Asia, the Mediterranean region, East Africa, and parts of Latin America), who felt that in addition to work on protected areas and improved management, there was an urgent need to develop a programme of work on forest restoration in an effort to begin to counter the ongoing process of forest loss and degradation in many parts of the world.

With an increasing focus on implementing forest conservation in landscapes, Forests for Life is now actively working to integrate the approaches and efforts toward achieving its targets—protected areas, improved forest management and restoration—within priority landscapes that have been identified within WWF Global 200 ecoregions.

Through the forest restoration component of Forests for Life, WWF is working with governments, international organisations, indigenous peoples, and other communities, as well as the private sector on the following activities:

- Developing and implementing a portfolio of forest landscape restoration projects/programmes (see http://www.panda.org/forests/restoration/) within priority landscapes
- Assisting others, and building local capacity to plan and implement forest restoration interventions within the broader landscape context
- Developing suitable monitoring tools and techniques to measure progress
- Promoting the use of a forest landscape restoration approach through both local collaboration and broader partnerships such as the Global Partnership on Forest Landscape Restoration (http://www.unep-wcmc.org/forest/restoration/globalpartnership/)
- Documenting, exchanging, and disseminating lessons learnt and experiences
- Highlighting the ways in which governments and the private sector, including plantation companies, can make their contribution to the restoration of forests and their full range of functions in degraded areas
- Working to eliminate/redirect economic, financial, and policy incentives that contribute to forest loss or degradation
- Identifying, researching, and catalysing potential investments and funding mechanisms that can support forest landscape restoration activities, e.g., carbon knowledge projects, and payments for environmental services.

In addition, many others including IUCN, the U.K. Forestry Commission, CIFOR SERI, the governments of El Salvador, Finland, Italy, Japan, Kenya, South Africa, Switzerland, and the United States, and restoration practitioners worldwide are committed to forest landscape restoration, and are making their own significant contributions to ensuring that future restoration efforts are planned and implemented within a landscape context and enhance both ecological integrity and human well-being.

In this challenging context it is crucial that we work together, developing strategic partnerships where required in order to ensure that we have more healthy forests that are able to support people and biodiversity into an uncertain future.

If we do this, and learn and adapt from the lessons and experiences along the way, then we can realise this vision, and we will be able to look back in 20 or 30 years and agree that the first decades of the 21st century really did mark the start of a global effort to successfully restore the world's damaged and degraded forest areas for future generations of biodiversity and people.

Appendix 1
Selection of Identified Ecological Research Needs Relating to Forest Restoration

1. Long-Term Impacts of Restoration on Forest Ecosystems

- Understanding of the long-term dynamics of different ecosystems to help develop realistic restoration targets
- Understanding the ability of different forest ecosystems to recover quality over time and particularly about the likely speed of recovery and the length of time after degradation when a forest can still recover (linked, for instance, to survival time of buried seed populations), all of which are critical for determining whether natural regeneration will suffice or more active efforts are required
- Measuring the sustainability of different restoration efforts, from ecological, social, and economic viewpoints
- Identifying the opportunities for manipulating natural succession to favour desired outcomes
- Understanding what could enhance natural succession after land abandonment

2. Climate Change and Adaptation

- Implementation of field projects to test and if appropriate develop restoration's role in mitigating as well as in building resilience to climate change
- Creative partnerships to analyse climate impacts and proposed restoration activities

3. Knowledge of Species

- Understanding the role that individual species and microhabitats have in the restoration of ecosystem processes
- Clarifying the potential of indigenous species in restoration where planting is necessary, including information on genetics, propagation techniques, the dynamics of ecological succession, the relationships between different species, the performance of indigenous species in plantation conditions, and the production of specific species in nurseries
- Disseminating information on where to obtain seed of indigenous species, how to store the seeds, how to raise seedlings, and how to establish these seedlings in the field

4. Plantations

- Developing user-friendly and location-specific silvicultural guidelines for plantations with indigenous species to increase their adoption by local farmers
- Gathering more information on the long-term dynamics of tree regeneration in plantations (to date, most studies have focussed on young plantations)
- Enhancing understanding of the role and limitations of plantations in landscapes

5. Linkages and Connectivity

- Understanding the role of corridors and ecological stepping stones and in particular how to make these most effective, conditions in which they will and will not work, challenges, problems to avoid, information about distances species will disperse over unsuitable habitat, use of corridors by invasive or pest species
- Developing greater experience on issues related to connectivity of forests across landscapes; for example, connectivity can be at least obtained through the use of lines or even isolated trees in the landscape, serving to buffer plantation areas, changing the "shape" of the plantation, etc.

6. Fires

- Increasing understanding of natural fire regimes including the forest structure needed to avoid high-intensity destructive fires and the associated management implications
- Developing cost-effective fire control measures with minimal biodiversity impacts

7. Invasive Species

- Improving methods for the control of invasive species
- Developing a comprehensive solution for dealing with invasive alien species as part of forest restoration

8. Artificial and Natural Disturbance

- Drawing up codes of practice and perhaps principles for artificial disturbance
- Developing and disseminating methods of enriching degraded or regrowth forests
- Developing enrichment planting guidelines that are species- and site-specific

9. Water and Forests

- Developing tools and methodologies for calculating net gains of different restoration and management actions from the perspective of water supply
- Improving understanding of watershed-scale processes

10. Links Between Site Conditions and Species

- Clarifying species-site relationships—there is often surprisingly little knowledge of the distribution patterns and site requirements of most tropical tree species
- Quantifying better the influence of site conditions (precisely for each parameter) on species' development and growth and on communities' composition, and diversity, along with a better comprehension of the potential trajectories of the communities (i.e., rupture thresholds, lag of time response).

Index

A
abandoned land *see* land abandonment
access controls, 211
access rights, clarification, 235
adaptive management approach, 417
ADPM, 335
advocacy, 124, 139
afforestation, definition, 10
agriculture, shifting, 274
AGROCANP, 409
agroforestry, 247, 274–279, 406, 407–409
  "agriculture in stages", 410
  definition, 275
  future needs, 279
  overcoming impediments, 296
  techniques, 141
  tools, 278–279
Al Shouf Cedar Reserve, Lebanon, 187
Albatera, Spain, forest restoration, 316–317
Algeria, reforestation, 317–318
alley cropping, 275, 277–278
Altai Sayan, Russia, 122
Alternative Association of Producers (APA), 409
Amazon, coca in, 234
amenity, emphasis on, 104
Amur honeysuckle, 388
ancient woodland, definition, 112
Andresito, Argentina, 237, 253
animal dispersal, 357
anthropogenic disturbance control, 251–252
Appalachian region, 264

Area de Conservación Guanacaste (ACG), Costa Rica, 251–252
Argentina, Atlantic forest restoration, 75, 237–238, 253
"artificial negative selection", 286
Asian Development Bank (ADB), 139
Australia
  exclusion zones, 211
  fire control, 272
  linkage corridors, 292
  mining reclamation, 372–373
  monoculture plantations, 292–293
  Tasmania, southern forests, 205
avalanche control, 104

B
Bai Bang Pulp and Paper Mill, Vietnam, 409
Bandipur National Park, India, 111
barrier elimination, 254
BATNA, 129
bauxite mines, forest restoration, 292, 372–373
beetles, saproxylic, 186, 203
beneficial use laws, 79
Bialowieza forest, Poland, 204
bilateral donors, 139, 163
biodiversity
  conservation
    goals, 42
    payments for, 167, 169–170
    plantation management in, 382
  in even-aged plantations *see* even-aged plantations
  forest loss impact, 17–21

modelling tools, 104
reestablishment, 195, 247–248
reservoirs, 360
survey methods, 19–20
Biodiversity Conservation Network, 163
biological targets, 116–117
biological values, in plantations, 394–395
biomass, incorporation in soil, 351
bird species, habitat restoration for, 200
Bitterroot National Park, USA, 336
Borneo
  forest regeneration, 137, 187, 310
  log landings rehabilitation, 364
  rubber, 276
Brazil
  Atlantic forest
    forest loss, 19
    tree cover restoration, 252
  commercial plantations, 380–381
  forest rehabilitation, 405, 406, 409–410
  Plantar project, 172–173
  restoration after mining, 292, 373
bridging substitutes, 206
British Columbia, carbon sequestration payments, 168–169
buffer strips, 246
buffer zones, 35–36, 309–310
Bulgaria, forest policy change, 137–138
burning, prescribed, 186–187, 272

C

C-Plan, 119
California, giant forest restoration, 335
campaigning, 139
Canada
  carbon sequestration payments, 168–169
  eastern, deciduous hardwood restoration, 242–243
  Pacific Northwest forests, 205
capacity building, 127, 133, 421
carbon knowledge projects, 171–175
carbon market, 172, 174, 175
carbon sequestration, 32, 382
  estimation, 174
  payments for, 167, 168–169
carbon sinks, 171
Carrifran, 9
case studies, as policy change stimulus, 124
CATIE, 263, 264, 266
Catskill State Park, USA, 230
cattle grazing, 254
CEAM Foundation, 154
Cebu, Philippines, 407–408
CEDISA, 408
CELOS system, 362
Central America
  and Kyoto protocol modification, 123
  shade-grown coffee, 276–277
Central Truong Son initiative, Vietnam, 69, 153–154, 157–158
Centre for International Forestry Research (CIFOR), 405
  Co-learn tools, 411
  institutional agreement indicators, 412
  Review of Forest Rehabilitation Initiatives, 405
  see also rehabilitation, sustainable
  socioeconomic impact indicators, 412
Centre for Tropical Forest Science (CTFS), 111
change drivers, 103
Chesapeake Bay watershed, USA, 309–310
Chiapas, Mexico, 358

Chile, temperate forest restoration, 324–325
China
  forest ownership policy, 86
  forest rehabilitation, 405, 406, 407, 410–411
  Grain-for-Green programme, 80
  mobile dune stabilisation, 352–353
  restoration benefits and incentives, 87–88
  restoration drivers, 91
  slope stabilisation, 352
CIFOR *see* Centre for International Forestry Research
Clean Development Mechanism (CDM), 172, 174
climate change
  and invasive alien species, 349
  link to CO_2 emissions, 171
  research needs, 424
  restoration in face of, 31–36
  threat to biodiversity, 31
Climate, Community, and Biodiversity (CCB) standards, 174
closures, 254
cloud forest, 229, 303–305
CO_2Fix, 174
coal mines, forest restoration, 373–374
cocoa, 276
codes of practice, 124
coffee, shade-grown, 276–277
Colombia, biodiversity conservation payments, 169
commercial plantations, in forest landscape restoration, 379–382
Common Agricultural Policy (CAP), forestry-related incentives, 80, 82–83
communications
  about forest landscape restoration, 176–180
  messages for specific audiences, 177
  after storms, 343
  effective, 133–134
  proactive, 179
  rapid-response, 179–180
  tools, 139
  via Web sites, 180

communities, compensating, 141
community-based cost-benefit analysis, 28
community-based fire management (CBFiM), 337
community-based forest management (CBFM), 407–408
company practices, changing, 139
conceptual modelling, 76
conflict management, 126–135
  analytical tools, 132–133
  building blocks, 127
  capacity building, 127, 133, 421
  creative thinking, 134
  effective communications, 133–134
  examples, 130
  types of conflict, 126–127
  see also negotiation
connectivity
  in plantation biodiversity restoration, 389
  research needs, 425
  strategy, 47
  see also fragmentation
consensus building workshops, 62
conservation
  by design, 55
  landscapes *see* landscape(s)
Conservation Measures Partnership (CMP), 147
conservationists, training, 422
cork oak forests, 217–218
Coronado National Forest, Arizona, USA, 210
Corrimony, Scotland, UK, 242
cost-benefit analyses, 418
  alluvial forests, 311–312
  community-based, 28
  extended, 62
Costa Rica
  anthropogenic disturbance control, 251–252, 259
  biodiversity conservation payments, 169
  degraded pasture restoration, 264–265
  forest regeneration, 210, 287
  habitat linking, 54
  mixed plantations, 386–387
  thinning in teak plantations, 387
  watershed protection payments, 168, 231
Côte d'Ivoire, cocoa, 276

critical thresholds, for species, 17
cultural keystone species (CKS), 234
cultural values, restoring landscape for, 233–236

D

dams, 308
Dana Nature Reserve, Jordan, 209
deadwood
  assessment, 205
  future needs, 206–207
  habitats provided by, 186, 204
  importance, 203
  restoration, 186, 199, 203–207
    artificial, 201–202, 206
  zoning, 206
decision support tools, future needs, 57, 420
deforestation
  definition, 23
  see also forest loss and degradation
degradation
  causes, 257
  definition, 23
  removing cause of, 243
  vs. restoration, 101–102
  see also forest loss and degradation
Denmark, arable land afforestation, 265
designer landscapes, 103–104
development trajectories, 103
diagnostic sampling, 366
direct planting, 367
direct seeding, 244–245
dispersers, management of, 254
disturbance(s)
  natural, 299
  patterns, influencing, 188
  research needs, 425
  using, 244
diversity nuclei/islands, 252, 254
donor engagement, 177
drivers of change, 103
dry tropical forests *see* tropical dry forests
Dyfi estuary, Wales, UK, 186, 189–190

E

Earth Conservation Toolbox, 55
East Kalimantan, Indonesia, 334–335
ecolabelling, 167
ecological attributes, vital, 153
ecological integrity, 5
  definition, 18
ecological processes, 47
ecological reconstruction, 245–246
ecological restoration, definition, 9
ecological succession *see* succession
economic analysis, 104, 124
economic incentives, 124
ecoregion(s)
  definition, 4
  Global 200, 42, 51, 422
  terrestrial, 42, 43
ecoregion conservation (ERC), 41–49
  determining area to restore, 48
  goals, 42
    restoration and, 44–48
  tools available, 49
ecoregional planning tools, 54–55
ecosystem(s)
  definition, 192
  long-term impacts of restoration on, 425
ecosystem consumption, management, 258
ecosystem fragmentation, 35, 292
  see also connectivity
ecosystem processes, 192
  restoration, 192–196
ecosystem service payment schemes, 28
ecosystem values, evaluation, 359
Ecuador
  payment for watershed services scheme, 162–163
  water management, 229–230
edge effects, 35
egalitarianism, 87
empowerment, 419
endangered local species
  saving, 263
  see also native species
engagement, 419
Enhanced 5-S Project Management Process, 147
enrichment planting, 245, 260, 295–296, 364, 367
environmental change, planning for, 47–48
environmental education programmes, 255
environmental externalities, persistence, 79
environmental values, in plantations, 395
equity
  intergenerational, 86
  issues in community-owned forests, 87
ERDAS, 119
erosion
  control, 69, 299, 350–355, 375
    future needs, 355
    tools, 353–355
  hill slope, 350
  in Iceland, 193, 194
  mass movement, 351–352
  models, 374–375
  wind, 351
ESRI, 119
Ethiopia, user rights for forest restoration, 88–89
ethnobotanical surveys, 236
European Union
  afforestation policies, 80, 82–83
  forest reserves, 111
  grazing in woodlands, 123
  subsidies after storms, 342, 343
evaluation *see* monitoring
even-aged plantations, 384
  biodiversity restoration in, 384–390
    factors influencing natural regeneration, 388–389
    future needs, 389–390
    planting to improve microclimatic conditions, 388
    seed dispersal agent attraction, 388
  factors altering biodiversity, 385–386
evolutionary processes, 47
exclusion zones, 211

F

Fagerön, Sweden, managed forests, 186
Fair-Trade Labelling Organisation (FLO), certification, 220
fallow, improved, 277
farmers
  market information for, 296
  species preferences, 264, 390
  training, 421
FARSITE model, 271, 272

Index

fencing, 260
financing, 161–165, 255, 404
  domestic public sources, 163
  international systems of payments, 164
  payment for goods and services, 164
  private for-profit sources, 164
  private not-for-profit sources, 163–164
  sustainable, 420
Finland
  boreal forest restoration, 327–328
  deadwood requirements, 199
  prescribed burning, 186–187, 327–328
  protected area interventions, 210
  southern region restoration policy, 204–205
  species' transfers, 200–201
fire
  as degradation factor, 334
  historical account, 331
  impacts, 332–333
  in the landscape, 331–332
  as natural disturbance, 333–334
  research needs, 425
  restoration after, 333–338
    potential adverse impacts, 336
    tools, 336–337
  as tool, 334
fire-dependent specialist species, 199, 201
fire management, 141, 201, 337, 396
fire risk, 82
  management strategies, 269–270
firebreaks, 269–273
  widths, 270, 271
floodplain forests
  characteristics, 306–307
  restoration, 306–312
    assessment, 310
    bedload transport, 307–308
    examples of measures, 308–309
    forest structure, 308
    future needs, 311–312
    hydrological connections, 307
    integrated river basin management, 310–311
    monitoring, 310
    scales, 307
focal species, 45
focus groups, 61
fodder harvest, 223
FONAFIFO, 168
Fontainebleau Forest, 204, 210, 340–341
forcefield analyses, 132
forest authenticity, 18
  assessment of levels, 187, 188
Forest Biodiversity Indicators Project, 148
forest certification, 389
  NTFPs and, 220
forest dependence
  degree of, 85
  poverty and, 22, 26
forest dynamics plots, 111
forest fires, mimicking see fire management
forest fragments, 113, 205, 301
forest landscape restoration (FLR), 8
  active vs. passive, 95
  after fire see fire, restoration after
  background, 3–4
  balancing needs, 6, 404
  broader approach, 4–6
  capacity, 97
  challenges based on experience to date, 94–98
  commercial plantations in, 379–382
  communications about see communications
  definition, 5, 10–11
  end point, 96
  framework, 417–422
  funding see financing
  goals, 94–95, 101–105, 109, 419
  growing recognition of need, 401–402
  guidelines, 12
  integration with protection and management, 402
  key elements, 11
  lessons learnt, 415–417
  planning see restoration planning
  practical interventions see tactical interventions
  as a process, 402
  process of, 53
  reasons for landscape scale, 6, 52
  as resilience/adaptation strategy, 35–36
  resources, 96
  social impact, guiding questions, 26–27
  suite of responses required, 402–403
  support needed, 404
  trade-offs in see trade-offs
  valuation of goods and services, 95–96, 139–140, 170
forest loss and degradation
  addressing underlying causes, 418
  impact assessment, 418
  impact on biodiversity, 17–21
  impact on human well-being, 22–29
  examples, 25, 27–28
forest ownership
  communal, 86–87
  definitions, 84–85
  and forest restoration, 84–92
  future needs, 91–92
  tools to address issues, 90–91
  and goods and services rights, 86
  stability, 86
forest plantations, definition, 379, 384
forest quality
  assessment, 20, 187, 188
  restoration, 185–189
Forest Stewardship Council (FSC), 164
  certification, 220–221
forestry officers, training, 421
Forests for Life Programme, 422
Forests of the Lower Mekong ecoregion, Indochina, 44
"founder effect", 244
fragmentation, 35, 292
  see also connectivity
"framework species" approach, 245, 252–253, 289
France
  badlands restoration, 152–153, 265
  deadwood, 204
  floodplain forest restoration, 309
  forest management, 177–178
  Japanese knotweed invasion, 210

lack of ecological monitoring, 69
restoration after storm, 341–342
storm disturbance data, 340–341
frontier forest
analysis, 20
definition, 112
fuel management, 271
vs. fire suppression, 272
fuelwood, 223
forest restoration for, 223–226
plantation eras, 224–226
Fundación Vida Silvestre Argentina (FVSA), 75, 237

G
gap analysis, 57, 113
gap planting, 364
gene flow, 47
genetic diversity, maintenance, 36
genetic selection, 263, 265–266
geographic information system (GIS) tools, 119
in conservation/restoration planning, 49, 325, 374
in fire risk analysis, 271
in suitability modelling, 117–118
in threat assessment, 76–77
Ghana, collaborative forest management, 27
Gifts to the Earth tool, 139
Glen Affric, Scotland, UK, 323–324
Global 200 ecoregions, 42, 51, 422
global change issues, and invasive alien species, 349
Global Environmental Facility (GEF), 164
Global Invasive Species Programme (GISP), 347
Global Partnership on Forest Landscape Restoration, 422
global warming, 32, 287
see also climate change
goods and services
payment for, 164
valuation, 95–96, 139–140, 170
government incentives, 78–81

government policies
changing, 138
and erosion control, 355
grazing management, 353–354
green markets, facilitating access to, 29
Greenhouse Emissions Reduction Trading (GERT), 169
GTZ
property legislation principles, 91
Sustainable Forest Management Project, 334
Guanacaste National Park, Costa Rica, 210, 259, 287
Guangdong, China, 410–411
Guatemala, montane forest restoration, 299–300
Guinea, forest restoration, 53–54
Gunung Kidal, Indonesia, 410

H
habitat
loss, 386
modelling, 116
provided by deadwood, 186, 204
reconnection, 46, 54
Hawaii
alien grass control, 346–347
native forests, 195, 205
hedgerow intercropping, 275, 277–278
"hidden forest harvest", 219
high conservation value forests (HCVF), 20, 235
high conservation values (HCVs), 235
Hmong people, and land rights, 88
home gardens, 235
multistorey, 276
homogeneous monocultures, restoration, 201
human well-being
definition, 11, 23
forest loss impact, 22–29
examples, 25, 27–28
Hungary, mine site regeneration, 259
hurricanes, 299
hydrological models, 374–375

I
Iceland, substrate stability, 193, 194
IDRISI, 118, 119
IFOAM, certification, 220–221
impact, definition, 23
India
joint forest management, 27
Nilgiri Biosphere, 111
sacred forests, 234
indigenous species see native forests; native species
Indonesia
cloud forest conservation, 303–305
enrichment planting, 364
forest rehabilitation, 405, 406, 407, 410
plantation development incentives, 79
protection forests, 89–90
pulp plantations, 380, 381
rainforest rehabilitation, 334–335
Indonesian deer, 347
industrial plantations
best practice guide, 394–397
era of, 225
inoculation, 264, 289, 294
institutional arrangements, for rehabilitation projects, 407, 410–411
integrated approach, 417
Integrated Conservation and Development Projects (ICDPs), 403
intergenerational equity, 86
International Erosion Control Association, 355
International Institute of Rural Reconstruction, advice on land tenure issues, 91, 92
International Plant Protection Convention (1951), 347
International Tropical Timber Organisation (ITTO)
planted forest guidelines, 381–382
restoration guidelines, 90, 382, 412
invasive (alien) species (IASs), 345–346
control/removal, 346–349, 387–388

invasive (alien) species (IASs) (*cont.*)
  by planting native species, 253
  future needs, 347–349
  methods, 187, 189, 260
  research, 140, 348
  tools, 347
 impact, 195
 introduced intentionally, 346, 347
 introduced unintentionally, 346
 research needs, 425

J
Jari plantations, Brazil, 380–381
Jarrah forest, Australia, 372–373
Jordan, forest regeneration, 209

K
Kenya
 improved fallow, 277
 montane forest restoration, 299
 quarry restoration, 9, 123
 water supply protection, 230
keystone species, 195, 198
 cultural (CKS), 234
Kinabatangan River, Malaysia, 137, 187, 310
Kings Canyon National Park, USA, 335
KMYLB, 407–408
knowledge, dissemination and exchange, 421
Kyoto protocol, 123, 168–169, 172

L
La Selva Biological Station, Costa Rica, 264–265
Lafarge, quarry rehabilitation, 123–124
land abandonment, 356
 forest restoration after, 356–360
  active, 358–359
  passive, 358
  socioeconomic tools, 359
land care, 104
land mapping, 90, 117
land ownership *see* forest ownership
land tenure *see* tenure
land-use scenarios, 67
land value, mapping, 117

landscape(s)
 multifunctional, 6, 60, 216
 promotion, 95
 see also forest landscape restoration
landscape architecture, 104
landscape beauty, payment for, 167
landslides, 298–299
Latvia, forestry regulations, 122
learning by doing, 105
Lebanon, forest management, 187
liberation thinning, 366
line planting, 364
livelihood(s)
 analysis, 28–29, 278
 definition, 23
 needs, in rehabilitation projects, 406–407, 409–410, 412
lobbying, following storms, 340, 343
local participation, in rehabilitation projects, 406, 407–409, 411–412
log landings rehabilitation, 364
logging
 biodiversity impacts, 362
 monocyclic, 362
 polycyclic, 362
 reduced-impact (RIL), 363
 see also overlogged forests
Lombok, Indonesia, 303–305
LULUCF, 174

M
Madagascar
 choosing priority landscape, 97
 forest restoration, 74–75, 107–108, 288
 microenterprise development programmes, 141
 plantation projects, 10
 seed dispersal problems, 357
Malaysia
 forest reconnection, 187, 310
 log landings rehabilitation, 364
 native species silviculture, 293
 priority species identification, 98
 restoration methods research, 137
Mandena Conservation Zone, Madagascar, 74

mangrove restoration, 32–34, 47–48
mapping
 examples, 118–119
 future needs, 119
 in long-term modelling, 118
 of opportunities, 117–118
 to meet or set targets, 116–117
market pressure, 139
market research, 140, 413
marketing, of forest landscape restoration, 176–177
Mediterranean region
 forest degradation, 313–314
 forest restoration
  activities, 314–315
  after fires, 335–336
  examples, 315–318
  future needs, 319
  programme evaluation, 154
  tools, 318–319
 land tenure, 314
 NTFPs in, 217–218
 plantation management, 357–358
 reference forests, 111
 wildfires, 314
Meket district, Ethiopia, 88
METSO, 205
Mexico
 active restoration research, 358
 natural forest regeneration, 358
 pilot forest plan based on NTFPs, 220
 Scolel Té project, 173–174
 shade-grown coffee, 277
microenterprise development, 141
migration, 47
mine site regeneration, 259
 see also open-cast mining reclamation
mixed species plantations, 247, 266–267, 389
Model Code of Forest Harvesting Practices, 363
modelling tools, 420
Mombasa, Kenya, disused quarry rehabilitation, 9, 123
monitoring, 150–155, 420–421
 in adaptive management context, 145–148
 common mistakes, 147
 framework for, 152

future needs, 155, 420–421
indicator selection, 151–152
as key to success, 403
long-term, 96, 118
as management tool, 103
of plantations, 397
pressures, 288
tools, 154–155
vital attributes, 153
monoculture plantations, 246, 292–293
monocultures, mosaics of, 246
Morocco, forest restoration, 318
Mount Kenya national park, 299
mountain gorillas, 19
Mudumalai Wildlife Sanctuary, India, 111
multicriteria evaluation (MCE), 62, 115, 117–118
multidisciplinary teams, 420
multifunctionality, 6, 60, 216
promotion, 95
multilateral donors, 139, 163
multipurpose tree, 275
mycorrhizae inoculation, 264, 289, 294

N

Nairobi, Kenya, water supply, 230
national level surveys, 19–20
native forests
definition, 112
restoration, 186, 190–191, 195
native species
endangered, saving, 263
issues related to use, 263–264
planting, 253
silviculture, 293
natural communities
representation, 44–45
seral stages, 45
natural regeneration stimulation, 250–255, 367
anthropogenic disturbance control, 251–252
diversity nuclei use, 252
"framework species" method, 252–253
future needs, 254–255
invasive species elimination, 253
limiting factors, 251
tools, 254
vegetation as regeneration facilitators, 253

natural succession *see* succession, natural
naturalness
assessment, 210–211
components, 185–186
Neem tree, 235
negotiation
alternative to, 129
cultural considerations, 130
need for, 418
phases, 132
principles, 128, 131
process, 130–132
skills, 131
of trade-offs, 61–62, 279
Nepal, community forestry, 27
New Caledonia
forest loss, 18
invasive species control, 347
tropical dry forests programme, 68–69, 97–98, 140, 287–288
New York City, water supply, 230
New York State, salvage logging ban, 341
Nicaragua, biodiversity conservation payments, 169
Niger, watershed restoration, 353
nontimber forest products (NTFPs)
community-based income-generating systems based on, 220
definition, 215
environmental values, 216
and forest certification, 220
impact of loss of, 26
legal frameworks for, 221
in national forestry curricula, 221
as response to poverty, 216–217
restoration guidelines, 219
socioeconomic benefits, 215, 216
valuing in rural development, 219
Novo Paraíso, Brazil, 409–410
nurseries
design, 141
seed availability in, 264

O

Oaxaca, Mexico, 358
obstructions, above-ground, 354

old-growth, definition, 112
open-cast mining reclamation, 9–10, 264, 292, 370–375
conceptual framework, 371
future needs, 375
laws, 375
planning, 371
problems of mine soils, 372
tools, 374–375
opportunity costs, 86, 104
Oregon, USA, H.J. Andrews Experimental Forest, 110–111
organic matter addition, 195
original forests, definitions, 112
outgrower schemes, 162
overland flow, 350
overlogged forests
definition, 361
restoration, 363–367
area protection, 365
future needs, 367
logging practice improvements, 363
planning, 365–366
reasons, 363
silvicultural interventions, 366–367
overseas development assistance (ODA), 162, 163
overstorey removal, 366
ownership, forest *see* forest ownership

P

PALNET system, 421
Paluarco river, Ecuador, 162–163
Panama, reforestation in catchments, 230
participatory appraisal, 132
participatory rural appraisal (PRA), 90–91, 278
PASOLAC, 27
payment for environmental services (PES), 162, 166–170, 231
valuation tools, 170
people first era, 225–226
Peru
Croton restoration, 218–219
forest rehabilitation, 405, 406, 408–409
pests, 346
control, 396

Philippines, forest rehabilitation, 405, 406, 407–408
Plan Vivo system, 174
plant ecology, 266
Plantar project, 172–173
plantation companies, training, 396–397, 421
plantation trees, as nurse plants, 259
plantations
  best practice guide, 394–397
  commercial, in forest landscape restoration, 379–382
  even-aged *see* even-aged plantations
  locating, 393
  managing, 393–397
  mixed species, 247, 266–267, 389
  monoculture, 246, 292–293
  monospecific, 384
  research needs, 424
  rubber, 379
  sustainability elements, 392–393
  tree species selection, 262–267
    future needs, 267
    goals, 263
    issues related to native species use, 263–264
    tools, 265–267
Poland, Bialowieza forest, 204
policy changes, 402, 403
policy incentives
  perverse, 78–81
  redirection of, 81
policy interventions, 121–125
  tools, 124
political environment, supportive, 418–419
pollen analysis, 113
polyacrylamides (PAMs), 355
population viability analysis (PVA), 45–46
Portugal, restoration after fires, 335–336
poverty
  avoidance/mitigation, 26
  degrees of, 24
  elimination, 26
  and forest dependence, 22, 26
  mapping and assessment, 104
  NTFPs as response to, 216–217
predator–prey dynamics, 47
pressures, monitoring, 288

Prestige oil spill, 178
primary woodland, definition, 112
prioritisation, 418
  tools, future needs, 57
priority landscapes, 42
  identification, 67
  implementing conservation in, 55
problem trees, 132
process management, 128, 129
PROCYMAF project, 28
property
  definition, 84–85
  rights, problems, 79
  types, 85
protect–manage–restore approach, 44, 52–53, 55
  stages, 56–57
protected areas
  categories, 211
  restoration in, 208–212
  threats, 208
  zoning, 211
Puerto Rico
  restoration via natural succession, 292
  substrate stability, 193–194
  tree plantations, 259, 386

Q

quality, forest *see* forest quality
Quintana Roo, Mexico, pilot forest plan, 220
Quito, Ecuador, water supply, 229

R

racks, installation of, 254
Rainforest Alliance, Smartwood Programme, 221
range maps, 117
Rapid Ecological Assessment methodology, 20
rapid rural appraisal (RRA), 90–91, 278
rattan, 218
REACTION programme, 154, 319
reclamation *see* open-cast mining reclamation
reduced-impact logging (RIL), 363
reference forests/landscapes, 55, 103, 109–113, 258
  tools, 112–113
reforestation, definition, 10

"regeneration nuclei", 251
rehabilitation
  definition, 9
  sustainable, 405–413
    future needs, 413
    institutional arrangements, 407, 410–411
    lessons from past projects, 406–407
    local participation, 406, 407–409, 411–412
    socioeconomic needs, 406–407, 409–410, 412
    tools, 411–412
relics, 366
representation, natural community, 44–45
resilience-building, and forest restoration and protection, 33
restoration databases, 155
restoration planning
  framework, 66–68
  future needs, 70
  goals and targets, 94–95, 101–105, 109, 419
  multiple scales, 419
  need for, 65–66
  tools, 69–70
restoration trajectories
  identification, 68
  reappraisal, 68
Rhone River, 309
rills, 350
Rinjani National Park, Indonesia, 303–305
Rio Cumbaza Basin, Peru, 408–409
RISEMP, 169
risk, sources of, 26
river basin management, integrated, 310–311
rubber, 276
  plantations, 379
runoff control, 375
Rural Development Regulation (RDR), 82
RUSLE model, 375
Russia, woodland certification, 122

S

Sabah, Malaysia
  forest regeneration, 137, 187, 310

log landings rehabilitation, 364
sacred groves/forests/gardens, 234
safety net, forests as, 24
Saignon, 152
salvage logging, 342
  banning, 341
SAPARD, 80, 82
Saracá-Taquera National Forest, Brazil, 373
scattered tree plantings, 245
scenarios, 62, 102–103
  modelling tools, 102
Scolel Té project, Mexico, 173–174
Scotland
  commercial plantations, 380, 381
  natural regeneration with grazing, 242
  pine forest restoration, 323–324
SEAGA, 91
secondary forests, 246, 276
  restoration potential, 321–322
seed
  availability, 264
  collection, 141, 294
  dispersal, 357, 388
seeding, direct, 244–245
Sequoia National Park, USA, 335
Shaanxi Province, China, 352–353
shifting agriculture, 274
Sichuan Province, China, 352
Sierra de las Minas, Guatemala, 299–300
Sierra Espuña, Spain, reforestation, 315–316
SilvaVoc, 12
silvopastoral systems, 169
SIMILE, 102
site-level restoration, 241–248
  approach determination, 241–242
  degrading influence reduction, 243
  future needs, 248
  management considerations, 247–248
  reforestation for productivity and biodiversity, 246–247
  tree cover initiation/ improvement, 244–246
site-scale survey methods, 20

SITES/Marxan, 119
skid trails rehabilitation, 364
Slovakia, Tatra National Park, 341
Smartwood Programme, 221
social values, 394
  see also cultural values; socioeconomic needs
Society for Ecological Restoration International (SERI), 421
Socio-economic and Gender Analysis (SEAGA), 91
socioeconomic needs, in rehabilitation projects, 406–407, 409–410, 412
socioeconomic research, 140
socioeconomic targets, 117
Soil Association, Woodmark Programme, 221
soil conditioners, 355
soil microcarbon analysis, 113
soil nutrient reduction, 195
soil protection, 351, 354
soil remediation, 372, 375
soil stabilisation, 266, 351
soil surface manipulations, 351, 354
Song Thanh Nature Reserve, Vietnam, 75, 122, 293
SOS Sahel, 89
South Africa
  outgrower schemes, 162
  toxic conditions amelioration, 194
South Wales coalfield, 374
Southeast Asia, rattan production, 218
Spain
  firebreaks, 271–272
  mining reclamation, 373–374
  natural regeneration stimulation, 253
  Prestige oil spill, 178
  reforestation, 314–316
spatial modelling, 325
species
  knowledge of, research needs, 424
  transfers of, 200–201
species-based targets, 117
species-site relationships, 295, 425
Sri Lanka, silvicultural treatment guidelines, 390

staff training, in plantations, 396–397, 421
stakeholder(s)
  external, 60
  primary, 60
  in scenario development, 102
  secondary, 60
stakeholder analysis, 91, 132
STEEP, 132
STELLA, 102, 303
Stockholm, Sweden, water supply, 230
storm disturbance
  forest restoration after, 339–343
  key ideas, 340–341
Stradbroke Island, Queensland, Australia, 211
subsidies, government, 79
substrate fertility, 194
substrate stability, 193, 194–195
succession, 192
  direction/manipulation, 194, 195, 244, 257–260
  tools, 259–260
  dynamics of, 254–255
  minimal intervention design, 258–259
  natural
    causes halting, 257
    stimulation, 244
    understanding, 257–258
suitability modelling, 115, 117–118
Sumatra, Indonesia, pulp plantations, 381
surveys, stakeholder, 61–62
sustainability analysis, 132
Sustainable Forest Market Transformation Initiative (SFMTI), 163
sustainable rehabilitation *see* rehabilitation, sustainable
Sweden
  deadwood microhabitat re-creation, 186
  water quality protection, 230
Switzerland, continuous cover forestry, 53
SWOT, 132
systems approach, 417

T

tactical interventions, 136–142
Tanzania, agroforestry, 243

target species
  categories, 197–198
  as indicators of successful restoration, 198–199
  restoration for, 197–202
    future needs, 202
    planning, 200–201
    stand-level restoration methods, 201–202
targets
  biological, 116–117
  socioeconomic, 117
Tasmania, southern forests, 205
Tatra National Park, Slovakia, 341
Tebang Pilih system, 362
temperate forests
  characteristics, 320–321
  ecological attributes, 321, 322
  restoration, 320–325
    future needs, 325
    issues, 321–323
    tools, 325
tenure
  clarification, 235
  customary, 84, 85
  mapping, 117
  rights of, 29
  security of, 86
Terai Arc, Nepal, 46, 47
Thailand
  "framework species" approach, 252–253
  land rights, 88
thinning, 260, 292, 387, 389
  liberation, 366
threat(s)
  direct, 73–74
  examples, 138
  indirect, 73, 74
  potential, 73–74
  removal of, 138
threat assessment
  future needs, 77
  information needed, 73
  tools, 76–77
threat mapping, 76
threat matrices, 76
threshold barriers, 257–258
tigers, 46
timber, production objectives, 104
timber stand improvement (TSI), 366
Tonda de Tamajón woodland, Spain, 253

toxic conditions amelioration, 194, 195
tracking tools, for landscapes, 105
trade-offs, 59–62, 248
  negotiation, 61–62, 279
  types, 60–61
  win–win situations, 59
training
  in restoration techniques, 140–141
  tailored, 421–422
transects, 90, 278
tree crops, 104
  and forest restoration, 276–277
Trombetas, Brazil, 9, 373
tropical dry forests (TDF)
  attractiveness to people, 286
  characteristics, 285–286
  restoration
    active, 289
    Guanacaste National Park, Costa Rica, 210, 259, 287
    monitoring pressures, 288
    New Caledonia, 68–69, 97–98, 140, 287–288
    passive, 288–289
    reasons for, 286–287
  soil fertility, 289
tropical moist forests
  restoration, 291–296
    choice of method, 293–294
    choice of species, 294
    fostering animal diversity, 295
    future needs, 295–296
    obtaining seed, 294
    production-biodiversity trade-off, 295
    raising seedlings, 294
tropical montane forests
  characteristics, 298
  overcoming natural succession barriers, 300
  restoration, 298–301
    choice of species, 300–301
    in face of natural disturbance, 299
    remnant forest role, 301
    socioeconomic rationale, 298–299
Tunisia, access to NTFPs, 220

U
Uganda, forest loss, 19
umbrella species, 198

underplanting *see* enrichment planting
understorey development encouragement, 247
United Kingdom
  plantations, 54, 381
  see also Scotland; Wales
United States
  alien grass control, 346–347
  buffer zone restoration, 309–310
  fire control, 272
  giant forest restoration, 335
  Hawaiian forests, 195, 205
  H.J. Andrews Experimental Forest, 110–111
  honeysuckle control, 388
  longleaf pine ecosystems, 146–147
  mine spoil restoration, 264
  salvage logging ban, 341
  water supply protection, 230
  wilderness values restoration, 210
  wildfires, 336
urban/forest interface, fire risk, 270–271
urban frontier, proximity to, 48
Utrillas coalfield, Spain, 373–374

V
Valdivian ecoregion, Chile, 324
vegetation, as regeneration facilitators, 253
VENSIM, 102
viable populations, of species, 45–47
Vietnam
  forest rehabilitation, 405, 406, 407, 409, 411
  integrated restoration approach, 69
  land rights, 88
  mangrove restoration, 34
  participatory monitoring system, 153–154, 157–158
  pressures on remaining forests, 97
  reforestation programme, 122–123, 293
  three-dimensional model of threats, 75–76
vision(s)
  development, 102–103
  fine-tuning tools, 104

working together toward, 422–423
voice, development of, 28
vulnerability
  household, 23–24
  to climate change, 34–35

W

Wales
  commercial plantations, 381
  mining reclamation, 374
  native forest restoration, 186, 190–191
Walomerah protection forest, Indonesia, 89
water
  quality and quantity, 228–231
  research needs, 425
  scarcity, 228
Water Framework Directive, 311
watershed protection, payments for, 167, 168, 231
watershed values, 231
weed control, 396
well-being *see* human well-being
Western Europe, forest loss, 18–19
wetland, restoration, 189–190
wilderness
  assessment, 210–211
  re-creation, 209
wildfires
  in Mediterranean region, 314
  in United States, 336
wildwood, definition, 112
wind
  erosion by, 351
  resistance to, 340–341
windbreaks, 301
wood harvesting methods, 354
woodlot era, 225
Woodmark Programme, 221
WWF
  challenges based on experience to date, 94–98
  and forest management in France, 178
  Forests for Life Programme, 422
  lessons from experience to date, 401–404

Y

yerba mate, 253
Ynyshir bird reserve, Wales, UK, 186

Z

Zambia, improved fallow, 277